Methods in Enzymology

Volume XLVIII
ENZYME STRUCTURE
Part F

METHODS IN ENZYMOLOGY

EDITORS-IN-CHIEF

Sidney P. Colowick   Nathan O. Kaplan

Methods in Enzymology

Volume XLVIII

Enzyme Structure

Part F

EDITED BY

C. H. W. Hirs

DIVISION OF BIOLOGICAL SCIENCES
INDIANA UNIVERSITY
BLOOMINGTON, INDIANA

Serge N. Timasheff

GRADUATE DEPARTMENT OF BIOCHEMISTRY
BRANDEIS UNIVERSITY
WALTHAM, MASSACHUSETTS

1978

ACADEMIC PRESS   New York   San Francisco   London
A Subsidiary of Harcourt Brace Jovanovich, Publishers

ACADEMIC PRESS, INC.
111 Fifth Avenue, New York, New York 10003

United Kingdom Edition published by
ACADEMIC PRESS, INC. (LONDON) LTD.
24/28 Oval Road, London NW1 7DX

Library of Congress Cataloging in Publication Data

Main entry under title:

Enzyme structure.

(Methods in enzymology, v. 11, 25–26, 48)
    Part B—   edited by C. H. W. Hirs and S. N. Tim-
asheff.
    Includes bibliographical references.
    1. Enzymes––Analysis. I. Hirs, Christophe Henri
Werner, Date   ed.
QP601.M49 vol. 11, etc.   547'.758   79–26910
ISBN 0–12–181948–5 (v. 48)

Table of Contents

Section I. Molecular Weight Determinations and Related Procedures

Section II. Interactions

Contributors to Volume XLVIII

Article numbers are in parentheses following the names of contributors.
Affiliations listed are current.

E. T. ADAMS, JR. (5), *Chemistry Department, College of Science, Texas A & M University, College Station, Texas*

KIRK C. AUNE (7), *Department of Biochemistry, Baylor College of Medicine, Houston, Texas*

ALFRED D. BARKSDALE (16), *Department of Laboratory Medicine and Pathology, University of Minnesota, Minneapolis, Minnesota*

VICTOR A. BLOOMFIELD (19), *Department of Biochemistry, College of Biological Sciences, University of Minnesota, St. Paul, Minnesota*

JOHN R. CANN (11, 12, 14), *Department of Biophysics and Genetics, University of Colorado Medical Center, Denver, Colorado*

CHARLES R. CANTOR (17), *Departments of Chemistry and Biological Sciences, Columbia University, New York, New York*

DAVID J. COX (10), *Department of Biochemistry, Kansas State University, Manhattan, Kansas*

EDWARD F. CRAWFORD (5), *Chemistry Department, College of Science, Texas A & M University, College Station, Texas*

T. H. CROUCH (4), *Department of Biochemistry, School of Medicine, University of Virginia, Charlottesville, Virginia*

F. W. DAHLQUIST (13), *Institute of Molecular Biology, and Department of Chemistry, University of Oregon, Eugene, Oregon*

J. DANDLIKER (18), *Department of Biochemistry, Scripps Clinic and Research Foundation, La Jolla, California*

W. B. DANDLIKER (18), *Department of Biochemistry, Scripps Clinic and Research Foundation, La Jolla, California*

T. E. DORRIER (6), *Department of Biochemistry, School of Medicine, University of Virginia, Charlottesville, Virginia*

ROBERT H. FAIRCLOUGH (17), *Department of Chemistry, Columbia University, New York, New York*

GEOFFREY A. GILBERT (9), *Department of Biochemistry, University of Birmingham, Birmingham, England*

LILO M. GILBERT (9), *Department of Biochemistry, University of Birmingham, Birmingham, England*

A. N. HICKS (18), *Department of Biochemistry, Scripps Clinic and Research Foundation, La Jolla, California*

MARGARET J. HUNTER (3), *Biophysics Research Division, University of Michigan, Ann Arbor, Michigan*

GERSON KEGELES (12, 15), *Section of Biochemistry and Biophysics, Department of Biological Sciences, The College of Liberal Arts and Sciences, The University of Connecticut, Storrs, Connecticut*

R. J. KELLY (18), *Department of Biochemistry, Scripps Clinic and Research Foundation, La Jolla, California*

D. W. KUPKE (4, 6), *Department of Biochemistry, School of Medicine, University of Virginia, Charlottesville, Virginia*

S. A. LEVISON (18), *Department of Biochemistry, Scripps Clinic and Research Foundation, La Jolla, California*

TONG K. LIM (19), *Department of Biochemistry, College of Biological Sciences, University of Minnesota, St. Paul, Minnesota*

THOR B. NIELSEN (1), *Section on Membrane Regulation, Laboratory of Nutrition and Enzymology, National Institute of Arthritis, Metabolism, and Digestive Diseases, National Institutes of Health, Bethesda, Maryland*

JACQUELINE A. REYNOLDS (1, 2), *Department of Biochemistry, Duke University Medical Center, Durham, North Carolina*

ANDREAS ROSENBERG (16), *Department of Laboratory Medicine and Pathology,*

University of Minnesota, Minneapolis, Minnesota

JOHN C. H. STEELE, JR. (2), *Department of Biochemistry, Duke University Medical Center, Durham, North Carolina*

CHARLES TANFORD (2), *Department of Biochemistry, Duke University Medical Center, Durham, North Carolina*

PETER J. WAN (5), *The Oil and Seed Products Division, Food Protein Research and Development Laboratory, Texas Engineering Experiment Station, College Station, Texas*

JOHN U. WHITE (18), *White Development Corporation, Stamford, Connecticut*

ROBLEY C. WILLIAMS, JR. (8), *Department of Molecular Biology, Vanderbilt University, Nashville, Tennessee*

Preface

"Enzyme Structure," the eleventh volume in this series, was published eleven years ago. Three supplements, Parts B, C, and D, appeared in 1972 and 1973 and served to update and expand the methodology covered in the original volume. Part E, which appeared recently, is concerned primarily with chemical techniques. This volume and Part G, which is now in press, deal in detail with physical methods. It is hoped that they will give a broad coverage of techniques currently available for the study of enzyme conformation and interactions.

As in the past, these volumes present not only techniques that are currently widely available but some which are only beginning to make an impact and some for which no commercial standard equipment is as yet available. In the latter cases, an attempt has been made to guide the reader in assembling his own equipment from individual components and to help him find the necessary information in the research literature.

In the coverage of physical techniques, we have departed somewhat in scope from the traditional format of the series. Since, at the termination of an experiment, physical techniques frequently require much more interpretation than do organic ones, we consider that brief sections on the theoretical principles involved are highly desirable as are sections on theoretical and mathematical approaches to data evaluation and on assumptions and, consequently, limitations involved in the applications of the various methods.

The organization of the material is the same as in Parts C and D, with Part F being devoted primarily to techniques related to molecular weight measurements, associations, and interactions, while Part G contains the sections on conformational analysis and on optical and resonance spectroscopy.

We wish to acknowledge with pleasure and gratitude the generous cooperation of the contributors to this volume. Their suggestions during its planning and preparation have been particularly valuable. Academic Press has provided inestimable help in the assembly of this volume. We thank them for their many courtesies.

C. H. W. HIRS
SERGE N. TIMASHEFF

METHODS IN ENZYMOLOGY

EDITED BY

Sidney P. Colowick and Nathan O. Kaplan

VANDERBILT UNIVERSITY
SCHOOL OF MEDICINE
NASHVILLE, TENNESSEE

DEPARTMENT OF CHEMISTRY
UNIVERSITY OF CALIFORNIA
AT SAN DIEGO
LA JOLLA, CALIFORNIA

METHODS IN ENZYMOLOGY

EDITORS-IN-CHIEF

Sidney P. Colowick   Nathan O. Kaplan

Section I

Molecular Weight Determinations and Related Procedures

[1] Measurements of Molecular Weights by Gel Electrophoresis[1]

By THOR B. NIELSEN and JACQUELINE A. REYNOLDS

A charged particle in an electric field will acquire a steady-state velocity that is proportional to the field strength and the magnitude of the charge and inversely proportional to the frictional coefficient of the particle. This last property reflects the molecular dimensions of the particle and the viscosity of the medium in which it is suspended. This movement of ions in an electric field is termed electrophoresis, and the proportionality constant between the steady-state velocity and the field strength is commonly called the electrophoretic mobility.

$$u = \epsilon q / f \qquad (1)$$

where u is the steady-state velocity, ϵ is the electric field strength, q is the charge, and f is the frictional coefficient. The electrophoretic mobility, U, is q/f.

Theoretical treatments of electrophoretic mobility in solution are fraught with difficulties, which are discussed in detail in a number of other sources.[2-4] The important considerations in the use of any electrophoretic technique as a means of characterizing charged particles (including macroions, such as proteins and polynucleic acids) can be summarized as follows:

1. The molecular parameters of charge and size are reflected in the electrophoretic mobility.

2. A frictional effect due to the viscosity of the medium is also reflected in the electrophoretic mobility.

3. There is no *rigorous* theory relating electrophoretic mobility to molecular dimensions of a macroion (i.e., charged macromolecules with which we are primarily concerned in this chapter).

Following the classic studies of Tiselius[5] on the electrophoretic properties of human plasma proteins using the free-boundary procedure, a number of investigators introduced solid-support materials to the solution medium in an attempt to stabilize the bands of macroions observed during

[1] This work was supported in part by National Institutes of Health Grants HL-14882 and NS-12213.
[2] C. Tanford, "Physical Chemistry of Macromolecules," p. 412. Wiley, New York, 1961.
[3] F. Booth, *Proc. R. Soc. London Ser. A* **203,** 514 (1950).
[4] J. Th. G. Overbeek, *Kolloid Beih.* **54,** 287 (1943).
[5] A. Tiselius, *Biochem. J.* **31,** 1464 (1937).

electrophoresis (see Sober *et al.*[6] for a review of this field.) The most common support material in use today is polyacrylamide, the application of which was originally investigated by Raymond and Weintraub.[7] The electrophoretic mobility of a macroion in a polyacrylamide gel containing an appropriate solvent is a function of the same parameters itemized above, but it is also dependent on the pore size of the gel that acts as a molecular sieve. Thus, in addition to the retardation effect of the solvent viscosity, one must also consider the retardation (relative to solution mobility) of the gel pores themselves. Equation (1) would then be written as follows:

$$u = \epsilon q/f \; \phi(r) \tag{2}$$

where $\phi(r)$ represents the retardation effect of the gel pores. A recent review by Rodbard[8] discusses the various theoretical approaches to the electrophoretic mobility of a macroion in a porous matrix.

This chapter deals with the use of polyacrylamide gel electrophoresis as an analytical tool for separation and qualitative characterization of charged macromolecules. Restrictions on the use of this procedure for the estimation of molecular weights are emphasized, particularly with respect to the now widespread use of polyacrylamide gel electrophoresis in the presence of denaturing detergents.

Polyacrylamide Gel Electrophoresis in the Absence of Detergents

If the migration of a variety of proteins subjected to polyacrylamide gel electrophoresis are compared under conditions where the solvent medium, field strength, and pore size are identical, the relative mobilities are proportional to the relative values of q/f [Eq. (2)]. No absolute molecular data can be obtained from such an experiment since there are two unknowns—the net charge on the protein and the frictional coefficient. However, this procedure has become a powerful tool for separating and categorizing protein components in a mixture so long as the following reservations are kept in mind. (1) Different proteins can have the same q/f, and thus the observation of a single band is not a sufficient criterion for purity. (2) Relative mobilities are not necessarily an indication of relative molecular weights.

[6] H. A. Sober, R. W. Hartley, Jr., W. R. Carroll, and E. A. Peterson, *in* "The Proteins" (H. Neurath, ed.), Vol. III, p. 31. Academic Press, New York, 1965.
[7] S. Raymond and L. Weintraub, *Science* **130,** 711 (1959).
[8] D. Rodbard, *in* "Methods in Protein Separation: A Modern Survey" (N. Catsimpoolas; ed.), p. 145. Plenum, New York, 1976.

Polyacrylamide Gel Electrophoresis in the Presence of Denaturing
  Detergents

Protein chemists have been aware for many years that anionic and
cationic detergents with alkyl chains longer than ten or twelve carbon
atoms bind to most water-soluble proteins and induce a concomitant conformational change.[9] The most extensively studied of these detergents is
sodium dodecyl sulfate (SDS) which dissociates a large number of water-
soluble proteins to their constituent polypeptides in the presence of reducing agent and binds to these polypeptides at a level of approximately 1.4 g
of SDS per gram of protein. The polypeptides undergo a binding-induced
conformational change such that the Stokes radius of the polypeptide–
detergent complex is proportional to the molecular weight of the
polypeptide raised to the 0.73 power.[10,11] Characterization of the SDS–
polypeptide complexes has been carried out primarily in sodium phosphate buffer at pH 7.2. The dependence of binding and Stokes radius on
pH, nature of buffer ions, and addition of urea has not been thoroughly
investigated. Table I lists data presently available for complexes formed
between water-soluble, reduced proteins and SDS.

It is clear from the previous discussion that if the ratio of charge to
frictional coefficient is a unique function of the molecular weight of a
group of polypeptides, the relative electrophoretic mobilities in solution
or in polyacrylamide gel electrophoresis will also be a unique function of
the molecular weight. It is on this premise that the use of polyacrylamide
gel electrophoresis in SDS for molecular weight determinations is based.
This is a valid technique only if the following criteria are met:

1. All polypeptides (standards and unknowns) must bind the same
amount of SDS on a gram per gram basis, thus making q a unique function
of the molecular weight.

2. All polypeptides (standards and unknowns) must assume the same
unique conformation in the SDS complex, i.e., the hydrodynamic radius
must be proportional to $M_r^{0.73}$, thus making the molecular contribution to f
a unique function of the molecular weight.

3. Standards and unknowns must be subjected to the same electric
field strength, medium viscosity, and polyacrylamide pore size.

These are stringent requirements and cannot be circumvented by any
experimental manipulations. Since the only rigorous data on the
physicochemical properties of polypeptide–SDS complexes have been

[9] J. Steinhardt and J. A. Reynolds, "Multiple Equilibria in Proteins," p. 234. Academic
  Press, New York, 1970.
[10] J. A. Reynolds and C. Tanford, *J. Biol. Chem.* **245,** 5161 (1970).
[11] W. W. Fish, J. A. Reynolds, and C. Tanford, *J. Biol. Chem.* **245,** 5166 (1970).

TABLE I

PROPERTIES OF PROTEIN–SDS COMPLEXES

Protein	M_r	R_s (Å)	Method[a]	SDS/ protein (g/g)	Ionic strength[b]
Cytochrome c	11,700	26.2[c]	GF	—	—
Lysozyme	14,400	27.3[c]	[η]	1.39	0.26[d]
Hemoglobin	16,700	27.7[c]	[η]	1.38	0.17[e]
				1.40	0.26[d]
AII	17,400[l]	28.0[f]	S	1.40	0.033[g]
β-Lactoglobulin	18,400	31.5[c]	GF	1.19	0.17[e]
Chymotrypsinogen	25,700	40.1[c]	[η]	1.40	0.26[d]
AI	28,400	44.0[f]	S	1.40	0.033[g]
Pepsin	35,500	53.0[h]	GF	1.20	0.26[h]
Lactate dehydrogenase	35,000	54.0[i]	GF	—	—
Ovalbumin	45,000	61.3[c]	[η]	1.46	0.26[d]
		63.0[j]	S	1.42	0.17[e]
Alkaline phosphatase	43,000	61.3[c]	[η]	—	—
IgG (heavy chain)	49,500	66.5[c]	GF	—	—
Catalase	60,000	—	—	1.38	0.17[e]
Bovine serum albumin	69,000	83.5[c]	[η]	1.35	0.005[d]
Xanthine oxidase	147,000	144.0[j]	S	—	—
Myosin (heavy chain)	194,000	170.0[j]	GF	1.41	0.26[d]
Spectrin	210,000	184.0[k]	S	2.1	0.055[k]

[a] GF, gel filtration; [η], intrinsic viscosity; S, sedimentation velocity.

[b] All measurements were carried out in sodium phosphate buffer, pH 7.2, at the indicated ionic strength with the exceptions noted here. AI and AII (human serum high-density lipoprotein) were determined at pH 8.2. Spectrin was determined in 50 mM NaCl, 0.1 M EDTA, 10 mM Tris-Cl, pH 8.0.

[c] J. A. Reynolds and C. Tanford, *J. Biol. Chem.* **245**, 5161 (1970).

[d] J. A. Reynolds and C. Tanford, *Proc. Natl. Acad. Sci. U.S.A.* **66**, 1002 (1970).

[e] R. Pitt-Rivers and F. S. Impiombato, *Biochem. J.* **109**, 825 (1968).

[f] S. Makino, C. Tanford, and J. A. Reynolds, *J. Biol. Chem.* **249**, 7379 (1974).

[g] J. A. Reynolds and R. H. Simon, *J. Biol. Chem.* **249**, 3937 (1974).

[h] G. Whitmyre and C. Tanford, unpublished data (1976).

[i] W. W. Fish, J. A. Reynolds, and C. Tanford, *J. Biol. Chem.* **245**, 5166 (1970).

[j] J. A. Reynolds, unpublished data (1976).

[k] N. M. Schechter, Ph.D. dissertation, Duke University, Durham, North Carolina, 1976.

[l] Molecular weight of disulfide-bonded dimer.

obtained in the absence of other denaturing agents, such as urea, and in a limited number of buffer systems (Table I), appropriate solutions for electrophoretic measurements are limited. Use of different pH or different buffer ions requires determinations of the binding capacity and Stokes radii for standard polypeptides under the altered conditions.

In particular the use of urea in SDS–polyacrylamide gel electrophoresis is unsound. Each of these reagents induces a different type of conformational change in proteins, and therefore, one is not certain that all proteins assume an identical conformation in this mixed medium. Neither can one be assured that detergent binding is unaltered in the presence of urea.

Polypeptides with large side-chain branches (such as carbohydrate moieties or intact disulfide bonds) and intrinsic membrane proteins rarely meet the criteria listed above.[12]

In principle, long alkyl chain cationic detergents could replace SDS in polyacrylamide gel electrophoresis. However, there is a limited amount of data available regarding the interaction of cationic detergents with proteins and only tetradecyltrimethylammonium chloride (TMAC) has been investigated in any detail.[13] The behavior of a number of proteins in this detergent is similar to that in SDS, and the results are summarized as follows:

1. TMAC binds cooperatively to a number of reduced polypeptides at a 10-fold higher detergent concentration than is required for saturation of binding sites by SDS.

2. The accompanying conformational change produces polypeptide–detergent complexes somewhat smaller than those in SDS, but the relationship between Stokes radius and molecular weight appears to be the same (i.e., $R_s \sim M_r^{0.73}$).

The restrictions discussed previously for the determination of molecular weights by polyacrylamide gel electrophoresis in SDS apply to the use of TMAC as the denaturing agent.

Experimental Procedures

Polyacrylamide Gels

Detailed procedures for the preparation of polyacrylamide gels have been published elsewhere.[14,15] Commercially prepared gels can be purchased from Bio-Rad Laboratories and must be preelectrophoresed before use. In the authors' experience, these gels are particularly suitable for procedures involving extraction of polypeptides following elec-

[12] C. Tanford and J. A. Reynolds, *Biochim. Biophys. Acta* **457,** 133 (1976).
[13] Y. Nozaki, J. A. Reynolds, and C. Tanford, *J. Biol. Chem.* **249,** 4452 (1974).
[14] K. Weber, J. Pringle, and M. Osborn, this series Vol. 26, p. 3.
[15] J. V. Maizel, *in* "Methods in Virology," (K. Maramarosch and H. Koprowski, eds.), Vol. 5, p. 179. Academic Press, New York, 1971.

trophoretic separation, since the concentration of water-soluble contaminants is extremely low.[16]

Solutions for Polyacrylamide Gel Electrophoresis

Buffer solutions containing SDS for gel electrophoretic separation of polypeptides must be at an ionic strength sufficiently low to allow saturation of the detergent binding sites by monomeric SDS. Previous studies have demonstrated that the equilibrium concentration of SDS monomers required for saturation ranges from 0.8 to 2 mM.[10,17] Since the critical micelle concentration of SDS is a function of ionic strength,[18] buffers for SDS–polyacrylamide gel electrophoresis should not exceed 0.26 ionic strength and should have the same composition as those used for characterization of detergent binding to standard proteins (Table I). Discontinuous buffer systems[19-21] have been proposed that appear to produce sharper polypeptide bands in SDS–polyacrylamide gel electrophoresis. The pitfalls in using such a system are obvious from the previous discussion. Alterations in ionic strength and pH as well as buffer ions may result in changes in both SDS binding and conformation of the resultant complex. In the absence of rigorously obtained data on these potential effects, molecular weights cannot be deduced from apparent electrophoretic mobilities in these systems.

Removal of Other Detergents before SDS– Polyacrylamide Gel Electrophoresis

In the course of isolation and purification of membrane proteins, a number of different detergents are often used. Recognizing the high probability that intrinsic membrane proteins will not obey the criteria for molecular weight determinations in SDS–polyacrylamide gel electrophoresis, many investigators use this technique for qualitative cataloguing of the number of separable species. The experimental difficulty lies in replacement of other degergents by SDS. If the detergent in question has a high critical micelle concentration, it can be readily exchanged by dialysis. However, the most commonly used detergents in membrane studies are nonionics with relatively low critical micelle con-

[16] S. J. Friedberg and J. A. Reynolds, *J. Biol. Chem.* **251**, 4005 (1976).
[17] T. Takagi, K. Tsujii, and K. Shirahama, *J. Biochem.* (Tokyo) **77**, 939 (1975).
[18] M. F. Emerson and A. Holtzer, *J. Phys. Chem.* **71**, 1898 (1967).
[19] U. K. Laemmli, *Nature (London)* **227**, 680 (1970).
[20] A. R. Ugel, A. Chrambach, and D. Rodbard, *Anal. Biochem.* **43**, 410 (1971).
[21] D. Neville, *J. Biol. Chem.* **246**, 6328 (1971).

TABLE II
STACKING AND SEPARATING POLYACRYLAMIDE GELS[a]

Gel	Buffer[b] (ml)	Acrylamide[c] (ml)	TEMED[d] (μl)	Persulfate[e] (ml)	H_2O (ml)
Stacking	10	4.5	30	2	4.5
Separating	20	18.0	60	2	—

[a] L. J. Rizzolo, unpublished work 1976.
[b] Gel buffer: 50 mM sodium phosphate, pH 7.2: 19.3 g of $Na_2HPO_4 \cdot 7 H_2O$, 3.9 g of $NaH_2PO \cdot H_2O$, 2 g of SDS in 1 liter of H_2O (diluted 1 : 2 for running buffer).
[c] 7.5% separating gel and 3.7% stacking gel: 16.6 g of acrylamide, 0.45 g methylene bisacrylamide made up to 100 ml with H_2O. 5% separating gel and 2.5% stacking gel: 11.1 g of acrylamide, 0.30 g of methylene bisacrylamide made up to 100 ml with H_2O.
[d] N,N,N',N'-Tetramethyl-1,2-diaminoethane.
[e] Ammonium persulfate, 0.188 g in 50 ml of H_2O.

centrations and are often tightly bound to membrane proteins.[12,22] It is frequently possible to exchange these detergents with SDS by one of the following procedures.

A large excess of SDS micelles is added to the sample and competes with the polypeptides for the nonionic detergent. The sample is subjected to polyacrylamide gel electrophoresis in the customary manner except that a low-percentage polyacrylamide stacking gel is used above the separating gel (Table II). The mixed SDS-nonionic detergent micelles rapidly migrate away from the polypeptide–SDS complexes, which tend to concentrate at the interface between the stacking and separating gels. This method has provided reproducible relative electrophoretic mobilities for those polypeptide–SDS complexes that are large relative to the mixed micelles.

Alternatively, a gel with a gradient in polyacrylamide concentration can be used without a stacking gel (Table III). A large excess of SDS micelles must also be present in the sample. The principle in each of these procedures is to remove nonionic detergent by competitive binding of SDS to the polypeptide and by competitive formation of mixed detergent micelles. The gel systems are designed to remove the mixed micelles from the solution of polypeptide–SDS complexes by maximizing the differences in the rates of migration.

It must be emphasized again that these procedures do *not* provide a measure of polypeptide molecular weight unless the three criteria stated in Section III are met.

[22] A. Helenius and K. Simon, *Biochim. Biophys. Acta* **415**, 29 (1975).

TABLE III
GRADIENT POLYACRYLAMIDE GELS

Solution[a]	Buffer[b] (ml)	Acrylamide[c] (ml)	TEMED[d] (μl)	Sucrose (g)	Persulfate[e] (ml)
A	10	9	10	2	1
B	10	9	30	—	1
C	5	4.5	30	—	1

[a] The total volume of solutions A and B should equal the volume of the gel to be made. The denser solution, A, is placed in the mixing chamber, and the less dense solution, B, in the inflow chamber of a gradient forming device. The highest-density portion of the gradient flows to the bottom of the slab space or tube. As A and B mix, the density and the acrylamide concentration decrease to form the continuous gradient. After the gradient gel is formed, 3 ml of C are introduced around the comb of a slab gel apparatus to form sample wells. For cylindrial gels, overlay the gradient with H_2O.
[b] Gel buffer: 50 mM sodium phosphate, pH 7.2: 19.3 g of $Na_2HPO_4 \cdot 7H_2O$, 3.9 g of $NaH_2PO_4 \cdot H_2O$, 2 g of SDS in 1 liter of H_2O (diluted 1:2 for running buffer).
[c] A: 22.2 g of acrylamide, 0.6 g of methylene bisacrylamide made up to 100 ml H_2O; B: 11.1 g of acrylamide, 0.3 g of methylene bisacrylamide made up to 100 ml H_2O; C: use stock solution B.
[d] N,N,N',N'-Tetramethyl-1,2-diaminoethane.
[e] Ammonium persulfate, 0.188 g, in 50 ml of H_2O.

Detection of Anomalous Behavior of Polypeptide–SDS Complexes

Under specific conditions (see Table I) a number of water-soluble proteins subjected to SDS–polyacrylamide gel electrophoresis in the presence of reducing agent have q/f ratios that are a unique function of the molecular weight. The pore size of the polyacrylamide gel is reflected in the term, $\phi(r)$, in Eq. (2), so it is apparent that alterations in pore size will change the mathematical function describing the relationship between migration rate and molecular weight. Any polypeptide–SDS complex that has a q/f ratio that is *not* the same unique function of molecular weight as the standards will have different *apparent* molecular weights as the pore size is varied, since $\phi(r)$ is some function of a hydrodynamic radius of the complex. Therefore, a comparison of the migration rate of an unknown complex to a group of standard polypeptide–SDS complexes as a function of polyacrylamide concentration will often demonstrate anomalous binding or conformation in the unknown.[23]

[23] G. Banker and C. Cotman, *J. Biol. Chem.* **247**, 5856 (1972).

[2] Determination of Partial Specific Volumes for Lipid-Associated Proteins[1]

By JOHN C. H. STEELE, JR., CHARLES TANFORD, and JACQUELINE A. REYNOLDS

The study of lipid-associated proteins, as found in membranes and serum lipoproteins, has become an active field of research during the past decade. Yet careful physicochemical characterization of these polypeptides has not been extensively undertaken; in part, this is due to the insolubility of most such proteins in simple aqueous solutions, necessitating the use of a solubilizing agent to maintain the protein in solution. For reasons detailed elsewhere,[2] detergents are generally the most useful compounds in this regard, promoting solubility while permitting the detailed characterization of the polypeptide moiety.

However, the addition of solubilizing agents to the protein–water (or buffer) system precludes the application of physicochemical techniques based on two-component systems. Thus, most workers have relied on empirical techniques to determine properties such as the molecular weight of lipid-associated proteins. The most widely used method for this purpose is polyacrylamide gel electrophoresis in sodium dodecyl sulfate (SDS) or in SDS/urea buffer systems. While this may be quite useful in preliminary characterization of membranes or lipoproteins, especially when dealing with mixtures of proteins (e.g., as obtained on initial solubilization), the molecular weights obtained by SDS–polyacrylamide gel electrophoresis must be considered as apparent. This results from the basic assumptions made in SDS–polyacrylamide gel electrophoresis— namely, that the polypeptides being studied interact with SDS in a fashion identical with that of the water-soluble proteins that are used as standards for molecular weight calibration. This assumption is often unjustified. Similarly, molecular weights of polypeptides obtained from the determination of elution position in gel filtration chromatography in denaturing agents, such as guanidine hydrochloride (GuHCl) or SDS, are based on comparable assumptions that are often not valid for membrane proteins or serum lipoproteins.[2]

For these reasons, the rigorous measurement of the molecular weight of a lipid-associated polypeptide is usually possible only by the use of

[1] This work was supported in part by National Institutes of Health grants HL-14882 and NS-12213. J. C. H. S. was supported by a U.S. Public Health Service predoctoral fellowship, 5-TO-5-GM-01678.
[2] C. Tanford and J. A. Reynolds, *Biochim. Biophys. Acta* **457**, 133 (1976).

sedimentation equilibrium in a solvent that maintains the protein in a soluble, monodisperse state. If the protein in question is completely disaggregated by a detergent such as SDS, the molecular weight of the monomeric polypeptide can be determined rigorously by ultracentrifugal techniques. Characterization of a lipid-associated protein in its "native" state (which may be oligomeric) can be carried out using sedimentation equilibrium in the presence of bound, nondenaturing detergent. Frequently, residual amounts of tightly bound lipid will also be present which are required for maintenance of structure and function.

The use of sedimentation methods (equilibrium and velocity) to study lipid-associated proteins in soluble form requires that the effect of any detergents and/or lipids bound to the protein be considered. As shown below, this necessitates a knowledge of the partial specific volumes of these components and/or the protein–lipid–detergent complex. This information may also be necessary for interpretation of data obtained by other techniques (e.g., light scattering[3] and small-angle X-ray scattering[4]). This chapter discusses the theory and methods relating to the use of partial specific volumes in sedimentation studies and tabulates experimentally determined values of this quantity for a variety of detergents and lipids.

Definitions

The partial specific volume, $\bar{v}_i$, of a component, i, is the change in volume of a solution produced by adding 1 g of component i to that solution, the addition being carried out at constant temperature (T), pressure (P), and mass (w) of all other components of the system. Mathematically, this is expressed as

$$\bar{v}_i = (\partial V/\partial w_i)_{T,P,w_j} \tag{1}$$

where V is the total volume of the solution and the subscript j refers to all components of the system other than i.

If the differentiation is carried out at constant chemical potential, μ_j, of all components, j, other than i, the effective partial specific volume ϕ_i' is obtained.

$$\phi_i' = (\partial V/\partial w_i)_{T,P,\mu_j} \tag{2}$$

This quantity, as shown by Casassa and Eisenberg,[5] is applicable when

[3] E. P. Pittz, J. C. Lee, B. Bablouzian, R. Townend, and S. N. Timasheff, see this series Vol. 27, p. 209.

[4] H. Pessen, T. F. Kumosinski, and S. N. Timasheff, see this series Vol. 27, p. 151.

[5] E. F. Casassa and H. Eisenberg, *Advn. Protein Chem.* **19**, 287 (1964).

dealing with systems in which i is a nondiffusible component at equilibrium across a semipermeable membrane with diffusible components, j. Thus, the effective partial specific volume contains contributions from component i and any other components which are preferentially bound to i.

Theory

Sedimentation equilibrium and sedimentation velocity studies of multicomponent systems are treated by the method of Casassa and Eisenberg.[5] The application of this theory has been discussed in detail elsewhere.[2,6] Assuming thermodynamic ideality, the directly measured quantity in sedimentation equilibrium studies is $M_p(1 - \phi'\rho)$ where M_p is the molecular weight of the protein, ρ is the density of the solution at zero protein concentration, and ϕ' is the effective partial specific volume of the protein plus all bound components.

The term $(1 - \phi'\rho)$ can be evaluated directly if a sufficient quantity of purified protein is available. When this is not the case, it is possible to approximate it closely by Eq. (3).

$$(1 - \phi'\rho) = (1 - \bar{v}_p\rho) + \delta_d(1 - \bar{v}_d\rho) + \Sigma_j\delta_j(1 - \bar{v}_j\rho) \tag{3}$$

$\bar{v}_p$ is the true partial specific volume of the protein, $\bar{v}_d$ is the partial specific volume of bound detergent (which is assumed to be the same as that of the detergent micelle), $\bar{v}_j$ is the partial specific volume of any other bound component (e.g., residual lipid), and δ is the amount bound in grams per gram of protein. (In solutions with high concentrations of a third component, e.g., guanidine or sucrose, molecules of other components, such as water, may be displaced from the protein, necessitating the inclusion of negative contributions for these components in the last term of Eq. (3). However, in dilute solutions of solubilizing agents, these negative contributions are generally negligible.) Determination of these quantities is described below under Methods.

The measurement of δ for nonionic detergents is often not feasible because of the lack of accurate assays or the nonavailability of radioactive forms. In such cases it is possible to adjust ρ to $1/\bar{v}_d$ or to extrapolate to this density from measurements of $M_p(1 - \phi'\rho)$ made at several densities (using D_2O-H_2O). At $\rho = 1/\bar{v}_d$ the quantity $(1 - \bar{v}_d\rho)$ is zero and a knowledge of δ is unnecessary.[7]

Conversely, it is possible to consider the protein–lipid–detergent complex as a single component and to formulate the sedimentation results in

[6] C. Tanford, Y. Nozaki, J. A. Reynolds, and S. Makino, *Biochemistry* **13**, 2369 (1974).
[7] J. A. Reynolds and C. Tanford, *Proc. Natl. Acad. Sci. U.S.A.* **73**, 4467 (1976).

terms of a two-component system. The quantity determined experimentally in sedimentation equilibrium is in this case $M_c(1 - \bar{v}_c\rho)$ where

$$M_c = M_p(1 + \delta_d + \Sigma\delta_j) \tag{4}$$

and

$$\bar{v}_c = (\bar{v}_p + \delta_d\bar{v}_d - \Sigma\delta_j\bar{v}_j)/(1 + \delta_d + \Sigma\delta_j) \tag{5}$$

From sedimentation equilibrium data at two or more solvent densities,[8] it is possible to obtain M_c and $\bar{v}_c$ directly if the amount of bound components does not vary with the density of the solvent. M_p can then be determined using Eq. (4) if δ_d and δ_j are known.

Theoretically, it is possible to obtain a value for $M_p(1 - \phi'\rho)$ from sedimentation velocity experiments if the frictional coefficient of the hydrodynamic particle can be determined independently.[6] However, experience in the authors' laboratory has shown that the Stokes radius measured by gel chromatography is not always reliable, especially when the particle is highly asymmetric.[9] Therefore, until precise and simple methods of determining the frictional coefficient are available, the accuracy of polypeptide molecular weights obtained in this manner may be questioned, especially for large polypeptides with inherent or induced (e.g., by binding of SDS) asymmetry.

Finally, the use of techniques such as sedimentation velocity in sucrose density gradients[10] to determine molecular weights, while appealing from the standpoints of experimental ease and less rigid purity requirements, is fraught with interpretive difficulties.[2,6] The accuracy of molecular weights obtained in this manner is highly suspect.

Methods

Density Measurements

The determination of the partial specific volume $(\bar{v})$ of a component in solution involves the measurement of the densities of a series of solutions containing varying concentrations of that component (see next section). In order to obtain accurate values for $\bar{v}$, it is necessary to measure densities with a precision of 2×10^{-6} g/ml. Although several techniques have been devised for density determinations with this precision[11] (see also refer-

[8] S. J. Edelstein and H. K. Schachman, *J. Biol. Chem.* **242,** 306 (1967).

[9] Y. Nozaki, N. M. Schechter, J. A. Reynolds, and C. Tanford, *Biochemistry* **15,** 3884 (1976).

[10] R. G. Martin and B. N. Ames, *J. Biol. Chem.* **236,** 1372 (1961).

[11] D. W. Kupke, *in* "Physical Principles and Techniques of Protein Chemistry" (S. J. Leach, ed.), Part C, p. 1. Academic Press, New York, 1973.

ences given in tables for specific methods), the most common procedure currently in use is the mechanical oscillator densimeter. The theory, construction, and operation of this instrument have been discussed in detail elsewhere.[12] Pycnometers[11,13] generally cannot provide the necessary accuracy. However, $\bar{v}$ values determined by pycnometry are included in the tables when measurements by more accurate methods have not been made.

Whatever technique is used, several factors should be kept in mind. Precise temperature control is necessary; a variation of 0.1° in the temperature range (20°–25°) of most measurements produces a change in the density of water of about 2×10^{-5} g/ml.[14] Thus, temperature control of ±0.01° is required. Meticulous care must be used in filling the sample compartment of the instrument to avoid the introduction of air bubbles. Finally, the possibility of adsorption of the component being measured to the glass surfaces in the instrument should be kept in mind.

Calculation of Partial Specific Volumes from Densities

The partial specific volume of a solute is obtained from a series of accurate density measurements at varying solute concentrations (including zero). At any given concentration, c, the *apparent* partial specific volume is

$$v = [1 - (\rho - \rho_0)/c]/\rho_0 \tag{6}$$

where ρ is the solution density and ρ_0 the solvent density in grams per milliliter. If this value of v is constant over a range of concentrations, it can be taken as the true partial specific volume, $\bar{v}$, of the component in that concentration range.

Alternatively, the limiting slope as the component concentration approaches zero (or the critical micelle concentration, cmc, when detergent micelles are the component) of a plot of ρ vs c yields $\bar{v}$ from

$$\lim_{c \to 0} (\partial\rho/\partial c) = 1 - \bar{v}\rho_0 \tag{7}$$

The apparent effective partial specific volume and the effective partial specific volume, ϕ', of a protein–lipid–detergent complex are evaluated analogously.

[12] O. Kratky, H. Leopold, and H. Stabinger, see this series Vol. 27, p. 98.
[13] N. Bauer and S. Z. Lewin, *in* "Physical Methods of Organic Chemistry" (A. Weissberger, ed.), 3rd ed., Part I, p. 131. Wiley (Interscience), New York, 1959.
[14] "Handbook of Chemistry and Physics" (R. C. Weast, ed.), 49th ed., p. F-4. Chemical Rubber Co., Cleveland, Ohio, 1968.

*Determination of the Effective Partial Specific Volume of the
    Protein–Lipid–Detergent Complex*

If adequate amounts of the complex are available, the effective partial
specific volume, ϕ', can be determined directly as shown above. In this
case, however, the solvent has been equilibrated with the solution of
complex by equilibrium dialysis. If the detergent concentration is below
the critical micelle concentration, equilibrium dialysis of the complex
against a large volume of the desired solvent provides the simplest means
of preparing the necessary solutions. If the detergent concentration is
above the cmc, equilibrium across a semipermeable membrane is generally reached very slowly. In this case, gel filtration on a column equilibrated with the solvent can be used. It is recommended that the eluted
complex be dialyzed against the column buffer to correct any small concentration changes in diffusible buffer ions that may result from evaporation of the eluate during fraction collecting.

Determination of the Partial Specific Volume of the Apoprotein

With sufficient amounts of material, it is possible to directly measure
the partial specific volume of the apoprotein. If the apoprotein is water
soluble, measurements can be done in aqueous solution; generally this is
not the case, and it is necessary to include a solubilizing agent *at constant
mass* in the solvent and protein solutions—i.e., the detergent or lipid is
one of the components *j* of Eq. (1).

Usually such measurements will not be feasible owing to insufficient
amounts of protein; in such cases a reliable estimate[15] of $\bar{v}$ can be obtained
from the amino acid composition using the procedure of Cohn and Edsall.[16] This involves calculating the weight percentage, w_r, of each amino
acid residue and the corresponding volume percentage, $w_r\bar{v}_r$, where $\bar{v}_r$ is
the calculated partial specific volume of that residue:

$$\bar{v} = (\Sigma w_r \bar{v}_r)/(\Sigma w_r) \tag{8}$$

Determination of the Partial Specific Volume of the Detergent or Lipid

The partial specific volume of detergents and lipids can be obtained as
outlined above from density measurements as a function of solute concentration. It should be noted that $\bar{v}$ in general will be different for these
compounds above and below the cmc. When adsorption of such am-

---

[15] T. L. McMeekin and K. Marshall, *Science* **116**, 142 (1952).
[16] E. J. Cohn and J. T. Edsall, *in* "Proteins, Amino Acids and Peptides as Ions and Dipolar
    Ions" (E. J. Cohn and J. T. Edsall, eds.), p. 375. Reinhold, New York, 1943.

phiphiles is a serious problem, other techniques can be used. A graph of sedimentation coefficient vs density can be constructed, and interpolation to a sedimentation coefficient of zero provides $\bar{v}$ as the reciprocal of that density.[17] Alternatively, sedimentation equilibrium in sucrose density gradients has been used to estimate $\bar{v}$ as the reciprocal of the density at which the amphiphile bands.[18,19] The use of sucrose enables one to achieve densities not obtainable with D_2O and therefore circumvents the need for a long extrapolation to measure $\bar{v}$ when the partial specific volume is not close to unity. However, this method suffers from a potentially serious problem in that the hydrodynamic particle contains bound H_2O, and in a solvent system containing sucrose the $(1 - \bar{v}_{H_2O}\rho_{solv.})$ term is not equal to zero. In the case of sodium taurodeoxycholate, pycnometry indicated a $\bar{v} = 0.76$ ml/g whereas sedimentation velocity measurements in sucrose gradients implied that $\bar{v} = 0.93$ ml/g.[20]

When using detergents that are a mixture of molecular species (e.g., most commercial nonionic detergents), the possibility exists that the protein may preferentially bind one (or a few) of these species. In such a case the $\bar{v}_d$ of the bound detergent is not accurately determined by the measured $\bar{v}_d$ of the total mixture. If protein–detergent complexes with different amounts of bound detergent are available, $M_p(1 - \phi'\rho)$ can be measured at several densities in H_2O–D_2O mixtures. A plot of this quantity as a function of density for two different levels of binding will give two intersecting lines. The density at the point of intersection is the reciprocal of the partial specific volume of the bound detergent species. An accurate value of $\bar{v}$ can be obtained in this manner only if the amount of detergent bound is significantly different and is relatively insensitive to small changes in the free detergent concentration. It is assumed that any changes in binding, partial specific volumes, molecular weights, etc., in altering the H_2O–D_2O composition are negligible or are taken into account in the calculations. This method should also provide a means of justifying the basic assumption (see section Theory, above) that the $\bar{v}_d$ of the detergent when bound to the protein closely approximates that of the pure micelle.

Calculation of Partial Specific Volumes from Homologous Amphiphiles

When it is not feasible to directly measure the partial specific volume of a particular amphiphile, it is possible to estimate its value, if the v of a

[17] C. Huang and J. P. Charlton, J. Biol. Chem. 246, 2555 (1971).

[18] M. E. Haberland and J. A. Reynolds, Proc. Natl. Acad. Sci. U.S.A. 70, 2313 (1973).

[19] M. E. Haberland and J. A. Reynolds, J. Biol. Chem. 250, 6636 (1975).

[20] T. C. Laurent and H. Persson, Biochim. Biophys. Acta 106, 616 (1965).

TABLE I
IONIC DETERGENTS

Class	Compound	$\bar{v}$	T	Solvent	Method	Comment
1. Alkyl sulfates[a]	NaC_8SO_4	0.766	25	N	NS	NED[d]
	$NaC_{10}SO_4$	0.816	25	W	CD	PSV[e]
	$NaC_{12}SO_4$	0.854	25	W	CD	PSV[e]
		0.863	25	S	DM	PSV[f]
	$NaC_{14}SO_4$	0.890	26	N	DL	PSV[g]
2. Alkyl sulfonates[a]	$NaC_{12}SO_3$	0.876	31.5	N	DL	PSV[g]
	$NaC_{14}SO_3$	0.923	39.5	N	DL	PSV[g]
3. Trimethylammonium chloride[b]	$C_{12}TMACl$	1.107	23	W	PY	PSV[h]
		1.083	23	S	PY	PSV[h]
	$C_{14}TMACl$	1.113	23	W	PY	PSV[h]
		1.087	23	S	PY	PSV[h]
		1.110	25	B	DM	PSV[i]
	$C_{16}TMACl$	1.11	27	N	PY	NED[j]
4. Trimethylammonium bromide[b]	C_8TMABr	0.902	25	W	CD	PSV[e]
	$C_{10}TMABr$	0.936	25	W	CD	PSV[e]
	$C_{12}TMABr$	0.958	25	W	CD	PSV[e]
		0.97	25	B	CT	—[i]
	$C_{14}TMABr$	0.985	25	W	CD	PSV[e]
	$C_{16}TMABr$	1.003	25	W	CD	PSV[e]
		0.995	25	B	CT	—[i]
5. Dimethylalkylammoniopropane sulfonates[c]	$DC_{10}APS$	0.920	30	W	MF	PSV[k]
	$DC_{12}APS$	0.957	30	W	MF	PSV[k]
	$DC_{16}APS$	0.986	30	W	MF	PSV[k]

[a] The symbols NaC_nSO_4 and NaC_nSO_3 indicate the compounds $CH_3(CH_2)_{n-1}SO_4^- Na^+$ and $CH_3(CH_2)_{n-1}SO_3^- Na^+$, respectively.

[b] The symbols C_nTMACl and C_nTMABr indicate the compounds $CH_3(CH_2)_{n-1}N(CH_3)_3^+Cl^-$ and $CH_3(CH_2)_{n-1}N(CH_3)_3^+Br^-$, respectively.

[c] The symbol DC_nAPS indicates the compound $CH_3(CH_2)_{n-1}N^+(CH_3)_2CH_2CH_2CH_2SO_3^-$.

[d] M. Tanaka, S. Kaneshina, K. Shin-no, T. Okajima, and T. Tomida, *J. Colloid Interface Sci.* **46**, 132 (1974).

[e] J. M. Corkill, J. F. Goodman, and T. Walker, *Trans. Faraday Soc.* **63**, 768 (1967).

[f] J. A. Reynolds, unpublished work (1976).

[g] K. Shinoda and T. Soda, *J. Phys. Chem.* **67**, 2072 (1963).

[h] Calculated from density data of L. M. Kushner, W. D. Hubbard, and R. A. Parker, *J. Res. Natl. Bur. Stand.* **59**, 113 (1957).

[i] C. Tanford, Y. Nozaki, J. A. Reynolds, and S. Makino, *Biochemistry* **13**, 2369 (1974).

[j] F. Reiss-Husson and V. Luzzati, *J. Phys. Chem.* **68**, 3504 (1964).

[k] L. Benjamin, *J. Phys. Chem.* **70**, 3790 (1966).

TABLE II
Nonionic Detergents

Class	Compound	$\bar{v}$	T	Solvent	Method	Comment
1. Polyoxyethylene alkylphenols[a]	$C_8\phi E_{9.7}{}^g$	0.908	25	W	DM	PSV[j]
	$C_8\phi E_{10.3}{}^g$	0.914	25	W	PY	PSV[k]
	$C_8\phi E_{20}$	0.874	25	N	PY	NED[l]
	$C_9\phi E_9{}^g$	0.91	—	N	PY	NED[m]
	$C_9\phi E_{9-10}{}^g$	0.922	25	W	DM	PSV[j]
	$C_9\phi E_{9.5}{}^g$	0.910	25	W	PY	PSV[k]
	$C_9\phi E_{13}{}^g$	0.91	—	N	PY	NED[m]
	$C_9\phi E_{15}$	0.972	25	W	PY	NED[n]
		0.965	25	S	PY	NED[n]
	$C_9\phi E_{50}$	0.957	25	W	PY	NED[n]
2. Polyoxyethylene alcohols[b]	C_4E_6	0.907	20	W	PY	PSV[o]
	C_6E_6	0.933	20	W	PY	PSV[o]
	C_8E_6	0.963	20	W	PY	PSV[o]
	$C_{12}E_6$	0.989	25	W	DM	PSV[p]
	$C_{12}E_8$	0.973	25	W	DM	PSV[p]
	$C_{12}E_{9.5}{}^h$	0.958	25	W	DM	PSV[p]
	$C_{12}E_{14}$	0.981	25	W	PY	NED[n]
	$C_{12}E_{28}$	0.962	26.3	N	NS	PSV[q]
	$C_{16}E_{10}{}^h$	0.955	25	W	DM	PSV[j]
	$C_{16}E_{20}{}^h$	0.919	25	W	DM	PSV[j]
	$C_{16,18}E_{16.5}{}^h$	0.929	25	W	DM	PSV[p]
	$C_{18}E_{10}{}^h$	0.973	25	W	DM	PSV[j]
	$C_{18}E_{14}$	0.985	25	W	PY	NED[n]
	$C_{18}E_{100}$	0.960	25	W	PY	NED[n]
3. Polyoxyethylene monoacyl sorbitans[c]	$C_{12}SorbE_{20}{}^i$	0.869	25	W	DM	PSV[j]
	$C_{18}SorbE_{20}{}^i$	0.896	25	W	DM	PSV[j]
	$C_{18}SorbE_x{}^i$	0.92	—	B	NS	ASV[r]
4. Dimethylalkyl amine oxides[d]	DC_8AO	1.035	30	W	MF	PSV[s]
	DC_9AO	1.095	30	W	MF	PSV[s]
	$DC_{10}AO$	1.106	30	W	MF	PSV[s]
	$DC_{11}AO$	1.119	30	W	MF	PSV[s]
	$DC_{12}AO$	1.112	30	W	MF	PSV[s]
		1.122	21	B	DM	PSV[t]
5. Dimethylalkyl phosphine oxides[e]	DC_8PO	1.088	30	W	MF	PSV[s]
	$DC_{10}PO$	1.118	30	W	MF	PSV[s]
	$DC_{12}PO$	1.190	30	W	MF	PSV[s]

TABLE II *(Continued)*

Class	Compound	$\bar{v}$	T	Solvent	Method	Comment
6. Alkylsulfinyl alkanols[f]	C_6SOC_2OH	0.952	25	W	CD	PSV[n]
	C_6SOC_3OH	0.963	25	W	CD	PSV[n]
	C_6SOC_4OH	0.974	25	W	CD	PSV[n]
	C_8SOC_2OH	0.991	25	W	CD	PSV[n]
	C_8SOC_3OH	0.998	25	W	CD	PSV[n]
	C_8SOC_4OH	1.004	25	W	CD	PSV[n]

[a] The symbols $C_8\phi E_n$ and $C_9\phi E_n$ indicate, respectively, C_8H_{17}—⟨ ⟩—$(OCH_2CH_2)_nOH$

and C_9H_{19}—⟨ ⟩—$(OCH_2CH_2)_nOH$. For detailed structural formula see C. R. Enyeart, *in* "Non-Ionic Surfactants" (M. J. Schick, ed.), p. 44. Dekker, New York, 1967. The value of n generally represents the average number of oxyethylene groups in a distribution that may vary from lot to lot.

[b] The symbol C_mE_n indicates the compound $CH_3(CH_2)_{m-1}(OCH_2CH_2)_nOH$. An integral value of n in this group generally represents a discrete number of polyoxyethylene groups; see specific references in this regard.

[c] The symbols $C_{12}SorbE_n$ and $C_{18}SorbE_n$ indicate, respectively, the monolaurate and monoleate esters of polyoxyethylene sorbitan, with n oxyethylene groups. (When n is not specified in the specific reference, x is used to indicate this.) For detailed structural formula, see F. R. Benson, *in* "Non-Ionic Surfactants" (M. J. Schick, ed.), p. 247. Dekker, New York, 1967.

[d] The symbol DC_nAO indicates the compound $CH_3(CH_2)_{n-1}N(CH_3)_2O$.

[e] The symbol DC_nPO indicates the compound $CH_3(CH_2)_{n-1}P(CH_3)_2O$.

[f] The symbol C_mSOC_nOH indicates the compound $CH_3(CH_2)_{m-1}S(=O)(CH_2)_nOH$.

[g] The following trade names are in common use: Triton X-100 ($C_8\phi E_{9.7}$), Ipegal CO-710 ($C_8\phi E_{10.3}$), Arkopal-9 ($C_9\phi E_9$), Triton N-101 ($C_9\phi E_{9-10}$), Surfonic N-95 ($C_9\phi E_{9.5}$), and Arkopal-13 ($C_9\phi E_{13}$).

[h] The following trade names are in common use: Lubrol-PX ($C_{12}E_{9.5}$), Brij-56 ($C_{16}E_{10}$), Brij-58 ($C_{16}E_{20}$), Lubrol-WX ($C_{16.18}E_{16.5}$), and Brij-96 ($C_{18}E_{10}$).

[i] The following trade names are in common use: Tween-20 ($C_{12}SorbE_{20}$), Tween-80 ($C_{18}SorbE_{20}$), and Emasol-4130 ($C_{18}SorbE_x$).

[j] C. Tanford, Y. Nozaki, J. A. Reynolds, and S. Makino, *Biochemistry* **13**, 2369 (1974).

[k] C. W. Dwiggins, Jr., R. J. Bolen, and H. N. Dunning, *J. Phys. Chem.* **64**, 1175 (1960).

[l] S. Ikeda and K. Kakiuchi, *J. Colloid Interface Sci.* **23**, 134 (1967).

[m] F. Husson, H. Mustacchi, and V. Luzzati, *Acta Crystallog.* **13**, 668 (1960).

[n] M. J. Schick, S. M. Atlas, and F. R. Eirich, *J. Phys. Chem.* **66**, 1326 (1962).

[o] A. T. Florence, *J. Pharm. Pharmacol.* **18**, 384 (1966).

[p] C. Tanford and J. A. Reynolds, *Biochim. Biophys. Acta* **457**, 133 (1976).

[q] H. Schott, *J. Colloid Interface Sci.* **24**, 193 (1967).

[r] B. Love and G. Cornick, *Chem. Phys. Lipids* **4**, 191 (1970).

[s] L. Benjamin, *J. Phys. Chem.* **70**, 3790 (1966).

[t] C. Sardet, A. Tardieu, and V. Luzzati, *J. Mol. Biol.* **105**, 383 (1976).

[u] J. M. Corkill, J. F. Goodman, and T. Walker, *Trans. Faraday Soc.* **63**, 768 (1967).

TABLE III
SODIUM SALTS OF BILE ACIDS

Class	Compound[a]	$\bar{v}$	T	Solvent	Method	Comment
1. Dihydroxy compounds	Deoxycholate	0.778	25	W	DM	PSV[b]
		0.76	22	S	PY	ASV[c]
	Glycodeoxycholate	0.77	22	W,S	PY	ASV[c]
	Taurodeoxycholate	0.76	22	S	PY	ASV[c]
	Chenodeoxycholate	0.78	22	S	PY	ASV[c]
	Glycochenodeoxycholate	0.77	22	S	PY	ASV[c]
	Taurochenodeoxycholate	0.76	22	S	PY	ASV[c]
	Fusidate	0.774	27.3	S	PY	ASV[d]
2. Trihydroxy compounds	Cholate	0.75	22	W,S	PY	ASV[c]
		0.771	25	B	DM	PSV[e]
	Glycocholate	0.75	22	W	PY	ASV[c]
	Taurocholate	0.75	22	S	PY	ASV[c]
3. Triketo compounds	Dehydrocholate	0.77	22	S	PY	ASV[c]

[a] Trivial names; chemical formulas are given by Small.[c]
[b] C. Tanford, Y. Nozaki, J. A. Reynolds, and S. Makino, *Biochemistry* **13**, 2369 (1974).
[c] D. M. Small, *Adv. Chem. Ser.* **84**, 31 (1968).
[d] M. C. Carey and D. M. Small, *J. Lipid Res.* **12**, 604 (1971).
[e] D. M. McCaslin, personal communication (1976).

homologous compound is known, by using Traube's rule of group additivity.[21] For instance, the addition of a methylene group to an alkyl chain increases the partial *molal* volume ($\bar{v}$ multiplied by the molecular weight) by 16.1 ml per mole. Partial molal volume increments for several groups and electrolytes have been published.[22–24]

Tabulation of Partial Specific Volumes of Amphiphiles

Table IV list values of partial specific volumes (ml/g) as measured or calculated for a wide variety of detergents or lipids at concentrations above their critical micelle concentrations. Values of apparent specific volumes, while not as precise as those of partial specific volumes, are listed in the expectation that an estimate of the $\bar{v}$ may be useful. In some

[21] J. Traube, *Samml. Chem. Chem. Tech. Vortr.* **4**, 255 (1899).
[22] H. Høiland and E. Vikingstad, *Acta Chem. Scand.* **A30**, 182 (1976).
[23] R. M. Noyes, *J. Am. Chem. Soc.* **86**, 971 (1964).
[24] F. J. Millero, *Chem. Rev.* **71**, 147 (1971).

TABLE IV

Lipids[a]

Class	Compound	$\bar{v}$	T	Solvent	Method	Comment
1. Monoacyl phospholipids	lyso C_{16}PC	0.949	20	G	SE	ASV[b]
	lyso EY PC	0.921	—	N	NS	NED[c]
		0.93	20	W	CT	—[d]
2. Diacyl phospholipids	EY PC	0.981	20	D	SV	PSV[e]
		0.988	20	S	SV	PSV[e]
	diC_6PC	0.865	25	B	DM	PSV[f]
	diC_7PC	0.888	25	B	DM	PSV[f]
	diC_{10}PC	0.927	25	B	CT	—[d]
	diC_{12}PC	0.945	25	B	CT	—[d]
	diC_{16}PC	0.976	25	B	CT	—[d]
	PG	1.015	20	W	CT	—[g]
	PE	0.965	20	W	CT	—[g]
	PS	0.93	20	W	CT	—[g]
	SM	1.005	20	W	CT	—[g]
3. Steroids	Cholesterol	0.949	20	G	SE	ASV[h]
	Digitonin	0.738	20	B	PY	ASV[i]
4. Acylcarnitines	C_{12}Carn	0.970	25	W	PY	PSV[j]
	C_{14}Carn	0.995	25	W	PY	PSV[j]
	C_{16}Carn	1.002	25	W	PY	PSV[j]
5. Monoglycerides	C_{10}Gly	0.93	25	W	PY	NED[k]
	C_{12}Gly	0.96	45	W	PY	NED[k]

[a] Symbols used are C_n:($CH_3(CH_2)_{n-1}$–), EY (egg yolk), PC (phosphatidylcholine), PG (phosphatidylglycerol), PE (phosphatidylethanolamine), PS (phosphatidylserine), SM (sphingomyelin), Carn (carnitine), Gly (glycerol). Lyso indicates one acyl chain and di indicates two acyl chains on a phospholipid.

[b] M. E. Haberland and J. A. Reynolds, *J. Biol Chem.* **250,** 6636 (1975).

[c] J. H. Perrin and L. Saunders, *Biochim. Biophys. Acta* **84,** 216 (1964).

[d] C. Tanford and J. A. Reynolds, *Biochim. Biophys. Acta* **457,** 133 (1976).

[e] C. Huang and J. P. Charlton, *J. Biol. Chem.* **246,** 2555 (1971).

[f] R. J. M. Tausk, J. van Esch, J. Karmiggelt, G. Voordouw, and J. Th. G. Overbeek, *Biophys. Chem.* **1,** 184 (1974).

[g] C. Tanford, Y. Nozaki, J. A. Reynolds, and S. Makino, *Biochemistry* **13,** 2369 (1974).

[h] M. E. Haberland and J. A. Reynolds, *Proc. Natl. Acad. Sci. U.S.A.* **70,** 2313 (1975).

[i] R. Hubbard, *J. Gen. Physiol.* **37,** 381 (1954).

[j] S. H. Yalkowsky and G. Zografi, *J. Colloid Interface Sci.* **34,** 525 (1970).

[k] F. Reiss-Husson, *J. Mol. Biol.* **25,** 363 (1967).

[l] While this manuscript was in press, the partial specific volume of dimyristoyl phosphatidylcholine (0.972 at 28°, 0.963 at 23° by SV in D_2O or by DM in buffer) was reported by K. C. Aune, J. G. Gallagher, A. M. Gotto, Jr., and J. D. Morriett, *Biochemistry* **16,** 2151 (1977). Also, the partial specific volumes of egg yolk phosphatidylcholine (0.984) and beef brain sphingomyelin (0.966) at 20° by DM in potassium chloride solution were given by Y. Barenholz, D. Gibbes, B. J. Litman, J. Goll, T. E. Thompson, and F. D. Carlson, *Biochemistry* **16,** 2806 (1977).

cases the measurement of a true $\bar{v}$ is not technically feasible. For brevity the following symbols are used:

Solvent:	B	buffer
	D	H_2O–D_2O mixture
	G	sucrose density gradient
	N	not specified (presumed water)
	S	NaCl solution, usually 0.1 M
	W	H_2O
Method:	CD	Cartesian diver
	CT	calculated using Traube's rule[25]
	DL	dilatometer
	DM	mechanical oscillator densimeter
	MF	magnetic float
	NS	not specified
	PY	pycnometer
	SE	sedimentation equilibrium
	SV	sedimentation velocity
Comment:	ASV	apparent specific volume[26]
	NED	no experimental details
	PSV	partial specific volume[27]

[25] Solvent and temperature listed are the same as for the homologous compound on which the calculation was based.

[26] Apparent specific volume measured at only one or a few concentrations.

[27] Apparent specific volume constant at several concentrations, or limiting slope at critical micelle concentration of density vs concentration used.

[3] Partial Specific Volume Measurements Using the Glass Diver

By MARGARET J. HUNTER

While some type of magnetic balancing system such as that developed by Beams et al.[1,2] currently offers the most rapid and accurate method for determining the density of small volumes (ca. 0.3 ml) of solutions at various temperatures and pressures, the required equipment is expensive and requires considerable expertise to maintain.

[1] J. W. Beams, C. W. Hulbert, W. E. Lotz, Jr., and R. Montague, Jr., *Rev. Sci. Instrum.* **26,** 1181 (1955).

[2] D. V. Ulrich, D. W. Kupke, and J. W. Beams, *Proc. Natl. Acad. Sci. U.S.A.* **52,** 349 (1964).

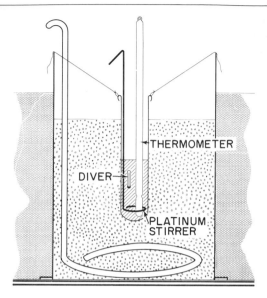

FIG. 1. Apparatus for density determinations. Reproduced from *J. Phys. Chem.* **70**, 3285 (1966); **71**, 3717 (1967) by permission from the American Chemical Society.

An alternative procedure, the flotation or isopycnic temperature method,[3] depends on the accurate determination of the temperature at which a calibrated small glass diver or float has the same density as (i.e., is isopycnic with) the solution in which it is immersed. Such a system has been applied to the determination of the specific volume of proteins in dilute solution[4,5] (0.5–4.5 g/100 g of solution). The sample volume was 3.0 ml. The apparatus required for these measurements is relatively inexpensive and simple to construct and maintain. The only major piece of equipment required is a system that will permit accurate temperature control over the range of 18°–35°.

Apparatus

The apparatus is shown in Fig. 1. The measurements were performed in a constant temperature bath that could be controlled to ±0.001°. A graduated 1-liter glass cylinder was placed in the bath and filled with distilled water until the level in the cylinder was 2–3 cm lower than the

[3] N. Bauer and S. Z. Lewin, *in* "Techniques of Chemistry" (A. Weissberger, ed.), incorporating 4th completely revised and augmented edition of "Techniques of Organic Chemistry," Volume 1, Part IV, p. 57. Interscience Publishers, Inc., New York, 1972.
[4] M. J. Hunter, *J. Phys. Chem.* **70**, 3285 (1966).
[5] M. J. Hunter, *J. Phys. Chem.* **71**, 3717 (1967).

water in the bath. A large glass stirrer was placed in the cylinder. Three glass hooks were sealed to the top of a Pyrex test tube (125 × 15 mm), and the test tube was suspended in the water in the cylinder by three fine nylon lines that were looped over the glass hooks and held to the cylinder by a heavy-duty rubber band, which had been placed around the top of the cylinder. The test tube could thus be lowered, raised, or straightened by manipulation of the nylon lines. Three milliliters of solution were placed in the test tube. A platinum stirrer and a small glass diver were placed in the solution and a precision thermometer (0–50°, graduated to 0.05°, and previously standardized against a calibrated calorimeter thermometer) was suspended in the solution so that the thermometer bulb and about 1 cm of the stem were immersed in the solution. The final level of the solution in the test tube was about 2 cm lower than that of the water in the cylinder.

The glass divers or floats were prepared by heat-sealing the ends of glass melting-point capillaries in such a manner that one end contained much more glass than the other. Such divers floated in a vertical position. The divers were about 1 cm long and weighed approximately 0.03 g. At temperatures below the isopycnic temperature, the diver rose; at temperatures above the isopycnic temperature, it sank.

Calibration of Divers

The divers were calibrated by determining their isopycnic temperatures (see below) in various standard KCl solutions of known density.[6,7] In general, the isopycnic temperature varies directly with the solute concentration and the density–temperature relationship can be expressed by an equation of the form

$$\rho = A - Bt$$

where ρ = density of the diver and t = temperature. A and B are constants.

A typical calibration curve is shown in Fig. 2. When a least-squares line was fitted to the experimental points, none of the experimental points deviated from the line by more than 5×10^{-6} g/ml in density and the variation appeared to be random. Sucrose density values determined by this method gave a density at 20° that agreed well with the accurate density data that are available for sucrose at this temperature.[2,8]

[6] T. W. Richards and J. W. Shipley, *J. Am. Chem. Soc.* **36**, 1 (1914).

[7] M. Randall and B. Longtin, *Ind. Eng. Chem., Anal. Ed.* **11**, 44 (1939).

[8] F. Plato, *Wiss. Abh.* **2**, 153 (1900); quoted by F. J. Bates and associates, *Natl. Bur. Stand.*, (U.S.), Washington, D.C., 1942.

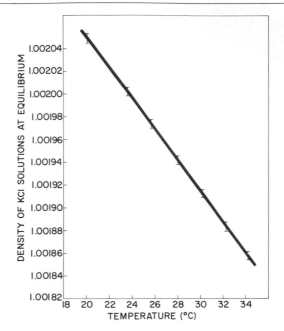

FIG. 2. Density of a diver at various temperatures. Reproduced from *J. Phys. Chem.* **70**, 3285 (1966); **71**, (1967) by permission from the American Chemical Society.

Method for the Determination of the Isopycnic Temperature

A diver was placed in the solution in the test tube and allowed to equilibrate, with occasional stirring, for 10 min. The temperature of the bath was then adjusted until the diver rose or sank very slowly, indicating that the temperature was close to the isopycnic temperature. The system was equilibrated with stirring at this temperature for 20 min.

After this time, the temperature of the bath was raised or lowered in increments of 0.01° until the isopycnic temperature had been passed and the direction of movement of the diver had been reversed. At least five observations were made at each temperature near the isopycnic temperature. The density of the solution was then obtained from the density–temperature correlation equation for that particular diver.

Discussion

The divers were sufficiently sensitive to change of 0.01° in temperature that the isopycnic temperature could be determined routinely with a reproducibility of ±0.01°. In aqueous solution this degree of accuracy corresponds to a change in density of about 2×10^{-6} g/ml.

Such reproducibility was not obtained outside the temperature range of 19°–35°. At temperatures below 19° the viscosity of the solution became great enough that the divers were very sluggish in their movements and did not respond adequately to small temperature changes. Above 35° it was very difficult to maintain a constant temperature in the test tube owing to the relatively large difference between the bath temperature and room temperature.

While the accuracy of the *solution* density data obtained by the determination of diver-solution isopycnic temperatures is more than adequate for most types of investigation, the method has one inherent disadvantage. For a particular diver, only one concentration of solute will give any particular isopycnic temperature. *Solute* density or specific volume data at various temperatures are thus obtained from solutions of different solute concentration. If the specific volume of the solute is concentration dependent, a series of apparent specific volumes (ϕ_2) will be obtained, and, unless solute specific volume–concentration correlation data as a function of temperature are available, the partial specific volume ($\bar{v}_2$) of the solute cannot be obtained. The necessary information can be supplied, however, if several divers of different density are used to determine the solute specific volume: for example, if four divers of different density are used, the specific volume of the solute at four solute concentrations can be determined at any particular temperature.

Since many solutes, including proteins, exhibit little concentration dependence in specific volume in dilute solution,[2,5,9,10] this intrinsic disadvantage of the method presents no real problem in the determination of protein partial specific volumes.

When the densities of bovine serum albumin solutions of different protein concentration were determined by measurement of their isopycnic temperatures with three divers of different density, three solution densities at 20° were obtained, one for each diver.[5] The concentration of albumin in the solutions varied from 0.5 to 4.5 g/100 g of solution, and the divers had densities of 0.999815, 1.001986, and 1.009866 at 20°. When the specific volume of the albumin was calculated from the densities at 20°, three values for the specific volume were obtained; 0.7346_8, 0.7346_8, and 0.7347_0. No significant variation in the apparent specific volume, ϕ_2, was thus observed, in agreement with other observations,[2,9,10] and the values represent the true partial specific volume ($\bar{v}_2$) of the protein. Since this is so, the temperature coefficient $(d\bar{v}_2/dt)$ of the partial specific volume can readily be obtained by measurement of the isopycnic temperatures of

[9] M. O. Dayhoff, G. E. Perlmann, and D. A. MacInnes, *J. Am. Chem. Soc.* **74**, 2515 (1952).
[10] P. A. Charlwood, *J. Am. Chem. Soc.* **79**, 776 (1957).

solutions of different protein concentration. Values of 3.66×10^{-4} ml/g deg and 3.33×10^{-4} ml/g deg were obtained for bovine plasma albumin and transferrin, respectively.

The reliability of the density measurements is adequate for the determination of $\bar{v}$ to within ± 0.003. However, since the accuracy of $\bar{v}$ is also dependent on the reliability of dry weight measurements, the actual error may be somewhat greater than this. The partial specific volumes of five

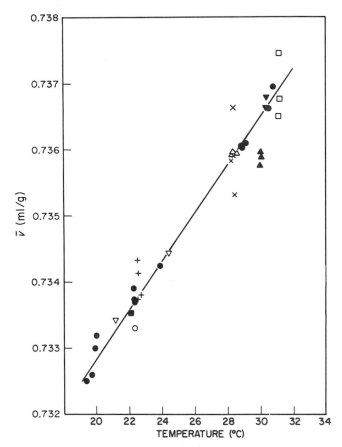

FIG. 3. Partial specific volume of bovine plasma albumin at various protein and KCl concentrations. Three divers were used in these determinations. ●, 0.50–4.50 g of albumin/100 g of solution; no KCl. ○, 0.20 g of KCl/100 g of H_2O; 4.00 g of albumin/100 g of solution. ▲, 0.56 g of KCl/100 g of H_2O; 1.00 g of albumin/100 g of solution. □, 0.79 g of KCl/100 g of H_2O; 0.58 g of albumin/100 g of solution. ▽, 1.68 g of KCl/100 g of H_2O; 0.67 g of albumin/100 g of solution. +, 0.19 g of KCl/100 g of H_2O; 0.34 g of albumin/100 g of solution. △, 0.26 g of KCl/100 g of H_2O; 1.50 g of albumin/100 g of solution. ×, 0.77 g of KCl/100 g of H_2O; 0.34 g of albumin/100 g of solution. ■, 0.80 g of KCl/100 g of H_2O; 2.60 g of albumin/100 g of solution. ▼, 1.67 g of KCl/100 g of H_2O; 1.40 g of albumin/100 g of solution. Reproduced from *J. Phys. Chem.* **70**, 3285 (1966); **71**, 3717 (1967) by permission from the American Chemical Society.

proteins, human and bovine serum albumin,[4,5] transferrin,[5] caeruloplasmin,[11] and ferredoxin[12] have been determined by this method.

The density measurements can equally well be performed in the presence of various small molecules, such as KCl (Fig. 3), provided adequate solute temperature–density–concentration correlation data are available.

The volume of solution required for these measurements could be greatly reduced if a small thermister were used to measure temperature and if the divers were slightly shorter. The 3 Figures are reproduced from J. Phys. Chem. 70, 3285, (1966); 71, 3717 (1967) by permission from the American Chemical Society.

[11] A. G. Morell, C. J. A. Van Den Hamer, and I. H. Scheinberg, J. Biol. Chem. 244, 3494 (1969).

[12] D. H. Petering, Ph.D. Thesis, University of Michigan, Ann Arbor, Michigan, 1969.

[4] Magnetic Suspension: Density-Volume, Viscosity, and Osmotic Pressure

By D. W. KUPKE and T. H. CROUCH

Improvements and additional applications in magnetic balancing for the study of proteins have appeared since this physical principle was last treated in this series.[1] In addition to density applications, the magnetic approach is now used also for viscosity and osmotic pressure. Viscosities of high accuracy can be obtained on a routine basis concurrently with the determination of the density on the same sample (~ 0.2 ml). Moreover, the viscosity method is enlarged by this approach because the change in viscosity with time is conveniently obtained and unusually small shearing stresses ($\sim 10^{-3}$ dyne, cm^{-2}) are applied without sacrifice of speed. The former attribute may be useful for the study of elongating proteins and of conformational changes, and the latter is applicable to systems that are sensitive to shearing. A newer application of magnetic balancing, while less developed, is in the measurement of very small osmotic pressures (to $\sim 10^{-2}$ cm H_2O). Osmotic pressure is theoretically the simplest and least ambiguous means for getting out the chemical potentials of macromolecular solutions; however, the sensitivity of the measurements has not been improved appreciably for routine use with the proteins since the inception of the method. This limitation in sensitivity still stands at $\sim 10^{-4}$ M in macromolecule concentration when results of high precision are required. Hence, conventional osmometry misses many of the systems of interest to enzymologists today. By the magnetic suspension approach, results at the 10^{-6} M range have been obtained with satisfactory precision. As currently practiced, this development is restricted to low-pressure studies (<10 cm H_2O or $<4 \times 10^{-4}$ M protein) and is not competitive with existing methods covering the higher pressure range (1 to 10^2 cm H_2O).

[1] D. W. Kupke and J. W. Beams, this series, Vol. 26, p. 74.

The chapter is divided into three parts, reflecting the three properties of protein solutions now being investigated with the magnetic suspension devices. Because the instrumentation for these techniques is not available in terms of self-contained units, identification of the components and descriptions for the assembly and use of the simplest designs are included.

Part I: Density-Volume

The theory and measurement aspects of the magnetic method for the rapid determination of the density of protein solutions are described in a previous volume of this series.[1] The applications, which were also discussed, considered the determination of the partial and the isopotential specific volumes, preferential interaction, composition analysis, volume change, and titration. Except for a general orientation to the magnetic approach, this section on density is restricted to newer developments in the volume change application and to instrumental aspects. The latter includes a brief description of the essential components for constructing the basic densimeter because a packaged instrument is not available commercially.

Survey of the Magnetic Method

The magnetic approach to density determinations, being amenable to a variety of purposes, has continued to receive development; a current list of the various types of densimeters in use is available.[2] For this orientation, the simplest kind of a magnetic densimeter may be envisioned (Fig. 1). Basically, a small (10–15 mg) magnetically soft, ferromagnetic cylinder (2 × 1 mm), suitably jacketed in glass, Kel-F, or polypropylene, is balanced at a reproducible height ($\pm 1 \mu$m) within a solution via a precisely controlled electromagnet. If the jacketed object (called a buoy or float) is to be more dense than the buoyant medium, the magnetic force emanates from above in order to oppose the force of gravity. The magnetic force in this case adds to that of buoyancy (Archimedes principle). If the buoys are to be less dense than the liquids, the electromagnet is placed below the liquid (as in Fig. 1), and the force adds to that of gravity (this design leaves the top of the cell containing the solution more accessible for biochemical experiments). In either design, the buoyant force generated by the liquid (or gaseous) medium is measured accurately in terms of the electric current generating the field that is required to balance the buoy at a given vertical distance from the electromagnet. At a reproducible height within the solution, the magnetic force, $M(dH/dz)$, is related to the difference in density, $(\rho_{(B)} - \rho)$, between buoy, B, and the solution by

$$M(dH/dz) = V_{(B)}g(\rho_{(B)} - \rho) \qquad (1)$$

[2] W. M. Haynes, *Rev. Sci. Instrum.* **48**, 39 (1977).

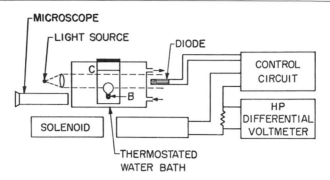

FIG. 1. Block diagram of the basic magnetic densimeter. B is the magnetically suspended buoy (or float); C is the glass cell for containing the liquid sample. From J. P. Senter, *Rev. Sci. Instrum.* **40,** 334 (1969).

where M is the magnetic moment of the induced magnet, H the intensity of the field along the vertical distance, z, on the magnetic axis, $V_{(B)}$ is the volume of the buoy, and g is the acceleration of gravity. [With liquids, the units of density are often expressed in grams per milliliter (g/ml) because the latter volume is conveniently defined in terms of a weight of pure water rather than in terms of a unit of length (e.g., cm³) which is difficult to maintain locally with the necessary accuracy.] The magnetic force according to Eq. (1) is positive when the buoy is lifted and negative when pulling the buoy down. The buoy is maintained in balance without visible fluctuation when observed through a telemicroscope by use of a servo-controlled circuitry acting in concert with the particular height-sensing device being utilized. By the use of solutions of known density, an instrument can be calibrated at a particular temperature so that the density of unknown solutions is obtained with high accuracy[3]; normal fluctuations in the atmospheric pressure limits the attainable precision to slightly better

[3] For the study of proteins in aqueous solutions ($\rho > 1$) in the room-temperature range, calibration solutions of gravimetrically prepared mixtures of sucrose and water are most often used. The densities of this system are available at weight fraction increments as small as 0.001 at some temperatures and are good (as an assigned standard) to $\sim 10^{-6}$ g/ml with smoothing: e.g., F. Plato, *Kaiserlichen Normal-Eichungs-Kommission, Wiss. Abh.* **2,** 153 (1900); quoted by F. J. Bates, *Natl. Bur. Stand. (U.S.) Circ.* **440;** p. 626 ff. (1942); cf. K. Scheel *in* "Landolt-Börnstein, Physikalish-Chemische Tabellen" (R. Börnstein and W. Meyerhoffer, eds.), p. 37, Springer-Verlag, Berlin and New York, 1905.

The authors have been using heat-dried, "Suprapur" cesium chloride (E. Merck, Darmstadt) as secondary standards in the higher density regions, where sucrose solutions become very viscous and require much more time and care in calibrating a buoy. These salt solutions are related to the sucrose density-composition tables of Plato because the available density data on cesium chloride on an absolute basis are inadequate in the higher density regions. An absolute table of sufficient scope and accuracy of aqueous cesium chloride would be very welcome; this system is much more ideal for density standards than the sucrose–water system.[1]

than 1×10^{-6} g/ml. The operator simply measures the current (across a standard resistance) required to balance the buoy at a prescribed height with respect to the cross hairs of the microscope. No other quantities, such as the volume and mass of the buoy need be determined. The total magnetic force in Eq. (1) is essentially proportional to the square of the current (in the simpler designs), hence, a parabolic fit of the current (or voltage) to the known densities of the calibration solutions suffices. (With certain other designs, the current is linear with the density; this is an advantage as $(\rho_{(B)} - \rho)$ increases.[1])

Since the volume of the sample solution is small (~ 0.25 ml), temperature equilibration is sufficiently rapid so that a measurement requires 5 min or less, including the insertion and removal of the liquid. With the simpler designs, a given calibrated buoy covers a density range of from 0.03 to 0.05 g/ml with good precision (from ± 1 to $\pm 2 \times 10^{-6}$ g/ml); thus, a given buoy may cover a span of protein concentration on the order of 100 mg/ml. [This range can be broadened substantially if the electromagnet is cooled for higher currents or if Helmholtz-like coils are used (see Part II).] If the temperature is maintained constant only to within 0.01°, the density can be determined on aqueous solutions with a precision of $\pm 2 \times 10^{-6}$ g/ml in the room-temperature range. Repetitive determinations on samples of a given solution have amply demonstrated this level of precision. Ordinarily, however, the precision of the weighings in preparing compound solutions limits the overall repeatability with separate solutions to $\pm 5 \times 10^{-6}$ g/ml unless great care is exercised.

As is evident, the magnetic method can be used for following the change in density (or volume) while a reaction is in progress by observing the change in voltage with time on a strip-chart recorder. In addition, the method is suitable for determining end points with incremental additions of a reagent (partial volume titrations) and it is convenient for temperature studies owing to the small volumes employed. The method has been utilized for following the change in density as a function of pressure (isothermal compressibilities).[4] A recently developed protocol for determining accurately the continuous volume changes by density on protein–small molecule interactions is described below.

Volume Change by Density

With the advent of rapid and sufficiently accurate density instruments, such as the magnetic densimeters, the determination of the change in volume, ΔV, of mixing and of transporting the protein and other reactants from one medium to another has become practical. For evaluating the

[4] P. F. Fahey, D. W. Kupke, and J. W. Beams, *Proc. Natl. Acad. Sci. U.S.A.* **63**, 548 (1969).

change in volume on mixing two solutions, A and B, of known densities, $\rho_{(A)}$ and $\rho_{(B)}$, weighed amounts, $m_{(A)}$ and $m_{(B)}$, of each solution are added together on the analytical balance to give $m_{(AB)}$; the density of the mixture, $\rho_{(AB)}$, is then determined, and ΔV is calculated by

$$\Delta V = m_{(AB)}/\rho_{(AB)} - [m_{(A)}/\rho_{(A)} + m_{(B)}/\rho_{(B)}] \tag{2}$$

where the first term on the right is the final volume and the term in brackets is the volume before mixing. With the magnetic densimeter, ~ 1 g of the mixture suffices for a triplicate determination. The masses and densities should be determined to better than 1 part in 10^5 so that ΔV is good to approximately 10^{-2} μl. If the densities of A and B differ appreciably (>0.05 g/ml), a correction for the difference in buoyancy in air of the 2 solutions may be significant (an average air density of 1.2 mg/ml at room temperatures is usually a sufficient value for this purpose). One means for reducing evaporation error during the weighings is to push a tightly fitting silicone stopper, with a small axial drill hole, into the neck of the weighing vessel such that a small-gauge needle on a gas-tight syringe can barely pass through the hole freely; a cap is placed on the vessel (or a plug in the hole) during the weighings following each addition.

This simple exercise may be particularly useful where ultracentrifugal analyses are used to study the equilibria in the case of association reactions, such as when protein or subunit A is mixed with protein or subunit B in the same solvent medium. In these cases the ΔV of mixing gives immediately the change in volume of the reaction owing to the association reaction; hence, the sign and magnitude of the volume change can be used to evaluate the effect on the equilibria by the pressure gradient in the ultracentrifuge cell. For this purpose, a sufficient concentration of protein is employed so that the association is virtually complete and the value of ΔV per gram of reactants is constant. A current example may illustrate what may be expected by this method (unpublished results by the authors). The 2 fragments of bovine serum albumin (BSA) of about equal size, called A and B, arising from peptic digestion,[5] are known to recombine in such a way that the enzymic activity of BSA in decomposing a Meisenheimer complex is restored.[6] Nearly equal volumes (~ 0.75 ml) of each purified fragment ($\sim 1.5\%$) in the same buffer medium were added one to another on the analytical balance with a weighing uncertainty of ± 5 μg. The densities, in triplicate, before and after mixing were evaluated to $\sim \pm 2 \times 10^{-6}$ g/ml. A volume decrease via Eq. (2) of -0.025 to -0.028 μl per 0.34 μmol was observed in repetitive experiments (figuring to about -80 ml per mole of the reassociated BSA complex). Thus, a measure of

[5] R. C. Feldhoff and T. Peters, Jr., *Biochemistry* **14**, 4508 (1975).
[6] R. P. Taylor and A. Silver, *J. Am. Chem. Soc.* **98**, 4650 (1976).

confidence to within a few nanoliters on very small volume changes may be achieved with practice in accurate weighings; the operation of the magnetic densimeter requires no special expertise, and the time involved for the triplicate density determinations on each sample of the experiment was less than 20 min.

When solvents are of different composition, the ΔV of mixing also includes contributions from mixing the solvent components. A typical example is the evaluation of the volume change upon transporting a protein from a native solvent to one containing a protein perturbant, such as a denaturant. In this case the densities of the 2 solvents containing no protein are also required. The difference in density, $\rho_{(A)} - \rho^0_{(A)} = \Delta\rho_{(A)}$, between protein solution A and solvent A, respectively, yields $\phi_{2(A)}$, the apparent specific volume of the protein (labeled component 2). Similarly, the density difference between protein solution B and solvent B (containing the perturbant), yields $\phi_{2(B)}$. These values for ϕ are calculated via the general expression

$$\phi_i = 1/\rho^0[1 - (\Delta\rho/c_i)_m]_{i \neq 1} \qquad (3)$$

where m refers to constant molality of all components except the protein (or other component i being varied which is not the major solvent component, such as water or buffer solution, labeled component 1). The concentration of the protein, c_2 (in grams per milliliter), in each calculation must be known and the density as a function of c_2 should be checked for linearity if this has not been established previously; [usually $(d\rho/dc_2)_m$ is constant for macromolecules at the present levels of precision exercised, but almost never so with small molecules.] The difference between $\phi_{2(B)}$ and $\phi_{2(A)}$ $(= \Delta\phi_2)$ gives immediately the change in volume per gram of protein, $\Delta V/g_2$, or the volume change upon transporting the protein to the perturbant medium independent of the other interactions. [The basis for this result lies in the definition of the apparent specific volume, which is the difference in total volume attending the addition of 1 g of a component $i \neq 1$. Thus, the volume of a solution, V, is defined in these terms by

$$V = \phi_i g_i + \sum_{j \neq i} \bar{v}_j^0 g_j \qquad (4)$$

where the second term on the right equals V^0, the volume of the solvent medium containing all components j. Note that V^0 is arbitrarily held constant (as if the solvent medium is ideal), therefore, the partial specific volumes, $\bar{v}$, of all components j retain the same values they had before any of component i was added. If $(d\phi_2/dg_2)_m$ is zero (i.e., ρ is linear with c_2), the apparent specific volume of the protein is equivalent to the partial specific volume $\bar{v}_2{}^0$ at vanishing concentrations since

$$\bar{v}_i = \phi_i + g_i \left(\frac{d\phi_i}{dg_i}\right)_m \qquad (5)$$

which is obtained by differentiating Eq. (4) with respect to g_i and holding the second term on the right constant as required by the definition of ϕ.[cf.7]

It may be instructive also to determine the change in volume upon adding the perturbant or other reactant (called component 3 for this purpose). In this exercise the apparent specific volume of component 3 is determined both in the presence and absence of the protein component. The difference, $\Delta\phi_3$, represents the increase or decrease in volume per gram of this reactant when transported from solvent to solvent + protein, and is independent of the other contributions to the volume. Equation (3) is used again for calculating each ϕ_3, where ρ^0 is the density of the appropriate solvent containing no component 3 in each case (e.g., water and water + protein). This approach may be useful as a titration procedure where the change in volume with the number of moles of a coenzyme, allosteric effector, metal ion, etc., is desired.

The power of the density method is more apparent by observing trends in ΔV and in $\bar{v}$ with composition. These trends are derived from smoothed curves generated from density-composition tables containing sufficient data entries to justify a satisfactory level of confidence over regions where transitions occur. Density-composition tables can be prepared over a broad range in the weight fraction of a component with a comparatively small investment of time and material. With the magnetic densimeter, about 1 ml of solution per triplicate determination allows for a rinsing between samples of a series (where $\Delta\rho < 10^{-3}$ g/ml). It can be shown that with only 2 density-composition tables, it is possible to extract curves depicting the changes in ΔV of mixing, the ΔV of transporting the protein and the ΔV of transporting the protein reactant as a function of the reactant or perturbant concentration. In addition, the same data can be used to display the variation in $\bar{v}$ of the reactant and of the solvating component (such as water) when the protein component is either present or absent; the differences, $\Delta\bar{v}_3$ and $\Delta\bar{v}_1$, resulting from the introduction of the protein are obtained over the entire composition range of interest.

The procedure for determining the variation in the various ΔV and $\bar{v}$ with concentration of a protein-perturbing substance for the case of a 3-component system is described. Two density-composition tables are prepared containing a suitable number of data entries to yield a smoothed

[7] D. W. Kupke, in "Physical Principles and Techniques of Protein Chemistry" (S. J. Leach, ed.), Part C, p. 1. Academic Press, New York, 1973.

curve of each via curve-fitting procedures. For the first table, the densities are determined on a series of gravimetrically prepared compositions comprising only components 1 and 3 (i.e., water or suitable solvating medium and the perturbant compound, respectively). This table, at the desired temperature, may be already available in the literature. The smoothed curve derived from this set of data is designated as curve 1. For the second table, the densities are determined on a series of compositions in which the protein (component 2) coexists with component 1 at a fixed mass ratio (g_2/g_1) or fixed weight fraction, W_2° $[= g_2/(g_1 + g_2)]$, as the proportion of component 3 is varied. The smoothed curve obtained from this set is designated as curve 2. The latter solutions are prepared gravimetrically by adding the stock solution containing only protein and component 1 of known composition to a preweighed amount of component 3. (If the latter component is very volatile, it may be more accurate to reverse these additions when the amount of that component occupies only a small fraction of the volume in the weighing vessel; the volume of this vessel should not greatly exceed that of the total components.) In the usual case, an appropriate volume of the stock solution is added by means of a gas-tight syringe through the axial hole into the vessel, such that the estimated weight based on the amount of component 3 (g_3) already in the vessel corresponds to the desired composition in terms of the ratio g_3/g_1. The proper weight of protein stock solution, m_p, at a given weight fraction, W_2°, is estimated for the desired ratio of g_3/g_1 by

$$g_3/g_1 = \frac{g_3}{m_p(1 - W_2^0)} \tag{6}$$

For the final composition, the observed weights are corrected for the difference in buoyancy in air since the density of the pure component 3 is frequently much different than that of the solvent or the stock solution; the buoyancy correction becomes more important as the weight fraction, $W_3 = [g_3/(g_1 + g_2 + g_3)]$, approaches 0.5. The value of W_2^0 of the stock solution is best obtained by dry weight analysis (see this volume [6]).

The data from the two density-composition tables are next subjected to polynomial fittings and error analysis. These curves may fit the form

$$\rho = \sum_{i=0}^{N} a_i W_3^i \tag{7}$$

where the number N of the coefficients a_i of W_3^i is chosen in each curve such that the variance is at the first minimum with respect to N. Usually, the standard error is smaller for the solvent series (curve 1) and fewer data points are required than for curve 2 because the variation is frequently more monotonic in the former. Interesting detail in the trends of ΔV and $\bar{v}$ have been observed if the standard error is maintained to within $\sim \pm 2 \times 10^{-5}$ g/ml. This level of precision can be exceeded with normal care. In the authors' experience, the above standard error was achieved even when several stock solutions, representing different preparations of a protein, were utilized in covering a large range in concentration of component 3—each stock solution being assayed for the value of W_2^0 by independent dry-weight determinations during a 2-year period. In cases where more than 1 stock solution is required for such a study, the concentrations, W_2^0, may be normalized to an assigned standard value, $W_2^{0(s)}$, on the assumption that the density at any molality of component 3 is essentially a linear function of c_2. Although such linearity seems to be the general observation with proteins, the assumption should be tested at selected concentrations of component 3 covering the span of values for c_2 actually used in the study. The standard density, $\rho^{(s)}$, is calculated from the experimentally determined density, ρ, the weight fraction, $W_2[= g_2/(g_1 + g_2 + g_3)]$, and the assigned or standard weight fraction, $W_2^{(s)}$ by

$$\rho^{(s)} = \rho^0[1 - (1 - \rho^0/\rho)W_2^{(s)}/W_2]^{-1} \tag{8}$$

where ρ^0 is the density taken from the identical value of g_3/g_1 in curve 1 containing no protein. The standard weight fraction quantity is given by

$$W_2^{(s)} = (g_2^{(s)}/g_1)(1 + g_3/g_1 + g_2^{(s)}/g_1)^{-1} \tag{9}$$

where $g_2^{(s)}$ is the mass of protein arbitrarily assigned to each gram of component 1. The volume differences and partial volumes that can be extracted from these 2 data curves are described below:

ΔV of Mixing

On occasions the total volume change of mixing different proportions of 2 compositions, A and B, along either curve 1 or curve 2 may be useful. The densities, $\rho_{(A)}$ and $\rho_{(B)}$, are calculated for the particular compositions by Eq. (7) and the appropriate coefficients a_i of W_3^i for insertion into Eq. (2) along with the chosen masses, $m_{(A)}$ and $m_{(B)}$. The value of $\rho_{(AB)}$ for the desired mixture is calculated similarly after ascertaining the weight fraction, $W_{3(AB)}$ of component 3 in the mixture by

$$W_{3(AB)} = \frac{W_{3(A)} m_{(A)} + W_{3(B)} m_{(B)}}{m_{(A)} + m_{(B)}} \tag{10}$$

If the effect of a perturbant on the protein system is essentially reversible, any number of solutions in any proportion along a regression curve, as derived from density-composition table No. 2, may be evaluated. The generally small differences in c_2 owing to the varying molalities of component 3 may be safely ignored because any nonlinearity in density versus c_2 would probably have only a minuscule effect. If $(d\rho/dc_2)_m$ is constant at any W_3, as is usually the case with proteins, such ΔV can be evaluated for any concentration of protein by generating curves at assigned values of $W_2^{0(s)}$ (i.e., hypothetical stock solutions) by Eqs. (8) and (10) after converting $W_2^{0(s)}$ to the weight fraction $W_2^{(s)}$ [Eq. (9)] in the mixture. It follows also, that solutions from curve 1 ($W_2^0 = 0$) may then be mixed with any solutions from curve 2 without generating additional curves.

ΔV for Transport of Protein

The volume change, $\Delta\phi_2$, attending the transport of a protein from one solvent medium to that of another, wherein other contributions to ΔV are eliminated, is usually ascribed to a change in conformation. With regression curves available, such as curves 1 and 2 discussed above, the values of ϕ_2 are readily generated as a smooth curve over the entire concentration range of interest in component 3. The value of ϕ_2 at each common value of g_3/g_1 (or W_3^0) in the 2 curves is calculated with Eq. (3) after converting W_2 to c_2 by the general expression $c_i = \rho W_i$; ρ^0 in each calculation is the density from curve 1 corresponding to the value of g_3/g_1 in the protein solution of curve 2. Thus, the difference, $\Delta\phi_2$, between any pair of compositions along curve 2 is immediately available. The differences in ϕ_2 between zero and specified concentrations of component 3 are the ones usually discussed. Here, in principle, the mass of a protein in the stock solution is being transported from a gram of water or solvating medium to another gram of the solvent, which contains also a specified amount of component 3. The values of these $\Delta\phi_2$ are apparent from the example shown in Fig. 2 wherein ϕ_2 is described as a smooth function of component 3 concentration. In this case isoionic ribonuclease A in water was studied between 0 and 8 M guanidinium chloride (denoted as M_3). For this projection the concentration of component 3 is also presented in terms of W_3^0, the weight fraction of the guanidinium chloride with respect to water exclusive of the amount of protein present [i.e., $W_3^0 = g_3/(g_1 + g_3)$]. The latter concentration variable serves the purpose of relating all volume difference and partial volume data in terms of a common molality of component 3. [In this example, the maximum and minimum values from crest to shallow trough between ~2 and 4 M guanidinium chloride in Fig. 2 correspond to the transition region of a conformational change assigned by other investigators using other kinds of probes. The increasingly positive values of ϕ_2, above 4 M where the protein is maximally denatured in this medium, are more compatible with the notion of the protein salting

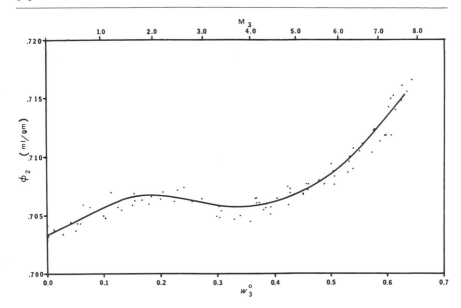

FIG. 2. Apparent specific volume, ϕ_2 of isoionic ribonuclease A vs weight fraction of guanidinium chloride with respect to water, W_3^0, at 20°. The points represent experimental values adjusted to those for a weight fraction value of ribonuclease in water of $W_2^0 = 0.0500$. The solid line represents the continuous values of ϕ_2 calculated by Eq. (3) from computer-drawn curves (of the form in Eq. 7) of the density values vs composition, W_3^0, of the solvent solution (curve 1 of the text) and of the protein solution (curve 2 of the text). M_3 is the molarity of guanidinium chloride in the protein solution. From T. H. Crouch and D. W. Kupke, *Biochemistry* **16,** 2586 (1977).

out the guanidinium salt than vice versa, according to independent solubility experiments.[8]]

ΔV for Transport of Perturbant

Rarely has the change in volume resulting from the transport of a perturbant or other small molecule in protein systems been studied. These ΔV are expected to be small, except initially, because the protein is usually a minor component in terms of the mole fraction. The quantitatively major interactions, as W_3^0 is increased, would tend to be with component 1, whose effects cancel out in the determination of $\Delta\phi_3$. By the use of smoothed density-composition curves (e.g., curve 1 and curve 2), however, the trends in the volume changes per gram of component 3 (i.e., $\Delta\phi_3$) may be of sufficient definition to provide a useful titration curve. In-

[8] T. H. Crouch and D. W. Kupke, *Biochemistry* **16,** 2586 (1977).

ferences drawn from such curves can lead to further experiments for focusing upon a segment of interest and in relating volume changes to results from other kinds of probes. In principle, the sign and magnitude of the difference in volume per gram of a reagent is ascertained as increasing amounts of it are moved from a given mass of the solvent to a like mass of this solvent to which a known mass of a protein has been added. An example of a titration curve of this kind is shown in Fig. 3, which was derived from the same density-composition curves as those that were utilized in Fig. 2 for estimating the volume changes upon transporting the protein. Although the volume differences per gram of the transported component 3 ($\Delta\phi_3$) are substantially smaller than those found when a gram of the protein was moved ($\Delta\phi_2$), the trends manifested with concentration

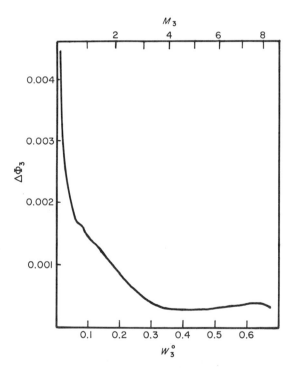

FIG. 3. Difference in the volume increase per gram of guanidinium chloride, $\Delta\phi_3$, versus W_3^0 (and M_3) upon adding this salt to a gram of water and to a gram of water + isoionic ribonuclease A ($W_2^0 = 0.0500$) at 20°. The smooth curve, obtained with Eq. (3), was derived from the same computer-drawn data curves 1 and 2, described in the text, as were used to generate the curve in Fig. 2. From T. H. Crouch and D. W. Kupke, *Biochemistry* **16**, 2586 (1977).

($W_3{}^0$ or M_3) are as reliable as the curves that were generated by smoothing the experimental values of the 2 density-composition tables.

Because titration curves of this kind are not in general use, a few explanatory remarks may be appropriate relative to Fig. 3. As noted, the volume changes accompanying the movement of the guanidinium salt from a gram of water to a gram of water + ribonuclease are positive at all concentrations of this salt to the limits of its solubility (0.675 g_3/g_1 at 20°). Obviously, the presence of the protein alters the normal interactions of the salt with water alone. The figure shows that the protein titrates the guanidinium salt such that the volume increase at all concentrations is greater than if the same amount of this salt were added at each point to water alone. (If the salt is added to a constant total volume of water and water + protein rather than to a constant mass of water, the curve is quite similar to that in Fig. 3 except that the actual values of $\Delta\phi_3$ are larger as $W_3{}^0$ increases; this is because the effect of the thermodynamic volume of the protein, $\phi_2 g_2$, is included in the volume change result and this contribution becomes more important as the interactions of protein with denaturant decrease.) The volume differences actually become more severe at concentrations far below that observed for the conformational transition (~ 2–$4 M$) of this protein. Clear evidence for interactions between guanidinium chloride and identifiable groups on ribonuclease at this low concentration range of the salt has been observed with nuclear magnetic resonance (NMR) spectroscopy[9]; included were interactions at 0.5–0.7 M denaturant corresponding to the wrinkle in Fig. 3 as well as other interactions through the major unfolding transition. Additional density experiments on very low concentrations of guanidinium chloride (0.1–2 mg/ml) showed that $\Delta\phi_3$ closely approached +0.1 ml/g_3 (~ 10 ml per mole) in the mole per mole range of the denaturant to protein; this behavior is analogous to the increase in volume observed when hydrocarbon is transported from water to a nonpolar phase. [A break in this titration curve (not shown in Fig. 3) occurred at ~ 3 mol of salt per mole of ribonuclease.[cf.9]] Reliable differences in the density can be obtained with the magnetic densimeter on micromolar increments of small-molecule reagents if their partial specific volumes are substantially different than that of water. For example, micromolar additions ($\sim 10^2$ μg/ml) of the guanidinium salt increases the density such that millivolt changes are observed that can be determined with good precision if the weighings are accurate. Also in Fig. 3, there is an almost constant residual volume increase at all denaturant concentrations above 4 M, where the unfolding of the protein is presumed to be complete. This small constant increase, however, may not reflect any significant

[9] F. W. Benz and G. C. K. Roberts, J. Mol. Biol. **91**, 367 (1975).

interaction of further added salt with protein because the protein is presumed to remove solvating water from the medium.[10]

Partial Specific Volumes

The total volume of a solution at constant temperature and pressure in terms of the partial specific volumes, $\bar{v}$, of all components reduces to

$$\sum_{i=1}^{N} \bar{v}_i c_i = 1 \tag{11}$$

when the masses of the components are cast as the concentrations, c. Hence, in a 3-component system ($N = 3$), if the partial specific volume of the protein and of the perturbing substance can be evaluated from the 2 density-composition tables which yielded the smoothed curves 1 and 2, then $\bar{v}_1$ for the solvent component is calculated by difference. Calculation of $\bar{v}_3$ for either curve 1 or curve 2 is straightforward, being a function of the slopes. In traditional form

$$(\partial \rho / \partial c_3)_m = (1 - \bar{v}_3 \rho)/(1 - \bar{v}_3 c_3) \tag{12}$$

In practice, it is more direct to utilize the weight fraction obtained from the gravimetric operations in preparing the solutions. Since $c_3 = \rho W_3$, the useful equation for obtaining $\bar{v}_3$ at any composition via the definition [Eq.

---

[10] The almost constant positive differences in the volume ($\triangle \phi_3 \approx +0.0003$ ml/g$_3$) above 4 M guanidinium chloride suggests that any additional increment of this salt at these concentrations interacts principally with the bulk-solvent phase rather than with the protein; that is, the upward curves of ϕ_3 versus component-3 concentration both in the presence and in the absence of ribonuclease were essentially parallel. The constant difference in the paired ϕ_3 values at each $W_3^0 > 0.35$ may be viewed in terms of the availability of water. If a gram of ribonuclease removes ~0.50 g of H_2O, at all $W_3^0 > 0.35$, the consequent increase in the effective concentration of any added guanidinium salt would raise the value of ϕ_3 by the amount shown (where $g_2/g_1 = 0.05263$). For example, at $W_3^0 = 0.4000$, the removal of 0.50 $g_1/g_2 \times 0.05263$ g_2/g_1 reduces the 1 g of water initially present to 0.9737 g_1 (as the amount available for interacting with any additional guanidinium chloride in the protein solution). The new, or effective, value of W_3^0 is 0.4064, which in the absence of the protein yields the higher value of ϕ_3 that was observed at $W_3^0 = 0.4000$ in the protein solution. [In this connection, the determined value of the preferential interaction parameter, ξ_3, was found to be $+0.10 \pm 0.02$ g_3/g_2 at $W_3^0 = 0.462$ as measured by the density and dialysis equilibrium method.[1] ξ_3 coupled with the value of 0.50 g_1/g_2 yields 0.53 g_3/g_2 as the amount of guanidinium chloride "bound" (or not of the bulk-solvent phase) by the relation $\xi_3 = \xi_3^* - \xi_1^*(c_3^0/c_1^0)$ (see derivation in Kupke[7]). Asterisks refer to the "bound," or nonbulk, solvent amounts of components 1 and 3 per gram of protein, and superscript zero refers to the protein-free concentrations of these diffusible components at dialysis equilibrium. The value, 0.53 $g_3/g_2 \approx 76$ mol of guanidinium chloride per mole of ribonuclease A at denaturing concentrations of this salt, is in approximate agreement with conclusions drawn from other kinds of studies: e.g., J. C. Lee and S. N. Timasheff, *Biochemistry* **13**, 257 (1974) and A. Salahuddin and C. Tanford, *Biochemistry* **9**, 1342 (1970).

(12)] is

$$\bar{v}_3 = \frac{\rho - (1 - W_3)(\partial\rho/\partial W_3)_m}{\rho^2} \qquad (13)$$

Thus, values for $\bar{v}_3$ along both density-composition curves 1 and 2 can be generated as smooth curves over the range of component-3 concentrations studied, and the differences, $\Delta\bar{v}_3$, at each value of $W_3{}^0$ (or g_3/g_1) represent the change in these partial quantities as a result of introducing the protein component. Since the protein is usually a minor component, these differences will tend to be very small except in the regions where the mass of component 3 approaches that of component 2. If ρ is found or known to be a linear function of c_2 at selected molalities of component 3 over the composition range of interest, then $\phi_2 = \bar{v}_2$ for these purposes [Eq. (5)]. Thus, the values of ϕ_2 (as in Fig. 2) combined with the corresponding values of $\bar{v}_3$ yields $\bar{v}_1$ at each composition by use of Eq. (11). Since component 1 is the major component in most studies, the differences in v_1 owing to the introduction of protein may become appreciable only after component 3 no longer interacts with component 2 or near the limit of solubility of one of these components. A sufficiently detailed study of the change in $\bar{v}_1$ over transition regions in the ΔV of transport or in the equimolar range of component 3 to component 2 has not been carried out to the authors' knowledge. The trends in $\bar{v}_1$ and in $\bar{v}_3$ with $W_3{}^0$ (Fig. 4) using the same density-composition data as for the curves in Figs. 2 and 3 suggest small differences in $\Delta\bar{v}_i$ $(i \neq 2)$ in certain regions, but these differences are unreliable in the absence of careful supplementary studies. It should be recognized also that the smoothed curves via the polynomial fittings described here are insensitive as $W_3{}^0 \rightarrow 0$; therefore, the differences in this region $(W_3{}^0 < 0.01)$ are not accurately reflected in the curves of Fig. 4a and 4b and a special density study on very low concentrations of the guanidinium salt were required for extrapolating these curves to $W_3{}^0 \rightarrow 0$. In this example, the value of $\bar{v}_3$ in water continued to decrease to an apparent limiting value, $\bar{v}_3{}^0$, of 0.692 ml/g_3, whereas in water containing isoionic ribonuclease, $\bar{v}_{3(R)}$, increased as $W_3{}^0 \rightarrow 0$ at various concentrations (2–6%) of the protein (R refers to ribonuclease). [At $W_2{}^0 = 0.0500$, the apparent limiting value of $\bar{v}_{3(R)}^0$ was ~0.80 ml/g_3; the error becomes increasingly large as $W_3{}^0 \rightarrow 0$, therefore, this limiting value is no better than ± 0.02 ml/g_3 owing to the steeper approach to zero concentration than with this salt in water alone.]

A Simple Magnetic Densimeter

A basic densimeter, such as diagrammed in Fig. 1 utilizing optical sensing, may be envisioned in terms of the following general items: (a)

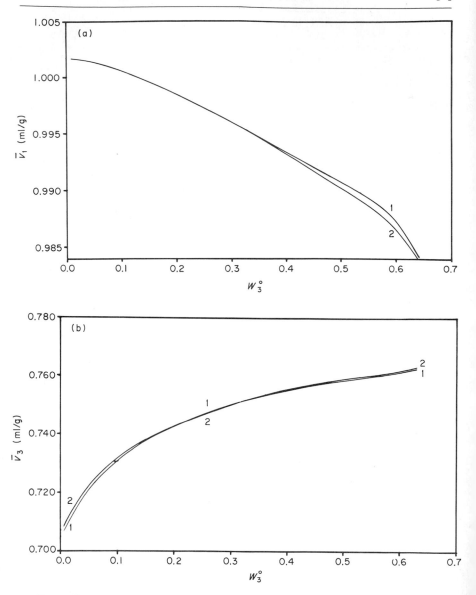

FIG. 4. Partial specific volumes $(\bar{v})$ of water and of guanidinium chloride as continuous functions of guanidinium chloride concentration with respect to water, W_3^0, at 20°. (a) Curve 1: water–guanidinium chloride; curve 2: water–isoionic ribonuclease A ($W_2^0 = 0.0500$)–guanidinium chloride. (b) Curve 1: water–guanidinium chloride; curve 2: water–isoionic ribonuclease A ($W_2^0 = 0.0500$)–guanidinium chloride. The curves in both (a) and (b) represent values of $\bar{v}_i$ calculated from the computer-drawn data curves 1 and 2 as described in the text. From T. H. Crouch and D. W. Kupke, *Biochemistry* **16,** 2586 (1977).

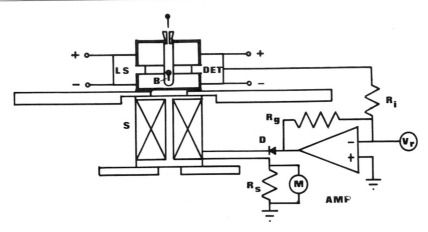

FIG. 5. Outline of circuitry for a basic magnetic densimeter (cf. Fig. 1). The cell containing buoy B is surrounded by a thermostated housing (heavy lines) with appropriate ports for optical sensing and for viewing.

solenoid, (b) cell and isothermal housing, (c) buoys, (d) temperature control units, (e) voltmeter (5½ digits), (f) telemicroscope (40×, nonmagnetic, (g) electronic and power components. Items (a) through (c) are fabricated according to the objectives of the user. (Details for the model in Fig. 1 are available[11]; however, we suggest a solid brass cell housing and a solenoid that can be fashioned to suit particular needs as described presently). Temperature control is also a local arrangement; reference is made to the analysis and choices of components given by McKie and Brandts.[12] The electronic and power components (item g above) referred to in Figs. 5 and 6 consist of the following: (a) operational power supply, (b) 5.2-V power supply, (c) field-effect transistor (FET), (d) phototransistor, (e) diode, (f) 6–12-V light source, (g) assorted resistors.

The overall diagram for the densimeter is shown in Fig. 5. Although optical sensing of the position of the buoy is utilized,[11] other sensing methods can be used because the amplifier (AMP) circuit remains the same and the instrument functions identically. As the buoy tends to rise or sink, the detector (DET) output voltage changes accordingly. This voltage is compared to a reference voltage (V_r) and the difference is amplified to drive the solenoid (S), which maintains the buoy at a constant height. In order to assure that the buoy is maintained at a constant vertical position relative to the solenoid, a 40× nonmagnetic telemicroscope (e.g., Gaertner Scientific Co., Chicago, Illinois) is used in positioning the image of the

[11] J. P. Senter, *Rev. Sci. Instrum.* **40**, 334 (1969).
[12] J. E. McKie and J. F. Brandts, this series, Vol. 26, p. 257.

top of the buoy onto the horizontal cross hairs by adjusting V_r. The telemicroscope is mounted on a horizontal, circular aluminum plate (~ 1 cm thick), which also supports the cell housing and solenoid. Since the buoy is at a constant vertical displacement from the solenoid, the current is a function of the fluid density only.[1] This current is measured as the voltage drop across a stable reference resistor, R_s (usually ~ 1 ohm and constructed of manganin wire around a plastic spool). The voltage drop across R_s is measured with sufficient accuracy with a 5½ digit, digital voltmeter or a Leeds and Northrop K-3 potentiometer (M). Ordinarily, about 0.1 mV corresponds to from 1 to 4×10^{-6} g/ml in the density, depending on the current; this is because the function relating density to current is parabolic. Position sensors other than optical types include marginal oscillators[13–15] and double transformers.[16–18] The optical method of sensing has been found by the authors to be the most convenient for general-purpose densimetry.

In Fig. 6, a detailed schematic diagram of Fig. 5 is shown. This is the system currently used by the authors for biochemical solutions in the room-temperature range and at atmospheric pressure. The detector functions in the following manner: As the buoy (B) tends to float, the amount of light incident on the HEP-P0001 phototransistor is reduced. This lowers the emitter voltage. The GE-2 field-effect transistor (FET) serves as a current amplifier with its output voltage in phase with its input voltage. Hence, as the buoy rises, the voltage output and, therefore, the current output of the detector is lowered. The detector current is subtracted from a reference current (the ±6.2-V sources) and is then amplified by a monopolar operational power supply (denoted BOP in Fig. 6), which inverts the sign of the input current difference. The output current from the amplifier is fed through a diode, D (to prevent inductive "kick," which may damage the power amplifier) to the solenoid and is measured across R_s with voltmeter M. Thus, as the buoy rises, the detector current is decreased and the solenoid current is increased to bring the buoy back to its predetermined position. The gain (controlled by the ratio R_g/R_i in Fig. 5) is such that no motion is detected in the telemicroscope; positioning can be reproduced to ±1 μm or less with experience (1 μm $\approx$ 0.5 to 2×10^{-6}

[13] J. W. Beams and A. M. Clarke, *Rev. Sci. Instrum.* **33**, 750 (1962).

[14] S. P. Almeida and T. H. Crouch, *Rev. Sci. Instrum.* **42**, 1344 (1971).

[15] M. G. Hodgins and J. W. Beams, *Rev. Sci. Instrum.* **42**, 1455 (1971).

[15a] J. W. Beams and M. G. Hodgins, *Bull. Am. Phys. Soc. Ser. II*, **15**, 189 (1970).

[16] D. W. Kupke, M. G. Hodgins, and J. W. Beams, *Proc. Natl. Acad. Sci. U.S.A.* **69**, 2258 (1972).

[17] S. C. Greer, M. R. Moldover, and R. Hocken, *Rev. Sci. Instrum.* **45**, 1462 (1974).

[18] W. M. Haynes, M. J. Hiza, and N. V. Frederick, *Rev. Sci. Instrum.* **47**, 1237 (1976).

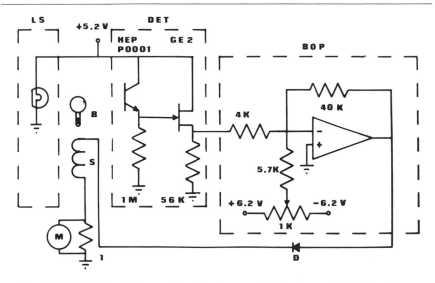

Fig. 6. Detailed circuit diagram for a basic magnetic densimeter. The dashed lines enclose the blocks shown in Fig. 5; V_r of Fig. 5 is detailed here as ± 6.2 V sources (contained in the operational power supply, BOP).

g/ml, depending on the current). With a double transformer that is operating properly, this error can be reduced. The authors currently use a Kepco BOP operational power supply (BOP in Fig. 6) which is bipolar and which also requires a diode of the appropriate current rating between amplifier and solenoid; the monopolar supply costs substantially less and there is no loss in performance.

   Since guidelines for the construction of solenoids are not found in the literature of densimetry, the following remarks may be helpful. The solenoid should be wound on a nonmagnetic cylinder which is electrically nonconducting. If a metal spool is used, which should be employed for good heat dissipation at excessive currents, the spool should be slotted radially from the center to its edge so that a shorted transformer secondary is not created. The solenoid used successfully by the authors is 5.7 cm in diameter by 5 cm long with a plastic core 0.65 cm in diameter. The size of the insulated copper wire (copper enamel) is determined by the voltage and current outputs of the control amplifier by applying Ohm's law. An equation approximating the relationship between solenoid size and the dc resistance, R, of the magnet (which determines the power dissipation for a given current) is

$$R \approx 2 \pi \sigma n \, \{l(r + \delta/2) + [l(l + 1)/2]\delta\} \qquad (14)$$

where l is the number of turns deep and n is the number of turns long, σ is the resistance per unit of length of the wire, δ is the diameter of the wire, and r is the radius of the solenoid core. We have found that with annealed slugs of HyMu-80 that are 2 mm long by 1 mm in diameter inside a buoy of proper density, a power input of 2.5 W is sufficient to cover a density range of at least 0.025 g/ml. If the power is increased to 10 W, the density range will be extended proportionately; heating at higher currents, however, becomes more of a problem, and water cooling of a solenoid (of these dimensions) may become necessary.

The magnetic field intensity H at a distance z from the solenoid top is given by

$$H = 2\pi J \left((z + L) \ln \frac{(r + D) + \sqrt{(z + L)^2 + (r + D)^2}}{r + \sqrt{(z + L)^2 + r^2}} \right.$$
$$\left. - z \ln \frac{(r + D) + \sqrt{z^2 + (r + D)^2}}{r + \sqrt{z^2 + r^2}} \right) \quad (15)$$

where J is the current density in the solenoid, L is the solenoid length, and D is the distance from r to the outside surface measured radially. The quantity J is the current per unit area in the solenoid. That is

$$J = (Inl)/LD \quad (16)$$

where I is the current through the solenoid. Thus, Eq. (15) becomes

$$H = 2\pi [(Inl)/LD] [X] \quad (17)$$

where the terms within the brackets of Eq. (15) are denoted by X. In general, it is better to construct a solenoid such that for a given resistance the magnetic field intensity is maximized. Using Eqs. (14) and (17) one can design a solenoid of any dimensions and of any wire size to meet spatial and electric current limitations for a particular purpose.

For the ferromagnetic core of the buoy, HyMu-80, a nickel–iron perm-alloy (Carpenter Steel Co.) or Moly Permalloy, containing 4% molybdenum (Allegheny-Ludlum Steel Corp.) have been used. These slugs are first machined into true cylinders ($\sim 2 \times 1$ mm) and are then annealed in a pure, dry hydrogen atmosphere at 1120°C (2050°F) on an Al_2O_3 bed in a ceramic tube for 3–4 hr. They are then allowed to cool at prescribed rates. For HyMu-80, the tube is allowed to cool to 650°, after which the rate of cooling is controlled at 167° per hour to 315° in an oven; the final cooling to room temperature is not controlled. Moly Permalloy slugs are cooled at a rate not to exceed 55° per hour to 593° and then at an uncontrolled rate to room temperature. Care should be taken to purge oxygen from the furnace prior to elevating the temperature and igniting the gas outflow. Annealing of Moly Permalloy is available commercially at small cost if several slugs are annealed in one run (e.g., Allegheny-Ludlum Steel Corp., Dept. 11, Brackenridge, PA 15014).

Sizes and shapes of buoys and the nature of the outer jacket can be varied depending on the application. The technique for constructing glass-jacketed buoys has been described in detail by Senter.[11] These buoys do not require frequent recalibration. Where skilled craftsmanship is unavailable, glass-jacketed buoys may be fashioned as follows: A symmetrical bulb, ~4 mm in diameter, is blown into one end of a length of glass capillary having an inside diameter that allows the slug to slide in freely. The bulb is cut off leaving ~6 mm of stem. The annealed ferromagnetic cylinder is inserted through the stem into the bulb along with enough sealing wax to achieve, approximately, a slightly higher density of the buoy than the value desired. The stem is then flame sealed and the sealed end is filed down, if necessary, to the prescribed weight. The metal cylinder is maneuvered into the closed stem via a hand magnet. The buoy is then centrifuged with a clinical centrifuge in a swinging-bucket tube containing hot cooking oil (~170°) to melt the wax. For this procedure, the stem of the buoy is aligned with the centrifugal field by inserting the stem into a well-centered drill hole in a suitable plug which fits the bottom of the centrifuge tube. The centrifugation is continued until the oil has cooled and the wax has hardened. By this means the mobile wax seats the slug so that the center of gravity of the buoy tends to be on the magnetic axis (ideally, the gravitational and magnetic vectors should be colinear). Asymmetries are observed as horizontal drifts of the buoy as the solenoid current is increased (i.e., as $\rho_{(B)} - \rho$ becomes greater); this effect, however, can be calibrated out if not too serious and, also, the instrument can be realigned for a given buoy. With experience, serviceable buoys covering the desired ranges in density (≤ 0.05 g/ml per buoy range) can be accumulated at very low cost. The authors have had initial success with another approach, which utilizes polypropylene cylinders. The symmetry of these solid cylinders can be controlled quite precisely by machining and also the material is less dense than water ($\rho \cong 0.90$ g/ml). A well-centered, axial hole is drilled at one end of the polypropylene cylinder to fit the ferromagnetic cylinder ($\rho \cong 8.7$ g/ml), which is then pushed inside. Sealing off the open end of the hole is accomplished by pressing in a tiny amount of glyptal or molten polypropylene. The densities of these buoys are adjusted by varying the dimensions of the plastic cylinders and/or by varying the length of the coaxial hole. Although this method is simpler and more precise than that for fabricating glass-jacketed buoys, the long-term behavior and frequency of calibrations of the polypropylene-jacketed buoys cannot be defined at this writing.

As noted at the outset, other more complicated sensing and support systems have been developed in the past. These sensing methods offer advantages where solutions are not transparent or are light sensitive; their complexity, however, makes them more difficult to construct, operate, and maintain. For buoys containing the magnetically soft, permalloy cores, only a single electromagnet is necessary; if permanent magnetic cores are used or a very large density span is to be encompassed by a single buoy, the 3-coil system (main solenoid and 2 opposing, gradient coils) should be employed (see the following section on Viscosity). Recent results with ceramic-type, permanent magnet cores in buoys demonstrate that the requirement for multicoiled support systems can be eliminated if the buoy contains a suitable permanent magnet.[2].

Part II. Viscosity

The magnetic viscometer–densimeter introduced by Beams and Hodgins[15a] has now been used in a number of studies.[16,19,20] In principle, the magnetic viscometer differs from the magnetic densimeter only in that the suspended buoy is caused to rotate by applying a torque via remote-drive coils situated around the solution cell. Thus, the viscometer functions somewhat like a very small, coaxial type of viscometer except that the cylindrical buoy (rotor) is completely immersed and independent of any mechanical support, such as a torsion wire. By this means, very small shear stresses (T) can be applied. Currently, the rate of shearing (G) is measured at a fixed power input which generates the rapidly rotating magnetic field for turning the buoy; the method, however, can be adapted to measure the torque or amount of shear stress required when maintaining a fixed rate of shearing.[21] The magnetic approach offers certain advantages not collectively available in any other kind of viscometer. These attributes along with tested and proposed applications and the limitations are outlined below. A succeeding section deals with a description of the components and operation of the simplest viscometer–densimeter (or viscodensimeter) for measuring rates of shearing at given applied stresses. These rates of shearing have been found to be remarkably proportional to the viscosity by calibrating with standard liquids; hence, the simplest design can be useful for determining relative viscosities routinely and accurately on small volumes (0.2 ml) without concern for the exact value

[19] C. H. MacGregor, C. A. Schnaitman, D. E. Normansell, and M. G. Hodgins, *J. Biol. Chem.* **249**, 5321 (1974).

[20] M. G. Hodgins, O. C. Hodgins, D. W. Kupke, and J. W. Beams, *Proc. Natl. Acad. Sci. U.S.A.* **72**, 3501 (1975).

[21] J. W. Beams and D. W. Kupke, *Proc. Natl. Acad. Sci. U.S.A.* (in press, October 1977).

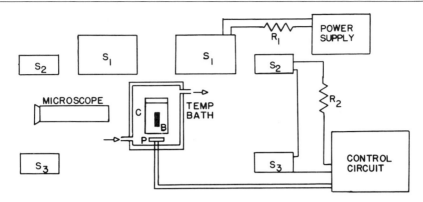

FIG. 7. Block diagram of the vertical support system for the rotor-buoy in the magnetic viscometer–densimeter. B is the rotatable buoy, C is the glass cell containing the sample solution, and P is the height-sensing coil. From J. W. Beams, *Rev. Sci. Instrum.* **40,** 167 (1969).

of the shear stress or the absolute viscosity. This design is also useful for observing the change in viscosity during slow reactions.

Characteristics and Applications

A schematic block diagram of the basic instrument as divided into the vertical support and the remote drive features is shown in Figs. 7 and 8. This design, described in more detail presently, incorporates the circuitry to measure accurately the supporting current for obtaining the density simultaneously.[15] For viscometric purposes, the rotor or buoy is usually more dense than the solutions to be used because part of the buoy must be electrically conducting. Hence, the main supporting solenoid (S_1) in Fig. 7 is placed above the cell for lifting (unlike that in the diagram of Fig. 1 for the basic densimeter). Since the difference in density between the suspended buoy and the solution is often too large for accurate density determinations, the main solenoid is used, in effect, to reduce this density difference by applying a preset amount of current to it. In order to determine accurate densities simultaneously, an auxiliary pair of solenoids (S_2 and S_3), similar to Helmholtz coils, are introduced which are of a configuration and placement such that the reduced density difference can be obtained very accurately.[22] The electric current in common to these linked solenoids, S_2, S_3, which maintain the desired vertical position of the buoy, is strictly proportional to the field gradient, dH/dz, because, in this design, the moment, M, of the buoy is held constant. Thus, the density is a

[22] J. W. Beams, *Rev. Sci. Instrum.* **40,** 167 (1969).

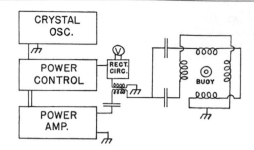

FIG. 8. Block diagram of the remote-drive circuit for the rotatable buoy in the viscometer–densimeter. From M. G. Hodgins and J. W. Beams, *Rev. Sci. Instrum.* **42**, 1455 (1971).

linear function of this current. The small drive coils in the plane of the buoy (Fig. 8) are fed a constant amount of power automatically via a feedback circuit between the power amplifier and a controlled oscillator. The power input to the buoy is so small that heating is negligible. For most experiments, the power to the buoy is such that the rotation rate is between 0.06 and 0.1 rad, sec^{-1} in water. The rotation is continued until the uncertainty in the rate is about 1 part in 10^4, depending on the timing procedure employed. The rotatable buoy itself is a small core ($\sim 2 \times 1$ mm) of an annealed permalloy (such as for the basic densimeter) encased in a hollow Kel-F cylindrical shell ($\sim 7 \times 2.8$ mm) to reduce the density of the total buoy to about 1.3 g/ml and to present an inert surface to the solutions. (A very thin band of nonmagnetic conducting material may be placed inside the upper part of the shell for better stability, but this is not always necessary.) Imperfection lines on the surface of the Kel-F shell as seen through a telemicroscope serve as reference marks for measuring the rate of rotation. The cell containing the solution is an accurately made glass cylinder of a diameter corresponding to the amount of gap desired between rotor surface and cell wall (usually about 1.4 mm). Other characteristic related to applications are described below.

Sample Size. A volume of 0.2 ml of a liquid sample fills the cell completely when containing the buoy; the solution is completely sealed from the environment. With such small volumes, temperature equilibration is rapid and much less protein is required than is customary for the series of measurements leading to a value of the intrinsic viscosity $[\eta]$. For example, ~ 15 mg of a nitrate reductase enzyme was utilized to obtain a value of $[\eta]$, which included triplicate determinations.[19]

Speed and Sensitivity. About 2–3 min. are required for a single determination, good to about 1 part in 10^3, after the solution has come to temperature. Usually, however, the rotation is continued in order to obtain a better time average, depending on the timing method employed. Thus, small differences in the viscosity between sample and reference liquid can be expanded (to the limits of the overall precision exercised) so that accurate values for the reduced specific viscosity, η_{sp}/c, may be obtained at lower protein concentrations than ordinarily for the extrapolation to a value of $[\eta]$. The viscometer is well suited for routine studies, such as the evaluation of $[\eta]$ of proteins in general. An example of a study on bovine serum albumin and on ribonuclease at increased levels of unfolding is shown in Fig. 9.

Low Shear Stress. Routinely, the applied stresses, T, are on the order of 10^{-3} dyne, cm^{-2}. This amount of stress is $\sim 10^4$ times smaller than those employed in conventional capillary viscometers. Such small stresses, while usually unnecessary for the globular proteins, may be of critical importance in systems that are sensitive to shearing, such as those containing highly asymmetric molecules or deformable macromolecular

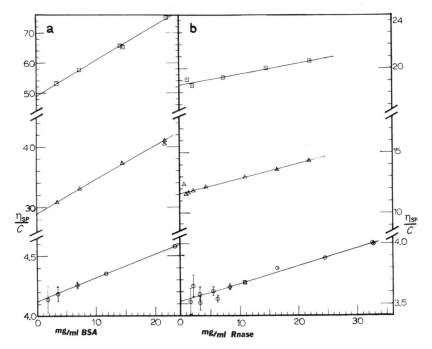

Fig. 9. Reduced viscosity versus protein concentration at 25°. (a) Bovine serum albumin (deionized). (b) Bovine pancreatic ribonuclease A (deionized).⊙, In 0.15 M KCl; △, in 6 M guanidinium chloride; ☐, in 6 M guanidinium chloride +0.1 M β-mercaptoethanol. Error bars reflect the uncertainty with hand timers on a fixed number of rotations (~ 300–400 sec per determination) at all dilutions ($T \sim 1.5 \times 10^{-3}$ dyne, cm^{-2}). From D. W. Kupke, M. G. Hodgins and J. W. Beams, *Proc. Natl. Acad. Sci. U.S.A.* **69**, 2258 (1972).

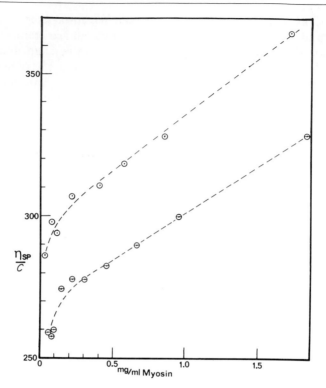

FIG. 10. Reduced viscosity versus concentration of rabbit muscle myosin at 10°. $\ominus$, 0.6 M KCL + 0.001 M dithiothreitol, pH 7.23; $\odot$, 0.5 M KCL + 0.2 M (K)PO$_4$ + 0.002 EDTA, pH 7.15. For protein preparation details, see references cited in footnote 15. Usually, 6 to 7 rotations were observed for each value of η_{sp}; the difference in total time between the most dilute solutions and the solvents was held to >6.8 sec (applied T ~ 1.5 × 10^{-3} dyne, cm^{-2}). From D. W. Kupke, M. G. Hodgins, and S. M. Mozersky, unpublished data, 1972.

complexes. (In a prototype instrument,[21] the stresses can be reduced much further.) Since dilute solutions can be studied under very low shear stresses, the method is amenable to some association reactions. Figure 10 is an example of 2 myosin preparations that were examined for curvature at low concentrations. These profiles support the careful observations that lead to the contention that myosin self-associates in a staggered parallel array.[23,24]

Kinetic Opportunities. When a reaction is in progress, a change in the viscosity of the solution at constant shear stress is reflected by a change in

[23] M. Burke and W. F. Harrington, *Nature (London)* **233**, 140 (1971); *Biochemistry* **11**, 1456 (1972).
[24] W. F. Harrington and M. Burke, *Biochemistry* **11**, 1448 (1972).

the observed rate of shearing G (i.e., $T = \eta G$). With reference marks suitably separated around the circumference of the buoy, several data entries can be made during a single rotation. (With a more recent design, the stress is monitored continuously while the rate of shearing is held constant at arbitrary levels[21]; this modification enlarges the potential for kinetic studies because the response to viscosity changes is reflected promptly.) An example of a kinetic experiment is shown in Fig. 11 on the rupture of the disulfide bonds of serum albumin in guanidinium chloride upon the addition of β-mercaptoethanol. These data were obtained with the original instrument in which only 1 to 2 data entries were obtainable per rotation of the buoy. Nonetheless, the nonsmooth character of the viscosity curve was reproducible and quite unlike the much smoother curve observed with ribonuclease in this medium.

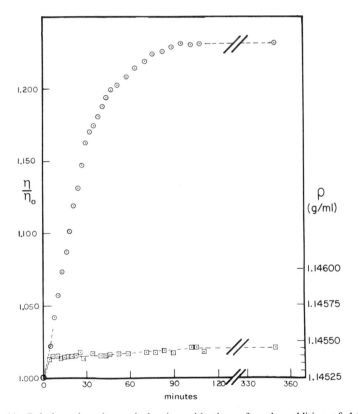

FIG. 11. Relative viscosity and density with time after the addition of 1.3 μl of β-mercaptoethanol to 200 μl of 14.1 mg/ml bovine serum albumin (deionized) in 6 M guanidinium chloride at 25°. ⊙, Viscosity; ⊡, density. From D. W. Kupke, M. G. Hodgins, and J. W. Beams, *Proc. Natl. Acad. Sci. U.S.A.* **69**, 2258 (1972).

Density. The density of a solution (and its change with time, Fig. 11) is gotten concurrently with the viscosity if the instrument is adapted for increasing the vertical support sensitivity, such as in Fig. 7. With the viscosity–density combination, it becomes feasible to study both the thermodynamic and hydrodynamic volumes of a protein simultaneously. For example, an investigation of an isometric virus confirmed other lines of evidence that a great deal of solvent is trapped within the structure. Values for the apparent specific volume and the intrinsic viscosity when coupled with preferential interaction studies by equilibrium dialysis and density showed that >0.4 g of solvent per gram of dry virus existed on the inside of the virion.[16]

Corrections. In general, the corrections applied in high-precision outflow types of viscometers are eliminated or reduced to very low levels. These include corrections for the density, kinetic energy and outflow end effects, surface tension, drainage, positioning and loading errors, and the less predictable effects at interfaces (the surface-to-volume ratio is far smaller in the magnetic method). In addition, by the magnetic method, the solution is in mechanical equilibrium, subject to only a very small and stable velocity gradient, so that chemical equilibrium is much less likely to be disturbed. As with capillary viscometry, however, filtering to remove lint, dust, and aggregates is very important. Since the constant force of gravity is not employed to create the velocity gradient, the power input system for applying torque to the rotor-buoy must be accurately controlled. This has been achieved to better than 1 part in 10^4, the latter uncertainty being the present level of precision experienced with repetitive determinations on a given sample (this imprecision, however, can be ascribed to the degree of temperature control exercised, which is not better than a variation of $0.005°$ per measurement). The symmetry and alignment of the buoy and the symmetry of the magnetic fields are also very important; for relative viscosity determinations, however, the effects of small imperfections are essentially calibrated out. The determination of absolute viscosities, which includes evaluation of the small end effects of the rotating cylinder, are considerably more convenient than with outflow viscometers. (The relevant equations and references for this purpose are given by Bradbury[25] and by Yang.[26])

Other. By incorporating broader shear stress options into the magnetic instrument, the viscous behavior of properly concentrated, shear-sensitive protein systems, such as occurs in local regions of living cells, may be studied. The bizarre nature of the viscosity at very low stresses following the extrusion of RNA from a virus has already been reported.[20]

[25] J. H. Bradbury, *in* "Physical Principles and Techniques of Protein Chemistry" (S. J. Leach, ed.), Part B, p. 99, Academic Press, New York, 1970.
[26] J. T. Yang, *Adv. Protein Chem.* **16,** 323 (1961).

At higher stresses these bizarre effect disappeared. Also, the magnetic approach would appear to be much more convenient for studying the viscosity as a function of temperature and compressibility (a pressure adaptation is currently under study by Dr. J. W. Beams). Finally, the potentially greater precision attainable for evaluating differences and changes in the reduced viscosity versus concentration may be useful for some purposes; this may result in improvements on the theory of viscous behavior at finite protein concentrations. In this connection, a comparison has been made with ribonuclease in water and in D_2O (Fig. 12). Since the molar volumes of these two waters are essentially the same, the shape and the degree of rigidity of the protein and the protein–protein and protein–solvent interactions must all be virtually identical in the 2 solvents according to present theory if the data in Fig. 12 are sufficiently accurate. That is, all slopes and intercepts are almost indistinguishable.

A Basic Viscodensimeter

The application of a constant torque to the magnetically suspended cylindrical buoy (Fig. 7) is achieved most conveniently by using 4 drive coils spaced around the thermostatted cell housing at 90° with respect to

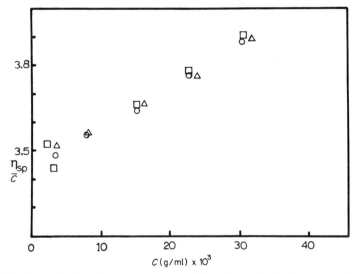

FIG. 12. Reduced viscosity versus concentration of isoionic bovine pancreatic ribonuclease A in water at 25°. ○, Ordinary water (distilled in Pyrex); △, D_2O (99.85 mol %); □, 50:50 ordinary water: D_2O. Corrections for water content of the lyophilized protein were performed after dry-weight analysis on samples taken at the time the protein was dissolved in the various waters. (The precision decreased with dilution of the protein, owing, perhaps, to the absence of any added salt.) From M. G. Hodgins and D. W. Kupke, unpublished data, 1971.

each other (Fig. 8). (A 3-coil drive design has proved to be more difficult.) An alternating current is passed through 2 coils (connected in series) such that each of 2 oppositely facing coils have the same current while the current to the other pair is 90° out of phase. This creates an induction motor with the buoy as a rotor. The power to these drive coils is regulated by sensing a magnetic-flux generated current through a small sensing transformer. As long as the voltage across the secondary of this transformer is kept constant, the root-mean-square current to the drive coils is constant. Figure 13 shows the network details relating to the block diagram of Fig. 8.

The drive circuit in Fig. 13 functions in the following manner. A crystal-controlled oscillator is used having an output frequency of $10^4 \pm 10$ kHz. The square-wave output of the oscillator is attenuated to about 1 V by a resistive voltage divider, and it is then fed into a single transistor amplifier. Here, a dc voltage is applied as determined by the voltage across the sensor transformer secondary; this voltage acts to raise or lower the square-wave amplitude according to the power output. This signal is amplified twice, and the square wave is converted to a sine wave by a band stop filter (L-C parallel) which shorts all higher harmonic frequencies of the square wave to ground. At this point the current to the drive coils passes through the sensor transformer primary. The signal

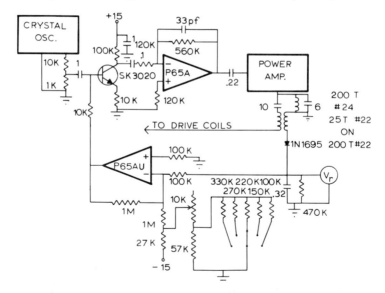

FIG. 13. Detailed network of the remote drive system currently employed for applying constant torque to the rotatable buoy in the magnetic viscometer–densimeter (cf. Fig. 8). By permission from M. G. Hodgins.

across this transformer secondary is rectified and monitored with a digital voltmeter (voltages being proportional to currents delivered to the drive coils). The resultant signal is compared to a reference voltage, which is amplified and used to adjust the square-wave amplitude into the transistor amplifier.

The drive coil current is split into two equal amplitude currents, and each is shifted 45° in phase from the initial signal in such a manner that they will differ from each other by 90°. The 2 capacitors used here are each a collection of paper-wound foil capacitors which add up to about 0.10 microfarad (μf) in one leg and 0.15 μf in the other leg (see Fig. 13). These capacitors are adjusted in order to give the proper phase shift while maintaining equal current amplitudes. The matched drive coils consist of 200 turns of American gauge wire (AGW) of No. 22 coated copper, which is wound on a plastic spool (1.25 cm long by 2.5 cm wide and a 0.63 cm core). These coils are mounted on a square bakelite frame with the in-phase coils facing each other. Plastic bolts are used to secure the coils because metal ones tend to generate heat. The signal amplitude is monitored in each leg of the drive circuit by a coil constructed with 50 turns of AGW No. 35 coated copper to yield dimensions of 0.38 cm long by 1.25 cm wide with a 0.63 cm core. This sensing coil is placed between the drive coil in each leg and the bakelite frame. These coils are used to sense the current amplitude in each leg of the drive circuit to assure that the currents are matched in amplitude. Signals from the sensing coils are separately rectified and filtered by the same type of diode and resistance capacitance configuration as that used in the sensor transformer secondary; the dc signals thus generated are monitored with a digital voltmeter. It should be noted that these rectifiers and filters are mounted in aluminum miniboxes on the support stand relatively close to the sensing coils themselves.

In the present instrument, the rotatable buoy is supported by the solenoid configuration referred to in Fig. 7. As noted in Part I, this type of support[22] has been used for obtaining densities with a single buoy over a wide range of densities, and it allows for a higher level of sensitivity than is ordinarily practiced with a single solenoid densimeter. (That is, the density is a parabolic function of the current in the latter kind of instrument, whereas it is a linear function of the measuring current via the configuration in Fig. 7.) The main solenoid (S_1) receives a very constant current (± 1 ppm) which is used to induce a constant magnetic moment, M, in the core of the buoy, while the opposing solenoids (S_2, S_3) are used to control the position of the buoy. The latter solenoids (spaced in a manner similar to opposing Helmholtz coils) control position by varying the magnetic field gradient, dH/dz, at the buoy core. The current for varying the field gradient is strictly linear with the density of the fluid at a particular vertical position of the buoy. This position is where the buoy slug is midway between S_2 and S_3. An inductive device is used for position sensing in this case because, when the buoy is rotating, slight irregularities make optical sensing difficult.

A detailed schematic diagram of the present magnetic support system for the viscometer is shown in Fig. 14. The system consists of the same basic functional units as the optical support system of the basic densimeter (Fig. 5) except that the inductance position detector is in a tuned circuit and the amplifiers are low-power devices which drive a pass transistor that regulates the current through solenoids S_2, S_3. The system functions in the following manner: The 160 $\mu\mu f$ capacitor, in parallel with coil Q, form the position-sensing network. The capacitor is tuned such that the resonance frequency of this configuration is slightly different than the frequency of the master oscillator (which operates at 4.55 MHz). Coil Q (27 turns of AGW No. 38 enameled copper with an air Core ~0.7 cm) is positioned at the base of the sample cell so that as the buoy moves vertically, the displacement of the permalloy core changes the inductance of Q and, thus, the impedance of the inductance-capacitance circuit. These changes are observed as changes in the root-mean-square voltage across Q. The output is in the form of a sine wave with a frequency of 4.55 MHz. The 1N1495 diode changes the high frequency alternating current to a direct current signal. The 6-V battery serves as the position-reference voltage source. The P65A amplifier subtracts the reference voltage from the position signal, and this difference is amplified and used to drive the

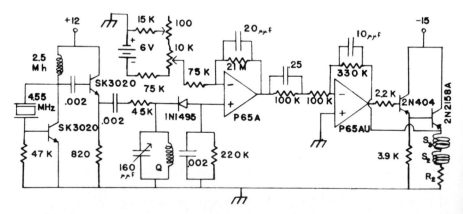

FIG. 14. Control circuit for maintaining vertical position of the buoy-rotor in the viscometer–densimeter (cf. Fig. 7). Adapted from M. G. Hodgins and J. W. Beams, *Rev. Sci. Instrum.* **42,** 1455 (1971).

pass transistor in series with the gradient support coils of solenoids S_2, S_3. These solenoid coils are physically arranged so that their magnetic fields at the plane midway between them cancel. The field gradients from these coils, however, add; i.e., only the field gradient (midway between the coils) varies with the current through the coils. When the center of the buoy core is positioned at the plane midway between S_2 and S_3, the magnetic force on the buoy becomes a linear function of the current through these coils and reference resistance R_2. The proper height of the buoy is found by calibrating the current with liquids of known density.

With soft magnetic material in the buoy core, the main solenoid, S_1 (or a permanent magnet), is used to induce a magnetic moment in the core (e.g., permalloy has a zero magnetic moment when no external magnetic field is present). This solenoid receives a constant current (to within 1 part in 10^6) to assure that only changes in the gradient coil circuit reflect fluid density changes. The overall solenoid configuration allows for greater sensitivity in determining densities because small changes in current through resistor R_2 can be amplified to reflect smaller density changes. This is because one is operating much closer to zero current than with the single-coil system. Thus, the overall effect of the main solenoid is to lower the apparent density of the buoy relative to the support circuit until it is close to the densities of the solutions to be studied. Hence, the current that provides the field gradient through S_2, S_3, and which is a linear function of the fluid density, can be held close to zero.

The principal complication with the 3-coil type of support is the great difficulty encountered in properly aligning the magnetic fields so that the rotating buoy does not precess. On the other hand, the alignment problem is greatly simplified for viscometry if only a single solenoid is employed; the instrument can then function as the basic densimeter with ordinary buoys to cover the lower density ranges when the drive coils are turned off. Alternatively, with the alignment being much less of a problem, several rotatable buoys of different density can be utilized for the simultaneous determination of density and viscosity.

Part III. Osmotic Pressure

The measurement of small osmotic pressures, utilizing magnetic suspension, was introduced recently by J. W. Beams and associates.[27] This approach provides a means of obtaining these "unencumbered" chemical potentials at the low molar concentrations of protein of current interest to

[27] J. W. Beams, M. G. Hodgins, and D. W. Kupke, *Proc. Natl. Acad. Sci. U.S.A.* **70**, 3785 (1973).

biochemists. By the use of the magnetic balancing principle it has been possible to make measurements of acceptable precision at concentrations as low as 10^{-6} M. Thus, the molecular weight of very large molecules ($>10^6$) and the dissociation behavior of proteins at low concentrations are being opened to the osmotic method.

The magnetic osmometer may be envisioned as a magnetic densimeter in which the magnetically suspended buoy can change its volume—hence its density—as a result of small pressure changes. In effect, the change in the density difference between buoy and solution is determined as a function of the pressure. The changes in pressure are too small in osmotic pressure experiments, ordinarily, to affect the density of the solution itself (e.g., the density of an aqueous solution changes only $\sim 10^{-6}$ g/ml per 0.023 atm, or ~ 25 cm of water pressure). The method makes use of Boyle's law in combination with the Archimedes principle, which is used in conventional magnetic densimetry. A hollow buoy containing trapped air is balanced magnetically at a preset height within the solution. Ignoring the volume of the buoy material for this didactic purpose, the fractional change in the pressure, $\Delta P/P$, would correspond to a similar fractional change in the volume of the buoy. Thus, at constant temperature and ~ 1 atm, a change in pressure corresponding to 0.1 mm of water ($\sim 10^{-5}$ atm) would change the volume of this hypothetical buoy on the order of 1 part in 10^5 and, hence, the density by a similar fraction. Such differences in the density are measurable to $\sim \pm 5\%$ if the temperature is controlled to $\pm 0.0003°$. On the other hand, 10^{-6} mol per liter of a nondiffusible solute yields an osmotic pressure, Π, of ~ 0.25 mm H_2O at 20° via the approximate relation, $\Pi \cong RT \cdot M_i$ where M_i is the molarity of the nondiffusible component and $R = 84.83$ liter, cm H_2O, degree^{-1}, mole^{-1}. Accordingly, the realizable precision at 10^{-6} M is increased to $\sim \pm 2\%$. In practice, the volume of the buoy material V (of mass m) is made as small as possible relative to the total volume $V_{(B)}$ and is designed to be of an average density, $\rho_{(B)}$, (including the trapped air), which is close to that of the solutions being studied ($\Delta\rho < 0.05$ g/ml). [The buoy may be a thin open cylinder (~ 0.003 cm thick) of soft ferromagnetic metal encased inside and outside between two snug-fitting Kel-F or polypropylene cups; on occasions, the metal is simply dipped in dissolved Lucite or glyptal and a thin plastic cover is affixed to the top end.] For this purpose, the buoy is made to float in the solutions of interest and a servo-controlled solenoid is placed below for pulling down and maintaining the buoy at a fixed height. In the two present designs, opposing Helmholtz-like coils are also utilized (as for the viscometer, see Part II) for covering greater ranges in the density difference with accuracy. (With the latter arrangement, an adjustable permanent magnet below the system on the magnetic axis can serve as the main force to balance the buoy, and the small current to the gradient coils serve as the measuring circuit.) In a recent modification, the buoy is balanced on the solvent side of the membrane and negative pressure differences are measured; in this way a more constant medium is exposed to the buoy for a given series of protein concentrations, thus eliminating possible adsorption and surface tension errors arising from the presence of variable amounts of the protein. The negative density difference $(\rho_{(B)} - \rho)$ is determined as a function of the current in a manner similar to that described in Part I. In the osmotic case, the difference is related to pressure changes by direct calibration at the time of a measurement; the mass and volume of the buoy, the density of the solution, and instrument constants are calibrated out. Because changes in atmospheric pressure would tend to swamp out the small pressure changes of interest, the system is connected to a constant-atmosphere tank, such as a large enclosed Dewar flask containing a small amount of the major solvent component (usually water). In actual practice, the equilibrium pressure difference and the calibration are determined rapidly (~ 1 min) to minimize the effect of any small temperature and pressure fluctuations.

Although the magnetic osmometer is undergoing continual development to provide for greater convenience and reliability, a brief outline of the simplest design, the principles involved, and an example of some of the results on an associating protein system are presented at this time. This is done in the hope that those having interest will be stimulated into attempting the method and, thus, to aid in closing the technological gap that has limited the application of the simple, but powerful, osmotic principle. In this connection, the application of nonideality theory to protein solutions in osmotic equilibrium have been reviewed recently.[28,29]

Theory of the Method

In magnetic osmometry the density of the buoy is the variable of interest. This density, $\rho_{(B)}$, is given by

$$\rho_{(B)} = m_{(B)}/V_{(B)} \tag{18}$$

where $m_{(B)}$ is the constant mass of the buoy and $V_{(B)}$ is the total volume. $V_{(B)}$ consists of a constant and a variable region such that

$$V_{(B)} = V + V^0 \tag{19}$$

[28] M. P. Tombs and A. R. Peacocke, "The Osmotic Pressure of Biological Macromolecules." Oxford Univ. Press, (Clarendon), London and New York, 1974.

[29] M. J. Kelly and D. W. Kupke, in "Physical Principles and Techniques of Protein Chemistry" (S. J. Leach, ed.), Part C, p. 77. Academic Press, New York, 1973.

where V is the constant volume of the buoy material and V^0 is the volume of the enclosed gas phase (air) at the constant or reference pressure, P^0, of the controlled atmosphere. Differentiating Eq. (18) with respect to the volume of the buoy gives

$$dp_{(B)} = - \frac{m_{(B)}}{V_{(B)}^2} dV_{(B)} = - \frac{m_{(B)}}{(V + V^0)^2} dV_{(B)} \qquad (20)$$

For a fractional change in the pressure, P/P^0, where P is another pressure which is close to P^0, the density of the buoy becomes

$$\rho_{(B)} = \frac{m_{(B)}}{V + (P^0/P)V^0} = \frac{Pm_{(B)}}{PV + P^0V^0} \qquad (21)$$

assuming that the ideal gas law applies for these small pressure differences (as has been shown by calibration). Differentiating Eq. (21) with respect to pressure and rearranging gives

$$dp_{(B)} = \frac{m_{(B)}dP}{P^0V^0[1 + (PV/P^0V^0)]^2} \cong \frac{m_{(B)}dP}{P^0V^0(1 + V/V^0)^2} \qquad (22)$$

where P/P^0 is essentially unity.

In the idealized case where $V \to 0$, or $V_{(B)} \approx V^0$, then $dp_{(B)} \approx dP$ since $\rho_{(B)}$ is nearly 1 g/ml and $P^0 \approx 1$ atm. Hence, the maximum useful sensitivity at $\Delta T \approx 0.0003°$ when $\Delta\rho_{(B)} = \pm 1 \times 10^{-6}$ g/ml is $\sim \pm 1 \times 10^{-6}$ atm or about $\pm 1 \times 10^{-3}$ cm H_2O. The volume V of the buoy material, however, is finite, ranging from 0.1 V^0 to ~ 3 V^0 in the various buoy designs employed. For the most unfavorable case ($V \approx 3$ V^0), the sensitivity reduces to

$$dp_{(B)} \approx 0.169 \, dP \qquad (23)$$

via Eq. (22) where $m_{(B)}$ is typically ~ 0.17 g (i.e., a change in density of $\sim 1.7 \times 10^{-4}$ g/ml per cm H_2O). The latter type of buoy, having thicker walls, can be fashioned symmetrically for better alignment with the magnetic axis.[30] Although the sensitivity drops about 6-fold from the ideal by the use of such "heavy" buoys, the density difference between buoy and solution can be maintained by magnetic suspension to better than 10^{-7} g/ml if the temperature is stable to within $0.0003°$ at constant pressure (atmospheric pressure fluctuations limit the precision in ordinary magnetic densimetry to $\sim 10^{-6}$ g/ml, regardless of temperature control to better than $\sim 0.005°$). In practice, the time taken for the final reading of the density difference and a calibration of the pressure upon releasing the developed pressure difference (~ 1 min) is such that temperature fluctuations of more than $0.0003°$ are not observed. Thus, the density difference has been found to remain constant to ± 0.2 μg/ml.

[30] T. H. Crouch, Dissertation, University of Virginia, Charlottesville, 1977.

If the volume of the liquid phase surrounding the buoy is truly constant (i.e., no bubbles or membrane distortions), the amount of fluid crossing the membrane during the approach to equilibrium may be calculated from the ideal gas law. That is,

$$(dV_g/dP)_{p^0} = V^0/P^0 \tag{24}$$

where V_g is the volume of the gas in the buoy at any pressure P. Substituting the gas volume of 0.063 ml for a typical buoy at the reference conditions V^0 when $P^0 = 1030$ cm H_2O (~ 1 atm), the derivative [Eq. (24)] is ~ 60 nl per cm H_2O or ~ 1.5 nl crosses the membrane when the nondiffusible solute is at 10^{-6} M ($\Pi \sim 0.025$ cm H_2O). Hence, the method holds the capability of being very rapid with respect to achieving equilibrium at small values of Π.

Methodology

Osmometers can conform to almost any geometry as long as the two liquid phases are in contact via a semipermeable membrane and a suitably quantitative manifestation of the pressure difference is observable. At present, the magnetic osmometers do not include the technology for changing the protein concentration without disassembling the cell compartment. Such an improvement would permit better evaluation of the slopes, $[d(\Pi/c_i)/dc_i]_{\mu,T}$, for studying the various factors contributing to nonideal behavior[31,32] (μ = constant chemical potential of the diffusible components, and i refers to the nondiffusible components). Although an improved model, with the buoy on the solvent side and with more convenient temperature control, has been effected,[30] the overall precision with the simpler, original design[27] has not been surpassed. The osmometer assembly of the latter, which amounts to a test tube containing the solvent, protein solution, membrane, and buoy, is shown in Fig. 15. Several such tubes are assembled and allowed to come to equilibrium at the desired temperature, after which each is inserted, in turn, into the magnetic suspension system for the measurements. A given tube is positioned reproducibly in an insulated, thermostable water bath with optical ports so that optical sensing and buoy positioning with the aid of a telemicroscope are facilitated. Alternatively, a thermostable brass block with appropriate portholes has been used for containing the tubes. The detailed circuitry for the optical height-sensing system of the buoy is to be found in the original paper[27]; it differs in some respects from that for the basic densimeter of Fig. 5, but the changes are not critical. This is because the

[31] G. Scatchard, *J. Am. Chem. Soc.* **68**, 2315 (1946).
[32] A. V. Güntelberg and K. U. Linderstrøm-Lang, *C. R. Trav. Lab. Carlsberg, Ser. Chim.* **27**, 1 (1949).

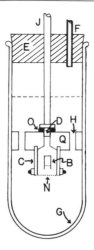

FIG. 15. Schematic diagram of the osmotic cell assembly. The assembled glass tube G (~15 × 1.8 cm) is inserted in a magnetic suspension system for measuring changes in density of the buoy B as a function of pressure. The cylindrical glass cell C (~2 × 0.6 cm), which contains B and the protein solution (~0.5 ml), is cemented into a nylon plug Q, which has an axial channel for filling. Membrane N is made to fit tightly over the bottom of C by slipping a latex band, 4–5 mm wide, over N and around C (the O-ring and gasket assembly, as shown, and perforated plastic plates against N to minimize ballooning were found to be unnecessary). The upper end of the channel in C is sealed by a neoprene ring O and a short nylon screw D, which is rigidly embedded into plastic rod J. J extends out of G through a tightly fitting stopper E. Q, which slides into G, contains a number of vertical holes H to permit free movement of the solvent. A small hole is drilled into G at about the desired height for Q, and a nylon screw (not shown) backed by an O ring protrudes through G and is aligned with a tapped hole in Q; this arrangement prevents any lateral movement of C when rod J is turned to open or close the liquid phase in C from the solvent phase. C is filled slowly by means of a gas-tight syringe into the channel until the solution runs out (at a temperature a little lower than that of the experiment); a hand magnet is used to prevent the buoy from floating and blocking off the channel during the filling. Solvent is then added into G through H until at Q; excess solution from C is blotted off, and J is slowly turned to close off C. Solvent is added to the dashed line and the stopper E is slid down along J until tightly embedded into G. For the measurements, the assembled osmometer is slipped into the densimeter cavity and is then attached to the insulated constant-atmosphere tank (see text) from glass tube F emerging from stopper E. A stopcock assembly between F and the tank (not shown) allows for opening the system to a manometer, the outside atmosphere, and/or to syringes for adjusting the pressure of the constant-atmosphere tank. It is best to submerge the tank and all leads from F in a constant-temperature bath; however, the system performed adequately when all exposed items were insulated. The tube of highest concentration in a series of assembled tubes may be kept connected to the measuring system for following the approach to equilibrium; this approach is monitored by the change in current across the reference resistance (Fig. 5 or Fig. 14) as observed with a digital voltmeter. At equilibrium, J is turned gently to release the developed pressure difference (for other details, see text). The pressure differences in the remaining tubes are then measured, in turn, quite rapidly. The protein solutions are then resealed from the solvent atmospheres, and the procedure is repeated to test for reproducibility. Adapted from J. W. Beams, M. G. Hodgins, and D. W. Kupke, *Proc. Natl. Acad. Sci. U.S.A.* **70**, 3785 (1973).

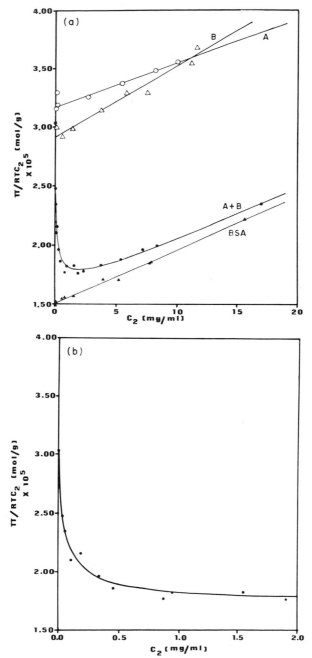

Fig. 16. Reduced osmotic pressure versus concentration of defatted and deionized bovine serum albumin (BSA) and of its peptic fragments A and B when separate and when mixed in equimolar proportions (see text) in 0.01 M bis-tris propane–0.10 M KCl, pH 8.6 at 20°. (a) Composite plot. (b) Expanded plot of the A + B complex at <2 mg/ml. From T. H. Crouch, doctoral dissertation, University of Virginia, 1977.

densities of the buoys can be made to be suitably close to those of the solutions and $\rho_{(B)}$ changes only a small amount with the pressure differences that develop. When equilibrium has been achieved, as indicated by no further change in the voltage which reflects the density difference between buoy and solution, the cell is opened to the solvent by way of the constant atmosphere tank (which may be an insulated container such as is used for storing liquid nitrogen). An amount of air is then added to the tank via a calibrated syringe to increase the pressure of the system a known amount until the density indicator is once again at the voltage corresponding to that observed at the end of the experiment. In this way the observed density difference is calibrated in terms of pressure within a minute of the final reading. The choice of a volume ratio of the tank to the syringe is arbitrary, depending on the precision by which the syringes can be read. With calibrated gas-tight syringes, the tank volume can be as small as 4 liters because 100 μl of air delivered by the syringe corresponds to $\sim$0.025 cm H_2O in the osmotic pressure ($\sim 10^{-6} M$ in protein concentration). It is necessary for the tank to contain water-saturated air at the temperature of the experiment. Calibration curves prepared over an extended range show that Boyle's law is obeyed in the buoy (for the precisions noted previously) up to pressure differences of about 10 cm H_2O.

The molecular weight of a virus (5.4×10^6) determined by this method has been reported,[33] and some results with an associating protein system are shown in Fig. 16. In this example, BSA was split into two nearly equal-sized fragments, A and B, by peptic digestion.[5] The enzymic activity of BSA in catalyzing the decomposition of a synthetic substrate (a Meisenheimer complex) is absent in either fragment, but is mostly regained when the fragments are mixed together.[6] The equilibrium constants taken from the curve for the A + B complex were in reasonable agreement with those derived from the activity analyses (1.8×10^6 liter/mole and 1.3×10^6 liter/mole, respectively). The molecular weights of the purified fragments obtained with the osmometer (A = 31,600; B = 34,400) and that of the purified BSA (66,400) are in close agreement with those deduced from the amino acid compositions.

Acknowledgments

It is to be noted that the ideas and the initial developments in magnetic suspension emanated from Professor J. W. Beams. We gratefully acknowledge the encouragement and counsel that he gave us in the preparation of this article. We also thank Dr. W. M. Haynes, U.S. National Bureau of Standards, for critically reviewing the manuscript.

Note added in proof: J. W. Beams died on July 23, 1977.

T. H. Crouch was supported by Grant 5-T32-GM07294 from the U.S. Public Health Service.

[33] D. W. Kupke, J. W. Beams, and M. G. Hodgins, *Fed. Proc., Fed. Am. Soc. Biol. Chem.* **33**, 1230 (1974).

[5] Membrane and Vapor Pressure Osmometry

By E. T. ADAMS, JR., PETER J. WAN, and EDWARD F. CRAWFORD

Mass transfer through a membrane (osmosis) is a very common phenomenon. Water, minerals, and other nutrients must be transferred through root membranes so that plants can grow. Osmotic phenomena are also evident when erythrocytes are crenated or hemolysed. The driving force of osmosis is due to the differences in the chemical potentials of solvents and other diffusible materials when two solutions are partitioned by a semipermeable membrane (SM) as shown in Fig. 1. The macromolecular solution is designated as phase β, and its counterpart is a solvent (or buffer or supporting electrolyte solution), designated as phase α. The membrane (SM) is permeable to the solvent and small diffusible molecules (such as KCl, buffer acids or bases, and buffer salts) but impermeable to the macromolecule. If the external pressure on both phases is the same (both are at p^0), then solvent (and other diffusible materials) will flow from phase α across the membrane to phase β until the liquid level in chamber β increases to a point where the hydrostatic pressure on phase β overcomes the tendency of solvent to flow across the membrane. This flow of solvent (and other diffusible molecules) across the semipermeable membrane is known as osmosis (derived from the Greek word for push or impulse). The flow of solvent across the membrane can be prevented by increasing

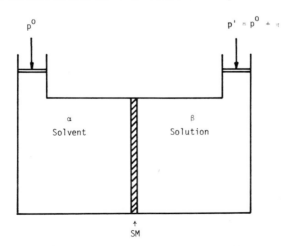

FIG. 1. Schematic drawing of an osmotic pressure cell. When the pressure exerted on phase β is $p' = p^0 + \pi$, no solvent will flow from phase α across the semipermeable membrane (SM) to phase β. The excess pressure π is known as the osmotic pressure.

the pressure on the solution in chamber β by an amount $p' = p^0 + \pi$. This excess pressure $\pi = p' - p^0$ is known as the osmotic pressure.[1,2] If the excess pressure is greater than π, then solvent will flow from phase β to phase α; this phenomenon is known as reverse osmosis, and, provided a suitable membrane could be found, it could be used to purify brackish water.[3]

Osmotic pressure, along with freezing point depression (cryoscopy), boiling point elevation (ebulliometry), vapor pressure lowering, dew point depression,[4,5] and elasto-osmometry,[6] is one of the colligative properties of solutions. Colligative properties are those properties that depend primarily on the number of molecules rather than on the nature of the molecules.[4] Here we shall be concerned with two of these colligative properties: membrane osmometry and vapor pressure osmometry. These techniques are useful for evaluating molecular weights or number average molecular weights, activity coefficients or osmotic pressure second virial coefficients for nonassociating solutes. For associating solutes one can use these techniques to establish the type of association present, if that is not known beforehand, as well as nonideal terms. Vapor pressure osmometry is most useful for studying lower molecular weight solutes: solutes having molecular weights between 0 and 20,000 (grams/mole).[7-9] Membrane osmometry, on the other hand, is more useful for the study of larger molecular weight solutes: macromolecules having molecular weights between 10,000 and 1,000,000.[7-9]

Osmometry, the measurement of osmotic pressure, is one of the oldest techniques for studying macromolecules and colloids. According to

[1] M. P. Tombs and A. R. Peacocke, "The Osmotic Pressure of Biological Macromolecules." Oxford Univ. Press, London and New York, 1974.
[2] M. J. Kelly and D. W. Kupke, Osmotic Pressure, in "Physical Principles and Techniques of Protein Chemistry." (S. J. Leach, ed.), Part C, p. 77. Academic Press, New York, 1973.
[3] S. Sourirajan, "Reverse Osmosis." Academic Press, New York, 1970.
[4] S. Glasstone, "Textbook of Physical Chemistry," 2nd ed. Van Nostrand-Reinhold, Princeton, New Jersey, (1946).
[5] Dew point depression is a technique used by physical chemists to measure vapor pressure lowering (see pp. 632 and 1266 of ref. 4). This technique has been used by McBain and co-workers to study the self-association of soaps and detergents [see J. W. McBain and C. S. Salmon, *J. Am. Chem. Soc.* **42**, 426 (1920); J. W. McBain and R. C. Williams, *J. Am. Chem. Soc.* **55**, 2250 (1933)]. A modern dew point depression apparatus is available from Wescor, Inc. Logan, Utah.
[6] S. Yamada, W. Prins, and J. J. Hermans, *J. Polym. Sci.* **A1**, 2335 (1963); H. J. M. A. Mieras and W. Prins, *Polymer* **5**, 177 (1964).
[7] F. W. Billmeyer, Jr., "Textbook of Polymer Science," 2nd ed. Wiley (Interscience), New York, 1971.
[8] J. L. Armstrong, *Appl. Polym. Symp.* **8**, 17 (1969).
[9] E. A. Collins, J. Bareš, and F. W. Billmeyer, Jr., "Experiments in Polymer Science." Wiley (Interscience), New York, 1973.

Glasstone,[4] the first recorded experiments of the phenomenon of osmosis are those of the Abbé Nollet in 1748. Nollet found that when alcohol and water were separated by an animal bladder membrane, water passed through into the alcohol causing an increase of pressure, but alcohol was not able to pass into water. Thomas Graham noted in 1854 that substances in the colloidal state cannot pass through certain membranes that are permeable to water.[4] Many other pioneers in colloid chemistry studied osmosis. The first theoretical treatments that put the phenomenon of osmotic pressure on a quantitative basis were the works of Van't Hoff[1,4,10] in 1888 and Pfeffer and De Vries[1,4,10] in 1889. Membrane osmometry was the earliest method used to study the molecular weight of macromolecules; it was used long before light scattering or ultracentrifugation. According to Tombs and Peacocke,[1] the first studies on the osmotic pressure of protein solutions were made by Starling[12] in 1899. This technique has attracted the attention of many famous physical chemists interested in studying proteins, such as Sørenson,[13] Adair,[14] Linderstrøm-Lang,[15,16] and Scatchard.[17] Membrane osmometry has also been used extensively in the study of synthetic polymers.[7,18] The number average molecular weights and osmotic pressure second virial coefficients obtained for many polymers can be found in the "Polymer Handbook."[19]

In multicomponent systems consisting of a solvent, one or more diffusible solutes, and a macromolecular solute, if the semipermeable membrane permits only the flow of solvent, but prevents the flow of the diffusible solute(s) and the macromolecular solute, then one would measure the total osmotic pressure. If the membrane allowed the passage of the diffusible solute(s) but restrained the macromolecular solute, then one would measure the colloid osmotic pressure.[1] No further distinction between these terms will be made here.

The advent of high speed membrane osmometry[1,2,8,9] has rekindled interest in osmotic pressure experiments. Banerjee and Lauffer[20] have

[10] J. H. Van't Hoff, *Phil. Mag.* **26**, 81 (1888).
[11] E. Pfeffer and J. DeVries, *Z. Physik. Chem. Stoechiom. Verwandschaftslehre* **3**, 103 (1889).
[12] E. H. Starling, *J. Physiol. (London)* **24**, 317 (1899).
[13] S. P. L. Sørenson, *C. R. Trav. Lab. Carlsberg* **12**, 1 (1917).
[14] G. S. Adair, *Proc. R. Soc. London, Ser. A*, **108**, 627 (1925); **120**, 595, 573 (1928).
[15] A. V. Gùntelberg and K. Linderstrøm-Lang, *C. R. Trav. Lab. Carlsberg, Ser. Chim.* **27**, 1 (1949).
[16] D. W. Kupke and K. Linderstrøm-Lang, *Biochim. Biophys. Acta* **13**, 153 (1954).
[17] G. Scatchard, *J. Am. Chem. Soc.* **68**, 2315 (1946). See also references in Table I.
[18] P. J. Flory, "Principles of Polymer Chemistry." Cornell Univ. Press, Ithaca, New York, 1953.
[19] J. Brandrup and E. H. Immergut, eds., "Polymer Handbook," 2nd ed. Wiley (Interscience), New York, 1975.
[20] K. Banerjee and M. A. Lauffer, *Biochemistry* **5**, 1957 (1966).

used this technique to characterize tobacco mosaic virus (TMV) protein subunits and follow the recombination of the subunits. T'so[21] and his associates have used vapor pressure osmometry to study the self-association of modified nucleic and nitrogen bases and to show that base-stacking interactions play an important role in the self-association of the nitrogen bases. We will give a brief outline of the theory of osmotic pressure and vapor pressure osmometry experiments and show how one can treat nonassociating and associating systems. Real data from our laboratory and other sources will be used to illustrate how some of the analyses are done. For a mixed association we will use a simulated example to illustrate the analysis. Except for the Tombs and Peacocke[1] monograph and the review by Kelly and Kupke,[2] most other reviews or references to osmotic pressure theory do not include self-associations or mixed associations. We will also give details of how to do osmotic pressure and vapor pressure osmometry experiments, and we shall discuss some experimental precautions. At the end of this section, we shall give a brief list of some important papers in this field that have appeared in the past few years.

There are a number of excellent references to osmometry, mostly membrane osmometry. The excellent laboratory manual of polymer chemistry by Collins, Bareš, and Billmeyer[9] gives some instructions for membrane osmometry and vapor pressure osmometry. The monograph by Tombs and Peacocke[1] is highly recommended, as it contains an extensive development of the theory, particularly with multicomponent systems; it gives details on how to do osmotic pressure experiments and calculate the results, and it has many references to current and older experimental results. The review by Kelly and Kupke[2] is very up to date, and it includes a discussion of associating solutes and how to analyze them. A very lucid and excellent presentation of the theory of membrane osmometry from a polymer chemist's point of view is presented by Overton;[22] this review discusses instrumentation and experimental methods. No discussion of associating solutes is included. There are many other reviews and treatments of osmotic pressure theory and practice; among them are articles of Guidotti,[23] Baldwin,[24] Edsall,[25] and Kupke.[26] Refer-

[21] P. O. P. Ts'o, *Ann. N. Y. Acad. Sci.* **153,** 785 (1969).

[22] J. R. Overton, *in* "Techniques of Chemistry" (A. Weissberger, ed.), Vol. 1, Part V, p. 309. Wiley (Interscience), New York, 1971.

[23] G. Guidotti, this series Vol. 27, p. 256.

[24] R. L. Baldwin, *in* "Comprehensive Biochemistry" (M. Florkin and E. H. Stotz, eds.), Vol. 7, Part 1, p. 184. Elsevier, Amsterdam, 1963.

[25] J. T. Edsall, *in* "The Proteins" (H. Neurath and K. Bailey, eds.), Vol. I, Part B, p. 578. Academic Press, New York, 1953.

[26] D. W. Kupke, *Adv. Protein Chem.* **15,** 57 (1960).

TABLE I
SOME REFERENCES TO MEMBRANE OSMOMETRY

Researchers	Reference[a]	Remarks
Edsall	(1)	Theory and technique of membrane osmometry. Tables IV and V give molecular weights of various proteins from osmotic pressure measurements. There are many references given to the older literature on membrane osmometry
Kreuzer	(2)	One of the first theoretical treatments of the analysis of self-associations from data obtained by colligative methods
Scatchard	(3)	Theory of osmometry for three-component and multicomponent systems
Scatchard, Batchelder, and Brown	(4)	Osmotic equilibrium in solutions of serum albumin and sodium chloride
Scatchard, Batchelder, Brown, and Zosa	(5)	Osmotic equilibrium in concentrated solutions of serum albumin
Güntelberg and Linderstrøm-Lang	(6)	Osmotic pressure studies of ovalbumin and plakalbumin. This paper also treats the theory of osmotic pressure
Steiner	(7)	The analysis of an ideal mixed association is described. Steiner shows how the number average degree of binding can be determined
Steiner	(8)	Analysis of ideal self-associations by various methods including light scattering and osmometry. Steiner shows how to obtain the number fraction of monomer from osmotic pressure experiments
Reiff and Yiengst	(9)	This paper reports on the development of an electronic membrane osmometer using a strain gauge to detect the osmotic pressure
Vink	(10)	Theory of osmotic measurements with solute-permeable membranes is developed
Brown	(11)	A very thorough study of the macromolecular properties of aqueous hydroxyethyl cellulose solutions. Osmotic pressure measurements done on a Fuoss–Meade block type osmometer are reported
Kupke	(12)	Ribonuclease and also insulin were studied in concentrated guanidinium hydrochloride so-

TABLE I (*Continued*)

Researchers	Reference[a]	Remarks
		lutions. Some of the problems in studying small macromolecules, like insulin, are discussed
Henley	(13)	A very thorough study of the macromolecular properties of cellulose in the solvent cadoxen (triethylenediamine cadmium hydroxide). Osmotic pressure studies were carried out in cadoxen solutions diluted 1:1 with water
Laurent and Ogston	(14)	Osmotic pressure studies on mixtures of human hyaluronic acid and human serum albumin
Brown, Henley, and Öhman	(15)	A very thorough study of the macromolecular properties, including osmotic pressure, of sodium carboxymethyl cellulose in 0.2 M NaCl and cadoxen solutions is reported
Lauffer	(16)	A theory of protein hydration is presented, which is applicable to the polymerization–depolymerization of tobacco mosaic virus protein
Adams	(17)	The first paper to show how the apparent weight (M_{wa}) average molecular weight and the apparent weight fraction (f_a) of monomer can be obtained from osmotic pressure experiments on nonideal self-associations
Preston, Davies, and Ogston	(18)	Osmotic pressures of mixed solutions of bovine hyaluronic acid and bovine serum albumin, as well as other studies, are reported
Lauffer	(19)	Osmotic pressure theory for hydrated proteins —an extension of the theory reported in ref. 16 of this table
Banerjee and Lauffer	(20)	Osmotic pressure studies on dilute solutions of tobacco mosaic virus (TMV) protein. Investigation of the beginning stages of the polymerization
Guidotti	(21)	Studies on the chemistry of hemoglobin, including osmotic pressure studies under conditions where association–dissociation equilibria are present
Lapanje and Tanford	(22)	Molecular weights, second virial coefficients, and unperturbed end-to-end distances of six

TABLE I (*Continued*)

Researchers	Reference[a]	Remarks
		proteins (ribonuclease, β-lactoglobulin, chymotrypsinogen, pepsinogen, aldolase, and serum albumin) in 6 M guanidinium hydrochloride solutions containing 0.1 M β-mercaptoethanol at 25° are reported
Castellino and Barker	(23)	Molecular weights and second virial coefficients are reported for seven native and dissociated (using guanidinium hydrochloride) proteins (serum albumin, ovalbumin, alcohol dehydrogenase, enolase, methemoglobin, lactate dehydrogenase, and aldolase) at 25°
Steiner	(24)	A very elegant way to analyze ideal mixed associations by colligative methods is presented
Diggle and Peacocke	(25)	Osmometric studies on the self-association of histone F2b at various pH are reported; this was also analyzed by sedimentation equilibrium. The monomer molecular weights of five histones in 6 M guanidinium hydrochloride solutions at 25° were measured by osmometry and sedimentation equilibrium
Harry and Steiner	(26)	A study on the monomer–dimer self-association of soybean trypsin and chymotrypsin inhibitor is presented
Adams, Pekar *et al.*	(27)	A method for analyzing nonideal mixed associations is reported
Jeffrey	(28)	Osmotic pressure and other studies on a low sulfur protein from wool are reported
Andrews and Reithel	(29)	Osmotic pressure experiments were carried out on jack bean urease under native and denaturing conditions
Steiner	(30)	This is a further extension of the theory for analyzing ideal mixed associations
Steiner	(31)	Here Steiner presents a theory for analyzing nonideal mixed associations (cf. refs. 24 and 30 of this table)
Vink	(32)	Precision measurements of osmotic pressure in concentrated, aqueous polymer (dextran, hydroxyethyl cellulose, polyvinylpyrrolidone and polyethylene oxide) are presented. Vink

TABLE I (*Continued*)

Researchers	Reference[a]	Remarks
		also reported on the design of an osmometer for this purpose
Bull and Breese	(33)	An osmometer capable of measuring colloid osmotic pressures of proteins up to 100 cm of mercury is described. Molecular weights and higher virial coefficients of three proteins (ovalbumin, bovine serum albumin, and bovine methemoglobin) are reported, as is an ingenious way of stirring the solutions in the osmometers
Vink	(34)	Precision measurements of osmotic pressure in concentrated, nonaqueous polymer solutions (cf. ref. 32 of this table) are described
Soucek and Adams	(35)	Osmometry and sedimentation equilibrium of a clinical dextran in aqueous solutions are reported; the molecular weight distribution of the dextran sample under ideal and nonideal conditions is also reported
Wan and Adams	(36)	Macromolecular characterization (viscometry, osmometry, sedimentation velocity, and sedimentation equilibrium) of two commercial dextran (M_w about 70,000) samples is reported

[a] Key to references:

1. J. T. Edsall *in* "The Proteins," 1st ed. (H. Neurath and K. Bailey, eds.), Vol. I, Part B, p. 578. Academic Press, New York, 1953.
2. J. Kreuzer, *Z. Phys. Chem.* **B53**, 213 (1943).
3. G. Scatchard, *J. Am. Chem. Soc.* **68**, 2315 (1946).
4. G. Scatchard, A. C. Batchelder, and A. Brown, *J. Am. Chem. Soc.* **68**, 2320 (1946).
5. G. Scatchard, A. C. Batchelder, A. Brown, and M. Zosa, *J. Am. Chem. Soc.* **68**, 2610 (1946).
6. A. V. Güntelberg and K. Linderstrøm-Lang, *C. R. Trav. Lab. Carlsberg, Ser. Chim.* **27**, 1 (1949).
7. R. F. Steiner, *Arch. Biochem. Biophys.* **47**, 56 (1953).
8. R. F. Steiner, *Arch. Biochem. Biophys.* **49**, 400 (1954).
9. T. R. Reiff and M. Yiengst, *J. Lab. Clin. Med.* **53**, 291 (1959).
10. H. Vink, *Ark. Kemi* **15**, 149 (1960); **19**, 15 (1962).
11. W. Brown, *Ark. Kemi* **18**, 227 (1961).
12. D. W. Kupke, *C. R. Trav. Lab. Carlsberg, Ser. Chim.* **32**, 107 (1961).
13. D. Henley, *Ark. Kemi* **18**, 327 (1962).
14. T. C. Laurent and A. G. Ogston, *Biochem. J.* **89**, 249 (1963).
15. W. Brown, D. Henley, and J. Öhman, *Ark. Kemi* **22**, 189 (1964). See also W. Brown, D. Henley, and J. Öhman, *Makromol. Chem.* **62** (1963).

ences to older reviews and theoretical treatments will be found in these articles, the Tombs and Peacocke[1] monograph, and the Kelly and Kupke[2] review. Some pertinent recent references to theoretical and experimental papers that involve membrane osmometry are listed in Table I; Table II gives a similar listing for vapor pressure osmometry.

Homogeneous, Nonassociating Solutes

Two-Component Systems[1,2,4,7,26]

For simplicity, a two-component system consisting of a solvent (component 1) and a neutral, nonassociating solute (component 2) whose mole fraction is X_2 will be considered first. It will be assumed that the solute is monodisperse. Note that it is improper to refer to the osmotic pressure of a solution or a solute as such, since osmotic pressure can only be detected and measured when a solution and solvent are separated by a semipermeable membrane (see Fig. 1). Let us now follow the course of events. Initially the pressure on the two phases is equal, i.e., in Fig. 1, $p^\alpha = p^\beta = p^0$ initially. Right away, solvent will flow across the membrane from phase α to phase β; the hydrostatic pressure in phase β will

16. M. A. Lauffer, *Biochemistry* **3**, 731 (1964).
17. E. T. Adams, Jr., *Biochemistry* **4**, 1655 (1965).
18. B. N. Preston, M. Davies, and A. G. Ogston, *Biochem J.* **96**, 449 (1965).
19. M. A. Lauffer, *Biochemistry* **5**, 1952 (1966).
20. K. Banerjee and M. A. Lauffer, *Biochemistry* **5**, 1957 (1966).
21. G. Guidotti, *J. Biol. Chem.* **242**, 3685, 3694 (1967).
22. S. Lapanje and C. Tanford, *J. Am. Chem. Soc.* **89**, 5030 (1967).
23. F. J. Castellino and R. Barker, *Biochemistry* **7**, 2207 (1968).
24. R. F. Steiner, *Biochemistry* **7**, 2201 (1968).
25. J. H. Diggle and A. R. Peacocke, *FEBS Lett.* **1**, 329 (1968); **18**, 138 (1971). See also pp. 131–135 of M. P. Tombs and A. R. Peacocke, "The Osmotic Pressure of Biological Macromolecules," Oxford Univ. Press, London and New York, 1974.
26. J. B. Harry and R. F. Steiner, *Biochemistry* **8**, 5060 (1970).
27. E. T. Adams, Jr., A. H. Pekar, D. A. Soucek, L.-H. Tang, G. Barlow, and J. L. Armstrong, *Biopolymers* **7**, 5 (1969).
28. P. D. Jeffrey, *Biochemistry* **8**, 5217 (1969).
29. A. T. deB. Andrews and F. J. Reithel, *Arch. Biochem. Biophys.* **141**, 538 (1970).
30. R. F. Steiner, *Biochemistry* **9**, 1375 (1970).
31. R. F. Steiner, *Biochemistry* **9**, 4268 (1970).
32. H. Vink, *Eur. Polym. J.* **7**, 1411 (1971).
33. H. B. Bull and K. Breese, *Arch. Biochem. Biophys.* **149**, 164 (1972).
34. H. Vink, *Eur. Polym. J.* **10**, 149 (1974).
35. D. A. Soucek and E. T. Adams, Jr., *J. Colloid Interface Sci.* **55**, 571 (1976).
36. P. J. Wan and E. T. Adams, Jr., *Biophys. Chem.* **5**, 207 (1976).

TABLE II
SOME REFERENCES TO VAPOR PRESSURE OSMOMETRY (VPO)

Researchers	Reference[a]	Remarks
Hill	(1)	The first paper on thermal osmometry. Two thermopiles covered with filter paper were put in a thermostatted chamber. Some solution was put on one of the filter papers and some solvent on the other. The vapor pressure lowering was detected by a change in temperature
Brady, Huff, and McBain	(2)	The first published work on vapor pressure osmometers equipped with thermistors. Osmotic coefficients of potassium laurate at 0° and 30° were determined. There was good agreement at 0° with osmotic coefficients obtained from freezing point depression of aqueous solutions
Huff, McBain, and Brady	(3)	Osmotic and activity coefficients of 19 detergents in aqueous solutions at 30° were determined. Additional experiments were carried out on 10 detergents solutions at 50°
Davies and Thomas	(4)	The indefinite self-association of amides in benzene solutions was studied. Davies and Thomas tried to analyze their data using ideal type I and type III indefinite self-associations
Burge	(5)	This was one of the first papers to report osmotic coefficients in aqueous solutions at 37° obtained from a commercial vapor pressure osmometer. The materials used were sucrose, NaCl, Na_2SO_4, $Al_2(SO_4)_3$, KH_2PO_4, and K_2HPO_4 at concentrations ranging from 0.01 m to 0.4 m
Lee, Wu, and Montgomery	(6)	Lee et al. used VPO to establish the molecular weight of an asparaginyl carbohydrate obtained from ovalbumin
T'so et al.	(7)	The base stacking of purine, 6 methylpurine, and 14 purine and pyrimidine nucleosides in aqueous solutions was studied by VPO and ${}^1H$ nuclear magnetic resonance. T'so et al. conclusively showed that base stacking exists by preparing derivatives that could not hydrogen bond; these derivatives associated more strongly in aqueous solutions

TABLE II (*Continued*)

Researchers	Reference[a]	Remarks
Elias and Bareiss	(8)	The self-association of polyethylene glycol in organic solvents at 25° was reported. Methods for analyzing ideal monomer-n-mer and type I indefinite self-associations were presented
Solie and Schellman	(9)	The self-association and also the mixed association of some nucleosides in aqueous solutions at 25° was examined by VPO and by sedimentation equilibrium experiments
Kertes and Markovits	(10)	Kertes and Markovits used the Bjerrum method for analyzing ideal self-associations of alkylammonium tetrahaloferrates in organic solvents. In addition they showed an alternative interpretation of the data, namely, as a nonspecific nonideality of the system in terms of activity and osmotic coefficients
Van Dam	(11)	This is a brief discussion of the theory of VPO; association equilibria were not considered
Armstrong	(12)	A comparison of two commercial vapor pressure osmometers is given here
Fontell	(13)	The osmotic coefficients of bile salts in aqueous solutions were reported. No supporting electrolytes were used. Fontell's data indicated self-association
Pörschke and Eggers	(14)	The self-association in aqueous solution of N-6,9-dimethyladenine and other related compounds was studied at four temperatures (10.5°–50°)
Elias	(15)	The self-association of nonionic detergents at three different temperatures was studied by light scattering, VPO, and sedimentation equilibrium experiments. Good agreement was obtained by all three methods
Lo, Escott et al.	(16)	The temperature-dependent self-association of dodecylammonium propionate in benzene and cyclohexane was examined. Methods for analyzing nonideal self-associations by VPO were presented

TABLE II (*Continued*)

Researchers	Reference[a]	Remarks
Plesiewicz, Stępień, Bolewska, and Wierzchowski	(17)	These workers studied the temperature-dependent self-association in aqueous solutions of 25 uracil derivatives. These derivatives were divided into three classes: alkylated uracils, 5- and 6-substituted 1,3-dimethyl uracils, and pyrimidine nucleosides. All the associations could be described as a sequential, indefinite self-association having all molar equilibrium constants equal

[a] Key to references:
1. A. V. Hill, *Proc. R. Soc. London, Ser. A,* **127,** 9 (1930).
2. A. P. Brady, H. Huff, and J. W. McBain, *J. Phys. Chem.* **55,** 304 (1951).
3. H. Huff, J. W. McBain, and A. P. Brady, *J. Phys. Chem.* **55,** 311 (1951).
4. M. Davies and D. K. Thomas, *J. Phys. Chem.* **60,** 763, 767 (1956).
5. D. E. Burge, *J. Phys. Chem.* **67,** 2590 (1963).
6. Y. C. Lee, Y. C. Wu, and R. Montgomery, *Biochem. J.* **91,** 9c (1964).
7. P. O. P. Ts'o, *Ann. N. Y. Acad. Sci.* **153,** 785 (1969) and references cited therein.
8. H.-G. Elias and R. Bareiss, *Chimia* **21,** 53 (1967).
9. T. N. Solie and J. A. Schellman, *J. Mol. Biol.* **33,** 61 (1968).
10. A. S. Kertes and G. Markovits, *J. Phys. Chem.* **72,** 4202 (1968). See also A. S. Kertes, O. Levy, and G. Y. Markovits, *J. Phys. Chem.* **74,** 3568 (1970) and G. Y. Markovits, O. Levy, and A. S. Kertes, *J. Colloid Interface Sci.* **47,** 424 (1974).
11. J. Van Dam, *in* "Characterization of Macromolecular Structure," pp. 336–342. Publ. No. 1573, National Academy of Sciences, Washington, D.C., 1968.
12. J. L. Armstrong, *Appl. Polym. Symp.* **8,** 17 (1969).
13. K. Fontell, *Kolloid-Z. Z. Polym.* **244,** 246 (1971).
14. D. Pörschke and F. Eggers, *Eur. J. Biochem.* **26,** 490 (1972).
15. H.-G. Elias, *J. Macromol. Sci., Chem.* **A7,** 601 (1973).
16. F. Y.-F. Lo, B. M. Escott, E. J. Fendler, E. T. Adams, Jr., R. D. Larsen, and P. W. Smith, *J. Phys. Chem.* **79,** 2609 (1975).
17. E. Plesiewicz, E. Stępień, K. Bolewska, and K. L. Wierzchowski, *Biophys. Chem.* **4,** 131 (1976).

increase exponentially with time, and then it will level off to a final equilibrium value. This is illustrated in Fig. 2. Note that solutions having different solute concentrations will show different final equilibrium pressures. Why does this flow of solvent occur?

When the solution was made up, the presence of the solute (component 2) lowered the chemical potential (μ_1) of the solvent. The pure solvent (phase α), on the other hand, has a higher chemical potential, which is designated as μ_1^0. This inequality in the solvent chemical potential ($\mu_1^\alpha > \mu_1^\beta$; note that $\mu_1^\alpha = \mu_1^0$) cause the solvent flow. Eventually the increased hydrostatic pressure on phase β will cause the chemical potential of the solvent in phase β to be the same as μ_1^0. How can the flow of solvent be prevented? The simplest way to do this is to increase the pressure on phase β so that $p^\beta = p^0 + \pi$. This excess pressure π, which prevents the flow of solvent from phase α to phase β, is called the osmotic pressure. What happens if $p^\beta > p^0 + \pi$? In this case solvent will flow

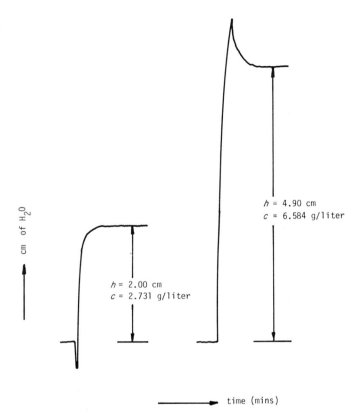

FIG. 2. Attainment of osmotic equilibrium. Here we have plotted pressure, in centimeters of solvent (H_2O), versus time. These drawings are redrawn from chart recorder tracings of osmotic pressure experiments by P. J. Wan (see Tables IV and V) on aqueous dextran solutions. These experiments were performed on Melabs (now Wescan) high speed membrane osmometers at 29.4°. The trace on the left was performed on a dextran solution whose concentration was 2.731 g/liter; here a value of $h = 2.00$ cm of H_2O was obtained. The trace on the right was obtained with a solution whose concentration was 6.584 g/liter; here a value of $h = 4.90$ cm of H_2O was observed. In each case the traces level off exponentially to the horizontal equilibrium value; this value, indicated by the vertical line labeled h, is used for the measurement of osmotic pressure. How the sample is introduced and how the solution drain valve is handled affect the transient pattern. The use of a rubber bulb to help introduce the sample can cause a downscale deflection, so that the pen tracing approaches equilibrium as shown in the left drawing.

from $\beta \rightarrow \alpha$; this phenomenon is known as reverse osmosis. If one can find a membrane impermeable to small solutes like NaCl or KCl, then reverse osmosis could be used to purify brackish water. There is an alternative way to prevent the flow of solvent from $\alpha \rightarrow \beta$; one can change the pressure on phase α so that $p^{\alpha} = p^0 - \pi$. The pressure on phase β will be $p^0 = p^{\beta}$, and Δp, the pressure difference, will be $\Delta p = p^{\beta} - p^{\alpha} = p^0 - (p^0 - \pi) = \pi$. Thus, as long as the difference in pressure between the two phases is π, no net solvent flow occurs across the membrane from $\beta \rightarrow \alpha$ (or from $\alpha \rightarrow \beta$).

How is π related to X_2? to the molecular weight of the solute, M_2? This will be shown in the discussion that follows.

The equilibrium conditions (no flow of solvent) are

$$T^{\alpha} = T^{\beta} \text{ (constant } T) \tag{1}$$

and

$$\mu_1^{\beta}(T, p, X_1) = \mu_1^{\alpha}(T, p^0, X_1 = 1) = \mu_1^0 \tag{2}$$

i.e., the chemical potential of the solvent (component 1) is the same in both phases. Note that the pressure p^{β} must equal $p^0 + \pi$ for $\mu_1^{\alpha} = \mu_1^{\beta}$.

The chemical potential of component i is defined by

$$\mu_i = \mu_i^0 + RT \ln a_i = \mu_i^0 + RT \ln f_i X_i \tag{3}$$

Here, μ_i^0 = standard state chemical potential, $a_i = f_i X_i$ = activity of component i, X_i = mole fraction of component i, f_i = activity coefficient of i (on the mole fraction scale), R = universal gas constant, and T = absolute temperature. The standard state for the solvent is chosen to be the pure solvent at T and p^0; the standard state for the solute is usually chosen so that $\lim_{X_2 \rightarrow 0} f_2 = 1$. There are several conventions for standard states, and these are discussed in texts on chemical thermodynamics.[27-29] An ideal solution is usually defined as one for which $f_i = 1$ and hence

$$\mu_i = \mu_i^0(T) + RT \ln X_i \text{ (ideal solution)} \tag{4}$$

Note that μ_i^0 is a function of temperature; thus μ_i^0 for the same solvent will be different at each temperature used.

The chemical potential is a function of temperature, pressure, and mole fraction of solute or solvent. Since $X_1 + X_2 = 1$, one need only specify one mole fraction. For a multicomponent system the chemical

[27] I. M. Klotz, "Chemical Thermodynamics." Benjamin, New York, 1964.

[28] K. Denbigh, "The Principles of Chemical Equilibrium," 3rd ed. Cambridge Univ. Press, London and New York, 1971.

[29] J. G. Kirkwood and I. Oppenheim, "Chemical Thermodynamics." McGraw-Hill, New York, 1961.

potential of any component will be a function of T, p, and n-1 mole fractions. Thus it is convenient to write

$$\mu_1 = \phi\,(T, p, X_2) \tag{5}$$

For each solution the quantities T and X_2 are held constant, but note that one uses several solutions having different values of X_2. For any one experiment the pressure is varied to make the chemical potential of the solvent the same on both sides of the membrane. To show this mathematically, one starts with a standard thermodynamic relation

$$(\partial\mu_1/\partial p)_{T,X2} = \overline{V}_1 \tag{6}$$

Here $\overline{V}_1$ is the partial molar volume of the solvent. The variables in Eq. (6) are separated and integrated as follows:

$$d\mu_1{}^\beta = \overline{V}_1\,dp \qquad \text{(const. } T, X_2)$$

$$\int_{p^0}^{p^0+\pi} d\mu_1{}^\beta = \int_{p^0}^{p^0+\pi} \overline{V}_1\,dp$$

This gives

$$\mu_1{}^\beta(T, X_2, p^0 + \pi) - \mu_1{}^\beta(T, X_2, p^0) = \pi\overline{V}_1 \tag{7}$$

if $\overline{V}_1$ is independent of pressure. Water has such a small isothermal compressibility that this is an excellent assumption. Even though organic solvents are compressible, the osmotic pressure encountered with macromolecular solutions in organic solvents is usually less than one atmosphere, so that this is still a good assumption in this case. When $p^\beta = p^0 + \pi$ there is no net flow of solvent across the membrane, so that

$$\mu_1{}^\beta\,(T, X_2, p^0 + \pi) = \mu_1{}^0\,(T, p^0) \tag{8}$$

This is the equilibrium condition. Thus Eq. (7) becomes[1,2,4,18,26,30]

$$\pi\overline{V}_1 = \mu_1{}^0\,(T, p^0) - \mu_1(T, p^0, X_2) \tag{9}$$

At this stage one substitutes Eq. (3) or Eq. (4) into Eq. (9) to obtain

$$\pi\overline{V}_1 = -RT\,\ln f_1 X_1 \tag{10}$$

or

$$\pi\overline{V}_1 = -RT\,\ln X_1 \text{ (ideal solution)} \tag{11}$$

Since $X_1 = 1 - X_2$ and since

$$\ln(1 - X_2) = X_2 - (X_2{}^2/2) - \cdots, X_2 < 1 \tag{12}$$

one obtains from Eq. (11)

$$\pi\overline{V}_1 = RT\,(X_2 + X_2{}^2/2 + \ldots) \cong RT\,X_2 \tag{13}$$

[30] C. Tanford, "Physical Chemistry of Macromolecules." Wiley, New York, 1961.

Suppose $M_2 = 10,000$ and one has a 10 g/liter aqueous solution, then we note that $n_2 = 0.001$ and $n_1 = 55.5$ (1 liter of water has 55.5 mol of water). Thus $X_2 = 0.001/55.501 \cong 0.001/55.5 = 1.8 \times 10^{-5}$. The mole fraction of component $2(X_2)$ in a solution containing 10 g/liter is quite small, so for large molecules one is justified in truncating the series in Eq. (12). For very dilute solutions[1,27]

$$X_2 = [P_2]\overline{V}_1^0/1000 = (c_2\overline{V}_1^0)/1000\, M_2 \qquad (14)$$

Here $[P_2]$ is the molar concentration of the solute, c_2 is its concentration in g/liter, and M_2 is its molecular weight. $\overline{V}_1^0$ is the partial molar volume of the solvent at infinite dilution; $\overline{V}_1^0 \cong \hat{V}_1^0$, the molar volume of the solvent. Assuming $\overline{V}_1/\overline{V}_1^0 \cong 1$, the substitution of Eq. (14) into Eq. (13) leads to

$$1000\,\pi/RT = [P_2] = c_2/M_2 \qquad (15)$$

Some authors prefer to retain the second term in the parentheses in Eq. (13); for this case, Eq. (15) becomes[30]

$$1000\,\pi/RT = c_2/M_2\,[1 + (c_2\overline{V}_1^0)/2000\,M_2 + \cdots] \qquad (15a)$$

For the nonideal case, one can use Tanford's[30] treatment of the chemical potential of the solvent. Using a statistical theory for the number of ways that a solute could be added to a fixed volume of solvent, Tanford[30] showed that

$$\mu_1 - \mu_1^0 = (-RT\overline{V}_1^0 c_2)/1000\,(1/M_2 + (Nuc_2)/2000\,M_2^2)$$
$$= -[(RT\overline{V}_1^0 c_2)/1000\,M_2]\,(1 + Bc_2/2) \qquad (16)$$

Here

$$B = (Nu)/(1000\,M_2) \qquad (17)$$

In Eq. (17) B is the second virial coefficient, N is Avogadro's number, and u is the excluded volume of the solute. The quantity u depends on the shape of the molecule. Comparison of Eqs. (9) and (16) leads to

$$1000\,\pi/RT = (c_2/M_2)(1 + Bc_2/2) = c_2/M_{2\,\text{app}} \qquad (18)$$

Here

$$1/M_{2\,\text{app}} = 1/M_2 + (B/2M_2)\,c_2 = 1/M_2 + B'c_2/2 \qquad (19)$$

For an ideal solution (on the c_2 or g/liter scale) $B = 0$, so that Eq. (18) reduces to Eq. (15). The simplest interpretation of nonideal behavior is to use the excluded volume (compare this with the covolume of the van der Waals' equation of state for a gas). In very simple terms the excluded volume attempts to account for the volume occupied by each solute molecule; no two solute molecules can occupy the same volume at the same time. If the solutes are considered as hard spheres, then the excluded volume, u, becomes[30]

$$u = (32/3)\pi r^3 = (8 M_2 v_2)/N \tag{20}$$

Here r is the radius of the solute molecule and v_2 is the specific volume of the solute. For information about excluded volumes for various shapes of solutes one should consult the Flory[18] and Tanford[30] monographs, as well as the papers by Ogston,[31,32] Ogston and Winzor,[33] Nichol, Jeffrey and Winzor,[34] and Nichol and Winzor.[35]

Another way to obtain the osmotic pressure equation for ideal or nonideal solutions is to use the Gibbs–Duhem equation,[1,36,37]

$$V dp = n_2 d\mu_2 \tag{21}$$

Since $\mu_1{}^\alpha = \mu_1{}^\beta$ at osmotic equilibrium, Eq. (21) relates changes in osmotic pressure to changes in the chemical potential of the solute. To develop the osmotic pressure equation, one first rearranges Eq. (21) to give

$$dp = (n_2/V) d\mu_2 = (c_2/1000 M_2) d\mu_2 \tag{22}$$

Then one uses the following relations:

$$\mu_2 = f(c_2, p, T) \tag{23}$$
$$\mu_2 = \mu_2{}^0 + RT \ln y_2 c_2 \tag{24}$$

and

$$\ln y_2 = BM_2 c_2 \tag{25}$$

Here y_2 is the activity coefficient of the solute on the c scale (g/liter) and B is a constant whose value depends on the temperature and the solute–solvent combination. At constant T, one notes that

$$d\mu_2 = (\partial\mu_2/\partial c_2)_{T,p} \, dc_2 + (\partial\mu_2/\partial p)_{T,c_2} \, dp \tag{26}$$
$$(\partial\mu_2/\partial p)_{T,c_2} = M_2\bar{v}_2 = \bar{V}_2 \tag{27}$$

and

$$(\partial\mu_2/\partial c_2)_{T,p} = RT[1/c_2 + (\partial \ln y_2/\partial c_2)_{T,p}] \tag{28}$$

[31] A. G. Ogston, *Trans. Faraday Soc.* **49**, 1481 (1953).
[32] A. G. Ogston, *J. Phys. Chem.* **74**, 668 (1970).
[33] A. G. Ogston and D. J. Winzor, *J. Phys. Chem.* **79**, 2496 (1975).
[34] L. W. Nichol, P. D. Jeffrey, and D. J. Winzor, *J. Phys. Chem.* **80**, 648 (1976).
[35] L. W. Nichol and D. J. Winzor, *J. Phys. Chem.* **80**, 1980 (1976).
[36] E. F. Casassa and H. Eisenberg, *Adv. Protein Chem.* **19**, 287 (1964).
[37] P. F. Curran and A. Katchalsky, "Nonequilibrium Thermodynamics in Biophysics." Harvard Univ. Press, Cambridge, Massachusetts, 1965.

The substitution of Eqs. (26–28) into Eq. (22) leads to

$$\frac{1000\,dp}{RT}\left(1 - \frac{\bar{v}_2 c_2}{1000}\right) = \frac{dc_2}{M_2} + BM_2 c_2 dc_2 \tag{29}$$

Division of both sides of Eq. (29) by $1 - (\bar{v}_2 c_2)/1000$ leads to

$$\frac{1000\,dp}{RT} = \frac{dc_2}{M_2} + \left(BM_2 + \frac{\bar{v}_2}{1000\,M_2}\right) c_2 dc_2 + \cdots \tag{30}$$

since

$$1/\left(1 - \frac{\bar{v}_2 c_2}{1000}\right) = 1 + \frac{\bar{v}_2 c_2}{1000} + \left(\frac{v_2 c_2}{1000}\right)^2 + \cdots \cong 1 + \frac{\bar{v}_2 c_2}{1000},\ \frac{\bar{v}_2 c_2}{1000} < 1$$

For most proteins $\bar{v}_2$ lies between 0.70 and 0.75 ml/g, so that the value of $\bar{v}_2 c_2/1000$ for a solution containing 10 g/liter (10 mg/ml) is $0.007 \ll 1$. The integration of Eq. (30) leads to

$$\int_{p^0}^{p^0+\pi} \frac{1000\,dp}{RT} = \int_0^{c_2} \left[\frac{1}{M_2} + \left(BM_2 + \frac{\bar{v}_2}{1000\,M_2}\right) c_2 + \cdots\right] dc_2$$

which becomes

$$\frac{1000\pi}{RT} = \frac{c_2}{M_2} + \left(BM_2 + \frac{\bar{v}_2}{1000\,M_2}\right)\frac{c_2{}^2}{2} + \cdots = \frac{c_2}{M_{2\,\mathrm{app}}} \tag{31}$$

If one assumes that the higher terms in Eq. (31) are negligible, then

$$\frac{1}{M_{2\,\mathrm{app}}} = \frac{1}{M_2} + \left(BM_2 + \frac{\bar{v}_2}{1000\,M_2}\right)\frac{c_2}{2} = \frac{1}{M_2} + \frac{B^* c_2}{2} \tag{32}$$

The limits of integration used to obtain Eq. (31) vary between p^0, the pressure exerted on the pure solvent ($c_2 = 0$), and $p^0 + \pi$, the pressure that must be exerted on the solution (phase β in Fig. 1) to prevent flow of solvent across the membrane.

It is also possible to obtain an expression for the osmotic pressure of nonideal solutions via the osmotic coefficient, g. The chemical potential of the solvent [see Eq. (3)] can also be written as[1,4,29]

$$\mu_1 = \mu_1{}^0 + g\,RT\,\ln X_1 \tag{33}$$

A comparison of Eqs. (3) and (33) leads to the relation between the activity coefficient, f_1, and the osmotic coefficient, g, namely,

$$f_1 = X_1{}^{g-1} \tag{34}$$

The osmotic coefficient, g, can also be defined as the ratio of the osmotic pressure of a real solution to the osmotic pressure it would exert if it were ideal, i.e.,[4]

$$g = \pi(\text{real})/\pi(\text{ideal}) \tag{35}$$

or

$$\pi(\text{real}) = g\pi(\text{ideal})$$

Thus for real solutions

$$\pi V_1^0/RT = gX_2 \tag{36}$$

For concentrations in grams per liter, one can expand g as a power series in c_2 to obtain [with the use of Eq. (14) also][29]

$$\frac{1000\pi}{RT} = \frac{c_2}{M_2} (1 + B_1'c_2 + B_2'c_2^2 + \cdots)$$

$$= \frac{c_2}{M_2} + \frac{Bc_2^2}{2} + \cdots = \frac{c_2}{M_{2\,\text{app}}} \tag{37}$$

which is the same as Eq. (18) or Eq. (31).

In this discussion we have used only one virial coefficient. Sometimes the solution may be more nonideal than we have considered, so that third or even fourth virial coefficients have to be included. For more details about this situation, the reader should refer to Flory[18] monograph or to the papers by Vink[38] or by Bull and Breese.[39]

Vapor Pressure Osmometry[7,9,40−42]

When a solution is made up, the presence of the solute or solutes causes the chemical potential of the solvent to be lower than it is in the pure state. Membrane osmometry is one way of detecting this chemical potential difference. Another way of detecting this difference is to measure the lowering of the vapor pressure of the solvent caused by the presence of the added solute or solutes. One of the most convenient ways of measuring vapor pressure lowering is to use vapor pressure osmometry.

[38] H. Vink, *Eur. Polym. J.* **7**, 1411 (1971); **10**, 149 (1974).
[39] H. B. Bull and K. Breese, *Arch. Biochem. Biophys.* **149**, 164 (1972).
[40] See, for example, "Operating and Service Manual for Vapor Pressure Osmometer 302B, Hewlett-Packard, Avondale Division, Avondale, PA, 1968; or the latest version of "Operating Instructions for the Vapor Pressure Osmometer," KG Dr.-Ing. Herbert Knauer and Co., GmBH West Berlin.
[41] J. L. Armstrong, *Appl. Polym. Symp.* **8**, 17 (1969).
[42] F. Y.-F Lo, B. M. Escott, E. J. Fendler, E. T. Adams, Jr., R. D. Larsen, and P. W. Smith, *J. Phys. Chem.* **79**, 2609 (1975).

In this technique a drop of solution is placed on one thermistor bead and a drop of solvent is placed on another thermistor bead. These thermistor beads are in a very precisely thermostatted chamber, which also contains a solvent reservoir. Because of the vapor pressure lowering, solvent will condense on the solution drop, diluting it and producing heat in this process, and this heat can be detected by the thermistors as a temperature difference. The two thermistors are part of a Wheatstone bridge circuit; one adjusts the slide wire resistance to balance the thermistors. Thus, the temperature difference, ΔT, is proportional to the resistance, r, or the microvolts imbalance, E, of the Wheatstone bridge circuit; and these quantities in turn are proportional to $\Delta p/p_0$, the vapor pressure lowering. More details about vapor pressure osmometry will be found in the book by Collins, Bareš, and Billmeyer,[9] the manufacturers' technical mannuals,[40] the review by Armstrong,[41] and the paper by Lo et al.[42]

Here is a brief outline of the theory. The vapor pressure lowering $\Delta p/p_0$, is proportional to X_2, the mole fraction of solute; thus[43]

$$(p_0 - p)/p_0 = \Delta p/p_0 = gX_2 \tag{38}$$

Comparison with Eq. (36) leads to

$$\Delta p/p_0 = \pi V_1^0/RT \tag{39}$$

This gives the relation between vapor pressure lowering and osmotic pressure. Now the Clapeyron equation is used to relate ΔT to $\Delta p/p_0$; thus[43]

$$(\Delta T)_{eq} = (RT^2/\Delta H_v) \, \Delta p/p_0 = (RT^2/\Delta H_v) \, gX_2 = K_{eq} \, gX_2 \tag{40}$$

Here ΔH_v is the molar latent heat of vaporization. Equation (40) describes the condition that would apply if equilibrium thermodynamic conditions existed. However, the vapor pressure osmometer is a steady-state instrument in which heat losses from the thermostatted chamber are balanced by a continuous condensation of solvent onto the solution. Therefore, the steady-state temperature differences, $(\Delta T)_{ss}$, can be expressed as

$$(\Delta T)_{ss} = K_{ss} \, gX_2 \tag{41}$$

Here K_{ss} is a proportionality constant between $(\Delta T)_{ss}$ and gX_2. In principle, one could evaluate K_{ss} and make vapor pressure osmometry an absolute method, but one would have to know such things as the size of the drop, the diffusion coefficient of the solvent in the vapor phase, and the latent heat of vaporization at the temperature of the chamber.[43] So it is easier to use an indirect method, and one calibrates the instrument using a

[43] J. Van Dam, in "Characterization of Macromolecular Structure." Publication No. 1573, National Academy of Sciences, Washington, D.C., 1968.

material of known molecular weight. With the use of Eqs. (38) and (41) one obtains

$$(\Delta T)_{ss} = K_{ss}(\pi V_1^0/RT) = \left(\frac{K_{ss}V_1^0}{RT}\right)\frac{c}{M_{2 \text{ app}}} = K'_{ss}(c/M_{2 \text{ app}}) \qquad (42)$$

Finally, one notes that $(\Delta T)_{ss}$ is proportional to the slide wire resistance, r, or to the microvolts imbalance, E, of the Wheatstone bridge circuit of the vapor pressure osmometer. Using E one obtains[42]

$$E/K_{vp} = c_2/M_{2 \text{ app}} = c_2/M_2 + Bc_2^2/2 \qquad (43)$$

The quantity K_{vp} is an apparatus constant whose value depends on the temperature and the solvent being used. The apparatus constant, K_{vp}, is determined by performing a calibration experiment at the desired temperature using a solution containing the desired solvent and a nonvolatile solute of known molecular weight. Benzil is a useful calibration standard with many organic solvents. Note that nonvolatile solutes must be used in vapor pressure osmometry. Also note that Eq. (43) gives the same result as does Eq. (37).

Three-Component, Ionizable Systems[1,2,30,36]

When one does osmotic pressure, light scattering, or sedimentation equilibrium experiments on a protein or a polyelectrolyte, one is usually doing measurements on a multicomponent system consisting of the protein or polyelectrolyte, supporting electrolyte like KCl or NaCl to control the ionic strength and buffers to control the pH. Clearly, we are not dealing with a two-component system. In aqueous solutions in the absence of a diffusible salt (component 3), component 2 (represented as PX_z) could ionize to give

$$PX_z + aq \rightarrow P^{z+} + z\ X^- \qquad (44)$$

Here z is the stoichiometric charge on the macroion P^{z+}; for incomplete ionization z would become the effective charge. Osmotic pressure experiments carried out on a solution of PX_z in water would give

$$\lim_{c_2 \to 0} \frac{\pi}{c_2 RT} = \frac{M_2}{|z| + 1} = M'_2 \qquad (45)$$

Here $|z|$ is the absolute value of z. If $|z| = 10$, for example, then $M'_2 = M_2/11$. This effect of reducing the molecular weight because of ionization is usually called the primary charge effect. It can be overcome by introducing sufficient diffusible salt (also known as supporting electrolyte) to shield the P^{z+} ion. Component 3, which is symbolically represented as BX, could be represented by NaCl, KCl, or some other electrolyte. While the presence of sufficient BX swamps out the primary charge effect, there are additional complications. There will be a residual charge effect due to the Donnan equilibrium.[1,2,30,36]

Since BX ionizes in water, i.e.,

$$BX + aq \rightarrow B^+ + X^- \tag{46}$$

the macroion P^{z+} will repel ion B^+ because both are positive ions. The ion P^{z+} cannot pass through the membrane (SM in Fig. 3), but the ion B^+ (and also the ion X^-) can pass through the membrane. Thus some of the B^+ ion will flow from phase β to phase α, and some of the X^- ion will be carried along with the B^+, since electroneutrality is maintained at all times. At

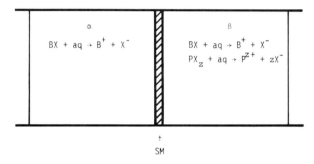

FIG. 3. Diagrammatic representation of the Donnan equilibrium. Here SM represents the semipermeable membrane. Note that H_2O and the ions B^+ and X^- can pass through the membrane, whereas the ion P^{z+} cannot. Thus at constant temperature the following conditions apply:

Initial conditions (T = constant)

Phase α		Phase β
μ_3'	$\neq$	μ_3
m_3'	$=$	m_3
$\mu_1(T,p^0)$	$\neq$	$\mu_1(T,p^0,m_2)$
$p^\alpha = p^0$		$p^\beta = p^0$

At osmotic equilibrium (T = constant)

Phase α		Phase β
μ_3'	$=$	μ_3
$\mu_1'\,(T,p^0)$	$=$	$\mu_1(T,p^0 + \pi, m_2)$
m_3'	$\neq$	m_3
$p^\alpha = p^0$		$p^\beta = p^0 + \pi$

The primes refer to components in phase α. Component 2, PX_z is in phase β only, since P^{z+} cannot pass through SM.

osmotic equilibrium one would note that there would be unequal concentrations of B^+ and X^- on both sides of the membrane. Careful measurements would show that

$$(m_{B^+})^\alpha > (m_{B^+})^\beta \tag{47}$$

and

$$(m_{X^-})^\alpha < (m_{X^-})^\beta \tag{48}$$

It has been shown for a three-component, ideal system that the osmotic pressure equation becomes[2,30]

$$\frac{1000\pi}{RT} = \frac{M_1}{\overline{V}_1}(m_2 + m_B - m_B' + m_X - m_X') \tag{49}$$

The primes refer to phase α in Fig. 3. At first this equation looks formidable, but one can use some consequences of the Donnan equilibrium to simplify these equations. At osmotic equilibrium

$$\mu_3' = \mu_3 \quad (T = \text{constant}) \tag{50}$$

Letting $(\mu_3^0)' = \mu_3^0$, it follows that

$$a_3' = a_3 \tag{51}$$

For phase α

$$a_3' = m_B' m_X' (\gamma_\pm')^2 = (m_3')^2 (\gamma_\pm')^2 \tag{52}$$

and

$$m_B' = m_X' = m_3' \tag{53}$$

because of electroneutrality. For phase β

$$a_3 = m_B m_X (\gamma_\pm)^2 \tag{54}$$

and the electroneutrality relation is

$$m_X = m_B + z m_2 \tag{55}$$

Assuming that $(\gamma_\pm')^2 \simeq (\gamma_\pm)^2$, then with the aid of Eqs. (51)–(55) one can show that[1,2,30]

$$(m_B - m_B') + (m_X - m_X') = \frac{Z^2 m_2^2}{4 m_3} + \cdots \tag{56}$$

so that Eq. (49) becomes

$$\frac{1000\pi}{RT} = \frac{M_1}{\overline{V}_1}\left(m_2 + \frac{Z^2 m_2^2}{4 m_3} + \cdots\right) \tag{57}$$

Neglecting the higher terms, Eq. (57) becomes[30]

$$\frac{1000\pi}{RT} = \frac{c_2}{M_2} + \frac{Z^2 \overline{V}_1 c_2{}^2}{4m_3 M_1 M_2{}^2} \tag{58}$$

for c_2 in grams per liter. Thus, there is a residual charge effect—the second term on the right-hand side of Eq. (58). However, this residual charge effect is much smaller than the primary charge effect. Furthermore, provided one can measure c_2 properly in a three-component system (more about this later), one notes that[30]

$$\lim_{c_2 \to 0} \frac{1000\pi}{c_2 RT} = \frac{1}{M_2} \tag{59}$$

Note that even though there are many more moles of component 3 (and hence of its ions) than there are of component 2 or the ion P^{z+}, we have been able to measure M_2 from osmotic pressure experiments.

What happens in the more general case where nonideal terms are included? Remember that the activity of component 2 will be affected by the concentration of diffusible solute components and vice versa. This problem has been discussed in great detail by Kelly and Kupke,[2] Edsall, Edelhoch et al.,[44] Casassa and Eisenberg,[36] and Tombs and Peacocke.[1] We can only give a brief sketch of how to deal with this problem.

This problem can be overcome by using a formulation for a nondiffusible (macromolecular) component based on a consideration of the distribution of species across a semipermeable membrane at osmotic equilibrium. This formulation is chosen so that the interactions between the nondiffusible and diffusible component vanish. Pioneering work was done in this area by Scatchard.[17] Subsequently, Casassa and Eisenberg[36] showed how one could obtain workable equations for three-component (or for multicomponent) systems that are formally identical to the osmotic pressure equation. For multicomponent systems the Scatchard formalism leads to this definition of the chemical potential of component J, namely,[1,17,36]

$$\mu_J = \mu_J{}^0 + RT \ln a_J = \mu_J{}^0 + RT \, \Sigma v_{iJ} \ln m_i + RT\beta_J \tag{60}$$

Here, $\mu_J{}^0$ is the chemical potential of component J in its standard state, a_J is the activity of component J, $RT\beta_J$ is the excess chemical potential of component J (a measure of the thermodynamic nonideality), v_{iJ} is the number of moles of species i (K^+ or Cl^-, for example) included in 1 mol of component J, and $m_i = \Sigma_J v_{iJ} m_J$.

[44] J. T. Edsall, H. Edelhoch, R. Lontie, and P. E. Morrison, J. Am. Chem. Soc. 72, 4641 (1950).

Since the concentrations of the components are taken as independent variables, it is required that the ν_{iJ} be taken in electrically neutral combinations—electroneutrality must apply. With the aid of Eq. (60) and the following equations:

$$\mu_{JK} = \left(\frac{\partial \ln a_J}{\partial m_K}\right)_{T,p,m_{q \neq k}} = \left(\frac{\partial \mu_K}{\partial m_J}\right)_{T,p,m_{q \neq J}} = \mu_{KJ} \tag{61}$$

$$a_{JK} = \left(\frac{\partial \ln a_J}{\partial m_K}\right)_{T,p,m_{q \neq k}} = \left(\frac{\partial \ln a_K}{\partial m_J}\right)_{T,p,m_{q \neq J}} \tag{62}$$

$$= \sum_i \frac{\nu_{iJ}\nu_{iK}}{m_i} + \beta_{JK} = a_{KJ}$$

and

$$a_{JK} = \mu_{JK}/RT \tag{63}$$

one can define the components in such a way that interactions between the diffusible components (such as KCl, NaCl, or buffer components) and the macromolecular or nondiffusible components vanish at osmotic equilibrium.[1,36] If this is done, component 2 is designated as 2* and $a_{23}^* = a_{32}^* = 0$. Then the following relation for a three-component system

$$1000 \frac{d\pi}{dc_2} = \frac{RT}{M_2} m_2 \left(a_{22} - \frac{a_{23}^2}{a_{33}}\right) \tag{64}$$

which is hard to deal with becomes

$$1000 \frac{d\pi}{dc_{2^*}} = \frac{RT}{M_{2^*}} m_{2^*} \left(a_{22^*} - \frac{a_{23}^{*2}}{a_{33}}\right) \tag{65}$$

$$= \frac{RT}{M_{2^*}} m_{2^*} a_{22^*}$$

$$= \frac{RT}{M_{2^*}} (1 + 2B^* M_{2^*} c_{2^*} + \cdots)$$

Here

$$B^* = \frac{V_m^0}{2(M_{2}^*)^2} \left[\sum_i \frac{(\nu_{2i}^*)^2}{m_i} + B_{22}^*\right] \tag{66}$$

This can be integrated to give[1,36]

$$1000\pi/RT = c_{2}^*/M_{2}^* + B^*(c_{2}^*)^2 = c_{2}^*/M_{2}^*{}_{app} \tag{67}$$

Note that Eq. (67) is formally identical to the equation for a two-component system [cf. Eqs. (32) or (37)].

What molecular weight does one measure this way? If the concentration determination is based on an elemental analysis (say percent N or S) and the percentage of the element in the macromolecule is known, then one obtains the molecular weight of the neutral macromolecular component. On the other hand, if the concentration determination is based on dry weights, and it is assumed that equal aliquots of the macromolecular and buffer solutions contain the same concentrations of supporting electrolyte or electrolytes, then one measures the molecular weight of a component 2* defined as

$$PX_z - (iz/2)BX$$

Here i is a number usually between 0 and 1 ($0 \leq i \leq 1$). For neutral nonionizing macromolecules $i = 0$. If the macromolecule ionizes completely then $i = 1$. For incomplete ionization $0 < i < 1$. It should be noted that the number of moles of 2 or 2* are the same, but that their weight concentrations will be different when 2* is defined as $PX_z - (iz/2)BX$.

In dealing with multicomponent systems, one should dialyze the material first to remove small molecular weight impurities, to obtain the right pH, and to obtain the right ionic strength. Several changes of solvent solution should be used. Concentration determinations should be performed on the solution and the solvent (or dialyzate) in dialysis equilibrium with the solution. For more details on multicomponent systems, especially the theory used in developing Eq. (67), one should consult the Casassa and Eisenberg[36] review or the Tombs and Peacocke[1] monograph.

An Example—The Molecular Weight and Second Virial Coefficient of Glycinin Subunits

Tombs[1] performed osmotic pressure experiments on glycinin, a soybean protein, in aqueous solutions containing 4 M guanidinium hydrochloride and 0.1 M Na_2SO_3. His experiments were done at 25°, and his experimental data are tabulated in Table III.

First one must convert h, the centimeters of H_2O, to π, the osmotic pressure, in atmospheres. To do this note that in CGS units $p = h\rho g$, where h is the height of the solvent, ρ is the density of solvent, and g is the acceleration of gravity. Thus we set

$$h\rho g \text{ (solvent)} = h\rho g \text{ (mercury)}$$

For 1 atmosphere, h (mercury) is 76 cm. Thus

$$h(\text{solvent}) = \frac{76\rho_{Hg}(\text{at } t)}{\rho \text{ solvent (at } t)}$$

At 25°, $h = (76)(13.5339)/(0.99707) = 1.032 \times 10^3$ cm of H_2O per atmo-

TABLE III

OSMOTIC PRESSURE DATA FOR GLYCININ IN 4 M GUANIDINIUM HYDROCHLORIDE[a]

c[b] (g/liter)	h[c] (cm of H_2O)	π[d] (atm)	π/c (atm-liter/g)
0.97	0.902	0.874×10^{-3}	0.902×10^{-3}
1.95	2.00	1.94	0.994
1.95	1.85	1.79	0.919
2.92	3.21	3.11	1.065
3.90	4.45	4.31	1.106
4.87	5.99	5.80	1.192
5.85	7.60	7.36	1.259
7.80	10.14	9.82	1.260
9.75	14.91	14.45	1.482
11.70	18.83	18.24	1.559

[a] These data are taken (by permission) from "The Osmotic Pressure of Biological Macromolecules" by M. P. Tombs and A. R. Peacocke, Oxford Univ. Press, London and New York, 1974.
[b] Total solute concentration (c).
[c] Osmotic pressure in centimeters of solvent measured on the osmometer (cf. Fig. 2).
[d] See text for details on how this is obtained.

sphere. Note that

$$\pi(\text{atm}) = \frac{h(\text{cm of } H_2O)}{1.023 \times 10^3 \text{ (cm/atm)}}$$

Figure 4 shows a plot of π/c vs c for glycinin in 4 M guanidinium hydrochloride (GuHCl) in 0.1 M Na_2SO_3. The inclined straight line indicates nonideal behavior, which may reflect in part the unfolding of the protein into a more random coillike configuration in the 4 M GuHCl–0.1 M Na_2SO_3 solution. The straight line through the data points was obtained by a linear regression (least squares) analysis, which gave

$$\text{Intercept} = \frac{RT}{M} = 0.859 \frac{(\text{ml-atm})}{\text{g}}$$

and

$$\text{Slope} = \frac{BRT}{2} = 6.09 \times 10^{-2} \frac{\text{ml-liter-atm}}{\text{g}^2}$$

Using $R = 82.05$ ml-atm/deg/mole and $T = 298.2°K$ (25°C), one obtains

$$M = 28.5 \times 10^3 \text{ g/mole}$$

and

$$B = 4.98 \times 10^{-6} \text{ (mole-liter/g}^2)$$

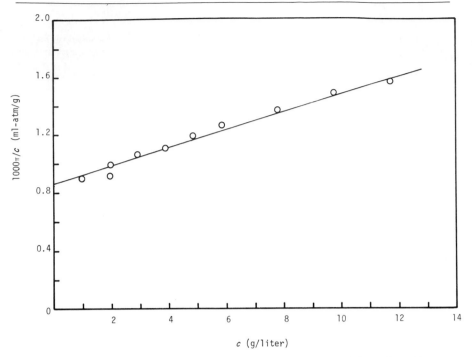

FIG. 4. Determination of M and B for a homogeneous, nonassociating solute. Here we have plotted $1000\pi/c$ vs. c for solutions of glycinin in 4 M guanidinium hydrochloride and 0.1 M Na$_2$SO$_3$. One obtains RT/M from the intercept and $BRT/2$ from the slope of this plot (see text). Redrawn from M. P. Tombs and A. R. Peacocke, "The Osmotic Pressure of Biological Macromolecules" (Oxford Univ. Press, London and New York, 1974), by permission of Dr. M. P. Tombs and the Oxford University Press).

Tombs and Peacocke[1] using a different method of fitting the data obtained

$$M = (27.81 \pm 2.80) \times 10^3 \text{ g/mole}$$

and

$$B = (4.98 \pm 1.12) \times 10^{-6} \text{ (mole-liter/g}^2)$$

The native protein has a molecular weight of 3.63×10^5 g/mole by sedimentation and diffusion studies; by light scattering the molecular weight was 3.45×10^5 g/mole. Using $M = 3.63 \times 10^5$ g/mole for the native protein, our estimate of the number of subunits is

$$\text{No. subunits} = (3.63 \times 10^5)/(28.5 \times 10^3) = 12.7 \text{ or } 13$$

Tombs and Peacocke reported the apparent number of subunits to be 14.

Heterogeneous, Nonassociating Solutes

Previously, we have been concerned with monodisperse, nonassociating solutes—solutes of only one size and chemical composition. Now our attention will be directed to polydisperse, nonassociating solutes; although the solute components are similar chemically, they will differ in size so that there will be a distribution of molecular weights. In some cases they may also show differences in degree of branching or in degree of substitution. Thus, one will no longer be dealing with a two-component system. Among the best known examples of polymers of biological or biochemical interest are gelatin, starch, glycogen, pectins, and bacterial capsular polysaccharides.

The solution in phase β of Fig. 1 consists of the solvent, component 1, and of solute components that cannot pass through the semipermeable membrane. These solute components are designated 2, 3, . . . , $q + 1$. The relation between the mole fractions of the components leads to

$$X_1 = 1 - X_2 - X_3 - \cdots - X_{q+1} = 1 - \sum_{i=2}^{q+1} X_i$$

The osmotic pressure equation can be developed from Eq. (35)

$$\pi(\text{real}) = g\pi(\text{ideal}) \tag{35}$$

The ideal osmotic pressure, $\pi(\text{ideal})$, can still be expressed by Eq. (11), since at osmotic equilibrium no solvent will flow across the semipermeable membrane (see Fig. 1). Thus

$$\pi = -g \frac{RT}{V} \ln X_1 \tag{68}$$

$$= -g \frac{RT}{V} \ln\left(1 - \sum_{i=2}^{q+1} X_i\right)$$

$$= g \frac{RT}{V} \sum_{i=2}^{q+1} X_i \qquad (\text{for } X_i \ll 1)$$

For dilute solutions

$$X_i = \frac{[P_i]V_1^0}{1000} \tag{69}$$

and

$$[P_i] = c_i/M_i = w_i/(VM_i) \tag{70}$$

so that (assuming $\overline{V}_1 = V_1^0$)

$$\pi = \frac{gRT}{1000} \Sigma[P_i] = g \frac{RT}{1000} \Sigma(c_i/M_i) = \frac{gRT}{1000} (c/M_n) \tag{71}$$

Here M_n is the number average molecular weight and is defined by

$$M_n = \frac{\Sigma n_i M_i}{\Sigma n_i} = \frac{\Sigma w_i}{\Sigma w_i/M_i} = \frac{\Sigma c_i}{\Sigma c_i/M_i} = \frac{c}{\Sigma c_i/M_i} \tag{72}$$

where n_i = number of moles of solute component i; $w_i = n_i M_i$ = weight of solute component i; M_i = molecular weight of solute component i; $w_i/V = c_i$ = concentration of component i; $w_i/(VM_i) = c_i/M_i$. If the solution is ideal, $g = 1$ and Eq. (71) gives M_n directly. If the solution is nonideal, then one can expand the osmotic coefficient, g, in integral powers of the solute concentration; thus

$$g = 1 + B_1'c + B_2'c^2 + \cdots \tag{73}$$

Using this relation in Eq. (71) leads to

$$1000\pi/RT = c/M_n\,(1 + B_1'c + B_2'c^2 + \cdots)$$
$$= c/M_n + B_1 c^2 + B_2 c^3 + \cdots = c/M_{n\,\mathrm{app}} \tag{74}$$

The virial coefficients, B_1, B_2, etc., are complicated functions of the q solute concentrations; they are also temperature dependent. In addition, the virial coefficients are dependent on the polymer–solvent combination.

The Gibbs–Duhem equation can also be used to obtain an osmotic pressure equation for macromolecular solutions. To do this one uses the following relations

$$dp = \sum_{i=2}^{q+1} \frac{n_i}{V}\,d\mu_i \ (\text{constant } T) \tag{75}$$

$$(\mu_1{}^\alpha = \mu_1{}^\beta)$$
$$\mu_i = f(p, T, c_2, c_3, \ldots, c_{q+1}) \tag{76}$$
$$= \mu_i{}^0 + RT \ln y_i c_i$$

$$\ln y_i = M_i \sum_{k=2}^{q+1} B_{ik} c_k \tag{77}$$

$$(\partial \mu_i/\partial p)_{T,c} = \overline{V}_1 = M_i \overline{V}_i \tag{78}$$

With these relations one obtains

$$1000\,dp\left(1 - \frac{\sum_i c_i \overline{V}_i}{1000}\right) = RT\left(\sum_i \frac{dc_i}{M_i} + \sum_i \sum_k B_{ik} c_i dc_k\right) + \cdots \tag{79}$$

or

$$\frac{1000\,dp}{RT} = d(c/M_n) + \sum_i \sum_k B_{ik} c_i dc_k + \sum_i \sum_k \frac{\overline{V}_i c_i}{1000} \frac{dc_k}{M_k} + \cdots \tag{80}$$

On the molal concentration scale it has been noted that $\mu_{ik} = \mu_{ki}$. These relations can be written as

$$\mu_{ik} = \frac{\partial \mu_i}{\partial m_k} = \frac{\partial \mu_i}{\partial c_k}\frac{\partial c_k}{\partial m_k} = \frac{\partial \mu_k}{\partial c_i}\frac{\partial c_i}{\partial m_i} = \frac{\partial \mu_k}{\partial m_i} = \mu_{ki} \tag{81}$$

Now note that

$$\frac{\partial \mu_j}{\partial c_q} = RTM_j B_{jq} \; (j = i \text{ or } k; q = i \text{ or } k) \tag{82}$$

$$c_q = 1000 \, m_q M_q / V_m \tag{83}$$

and

$$\left(\frac{\partial c_q}{\partial m_q}\right)_{T,c_{r \neq q}} = \frac{1000 \, M_q}{V_m}\left(1 - \frac{c_q \bar{v}_q}{1000}\right) \tag{84}$$

Here V_m is the volume (in milliliters) of solution containing 1 kg of solvent.

Since Eq. (77) for ln y_i is a Taylor's series about $c_r = 0$, we are only interested in the limiting values of the μ_{ik}; thus we note with aid of the above relations that

$$B_{ik} = B_{ki} \tag{85}$$

Hence one can use this to show that

$$\sum_i \sum_k B_{ik} c_i dc_k = \sum_i \sum_k B_{ik}\frac{d(c_i c_k)}{2} \tag{86}$$

Assuming for simplicity that $\bar{v}_i = \bar{v}_j = \bar{v}$, the integration of Eq. (79) (from $p = p^0$ to $p = p^0 + \pi$ on the left and 0 to c on the right) yields

$$\frac{1000\pi}{RT} = \frac{c}{M_n} + \sum_i \sum_k B_{ik}\frac{c_i c_k}{2} + \frac{\bar{v}c^2}{M_n} + \cdots \tag{87}$$

$$= \frac{c}{M_n} + \left[\sum_i \sum_k f_i f_k \left(B_{ik} + \frac{\dot{\bar{v}}}{M_k}\right)\right]\frac{c^2}{2} + \cdots$$

$$= \frac{c}{M_{n\,\text{app}}} = \frac{c}{M_n} + \frac{B_{os}c^2}{2}$$

Note that

$$\left(\sum_i c_i\right)\left(\sum_k \frac{dc_k}{M_k}\right) = cd\left(\frac{c}{M_n}\right) = \sum_i \sum_k \frac{c_i dc_k}{M_k} \tag{88}$$

This equation can be rearranged to give

$$1000/CRT(\pi/c) = 1/M_{\text{app}} = 1/M_n + (B_{os}c)/2 \tag{89}$$

Note carefully that this equation has the same form as the osmotic pressure equation for a homogeneous, nonassociating solute. Consequently, from a plot of $1000\pi/c$ vs. c, you cannot tell whether you are measuring M_{app} or $M_{n\ app}$; you have to know something about the history of the material in order to know which you are measuring. Or, you have to do light scattering or sedimentation equilibrium experiments and evaluate M_w or $M_{w\ app}$ (or their analogs if the refractive index increments and partial specific volume differ for each polymeric solute components) in order to know whether the material is homogeneous or heterogeneous. The same problem applies to light scattering, but it does not apply to sedimentation equilibrium experiments, since one can measure two apparent average molecular weights, $M_{w\ app}$ or $M_{z\ app}$, or their analogs. Also it is noted that this equation is formally identical with Eq. (37), and hence $B_1 = B_{os}/2$. For a polyelectrolyte solution, the components would be defined according to the Vrij–Overbeek[45] or Casassa-Eisenberg[36] conventions, and M_n and B_{os} or B_1 would refer to components defined according to these conventions. With vapor pressure osmometry one would obtain $M_{n\ app}$ from

$$E/c = K_{vp}/M_{na} \qquad (90)$$

which has the same form as does Eq. (43).

An Illustration—Evaluation of M_n and B_{os} for a Dextran T-70 Sample

In connection with some studies of molecular-weight distributions of some dextran samples similar to clinical dextrans, Wan and Adams[46] did some osmotic pressure experiments on these materials. Dextrans are slightly branched polysaccharides that are produced by bacteria such as *Leuconostoc mesenteroides*.[47-49] The original material has a high average molecular weight, and the sample we used was obtained by hydrolysis. It was nonionic and water soluble. A stock solution was made up in boiled, double-distilled water, and dilutions were made from the stock solution. Concentrations were determined by differential refractometry[50] at $\lambda = 546$

[45] A. Vrij and J. Th. G. Overbeek, *J. Colloid Sci.* **17**, 570 (1962).

[46] P. J. Wan and E. T. Adams, Jr., *Biophysical Chemistry* **5**, 207 (1976).

[47] Allene Jeanes, *in* "Encyclopedia of Polymer Science and Technology," 2nd ed. (H. F. Mark, N. G. Gaylord, and N. M. Mikales, eds.), Vol. 4, p. 805. Wiley (Interscience) New York, 1966.

[48] R. L. Whistler and C. L. Smart, "Polysaccharide Chemistry." Academic Press, New York, 1953.

[49] K. A. Granath, *J. Colloid Sci.* **13**, 308 (1958).

[50] E. P. Pittz, J. C. Lee, B. Bablouzian, R. Townend, and S. N. Timasheff, this series Vol. 27, p. 209.

nm using the refractive index increment reported in the Polymer Handbook.[19] The experiments were performed on a Melabs CSM 2 High Speed Membrane Osmometer using B19 cellulose acetate membrane (available from Schleicher and Schuell Co.). Table IV lists the experimental data. The values of h in centimeters of H_2O were obtained from a 10-inch, 1-mV chart recorder (Dohrmann Instruments). These data are plotted in Fig. 5 as plots of π/c vs c. From the intercepts one can obtain M_n, and from the slope B_{os}.

Note that one has the same type of plot for π/c vs c with a polydisperse, nonassociating solute, as one does with a homogeneous, nonassociating solute (cf. Fig. 5 to Fig. 4). The only way that one can tell whether one is measuring M and B or M_n and B_{os} is to know something about the history of the sample. Table V lists values of M_n and B_{os} for the three experiments reported in Table IV, as well as for an osmotic pressure experiment at 25° with another Dextran T-70 sample (Lot 7981). Figure 6 shows the differential molecular weight distribution, $f(M)$, vs molecular weight for Dextran T-70, Lot 693. The values obtained by the manufacturer (dashed line) using analytical gel chromatography and light scattering, and the values obtained by Wan and Adams,[46] after correction for nonideal behavior, by sedimentation equilibrium experiments are indi-

TABLE IV
OSMOTIC PRESSURE DATA FOR DEXTRAN T-70 (LOT 693)

T (°C)	c^a (g/liter)	h^b (cm H_2O)	$\pi \times 10^{3c}$ (atm)	$(\pi/c) \times 10^3$ (atm-liter)/g
16.0	1.481	0.93	0.90	0.608
	4.728	3.08	2.98	0.630
	6.597	4.41	4.27	0.647
	9.345	6.45	6.24	0.668
20.1	2.617	1.65	1.60	0.610
	4.918	3.31	3.20	0.651
	7.426	5.36	5.18	0.698
	10.011	7.61	7.36	0.735
25.0	2.614	1.80	1.74	0.666
	4.913	3.71	3.59	0.731
	7.417	5.88	5.70	0.768
	9.999	8.71	8.42	0.842

[a] Total solute concentration.

[b] Osmotic pressure in centimeters of H_2O as measured on the osmometer.

[c] The osmotic pressure in atmospheres was obtained in the same way as described for glycinin in Table III.

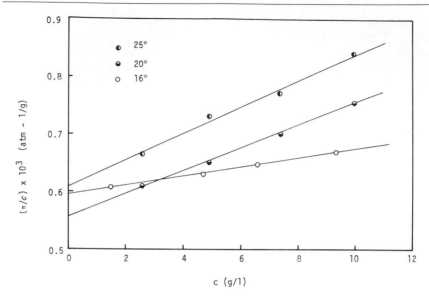

FIG. 5. Evaluation of M_n, the number average molecular weight, and B_{os}, the osmotic pressure second virial coefficient, for a heterogeneous, nonassociating solute. Here are shown plots of π/c vs c for aqueous dextran T-70 solutions at various temperatures. The intercept of each plot gives RT/M_n, and the slope of each plot give $RTB_{os}/2$ [see Eq. (89)]. Reproduced from Peter J. Wan and E. T. Adams, Jr., *Biophys. Chem.* **5**, 207 (1976), by permission of the authors and the North-Holland Publishing Co.

TABLE V

VALUES OF M_n AND B_{OS} AT VARIOUS TEMPERATURES FOR DEXTRAN T-70 SAMPLES[a]

Lot No.	t (°C)	$M_n \times 10^{-4b}$	$B_{os} \times 10^6$ (mole-l)/g²
7981	20.0	4.30	0.354
693	16.0	3.98	0.326
	20.1	4.24	0.713
	25.0	4.01	0.934

[a] This table is taken from P. J. Wan and E. T. Adams, Jr., *Biophys. Chem.* **5**, 207 (1976), by permission of the authors and the North-Holland Publishing Co.
[b] Average M_n (4.08 ± 0.11) × 10⁴ for Lot 693.

cated by the various symbols. We have indicated the values of M_n, M_w, and M_z on the abscissa of the plot of $f(M)$ vs M. A similar study with a different dextran sample will be found in the paper by Soucek and Adams.[51]

[51] D. A. Soucek and E. T. Adams, Jr., *J. Colloid Interface Sci.* **55**, 571 (1976).

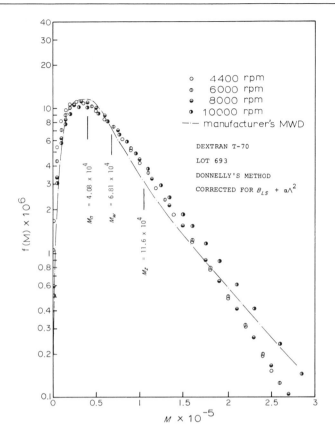

FIGURE 6. Plot of the differential molecular weight distribution, $f(M)$, vs molecular weight, M, for a dextran T-70 sample. The plot of $f(M)$ vs M obtained by the manufacturer (Pharmacia) using analytical gel chromatography and light scattering is shown by the dashed line. The plots of $f(M)$ vs M obtained from sedimentation equilibrium experiments at 20°, corrected for nonideality, are shown by the various symbols. Each symbol indicates a different speed. The values of M_n, M_w and M_z are indicated on the plot. These quantities are related to $f(M)$ as follows: $1/M_n = \int_0^\infty [f(M)/M]dM$; $M_w = \int_0^\infty Mf(M)dM$ and $M_wM_z = \int_0^\infty M^2f(M)dM$. Reproduced from P. J. Wan and E. T. Adams, Jr., *Biophys. Chem.* **5**, 207 (1976), by permission of the authors and the North-Holland Publishing Co.

Self-Associations

Background Information

Chemical equilibria of the type

$$nP_1 \rightleftarrows P_n, n = 2,3, \ldots, \tag{91}$$
$$nP_1 \rightleftarrows qP_2 + mP_3 + \ldots, \tag{92}$$

and related associations are known as self-associations. Here P represents the self-associating solute. The quantities P_1, P_2, . . . , P_n are referred to as the monomer (or unimer), dimer, . . . , and n-mer. Self-associations are widely encountered; many proteins, soaps and detergents, nucleosides, and nucleotides are known to undergo self-association.[52,53] Self-association may be important in biological regulation.[54]

The molecular weights of the various self-associating species have the following relation:

$$M_j = jM_1, j = 2, 3, \ldots \tag{93}$$

At constant temperature, the condition for chemical equilibrium is

$$n\mu_1 = \mu_n, n = 2, 3, \ldots \tag{94}$$

where μ_i ($i = 1, 2, \ldots$) is the molar chemical potential of associating species i. Because of the self-association, the number average molecular weight (M_{nc}) is concentration dependent; its values will range from M_1 at $c = 0$ to the molecular weight of the highest species present, unless the association is nonideal. Figures 7 and 8 show some plots of M_{nc} vs c obtained by Harry and Steiner[55] on soybean trypsin inhibitor under various solution conditions. For nonideal self-associations a plot of M_{na} vs c may go through a maximum and then decrease with increasing solute concentration.[42,56,57] Figure 9 gives a comparison of the type of plot that might be encountered with ideal and nonideal self-associations. The symbol M_{nc} is used to indicate that a self-association is present. The concentration dependence of M_{nc} or M_{na} introduces new problems in the analysis of the data. Note that[42,56]

$$\lim_{c \to 0} 1/M_{na} = 1/M_1 \tag{95}$$

and

$$\lim_{c \to 0} (d/dc)(1/M_{na}) = \lim_{c \to 0} (d/dc)(1/M_{nc}) + B/2 \tag{96}$$
$$= \frac{1}{2}[-(K_2/M_1) + B], \text{ if dimer is present}$$
$$= B/2, \text{ if dimer is absent}$$

Thus, one must find a different method of analysis, so that one can evaluate the type of association present if it is not known beforehand, as well as

[52] E. T. Adams, Jr., *Fractions*, No. 3, 1967

[53] E. T. Adams, Jr., W. E. Ferguson, P. J. Wan, J. L. Sarquis, and B. M. Escott, *Sepa. Sci.* **10**, 175 (1975).

[54] L. W. Nichol and D. J. Winzor, "Migration of Interacting Systems." Oxford Univ. Press, London and New York, 1972.

[55] J. B. Harry and R. F. Steiner, *Biochemistry* **8**, 5060 (1969).

[56] E. T. Adams, Jr., *Biochemistry* **4**, 1655 (1965).

[57] E. T. Adams, Jr. and D. L. Filmer, *Biochemistry* **5**, 2971 (1966).

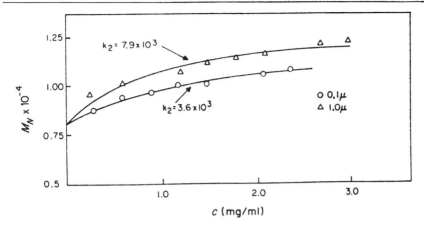

FIG. 7. Plot of $M_{n\ app}$ vs. c for a self-associating protein, the Bowman–Birk soybean trypsin inhibitor. These experiments were carried out by high speed membrane osmometry (Hewlett-Packard osmometer) at 25°, pH 7 and at two ionic strengths. The increase in M_{na} with increasing c (total solute concentration) is characteristic of a self-association. Note that the self-association is greater at an ionic strength (μ) of 1.0 than it is at 0.1 ionic strength, which indicates an electrostatic effect; at higher ionic strength the charge on the protein is shielded more, so that self-association is promoted. Reprinted (with the permission of Dr. R. F. Steiner and the American Chemical Society) from J. B. Harry and R. F. Steiner, *Biochemistry* **8,** 5060 (1969). Copyright by the American Chemical Society.

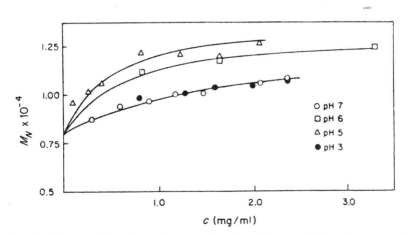

FIG. 8. Effect of pH on the self-association of the Bowman–Birk soybean trypsin inhibitor at ionic strength 0.1 and 25°. Reprinted (with permission of Dr. R. F. Steiner and the American Chemical Society) from J. B. Harry and R. F. Steiner, *Biochemistry* **8,** 5060 (1969). Copyright by the American Chemical Society.

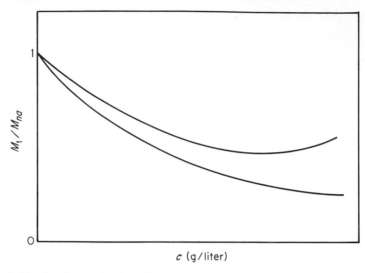

FIG. 9. Simulated example of a self-association. Here are shown plots of M_1/M_{na} vs c that might be encountered with an ideal self-association (lower curve) and a nonideal self-association (upper curve). The minimum shown in the upper curve indicates that a nonideal self-association with a positive BM_1 is present. If BM_1 is very small, then this effect may not be noticed unless experiments are carried out to fairly high concentrations.

the values of the equilibrium constant or constants (K_i) and the nonideal term (BM_1). If some simple assumptions are made regarding the associating species then the analysis turns out to be not too difficult. More details on this analysis can be found in the Fujita[58] monograph, the reviews by Adams,[52] Adams et al.,[53] or Williams et al.,[59] as well as the paper by Lo et al.[42]

The assumption usually made with regard to self-associating species are the following: (1) The natural logarithm of the activity coefficient of species i can be represented by[42,52,53,56]

$$\ln y_i = i B_* M_1 c \ (i = 1, 2, \ldots) \tag{97}$$

(2) The partial specific volumes of the self-associating species are equal, i.e., $\bar{v}_1 = \bar{v}_2 = \ldots = \bar{v}$. (3) The refractive index increments in volume per mass units (liter/g, etc.) are equal, i.e., $\psi_1 = \psi_2 = \cdots = \psi$, where $\psi = (\partial n / \partial c)_{T,P}$. Only the first assumption is needed in membrane or vapor pressure osmometry. Because one does not have to worry about assumptions 2 and 3, it follows that osmometry and related colligative methods

[58] H. Fujita, "Foundations of Ultracentrifugal Analysis," Wiley (Interscience), New York, 1975.

[59] H. Kim, R. C. Deonier, and J. W. Williams, Chem. Revs. 77, 659 (1977).

are the only methods that can give an unambiguous average molecular weight for heterogeneous, nonassociating solutes or for associating solutes.

Evaluation of M_{na}, M_{wa}, and $\ln f_a$.[42,52,53,58]

In order to analyze the self-association, one should first evaluate M_{na}, M_{wa}, and $\ln f_a$. As a consequence of Eq. (97), it follows that

$$c = c_1 + K_2 c_1^2 + K_3 c_1^3 + \cdots \tag{98}$$

where

$$K_2 = c_2/c_1^2, \quad K_3 = c_3/c_1^3, \text{ etc.} \tag{99}$$

Division of both sides of Eq. (98) by c leads to

$$1 = f_1 + f_2 + f_3 + \cdots \tag{100}$$

where $f_1 = c_i/c$ is the weight fraction of component i. The quantity M_{na} can be obtained from

$$1000\pi/RT = c/M_{na} \tag{101}$$

for membrane osmometry, or from

$$E/K_{vp} = c/M_{na} \tag{102}$$

from vapor pressure osmometry. Here M_{na} is defined by

$$1/M_{na} = 1/M_{nc} + Bc/2 \tag{103}$$

where

$$B = B_* + \bar{v}/1000 M_1 \tag{104}$$

and M_{nc}, the number average molecular weight is defined by

$$M_{nc} = \left(\sum_i n_i M_i \right) \bigg/ \sum_i n_i = c \bigg/ \left[\sum_i (c_i/M_i) \right]$$
$$= M_1 \left[c \bigg/ \sum_i (c_i/i) \right] = M_1 \bigg/ \sum_i (f_i/i) \tag{105}$$

Here n_i is the number of moles of associating species i. It follows from Eq. (105) that

$$\frac{M_1}{M_{nc}} = \frac{c_1 + K_2 c_1^2 + K_3 c_1^3 + \cdots}{c_1 + \dfrac{K_2 c_1^2}{2} + \dfrac{K_3 c_1^3}{3} + \cdots} = \frac{1 + K_2 c_1 + K_3 c_1^2 + \cdots}{1 + \dfrac{K_2 c_1}{2} + \dfrac{K_3 c_1^2}{3} + \cdots} \tag{106}$$

and one notes that

$$\frac{M_1}{M_{na}} = \frac{M_1}{M_{nc}} + \frac{BM_1 c}{2} \tag{107}$$

The apparent weight average molecular weight, M_{wa}, can be obtained from a plot of M_1/M_{na} vs c, since Adams[56] has shown that

$$\frac{M_1}{M_{wa}} = \frac{d}{dc}\left(\frac{cM_1}{M_{na}}\right) = \frac{M_1}{M_{na}} + c\,\frac{d}{dc}\left(\frac{M_1}{M_{na}}\right) = \frac{M_1}{M_{wc}} + BM_1c \quad (108)$$

Here M_{wc} is the weight average molecular weight, which is defined by

$$M_{wc} = \frac{\sum_i n_i M_i^2}{\sum_i n_i M_i} = \frac{\sum_i c_i M_i}{c} = \Sigma f_i M_i = M_1 \Sigma i f_i$$

$$= M_1\left(\frac{1 + 2K_2 c_1 + 3K_3 c_1^2 + \cdots}{1 + K_2 c_1 + K_3 c_1^2 + \cdots}\right) \quad (109)$$

The plot of M_1/M_{na} vs c can also be used in the evaluation of $\ln f_a$, where f_a is the apparent weight fraction of monomer. It has been shown that[56]

$$\ln f_a = \int_0^c \left(\frac{M_1}{M_{na}} - 1\right)\frac{dc}{c} + \left(\frac{M_1}{M_{na}} - 1\right)$$

$$= \int_0^c \left(\frac{M_1}{M_{wa}} - 1\right)\frac{dc}{c} = \ln f_1 + BM_1c \quad (110)$$

$$f_1 = c_1/c = \text{weight fraction of monomer}$$

In order to evaluate $\ln f_a$, it is necessary to make a plot of $[(M_1/M_{na}) - 1]/c$ vs c. Note that the intercept of this plot is $(-K_2 + BM_1)/2$, if dimer is present, or $BM_1/2$, if dimer is absent. Since this intercept is not known *a priori*, there may be complications in making the plot of $[(M_1/M_{na}) - 1]/c$ vs c, and one may prefer to evaluate by[42]

$$\ln(f_a/f_{a*}) = \int_{c^*}^c \left(\frac{M_1}{M_{na}} - 1\right)\frac{dc}{c} + \left(\frac{M_1}{M_{na}} - \frac{M_1}{M_{na*}}\right) \quad (111)$$

and use it for the analysis of self-associations. The choice of c_* is somewhat arbitrary, but it should be at a low concentration so that the value of $\ln (f_a/f_{a*})$ can be obtained over a greater concentration. The quantities M_1/M_{na} and M_1/M_{wa} can be combined in the following way to eliminate BM_1, the second virial coefficient.[42,53,58,60,61]

$$\xi = 2M_1/M_{na} - M_1/M_{wa} = 2M_1/M_{nc} - M_1/M_{wc} \quad (112)$$

The quantity ξ, as well as $\ln(f_a/f_{a*})$, can be used for the analysis of self-associations as shown below.

[60] P. W. Chun, S. J. Kim, J. D. Williams, W. T. Cope, L.-H. Tang, and E. T. Adams, Jr., *Biopolymers* **11**, 197 (1972).
[61] L.-H. Tang and E. T. Adams, Jr., *Arch. Biochem. Biophys.* **157**, 520 (1973).

Monomer-n-mer Self-Associations[42,54]

This association is described by Eq. (91). Here one notes that

$$c = c_1 + K_n c_1{}^n \tag{113}$$
$$f_n = 1 - f_1 \tag{114}$$

and

$$(1 - f_1)/f_1{}^n = K_n c^{n-1} \tag{115}$$

Thus a plot of $(1 - f_1)/f_1{}^n$ vs c^{n-1} will give a straight line going through the origin, if this model is correct, and the slope of this line will be K_n. Sometimes experimentally, error will cause the plot not to go through the origin; however, it should come close to the origin. If n is not known *a priori*, one has to try various values of n (2, 3, etc.), solve for f_1 for each choice, make the plots required by Eq. (115), and see whether one gets a straight line going through or close to the origin. If the plot shows curvature for all reasonable choices of n, then try some model other than a monomer-n-mer association.

The weight fraction of monomer, f_1, is obtained from ξ, since[42,53,60]

$$\xi = \frac{2M_1}{M_{na}} - \frac{M_1}{M_{wa}} = \frac{2M_1}{M_{nc}} - \frac{M_1}{M_{wc}} = \frac{2 + 2f_1(n-1)}{n} - \frac{1}{n + f_1(1-n)} \tag{116}$$

This has been shown to be a quadratic equation in f_1; thus[42,53,60]

$$f_1 = n/4(n-1)^2 \left\{ (n-1)(\xi + 2 - 2/n) - \left([(n-1)(\xi + 2 - 2/n)]^2 - (8/n)(n-1)^2(\xi n - 1) \right)^{1/2} \right\} \tag{117}$$

In order to obtain BM_1 note that

$$M_1/M_{na} - M_1/M_{nc} = (BM_1 c)/2 \tag{118}$$

and

$$M_1/M_{nc} = [1 + f_1(n-1)]/n \tag{119}$$

For larger values of BM_1 a plot of $M_1/M_{na} - M_1/M_{nc}$ vs c will give a straight line going through (or close to) the origin with a slope of $BM_1/2$. For very small values of BM_1 this plot may be too noisy. In this case one should set up an array of M_1/M_{na} vs c values and find the best value of BM_1 to fit the array. The criterion for best fit is that $\Sigma(\delta_i)^2$ is a minimum, where

$$\delta_i = [(M_1/M_{na})_{obsvd} - (M_1/M_{na})_{calcd}]_i \tag{120}$$

If $\ln(f_a/f_{a*})$ is used, then one notes that[42]

$$\ln(f_a/f_{a*}) = \ln(f_1/f_{1*}) + BM_1(c - c_*) \tag{121}$$

Here

$$f_1 = (nM_1/M_{nc} - 1)/(n - 1) = [n(M_1/M_{na} - BM_1c/2) - 1]/(n - 1) \quad (122)$$

The equation for f_{1*} is obtained by using M_1/M_{na*} and c_* in the equation above. To use $\ln(f_a/f_{a*})$ one has to set up an array of $\ln(f_a/f_{a*})$ vs c data, choose values of BM_1 and obtain values of $\epsilon_i = [\ln (f_a/f_{a*})_{Obsvd} - \ln(f_a/f_{a*})_{Calcd}]_i$, then look for sign changes in ϵ_i as BM_1 is varied. These choices can be narrowed by varying BM_1 over successively smaller intervals, and the apparent best choices of BM_1 can be used to calculate f_1 from Eq. (115). These values of f_1 can be used for the evaluation of K_n. Then these values of K_n and BM_1 are used to regenerate M_1/M_{na}. The best fit is the one which gives a minimum for $\Sigma_i(\delta_i)^2$. False solutions can be encountered which give negative values of K_1 and these can be discarded by inspection. It is obviously easier to use ξ for the analysis of a monomer-n-mer association. The analysis of other types of self-associations using the quantity $\ln(f_a/f_{a*})$ is done in a similar manner. β-Lactoglobulin A (βA) in 0.2 M glycine buffer at pH 2.46 is reported to undergo a monomer–dimer self-association between 10° and 30°; figures 3 and 4 of the paper by Tang and Adams[61] shows tests, based on the quantity ξ, for a monomer-n-mer association using $n = 2$, 3, and 4 for βA. Under similar solution conditions β-lactoglobulin C also undergoes a monomer–dimer association, and Figs. 3 and 4 of the paper by Sarquis and Adams[62] show similar tests for various monomer-n-mer associations. The enzyme lysozyme (or muramidase) was shown by Adams and Filmer[57] and also Deonier and Williams[63] to undergo a monomer–dimer self-association at pH 6.7; this has also been confirmed by Millthorpe, Jeffrey, and Nichol.[64] Elias[65] has studied the micellar (monomer-n-mer) association of aqueous solutions of n-octyl-β-D-glucoside and two nonylphenol (ethylene glycol)$_x$ ethers by light scattering, sedimentation equilibrium, and vapor pressure osmometry experiments; good agreement was obtained for the equilibrium constants as determined by the three methods.

Indefinite Self-Associations[42,53,66]

Sometimes it appears that a self-association continues without limit—it seems to be described by

$$n\ P_1 \rightleftarrows q\ P_2 + m\ P_3 + h\ P_4 + \cdots \quad (123)$$

[62] J. L. Sarquis and E. T. Adams, Jr., *Arch. Biochem. Biophys.* **163**, 442 (1974).

[63] R. C. Deonier and J. W. Williams, *Biochemistry* **9**, 4260 (1970).

[64] B. K. Millthorpe, P. D. Jeffrey, and L. W. Nichol, *Biophys. Chem.* **3**, 169 (1975).

[65] H.-G. Elias, *J. Macromol. Sci. Chem.* **A7**, 601 (1973).

[66] L.-H. Tang, D. R. Powell, B. M. Escott, and E. T. Adams, Jr., *Biophys. Chem.* **7**, 121 (1977).

This self-association and related self-associations are known as indefinite self-associations. They are actually made up of simultaneous equilibria of the type

$$P_1 + P_{(n-1)} \rightleftarrows P_n \ (n = 2, 3, \ldots) \tag{124}$$

For the ideal case, the equilibrium constant for this step becomes

$$K_{(n-1),n} = [P_n]/([P_1][P_{n-1}]) \tag{125}$$

Since this relation applies to all $P_{(n-1)}$ species, one notes that

$$[P_n] = K_{1,2}K_{2,3}K_{3,4} \ldots K_{(n-1),n}[P_1]^n \tag{126}$$

It would appear that if a large number of self-associating species are present, then one would have to try to evaluate several equilibrium constants. With the limited precision available to us at present, this is clearly impossible, and we are forced to make some simplifying assumptions in order to analyze these associations. In the analysis of indefinite self-associations it is convenient to use concentrations in grams per milliliter, which will be denoted by C for the total solute concentration or C_i for the concentration of species i. For the nonideal case it will be assumed that Eq. (97) applies, but here it will be written as

$$\ln y_i = i\hat{B}M_1C \ (i = 1, 2, \ldots) \tag{127}$$

Since $c = 1000C$, it follows that $\hat{B}M_1 = 1000 BM_1$. Please note that even if the experimental data are described adequately by an indefinite self-association, this does not mean that an unlimited number of associating species are present. It has been shown by Tobolsky and Thach[67] that a self-association, which could be described by 7 or 8 equilibrium constants, could also be described with just as good precision as an indefinite self-association with two equilibrium constants. While there are many varieties of indefinite self-associations, the analysis can be illustrated with four types that may be more commonly encountered.

Type I Indefinite Self-Association[42,52,53,58,66]

This is the simplest case to analyze. Here it is assumed that all species are present, and it is also assumed that the molar equilibrium constants for the association steps represented by Eq. (92) are equal. This means that ΔG^0 is constant for the addition of monomer to any preformed n-mer.

For this association it has been shown that the following relations

[67] A. V. Tobolsky and R. E. Thach, *J. Colloid Sci.* **17**, 410 (1962).

apply:

$$C = C_1 + 2kC_1^2 + 3k^2C_1^3 + 4k^3C_1^4 + \cdots$$

$$= C_1/(1 - kC_1)^2, \text{ if } kC_1 < 1, \quad (128)$$

$$k = 1000K/M_1 \text{ is the intrinsic equilibrium constant} \quad (129)$$

$$f_1 = C_1/C = (1 - kC_1)^2 \quad (130)$$

$$M_1/M_{na} = 1 - kC_1 + (\hat{B}M_1C)/2 = \sqrt{f_1} + (\hat{B}M_1C)/2 \quad (131)$$

and

$$M_1/M_{wa} = (1 - kC_1)/(1 + kC_1) + \hat{B}M_1C = \sqrt{f_1}/(2 - \sqrt{f_1}) + \hat{B}M_1C \quad (132)$$

The quantity ξ becomes

$$\xi = 2\sqrt{f_1} - \sqrt{f_1}/(2 - \sqrt{f_1}) \quad (133)$$

which is quadratic in $\sqrt{f_1}$. Thus $\sqrt{f_1}$ is obtained from

$$\sqrt{f_1} = 1/4 \{(\xi + 3) - [(\xi + 3)^2 - 16\xi]^{1/2}\} \quad (134)$$

Once the $\sqrt{f_1}$ is available, then one can obtain k from

$$(1 - \sqrt{f_1})/f_1 = kC \quad (135)$$

A plot of $(1 - \sqrt{f_1})/f_1$ vs C will give a straight line going through (or close to) the origin, if this model is correct; the slope of this line is k. Once k and f_1 are known, one can evaluate $\hat{B}M_1$ from Eq. (131), since

$$M_1/M_{na} - \sqrt{f_1} = (\hat{B}M_1C)/2 \quad (136)$$

In the event that $\ln(f_a/f_{a*})$ is used for the analysis [see Eqs. (111) and (121)], the appropriate form of f_1 to use is[66]

$$f_1 = (1 - kC_1)^2 = (M_1/M_{nc})^2 = [M_1/M_{na} - (\hat{B}M_1C)/2]^2 \quad (137)$$

The analysis is carried out in the same manner as described previously for the monomer-n-mer association.

Chun, Kim et al.[60] were able to show that the self-association data of Eisenberg and Tompkins,[68] obtained by light scattering, for bovine lactate dehydrogenase could be interpreted as a type I indefinite self-association. In a later study, using analytical gel chromatography, Chun, Kim et al.[69] confirmed this observation for the bovine lactate dehydrogenase. By vapor pressure osmometry, Ts'o and his associates[70] showed that in aqueous solutions purine underwent a type I indefinite self-association; Van Holde and Rossetti[71] confirmed this with sedimentation equilibrium exper-

[68] H. Eisenberg and G. Tompkins, *J. Mol. Biol.* **31**, 37 (1968).
[69] P. W. Chun, S. J. Kim, C. A. Stanley, and G. K. Ackers, *Biochemistry* **8**, 1625 (1969).
[70] P. O. P. Ts'o, *Ann. N.Y. Acad. Sci.* **153**, 785 (1969) and references cited therein.
[71] K. E. Van Holde and G. P. Rossetti, *Biochemistry* **6**, 2189 (1967).

iments. Plesiewicz, Stępień et al.[72] studied the base-stacking self-association of 25 uracil derivatives in aqueous solution at various temperatures using vapor pressure osmometry; they found that a type I indefinite self-association described their data better than a type III indefinite self-association.

Type II Indefinite Self-Association[47,66]

Here all odd species beyond the monomer are assumed to be absent; it is also assumed that all molar equilibrium constants are equal. Thus, the following relations apply.

$$C = C_1[1 + 2kC_1/(1 - k^2C_1^2)^2] = C_1[1 + 2X/(1 - X^2)^2], \text{ if } kC_1 < 1 \quad (138)$$
$$X = kC_1 = kCf_1 \quad (139)$$
$$k = 1000K/M_1 \quad (140)$$

$$\xi = \frac{2[1 + X/(1 - X^2)]}{1 + 2X/[(1 - X^2)^2]} - \frac{1 + 2X/[(1 - X^2)^2]}{1 + [4X(1 + X^2)]/[(1 - X^2)^3]} \quad (141)$$

One obtains $X = kC_1$ from Eq. (139) by using successive approximations $(0 \le kC_1 < 1)$. Then one can obtain f_1 from

$$f_1 = 1/[1 + 2X/(1 - X^2)^2] \quad (142)$$

Finally, a plot of X vs. Cf_1 [see Eq. (139)] will give a straight line going through (or close to) origin, if this model is correct; the slope of the straight line is k. If this plot fails, then the model is incorrect and one should try another model. In the event $\ln(f_a/f_{a\cdot})$ is used for the analysis, it can be shown that the appropriate form of f_1 to use is[66]

$$2f_1 = (3M_1/M_{nc} + M_{wc}/M_1 - 2) - [(2 - M_{wc}/M_1 - 3M_1/M_{nc})^2 - 4(M_{wc}/M_{nc} + 2M_1/M_{nc} - 2)]^{1/2} \quad (143)$$

Note that

$$M_1/M_{nc} = M_1/M_{na} - (\hat{B}M_1C)/2 \quad (144)$$
$$M_1/M_{wc} = M_1/M_{wa} - \hat{B}M_1C \quad (145)$$

and

$$M_{wc}/M_1 = 1/(M_1/M_{wc}) \quad (146)$$

Type III Indefinite Self-Association[47,66]

This is a variant of the type I indefinite self-association. Here it is assumed that $K_{12} \ne K_{23}$, K_{34}, etc., but it is also assumed that $K_{23} =$

[72] E. Plesiewicz, E. Stępień, K. Bolewska, and K. L. Wierzchowski, Biophys. Chem. 4, 131 (1976).

$K_{34} = \cdots = K$. For this association it has been shown that[66]

$$C = C_1 + 2k_{12}C_1^2 + 3k_{12}kC_1^3 + 4k_{12}k^2C_1^4 + \cdots \tag{147}$$
$$= C_1\{1 + k_{12}C_1[(2 - kC_1)/(1 - kC_1)^2]\}$$
$$= C_1\{1 + y[(2 - X)/(1 - X)^2]\} \text{ if } kC_1 < 1$$
$$k_{12} = 1000K_{12}/M_1 \tag{148}$$
$$k = 1000K/M_1 \tag{149}$$
$$y = k_{12}C_1 \tag{150}$$

and

$$X = kC_1 \tag{151}$$

The quantity ξ becomes[66]

$$\xi = \frac{2 + 2[y/(1 - X)]}{1 + y[(2 - X)/(1 - X)^2]} - \frac{1 + y[(2 - X)/(1 - X)^2]}{1 + y[(4 - 3X + X^2)/(1 - X)^3]} \tag{152}$$

This equation contains two unknowns, X and y. It can be dealt with using an iterative method.[66] First estimate a value of k_{12} and multiply Eq. (147) to obtain

$$k_{12}C = k_{12}C_1[1 + y(2 - X)/(1 - X)^2] \tag{153}$$
$$= y + y^2(2 - X)/(1 - X)^2$$

This equation is quadratic in y, i.e.,

$$y^2 + y[(1 - X)^2/(2 - X)] - k_{12}C[(1 - X)^2/(2 - X)] = 0 \tag{154}$$

The proper root to use (call it $g(X)$) is the one obtained using the positive square root of the discriminant. Thus

$$\xi = \frac{2[(1 - X)^2 + (1 - X)g(X)]}{(1 - X)^2 + (2 - X)g(X)} - \frac{(1 - X)^3 + (2 - X)(1 - X)g(X)}{(1 - X)^3 + (4 - 3X + X^2)g(X)} \tag{155}$$

Now solve Eq. (155) for X using successive approximations, remembering $0 < X < 1$. This is done most easily on a computer. If at this point the values of X required to solve Eq. (155) are greater than 1, then choose another value (say one half of the original estimate) and repeat the process. If apparent solutions of Eq. (155) are found for which $0 < X < 1$, then calculate values of $(\triangle\xi)_i$, where $(\triangle\xi)_i = (\xi_{\text{Obsvd}} - \xi_{\text{Calcd}})_i$. Pick the value of X that gives $|\triangle\xi|$ less than some prechosen limit (say 1×10^{-6}) that is near the limit of the precision of the computer. This process is repeated at various choices of C. Once the values of X are known for each value of C, then the values of y are also known [see Eq. (150)]. Once k_{12} is known, then k can be obtained from[66]

$$f_1 = C_1/C = 1/[1 + y(2 - X)/(1 - X)^2] \tag{156}$$

and

$$k = X/C_1 = X/Cf_1 \tag{157}$$

If one uses $\ln(f_a/f_{a*})$ for the analysis, then the appropriate form of f_1 to use is[66]

$$f_1 = [M_1/M_{nc}(2M_1/M_{wc} - M_{wc}/M_1 - 3) + 2]/[1 - M_{wc}/M_1] \tag{158}$$

Here M_1/M_{nc}, M_1/M_{wc}, and M_{wc}/M_1 are defined by Eqs. (144)–(146), respectively. The analysis is done in the same manner as described earlier for the monomer-n-mer analysis. A type III indefinite self-association has been found to be useful in describing self-association data in organic solvents for phenol and some alcohols[73,74] and also for some amides.[75]

Type IV Indefinite Self-Association[66]

This association is a variant of the type II indefinite self-association. For this case the relationships between the molar equilibrium constants are as follows: $K_{12} \neq K_{24}$, K_{26}, etc., but $K_{24} = K_{26} = \cdots = K$. The appropriate equations for C and ξ are

$$C = C_1 + 2k_{12}C_1^2 + 4k_{12}^2kC_1^4 + 6k_{12}^3k^2C_1^6 + \cdots \tag{159}$$
$$= C_1 \left[1 + \frac{2k_{12}C_1}{(1 - k_{12}kC_1^2)^2} \right]$$
$$= C_1 \left[1 + \frac{2k_{12}C_1}{(1 - k_*^2C_1^2)^2} \right]$$

and

$$\xi = \frac{2\left[1 + \dfrac{k_{12}C_1}{(1 - k_{12}kC_1^2)} \right]}{1 + \dfrac{2k_{12}C_1}{(1 - k_{12}kC_1^2)^2}} - \frac{1 + \dfrac{2k_{12}C_1}{(1 - k_{12}kC_1^2)^2}}{1 + \dfrac{4k_{12}C_1(1 + k_{12}kC_1^2)}{(1 - k_{12}kC_1^2)^3}} \tag{160}$$

provided $k_{12}C_1 < 1$ and $kC_1 < 1$. Here

$$k_{12} = 1000\, K_{12}/M_1 \tag{161}$$
$$k = 1000\, K/M_1 \tag{162}$$

and

$$k_*^2 = k\, k_{12} \tag{163}$$

Note that Eq. (160) has two unknowns: $X = k_{12}C_1$ and $y = kC_1$. One can solve for k_{12} and k by using an iterative method similar to that described for a type III indefinite self-association.

[73] E. G. Hoffman, Z. Physi. Chem. B53, 179 (1943).
[74] N. D. Coggeshall and E. L. Saier, J. Am. Chem. Soc. 73, 5414 (1951).
[75] M. Davies and D. K. Thomas, J. Phys. Chem. 60, 763, 767 (1956).

When $\ln(f_a/f_{a*})$ is used for the analysis, then f_1 is also given by Eq. (143); for further details consult the paper by Tang, Powell et al.[66]

Monomer-n-Mer-m-Mer Self-Associations

These discrete self-associations are described by equilibria of the type

$$n\,P_1 \rightleftarrows q\,P_n + h\,P_m(n = 2, 3, \ldots, m = 3, 4, \ldots, n \neq m) \qquad (164)$$

and related equilibria. One way to analyze these associations is to use Eq. (111) to evaluate $\ln(f_a/f_{a*})$ and then use it [see Eq. (121)] to analyze the self-association. For example, if a monomer–dimer–tetramer association were present, then the appropriate form of f_1 would be[76]

$$f_1 = \frac{1}{3}\left[\frac{8M_1}{M_{na}} - 6 - 4BM_1c + \frac{1}{M_1/M_{wa} - BM_1c}\right] \qquad (165)$$

Here

$$M_1/M_{wa} = 1/(4 - 3f_1 - 2f_2) + BM_1c \qquad (166)$$
$$M_1/M_{na} = (1 + 3f_1 + f_2)/4 + (BM_1c)/2 \qquad (167)$$
$$f_4 = 1 - f_1 - f_2 = K_4 c_1^4/c \qquad (168)$$

and

$$f_2 = K_2 c_1^2/c \qquad (169)$$

For other monomer-n-mer-m-mer associations, one would need to set down the appropriate equations (see Ferguson et al.[76] or Lo et al.[42] for details). One solves for BM_1 by setting up an array of $\ln(f_a/f_{a*})$ vs c values and solving for the best value of BM_1 as described earlier. Once BM_1 is known, then the f_1 values of each member of them are known, and the relations listed above can be used to find f_2 and f_4.

Rao and Kegeles[77] studied the self-association of α-chymotrypsin in aqueous solution at pH 6.2 by the Archibald method; their self-association data may well be described as a monomer–dimer–trimer association. Ferguson, et al.[76] found that the self-association of the disodium salt of adenosine 5'-triphosphate (ATP) could be described as a monomer–dimer–trimer association.

Other Methods for the Analysis of Self-Associations

Derechin[78-81] has used the multinomial theorem to describe the concentration dependence of any average molecular weight or its apparent

[76] W. E. Ferguson, C. M. Smith, E. T. Adams, Jr., and G. H. Barlow, Biophys. Chem. 1, 325 (1974).
[77] M. S. N. Rao and G. Kegeles, J. Am. Chem. Soc. 80, 5724 (1958).
[78] M. Derechin, Biochemistry 7, 3253 (1968).
[79] M. Derechin, Biochemistry 8, 921, 927 (1969).
[80] M. Derechin, Biochemistry 11, 1120, 4153 (1972).
[81] M. Derechin, Y. M. Rustrum, and E. A. Barnard, Biochemistry 11, 1792 (1972).

value. The coefficients of the power series can be expressed in terms of the equilibrium constant or constants and the second virial coefficient. As originally developed, this method requires a knowledge of limiting derivatives of any apparent average molecular weight, and this may be hard to obtain with strong self-association. Nevertheless, Derechin et al.[81] have applied this method to the monomer–dimer self-association of yeast hexokinases B and C in solution at pH 8 and above.

Lewis and Knott[82] have used nonlinear least squares in a study of the possibilities of discriminating between definite and indefinite modes of self-associations. Their tests were done with simulated data. They found that it might be very difficult to distinguish between a monomer–dimer–tetramer–octamer model and a type I indefinite self-association.

For ideal self-associations, Kreuzer[83] was the first to develop a method for analyzing monomer-n-mer and type I indefinite self-associations. By using equations for the equilibrium molarity (which is equal to c/M_{nc}) and the stoichiometric molarity (which is equal to c/M_1) he was able to obtain equations for the natural logarithm of the number fraction of monomer ($\ln z_1$), which can be shown to be identical to Steiner's[81] equation, i.e.,

$$\ln z_1 = \int_0^m (M_1/M_{nc} - 1)dm/m \qquad (170)$$

Here $z_1 = m_1/m$ is the number fraction of monomer, and m is the total equilibrium molality of the solute. Molarities could also be used here. At present one can only evaluate $\ln z_1$ for the ideal case. Harry and Steiner[55] have used this method to study the monomer–dimer self-association of soybean trypsin inhibitor under various solution conditions (see Figs. 7 and 8) using high speed membrane osmometry.

Kreuzer[83] also obtained an expression for the weight fraction of monomer, $\ln f_1$; this equation can be shown to be formally identical (for ideal self-associations) to the equation derived by Adams,[56] i.e.,

$$\ln f_1 = \int_0^c (M_1/M_{nc} - 1)dc/c + (M_1/M_{nc} - 1) \qquad (171)$$

The number fraction and weight fraction of monomer are interrelated, since

$$z_1 = f_1 (M_{nc}/M_1) \qquad (172)$$

Teller[85] has considered the case of discrete self-associations in which the equilibrium constants are equal. Šolc and Elias[86] have given a very

[82] M. S. Lewis and G. D. Knott, *Biophys. Chem.* **5**, 171 (1976).
[83] J. Kreuzer, *Z. Phys. Chem.* **B53**, 213 (1943).
[84] R. F. Steiner, *Arch. Biochem. Biophys.* **49**, 400 (1954).
[85] D. C. Teller, *Biochemistry* **9**, 4201 (1970).
[86] K. Šolc and H.-G. Elias, *J. Polym. Sci., Polym. Phys. Ed.* **11**, 1793 (1972).

TABLE VI
RAW DATA USED FOR STUDYING THE SELF-ASSOCIATION OF
DODECYLAMMONIUM PROPIONATE IN BENZENE AT 37°

w^a (g/kg)	E (μV)	$M_{na}{}^b$	c (g/liter)
2.963	141.1	355.3	2.550
3.704	162.5	385.3	3.188
4.444	188.4	398.8	3.825
5.926	227.2	440.9	5.100
8.889	304.8	493.0	7.651
11.852	375.6	533.4	10.201
14.815	445.0	562.8	12.751
17.772	510.4	588.8	15.301
20.741	573.0	611.9	17.852
23.704	626.0	640.1	20.402

[a] All solutions were made up by weight in volumetric flasks, hence w is the solute concentration in grams per kilogram and c is the solute concentration in grams per liter.

[b] Benzil was used as the calibration standard. At 37° in benzene $K_{vp} = 16.906 \times 10^3$. One obtains M_{na} from E and K_{vp} by noting that $M_{na} = K_{vp}/(E/w)$.

elegant treatment of a self-association involving a heterogeneous monomer; this type of association has been encountered with polyethylene glycols. For more details about self-associations one should consult the papers by Adams,[52] Van Holde et al.,[87] Roark and Yphantis,[88] the reviews by Adams et al.[53] or by Williams et al.[59] or the Fujita[58] monograph.

An Example–The Self-Association of Dodecylammonium Propionate in Benzene and Cyclohexane

Lo et al.[42] have studied the temperature-dependent self-association of the surfactant dodecylammonium propionate (DAP); in benzene and in cyclohexane DAP exhibits micellar catalysis, and the catalytic effect is stronger in cyclohexane than it is in benzene. In order to get a better picture of what was happening, it was decided to do some vapor pressure osmometry (VPO) experiments aimed at finding out what was the state of aggregation of DAP in these solvents. These experiments were carried out in a Hewlett-Packard Model 302B Vapor Pressure Osmometer, equipped with a variable temperature controller. Table VI shows some of the raw

[87] K. E. Van Holde, G. P. Rossetti, and R. D. Dyson, Ann. N.Y. Acad. Sci. 164, 279 (1969).

[88] D. E. Roark and D. A. Yphantis, Ann. N.Y. Acad. Sci. 164, 245 (1969).

data used in this analysis. In column 2 the quantity E is the microvolts imbalance in the Wheatstone bridge circuit of the vapor pressure osmometer. M_{na} is obtained from E [see Eq. (102)] as indicated in Table VI. The values of c and M_{na} were plotted as cM_1/M_{na} vs c, and these plots were smoothed using spline functions. A plot of M_1/M_{na} vs c was constructed from these smoothed data, and M_1/M_{wa} was evaluated from it [see Eq. (108)]. Figure 10 shows plots of M_1/M_{na} and M_1/M_{wa} vs c for DAP in cyclohexane at 27° and 50°. Several models were tested (see Lo et al.[42] for details) to see which type of self-association was present, including the four types of indefinite self-associations described here. It was found that a type I sequential, indefinite self-association gave the best description of the observed self-association.

Table VII lists some of the smoothed data used in these calculations.

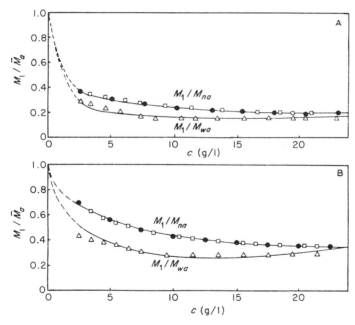

FIG. 10. Self-association of dodecylammonium propionate (DAP) in cyclohexane. Plots of M_1/M_{na} and M_1/M_{wa} [obtained from Eq. (108)] vs. c are shown for two temperatures: (A) 27°; (B) 50°. The filled circles indicate the experimental data; rectangles and triangles indicate values obtained by the free-spline technique for M_1/M_{na} and M_1/M_{wa}, respectively. The solid line indicates the curves obtained by the ship curve-fitting method. Reprinted (with the permission of Dr. E. T. Adams, Jr. and the American Chemical Society) from F. Y.-F. Lo, B. M. Escott, E. J. Fendler, E. T. Adams, Jr., R. D. Larsen, and P. W. Smith, *J. Phys. Chem.* **79**, 2609 (1976). Copyright by the American Chemical Society.

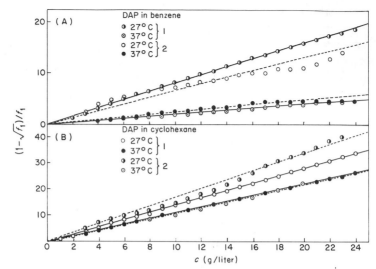

FIG. 11. Test for a sequential, indefinite self-association (type I): (A) Dodecylammonium propionate (DAP) in benzene at 27° and 37°; (B) DAP in cyclohexane at 27° and 37°. Two different ways were used to obtain f_1, the weight fraction of monomer, hence the two plots at each temperature. In both cases the straight line plot going through or close to the origin of $(1 - \sqrt{f_1})/f_1$ vs. c is characteristic of a type 1, indefinite self-association. Reprinted (with permission of Dr. E. T. Adams, Jr. and the American Chemical Society) from F. Y.-F. Lo, B. M. Escott, E. J. Fendler, E. T. Adams, Jr., R. D. Larsen, and P. W. Smith, *J. Phys. Chem.* **79,** 2609 (1975). Copyright by the American Chemical Society.

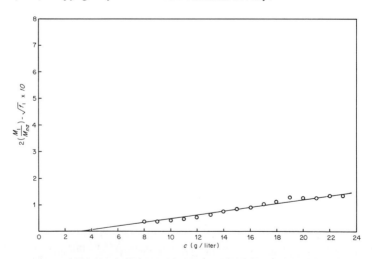

FIG. 12. Evaluation of $\hat{B}M_1$ for the self-association of dodecylammonium propionate (DAP) in cyclohexane at 43°. This plot is based on Eq. (136), and one can obtain $\hat{B}M_1$ from the slope of this plot. For this plot $\hat{B}M_1 = 7.14$ ml/g. Note that this plot does not go through the origin in this case, but nonetheless gives a reasonable straight line coming close to the origin. When $\hat{B}M_1$ is very small it may be better to use an array of data to find the best value of $\hat{B}M_1$ (see text).

TABLE VII

VAPOR PRESSURE OSMOMETRY DATA OBTAINED FOR THE SELF-ASSOCIATION
OF DODECYLAMMONIUM PROPIONATE IN BENZENE AT 37°

c (g/liter)	M_1/M_{na}[a]	$\xi = \dfrac{2M_1}{M_{na}} - \dfrac{M_1}{M_{wa}}$[b]	$(1 - \sqrt{f_1})^c/f_1$
3.35	0.6762	0.8837	0.5131
4.15	0.6299	0.8446	0.6930
5.35	0.5806	0.7559	1.2175
6.15	0.5574	0.7160	1.5274
7.35	0.5307	0.6724	1.9394
8.15	0.5165	0.6504	2.1828
9.34	0.4987	0.6246	2.5076
10.15	0.4885	0.6107	2.7008
11.35	0.4751	0.5936	2.9606
12.15	0.4670	0.5841	3.1164
13.35	0.4561	0.5720	3.3279
14.15	0.4493	0.5650	3.4558
15.35	0.4399	0.5559	3.6307
16.15	0.4340	0.5506	3.7372
17.35	0.4256	0.5436	3.8841
18.15	0.4203	0.5394	3.9741
19.35	0.4126	0.5338	4.0990
20.15	0.4077	0.5304	4.0761

[a] M_1/M_{na} was obtained from Eq. (102). For DAP, $M_1 = 253.4$.
[b] The spline-smoothed curve of M_1/M_{na} vs c (see Fig. 10) was used for the evaluation of M_1/M_{wa} [see Eq. (108)].
[c] Linear regression analysis of a plot of $(1 - \sqrt{f_1})/f_1$ vs c [see Eq. (135)] gave a slope of $k = 2.15 \times 10^2$ ml/g; the correlation coefficient was 0.976. For examples of plots of $(1 - \sqrt{f_1})/f_1$ vs c for DAP in benzene and cyclohexane, see Fig. 11.

Total solute concentration is in column 1, and corresponding values of M_1/M_{na} are listed in column 2. Values of M_1/M_{wa} were calculated from the spline-smoothed data and used to obtain ξ, whose values are tabulated in column 3. Column 4 lists values of $(1 - \sqrt{f_1})/f_1$; these were plotted against C [see Eq. (135)] to obtain k. Figure 11 shows some typical plots based on Eq. (135) for DAP in benzene and in cyclohexane. Figure 12 shows a plot, based on Eq. (136), that could be used for the evaluation of $\hat{B}M_1$.

Mixed Associations

Background Information

Associations between two reactants, A and B, of the type

$$n\text{A} + m\text{B} \rightleftarrows \text{A}_n\text{B}_m \ (n, m = 1, 2, \ldots) \tag{173}$$

or

$$\begin{cases} A + B \rightleftarrows AB \\ \quad 2A \rightleftarrows A_2 \end{cases} \tag{174}$$

and related equilibria are known as mixed associations. These associations can occur in a variety of ways, and as Eq. (165) indicates, both complex (AB) formation and self-associations can occur simultaneously. Mixed associations are frequently encountered in biochemistry. Two well known examples of mixed associations are the trypsin–trypsin inhibitor[89,90] reaction and the antigen–antibody[91-93] reaction. The trypsin–trypsin inhibitor[89,90] or the insulin–protamine[94] reactions may be a more complicated mixed association in which self-association occurs, since both insulin and trypsin are known to undergo self-association.

Although these equilibria have been studied by a variety of methods, relatively little has been reported on the study of mixed associations by sedimentation equilibrium, light scattering, or osmotic pressure experiments. Steiner has reported some very elegant theoretical and experimental studies on these associations by light scattering. He has also developed some elegant ways to analyze these associations by osmometry.[95-97]

Osmometry has its advantages since the theory is simpler, and one does not have to be as concerned about the associating species having unequal refractive index increments, density increments or partial specific volumes. We will illustrate the analysis first with the simplest mixed association, namely,

$$A + B \rightleftarrows AB \tag{175}$$

The condition for chemical equilibrium is

$$\mu_A + \mu_B = \mu_{AB} \tag{176}$$

where μ_i ($i = A$, B, or AB) is the molar chemical potential of constituent i. It will be assumed that the activity coefficients (y_i) of the associating

[89] R. F. Steiner, *Arch. Biochem. Biophys.* **49**, 71 (1954).

[90] R. F. Steiner, *Biopolymers* **9**, 1465 (1970).

[91] S. J. Singer, *in* "The Proteins," 2nd ed. (H. Neurath, ed.), Vol. 3, p. 270. Academic Press, New York, 1965.

[92] R. F. Steiner, *Arch. Biochem. Biophys.* **55**, 235 (1955).

[93] S. N. Timasheff, *in* "Electromagnetic Scattering" (M. Kerker, ed.), p. 337. Macmillan, New York, 1963.

[94] S. N. Timasheff and J. G. Kirkwood, *J. Am. Chem. Soc.* **75**, 3124 (1953).

[95] R. F. Steiner, *Biochemistry* **7**, 2201 (1968).

[96] R. F. Steiner, *Biochemistry* **9**, 1375 (1970).

[97] R. F. Steiner, *Biochemistry* **9**, 4268 (1970).

species obey the following relations[97-99]:

$$\ln y_A = M_A B_{AA} c_A^0 + M_A B_{AB} c_B^0 \tag{177}$$

$$\ln y_B = M_B B_{BA} c_A^0 + M_B B_{BB} c_B^0 \tag{178}$$

$$\ln y_{AB} = \ln y_A + \ln y_B \tag{179}$$

or

$$y_{AB}/y_A y_B = 1$$

Here

$$(\partial \ln y_i / \partial c_j^0)_{T, \, p, c_{k \neq j}(c_A = 0, c_B = 0)} = M_i B_{ij} \tag{180}$$

These assumptions are similar to those used with self-associations. Furthermore it is assumed that the relation[97-99]

$$B_{AB} = B_{BA} \tag{85}$$

still holds. We know that this is true if no association occurs, and since the $\ln y_i$ are written as a truncated Maclaurins series about zero solute concentration (where $c_A \cong c_A^0$ and $c_B \cong c_B^0$) it is assumed this relation applies to a mixed association. The total solute concentration becomes

$$c = c_A + c_B + K c_A c_B \tag{181}$$

where c_i ($i = A$ or B) denotes the chemical equilibrium concentration of species i. The total solute concentration can also be related to the initial concentrations, c_i^0 ($i = A$ or B); thus

$$c = c_A^0 + c_B^0 \tag{182}$$

It is convenient to define some mass balance equations that relate the c_i^0 to the c_i; thus

$$c_A^0 = c_A + \left(\frac{K c_A c_B}{M_{AB}} \right) M_A \tag{183}$$

$$c_B^0 = c_B + \left(\frac{K c_A c_B}{M_{AB}} \right) M_B \tag{184}$$

The quantity β will be defined as

$$\beta = c_A^0 / c_B^0 \tag{185}$$

Since one can vary the initial concentrations of A and B independently, it is evident that a solution of fixed total concentration (say 10 g/liter) can be made up in a multitude of ways. For each way the solution was made up, there would be a different value for an average molecular weight. In order to overcome this complication, it is convenient to prepare

[98] E. T. Adams, Jr., A. H. Pekar, D. A. Soucek, L.-H. Tang, G. H. Barlow, and J. L. Armstrong, *Biopolymers* **7**, 5 (1969).
[99] A. H. Pekar, P. J. Wan, and E. T. Adams, Jr., *Adv. Chem. Ser.* **125**, 260 (1973).

a master stock solution of A and also a master stock solution of B. If protein or other polyelectrolytes are involved, these master stock solutions would be dialyzed individually (in separate dialysis casings) against the same buffer or supporting electrolyte solution. A working stock would be prepared by blending the master stock solutions so that one has a predetermined value of β (say $\beta = 1$, 1.5, etc.), and a series of experiments would be performed on the working stock and dilutions made from it using the solvent, buffer, or supporting electrolyte solution as the diluent. This way β would remain constant for a series of experiments. At constant β it is evident that the number average (M_n) molecular weight or any other average molecular weight then becomes a function of the total solute concentration. Doing experiments in this way makes the analysis a lot easier.

With these relations and the use of the Gibbs–Duhem equation, one obtains[97]

$$1000\; dp/RT = d(c/M_n^{eq}) + \tfrac{1}{2} \sum_i \sum_j B_{ij} d(c_i^0 c_j^0)$$
$$+ (\bar{v}c/1000)\; d(c/M_n^{eq}) \quad (186)$$

Here it is assumed for simplicity that all partial specific volumes $(\bar{v})$ are equal. Integration of this equation between p^0 and $p^0 + \pi$ on the left and 0 and c (or 0 and $c_i^0 c_j^0$) on the right leads to

$$1000\pi/RT = c/M_n^{eq} + \tfrac{1}{2} \sum_i \sum_j B_{ij} c_i^0 c_j^0 + \bar{v}/1000 \int_0^c c\; d(c/M_n^{eq})$$
$$= c/M_{n\,app} \quad (187)$$

The quantity M_n^{eq} is the number average molecular weight; the superscript eq denotes that a mixed association is present. One defines M_n^{eq} by

$$M_n^{eq} = (\Sigma n_i M_i)/\Sigma n_i = c/(\Sigma c_i/M_i) \quad (188)$$

and $d(c/M_n^{eq})$ by

$$d(c/M_n^{eq}) = \Sigma\; dc_i/M_i \quad (189)$$

The last term on the right in Eq. (187) is quite small, so often it is ignored. Alternatively, from experiments at constant β over a range of concentrations, one could make a plot of $\bar{v}c/1000$ vs $c/M_{n\,app}'$ and estimate the term by numerical integration. Thus one can define $c/M_{n\,app}$ by

$$c/M_{n\,app} = 1000\pi/RT - \bar{v}/1000 \int_0^c c\; d(c/M_{n\,app}) \cong c/M_n^{eq}$$
$$+ \tfrac{1}{2} \sum_i \sum_j B_{ij} c_i^0 c_j^0 \quad (190)$$

Whether our convention for nonideal behavior is the best one or not is a moot point. It does have the virtue of simplicity, and it allows one to evaluate everything from the data. An alternative procedure has been advocated by Nichol and Winzor.[35] Because of interrelations such as $B_{AB} = B_{BA}$, etc., they would require six different virial coefficients (B_{ij}) to describe a nonideal A + B $\rightleftarrows$ AB association. While B_{AA} and B_{BB} could be evaluated from experiments on solutions containing only A or B, there would still be four virial coefficients to be evaluated. To estimate them, they would have to make educated guesses about the shape of the reactants and the AB complex; they would also need to know the amount of hydration in order to estimate the shape and size of these molecules. At best they have an educated guess, since it is difficult to measure the degree of hydration of a macromolecule. It should also be noted that with ionizable macromolecules, their nonideal treatment would include charge terms as well as the excluded volume term. If the charge items were much larger in magnitude so that they dominated the nonideal terms, then for an A + B $\rightleftarrows$ AB association $y_{AB}/y_A y_B = 1$. The same type of relation would apply to other mixed associations.

Ideal Mixed Associations

Case 1.

$$A + B \rightleftarrows AB^{95,98}$$

For the ideal case $\sum_i \sum_j B_{ij} c_i^0 c_j^0 = 0$ and $(c/M_{n\ app}) = (c/M_n^{eq})$. The following relations apply to this association[95,99]

$$\beta = c_A^0/c_B^0 \tag{185}$$

The total solute concentration, c, is

$$c = c_A + c_B + K c_A c_B \tag{181}$$

The mass balance equations are

$$c_A^0 = c_A + [(K c_A c_B)/M_{AB}]M_A \tag{183}$$
$$c_B^0 = c_B + [(K c_A c_B)/M_{AB}]M_B \tag{184}$$

The number average molecular weight, if no chemical equilibrium were present is M_n^0, and

$$c/M_n^0 = c_A^0/M_A + c_B^0/M_B \tag{191}$$

Insertion of the mass balance relations [Eqs. (183) and (184)] leads to

$$c/M_n^0 = c_A/M_A + c_B/M_B + (2K c_A c_B)/M_{AB} \tag{192}$$

The quantity c/M_n^{eq} is given by

$$c/M_n^{eq} = c_A/M_A + c_B/M_B + (Kc_Ac_B)/M_{AB} \qquad (193)$$

Thus

$$\Delta(c/M_n) \equiv c/M_n^0 - c/M_n^{eq} = (Kc_Ac_B)/M_{AB} \qquad (194)$$

This is a very neat result, and we can use this relation to show how one can test for the presence of an $A + B \rightleftarrows AB$ association. Rearrangement of the mass balance equations (183) and (184) and the use of Eq. (194) leads to

$$c_A = c_A^0 - \Delta(c/M_n)M_A$$

and

$$c_B = c_B^0 - \Delta(c/M_n)M_B$$

Insertion of these results into Eq. (194) leads to[78]

$$\Delta(c/M_n) = K/M_{AB}[c_A^0 - \Delta(c/M_n)M_A] \cdot [c_B^0 - \Delta(c/M_n)M_B] \qquad (195)$$

A plot of $\Delta(c/M_n)$ versus the product on the right-hand side of Eq. (195) will have a slope of K/M_{AB}, as illustrated in Fig. 13. Upward curvature in this plot could be due to the presence of other complexes, A_2B, AB_2, etc. Downward curvature in this plot could be due to nonideal effects.

Case 2.

$$2A + B \rightleftarrows A_2B^{98} \qquad (196)$$

Here we note that

$$c = c_A + c_B + Kc_A^2c_B \qquad (197)$$
$$C_A^0 = c_A + [(2Kc_Ac_B)/M_{A_2B}]M_A \qquad (198)$$
$$c_B^0 = c_B + [(Kc_Ac_B)/M_{A_2B}]M_B \qquad (199)$$
$$c/M_n^0 = c_A/M_A + c_B/M_B + (3Kc_A^2c_B)/M_{A_2B} \qquad (200)$$

For this case

$$c/M_n^{eq} = c_A/M_A + c_B/M_B + (Kc_A^2c_B)/M_{A_2B} \qquad (201)$$

and one notes that

$$\Delta(c/M_n) = (2Kc_A^2c_B)/M_{A_2B} \qquad (202)$$

With the aid of the mass balance equations, (198) and (199), one obtains

$$\Delta(c/M_n) = 2K/M_{A_2B}[c_A^0 - \Delta(c/M_n)M_A]^2 \cdot \{c_B^0 - [\Delta(c/M_n)M_B]/2\} \qquad (203)$$

One can construct a plot similar to that shown in Fig. 13 by plotting $\Delta(c/M_n)$ against the product on the right; the slope of this plot is $2K/M_{A_2B}$.

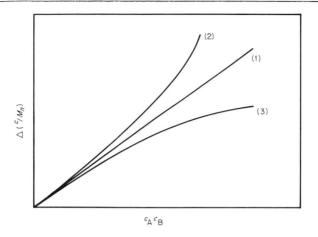

FIG. 13. Simulated mixed association. Here we have plotted $\Delta(c/M_n)$ vs. $c_A c_B$ [see Eq. (195)]. An ideal $A + B \rightleftarrows AB$ mixed association would give a straight line (plot 1). If higher species, such as A_2B, AB_2, etc., were present, then this plot would show upward curvature (plot 2). If only the AB complex were present and the system were nonideal, then the plot of $\Delta(c/M_n)$ vs $c_A c_B$ would show downward curvature (plot 3).

For the more general case

$$n A + m B \rightleftarrows A_n B_m \tag{173}$$

one notes that

$$\Delta(c/M_n) = (n + m - 1)K c_A{}^n c_B{}^m / M_{A_n B_m} \tag{204}$$

Case 3.

$$\begin{cases} 2A \rightleftarrows A_2 & {}^{95,98} \\ A + B \rightleftarrows AB \end{cases} \tag{174}$$

Here the following relations apply:

$$c = c_A + c_B + K_{A_2} c_A{}^2 + K_{AB} c_A c_B \tag{205}$$

$$c_A{}^0 = c_A + [(2K_{A_2} c_A{}^2)/2M_A]M_A + [(K_{AB} c_A c_B)/M_{AB}]M_A \tag{206}$$

$$= c_A + K_{A_2} c_A{}^2 + [(K_{AB} c_A c_B)/M_{AB}]M_A$$

$$c_B{}^0 = c_B + [(K_{AB} c_A c_B)/M_{AB}]M_B \tag{207}$$

At constant $\beta = c_A{}^0/c_B{}^0$ one obtains

$$\Delta(c/M_n) = (K_{A_2} c_A{}^2)/2M_A + (K_{AB} c_A c_B)/M_{AB} \tag{208}$$

If K_{A_2} is available from separate experiments on the self-association of A, then K_{AB} can be obtained from

$$\lim_{\substack{c\to 0 \\ \beta=\text{const}}} [\Delta(c/M_n)/(c_A^0 c_B^0)] = \lim_{\substack{c\to 0 \\ \beta=\text{const}}} [(\beta K_{A_2} c_A^2)/2(c_A^0)^2 M_A]$$

$$+ \lim_{\substack{c\to 0 \\ \beta=\text{const}}} [(K_{AB} c_A c_B)/(M_{AB} c_A^0 c_B^0)] = \beta K_{A_2}/2M_A + K_{AB}/M_{AB} \quad (209)$$

If experiments are conducted at two or more values of β, then one can obtain the values of K_{A_2} and K_{AB}. This illustrates the advantage of doing a series of experiments at constant β.

Steiner's Method[95-97]

In 1968 Steiner[95] published a very elegant paper which showed how one could obtain a relation between the number fractions, x_A and x_B, of A and B in mixed associations involving two reactants. His equation is

$$\ln x_A + \beta_m \ln x_B = \int_0^m (\alpha_n^{-1} - 1)dm/m$$
$$+ \ln 1/(1 + \beta_m) + \beta_m \ln [\beta_m/(1 + \beta_m)] \quad (210)$$

Here $x_A = (c_A/M_A)/(c/M_n^{eq}) = $ number fraction of A; $x_B = (c_B/M_B)/(c/M_n^{eq}) = $ number fraction of B; $m = c/M_n^{eq} = $ equilibrium molarity; $\alpha_n = M_n^0/M_n^{eq} = $ number average degree of associations; and $\beta_m = (c_B/M_B)/(c_A^0/M_A) = M_A/\beta M_B)$.

Since Eq. (210) does not include the number of moles of solvent, it is incorrect to refer to x_A and x_B as mole fractions; the correct term is number fractions. This relationship can be applied to any ideal mixed association involving two reactants A and B. For example, if the mixed associations is described by Eq. (166) and $\beta_m = 1$, the stoichiometry requires that $x_A = x_B$, so that Eq. (210) becomes

$$\ln x_A + \ln x_B = 2 \ln x_A = 2 \ln x_B = \int_0^m (\alpha_n^{-1} - 1)dm/m + 2 \ln(1/2) \quad (211)$$

The application of the Steiner method to other mixed associations is a bit more complicated, and the reader should refer to his papers[95-97] for more details about this ingenious method.

When species B is absent, $\beta_m = 0$. Then Eq. (211) becomes the same as Eq. (170) for the number fraction of monomer in a self-association.

Nonideal Mixed Associations[99]

When osmotic pressure experiments are performed on solutions containing only A or B, then one can obtain M_A or M_B, if they are unknown beforehand, as well as the second virial coefficients B_{AA} and B_{BB}. Then one uses Eq. (190) to obtain $(c/M_{n\ app*})$, which is defined by

$$c/M_{n\ app*} = c/M_{n\ app} - \tfrac{1}{2}[B_{AA}(c_A{}^0)^2 + B_{BB}(c_B{}^0)^2]$$
$$= c/M_n{}^{eq} + B_{AB}c_A{}^0 c_B{}^0 \quad (212)$$

The quantity $\Delta(c/M_{n\ app*}) = (c/M_n{}^0) - (c/M_{n\ app*})$ is then used for the evaluation of the equilibrium constant or constants and the nonideal term B_{AB}. This will be illustrated with some examples.

Case 1.

$$A + B \rightleftarrows AB \quad (175)$$

Here

$$\Delta(c/M_{n\ app*}) = c/M_n{}^0 - c/M_{n\ app*} = (Kc_A c_B)/M_{AB} - B_{AB}c_A{}^0 c_B{}^0 \quad (213)$$

and it follows that

$$\lim_{\substack{c\to 0 \\ \beta=\text{const}}} [\Delta(c/M_{n\ app*})/(c_A{}^0 c_B{}^0)] = K/M_{AB} - B_{AB} \quad (214)$$

Since this limit contains K and B_{AB}, how is K obtained? From the mass balance equations [see Eqs. (183) and (184)] one obtains

$$c_A{}^0 c_B{}^0 = c_A c_B[1 + K/M_{AB}(c_B M_A + c_A M_B) + (K^2 M_A M_B c_A c_B)/M_{AB}{}^2] \quad (215)$$

Thus,

$$\frac{\Delta(c/M_{n\ app*})}{c_A{}^0 c_B{}^0} = -B_{AB}$$
$$+ \frac{K}{M_{AB}[1 + K/M_{AB}(c_B M_A + c_A M_B) + (K^2 M_A M_B c_A c_B)/M_{AB}{}^2]} \quad (216)$$

and

$$\lim_{\substack{c\to 0 \\ \beta=\text{const}}} \frac{\partial}{\partial c} \frac{\Delta(c/M_{n\ app*})}{c_A{}^0 c_B{}^0} = \lim_{\substack{c\to 0 \\ \beta=\text{const}}} \frac{\partial}{\partial c_A{}^0} \frac{\Delta(c/M_{n\ app*})}{c_A{}^0 c_B{}^0} \frac{\partial c_A{}^0}{\partial c} \quad (217)$$
$$= \frac{-K^2}{M_{AB}{}^2}(M_A/\beta + M_B)\frac{\beta}{1+\beta}$$
$$= \frac{-K^2}{M_{AB}{}^2}\left(\frac{M_A + \beta M_B}{1+\beta}\right)$$

Instead of differentiating with respect to $c_A{}^0$, one could differentiate with respect to $c_B{}^0$ and obtain a similar expression. In order to get this result, the following relations have been used:

$$\beta = c_A^0/c_B^0 \tag{185}$$
$$c = c_A^0 + c_B^0 = c_A^0(\beta + 1)/\beta \tag{218}$$
$$\lim_{\substack{c \to 0 \\ \beta = \text{const}}} (\partial c_A/\partial c_A^0) = 1 \tag{219}$$

and

$$\lim_{\substack{c \to 0 \\ \beta = \text{const}}} (\partial c_B/\partial c_A^0) = 1/\beta \tag{220}$$

The following relations also apply and can also be used in the analysis:

$$\lim_{\substack{c \to 0 \\ \beta = \text{const}}} (\partial c_B/\partial c_B^0) = 1 \tag{221}$$

$$\lim_{\substack{c \to 0 \\ \beta = \text{const}}} (\partial c_A/\partial c_B^0) = \beta \tag{222}$$

and

$$(\partial c/\partial c_B^0)\beta = 1 + \beta \tag{223}$$

Case 2.

$$2A + B \rightleftarrows A_2B \tag{196}$$

Here one notes that

$$\Delta(c/M_{n\,\text{app*}}) = (Kc_A^2 c_B)/M_{A_2B} - B_{AB}c_A^0 c_B^0 \tag{224}$$

and that

$$\lim_{\substack{c \to 0 \\ \beta = \text{const}}} [\Delta(c/M_{n\,\text{app*}})/c_A^0 c_B^0] = -B_{AB} \tag{225}$$

Consequently, one can obtain

$$\Delta(c/M_n) = \Delta(c/M_{n\,\text{app}}) + B_{AB}c_A^0 c_B^0 \tag{226}$$

and this association then can be analyzed for K in the same way as would be done under ideal conditions.

Case 3.

$$\begin{cases} A + B \rightleftarrows AB \\ A + AB \rightleftarrows A_2B \end{cases} \tag{227}$$

Here one obtains

$$\lim_{\substack{c \to 0 \\ \beta = \text{const}}} \frac{\Delta(c/M_{n\,\text{app*}})}{c_A^0 c_B^0} = \frac{K_1}{M_{AB}} - B_{AB}$$

$$K_1 = \frac{c_{AB}}{c_A c_B} \tag{228}$$

Also note that

$$\lim_{\substack{c \to 0 \\ \beta=\text{const}}} \frac{\partial}{\partial c} \frac{\Delta(c/M_{n \text{ app*}})}{c_A{}^0 c_B{}^0} = \lim_{\substack{c \to 0 \\ \beta=\text{const}}} \frac{\partial}{\partial c_A{}^0} \frac{\Delta(c/M_{n \text{ app*}})}{c_A{}^0 c_B{}^0} \frac{\partial c_A{}^0}{\partial c} \qquad (229)$$

$$= \left[\frac{2K_2}{M_{A_2B}} - (K_1/M_{AB})^2 \frac{M_A + \beta M_B}{\beta} \right] \frac{\beta}{1+\beta}$$

By carrying out these experiments at two (or more) different values of β, one could obtain values of K_1 and K_2. Knowing these quantities one can readily obtain B_{AB} from Eq. (229).

The analysis of other nonideal mixed associations would be carried out in similar ways to the procedures described here. Once B_{AB} is known then one can calculate $\Delta(c/M_n)$ and recheck the calculations using the methods for the ideal case. Steiner[97] has developed a nonideal analog of Eq. (210), but it seems to be more difficult to apply it. The simplest procedure, at present, would be to estimate B_{AB} by the methods described here, then obtain $c/M_n{}^{\text{eq}}$ and apply the ideal form of Steiner's equation (Eq. 210), when one wants to use his method.

There are many, many other mixed associations besides the ones described here; some of them involve more than two reactants. Some of the methods used to study stability constants probably could be applied here. Many of these methods are summarized by Rossotti and Rossotti[100] in their monograph, and it should be consulted for further details.

A Simulated Example of a Mixed Association

Suppose that osmotic pressure experiments were carried out at $\beta = c_A{}^0/c_B{}^0 = 1$ on a mixture of two macromolecules A and B, for which $M_A = 40,000$ g/mole and $M_B = 60,000$ g/mole. It is thought that an A + B $\rightleftarrows$ AB association is occurring; for the AB complex $M_{AB} = 100,000$ g/mole. Given the following simulated data, which are listed in Table VIII, can we determine whether only an AB complex is formed?

The first column in Table VIII lists the values of the total solute concentration, c, at constant β; here $\beta = c_A{}^0/c_B{}^0 = 1$. It is assumed that separate experiments were performed on A and B, so that M_A and M_B are known; here it is assumed for simplicity that the solution is ideal. But this hypothesis can be tested as we will see. Values of $M_{n \text{ app*}}$ are tabulated in column 2; these would be evaluated using Eq. (212). The values of $c/M_n{}^0$ and $c/M_{n \text{ app*}}$ are listed in columns 3 and 4, respectively; they are used to evaluate $\Delta(c/M_{n \text{ app*}})$ (see Eq. 213), which is listed in column 5.

[100] F. J. C. Rossotti and H. Rossotti, "The Determination of Stability Constants." McGraw-Hill, New York, 1961.

TABLE VIII

SIMULATED OSMOTIC PRESSURE DATA FOR A MIXED ASSOCIATION

c (g/liter)	$M_{n\,app} \times 10^{-3}$ (g/mole)	$(c/M_n^0) \times 10^4$ (moles/liter)	$(c/M_{n\,app*}) \times 10^4$ (moles/liter)	$\Delta(c/M_{n\,app*}) \times 10^5$ (moles/liter)	c_A (g/liter)	c_B (g/liter)	$c_A c_B$ (g/liter)^2
12.0	55.38	2.500	2.167	3.332	4.672	4.001	18.692
11.0	54.98	2.292	2.007	2.909	4.336	3.755	16.282
10.0	54.55	2.083	1.833	2.502	3.999	3.499	13.993
8.0	53.60	1.667	1.493	1.741	3.304	2.955	9.763
6.0	52.54	1.250	1.142	1.080	2.568	2.352	6.040
5.0	51.94	1.042	0.9626	0.790	2.184	2.026	4.425
4.0	51.30	0.8333	0.7798	0.536	1.786	1.678	2.997
3.0	50.59	0.6250	0.5930	0.320	1.372	1.308	1.795
2.0	49.82	0.4167	0.4014	0.153	0.939	0.909	0.854
0	48.00[a]						

[a] $M_n^0 = 48{,}000 = c/(c_A^0/M_A + c_B^0/M_B) = (\beta + 1)/(\beta/M_A + 1/M_B)$. For $\beta = 1$, $c_A^0 = c_B^0$ and $c = 2c_A^0 = 2c_B^0$, since $\beta = c_A^0/c_B^0$.

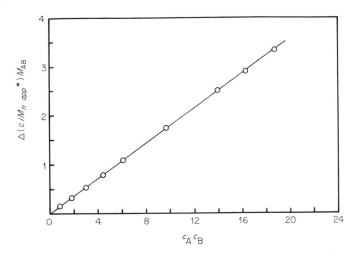

FIG. 14. Simulated example. Here we have plotted $\Delta(c/M_n)M_{AB}$ vs. $c_A c_B$ (see Eq. 195) using the data given in Table VIII. The straight-line plot indicates the presence of an ideal $A + B \rightleftarrows AB$ mixed association. The slope of this plot gives $K_{AB} = 0.1785$ liter/g.

The simplest procedure is to assume that the association is ideal, thus c_A and c_B, the equilibrium concentrations of A and B, respectively, can be calculated from the appropriate mass balance equations—in this case Eqs. (183) and (184). The values of c_A, c_B, and $c_A c_B$ are listed in columns 6, 7, and 8, respectively. Linear regression analysis of the plot of $[\Delta(c/M_{n\;app*})]M_{AB}$ vs $c_A c_B$ shown in Fig. 14 gave a straight line with a correlation coefficient of 0.99999 for 9 points used. The slope of this plot, based on Eq. (195), was $K = 0.1785$ liter/g. For the simulation we chose a value of $K = 0.1784$ liter/g. The intercept of the plot in Fig. 14 was 5×10^{-4}, which is essentially zero; this plot confirms that this is an ideal $A + B \rightleftarrows AB$ association. If other complexes had been present, such as A_2B, AB_2, then plot (2) in Fig. 13 would have shown upward curvature. On the other hand, if a nonideal $A + B \rightleftarrows AB$ mixed association were present, then the plot in Fig. 14 would have shown downward curvature, and one would have had to use the procedures described for this type of nonideal mixed association (see Fig. 13).

Equipment and Experimental Procedures

Membrane Osmometers

Two high speed membrane osmometers are currently available in the United States; these are the Wescan (formerly the Melabs High Speed

Membrane Osmometer; it is available from Wescan Instruments, Inc., 3018 Scott Blvd, Santa Clara, CA 95050) and the Knauer High Speed Membrane Osmometer (available from Utopia Instrument Co., P.O. Box 863, Caton Farm Road, Joliet, IL 60434). Figure 15 is a photograph of the Wescan osmometer, and Fig. 16 shows a photograph of the Knauer osmometer. The Hewlett-Packard (formerly the Mechrolab) high speed membrane osmometer is no longer being produced. Both the Wescan and the Knauer osmometer detect the flow of solvent electronically.

Figure 17 shows a schematic diagram of the Wescan osmometer cell. In the Wescan osmometer the solvent is on the underside of the membrane, and the solution is on the upperside. The flow of solvent through the membrane produces a slight negative pressure which causes the metal diaphragm (see Fig. 17) to move. The diaphragm is coupled to a strain gauge, and the movement of the diaphragm is detected by the strain gauge. The electrical signal from the strain gauge detecting circuit is sent to a 1 mV chart recorder. There are four pressure ranges for the Wescan osmometer (0–5, 0–10, 0–50, and 0–100 cm of solvent). One obtains the osmotic pressure as centimeters of solvent from the graph (see Fig. 2) on the chart recorder. For a 10-inch chart recorder, the 0–5 cm scale is the most sensitive scale available, although Kelly and Kupke[2] claim that the osmometer can be modified electronically to read the 0–1 cm range on the chart recorder.

In the Knauer osmometer, the solvent is below the membrane, which

FIG. 15. Photograph of the Wescan high speed membrane osmometer. Two models are available; Model 230 operates between 40° and 130° and Model 231 operates between 5° and 130°. (Courtesy of Dr. David Burge, Wescan Instruments, Inc., Santa Clara, CA 95050).

FIG. 16. Photograph of the Knauer high speed membrane osmometer. (Courtesy of Mr. J. L. Armstrong, Utopia Instrument Co., Joliet, IL 60434).

is rigidly held in place. The solution is above the membrane. Figure 18 shows a diagram of the osmometer cell. The flow of solvent across the membrane causes a metal diaphragm to move. This diaphragm is part of a capacitor, and the change in capacitance is used to detect solvent flow. This osmometer also requires a chart recorder, but in this case one needs

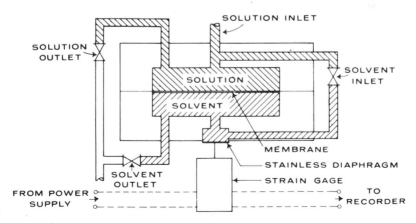

FIG. 17. Schematic diagram of the Wescan high speed osmometer cell. (Courtesy of Dr. D. E. Burge, Wescan Instruments, Inc., Santa Clara, CA 95050).

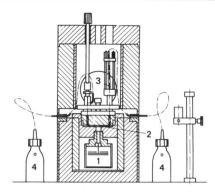

FIG. 18. Schematic diagram of the Knauer high-speed membrane osmometer cell. Measuring cell of the membrane osmometer: 1, capacitive pressure measuring system; 2, semipermeable membrane; 3, extractor for adjusting a constant liquid level; 4, extractor bottles. (Courtesy of Mr. J. L. Armstrong, Utopia Instrument Co., Joliet, IL 60434).

a recorder with a 10 mV full-scale range. The output from the capacitative detecting circuit is recorded as centimeters of solvent. Five pressure ranges from 2.5 to 40 cm of solvent are available. In both osmometers, the semipermeable membrane is held rigidly in place in the cell. The cell consists of two compartments, one for the solvent and one for the solution. Each compartment has precisely milled out grooves or channels that are interconnected; the ridges on the channels or grooves hold the membrane firmly in place when the cell is assembled and tightened.

The Knauer and Wescan osmometers have an advantage in that one can change the buffer solutions without having to replace the membrane. One simply flushes the osmometer cell with several changes of buffer. Also these instruments do not use a servo mechanism, nor does one have to use an air bubble to detect solvent flow and activate the servo mechanism, which is the case with the Hewlett-Packard osmometer. With the Hewlett-Packard osmometer, one has to disassemble the cell (the membrane chamber) to change the solvent or to replace the air bubble. Every time the osmometer cell is disassembled the membrane must be changed, since one cannot reposition it in the same orientation. When the cell is clamped, the grooves or ridges in the solution and solvent compartments will compress part of the membrane; in order to avoid additional problems, it is best to change membranes every time the cell is dismantled. Based on his experience, one of us (EFC) believes that the Hewlett-Packard osmometer may be better suited for work with organic solvents, whereas the Wescan osmometer may be better suited for aqueous solutions.

A block type, capillary osmometer (the Hellfritz osmometer) is available from Schleicher and Schuell, Inc. (543 Washington St., Keene, NH

03431). Two types are available: single- and double-chambered models. For each type, one can obtain membrane chambers made of glass or stainless steel. These osmometers can be used with aqueous or nonaqueous solutions. For further discussion on the operation of these block-type osmometers, one should refer to the review by Overton.[22]

A new, high-speed membrane osmometer, the IL 196 Weil Oncometer, is available from Instrumentation Laboratory, Inc., 113 Hartwell Ave., Lexington, MA 02173. This instrument is designed for measuring osmotic pressure of physiological fluids; it could be used for other applications as well. It requires 0.3 ml of sample, and it gives the colloid osmotic pressure (in mm Hg) in approximately two minutes. This osmometer operates at ambient temperature; at present there is no provision for temperature control of the osmometer cell.

Membranes

Membranes are commercially available from ArRo Laboratories, Inc. (P.O. Box 686, Caton Farm Road, Joliet, IL 60434), Schleicher and Schuell, Inc., and Wescan Instruments, Inc. Commonly used membranes are made of gel cellophane, deacetylated cellulose acetate, and cellulose acetate. The gel cellophane and deacetylated cellulose acetate membranes are used with organic solvents, whereas cellulose acetate membranes are used with aqueous solutions. Spectrum Medical Industries, Inc. (60916 Terminal Annex, Los Angeles, CA 90054) has announced some membranes in sheet form for dialysis; these come in various grades, one grade having a molecular weight cutoff as low as 2000.[101] Whether these membranes will be suitable for osmometry remains to be seen. For a more detailed discussion about membranes, their properties and their limitations, one should consult technical bulletins issued by ArRo Laboratories, Inc. or Schleicher and Schuell, Inc., the reviews by Overton[22] or by Kelly and Kupke[2] or the Tombs and Peacocke[1] monograph.

Before use, membranes should be conditioned. For aqueous solutions, the following procedure is suggested.

Pretreatment of Membranes for Use with Aqueous Solutions. (1) Membranes are generally stored in alcoholic solution (20% isopropanol) in the cold (a refrigerator or 4°–6° cold room). (2) The membrane should be rinsed thoroughly with distilled water and then with buffer, if a buffer solution is to be used. Precondition the membrane by soaking it in water (or buffer) for several hours (say overnight) in the refrigerator. (3) Degas the membranes by using a vacuum pump or water aspirator for 30 min or longer (until no visible air bubbles are formed on the surface of the membrane). Another procedure is to soak the membrane in warm water (or buffer) at constant temperature with *careful* mechanical agitation until no

---

[101] For more details, consult Spectrum Industries bulletin, *Molecular Products News*.

more air bubbles are produced from the membrane. Here the temperature of the water bath is chosen so that it is about 10° higher than the highest experimental temperature used. Try not to exceed 50°, since temperatures higher than this may cause deformation of the membrane. (4) After the membrane is degassed, it can be trimmed to the right diameter for the osmometer and installed in the cell compartment. (5) The solvent (water or buffer) and the solutions should be degassed, if possible. This may not be possible with easily denaturable proteins, so the solutions should be made up and dialyzed in degassed solvents (or buffers) in this case.

For their work with aqueous dextran solutions, Wan and Adams[46] used Schleicher and Schuell B19 membranes. The B19 and B20 membranes that have been used by us now have new catalog numbers, which are AC 62 (for the B-19) and AC 61 (for the B-20).

Pretreatment of Membranes for Use with Nonaqueous Solvents. (1) If the membrane is packed or stored in an aqueous alcoholic solution, and one wishes to work with nonaqueous solvents, then the membrane must be conditioned gradually and stepwise, like one prepares a histological sample. (2) For example, to change from water to toluene, one would use various water–isopropanol mixtures (25%, 50%, 75% v/v) and then 100% isopropanol. The membrane should be conditioned in each mixture, as well as the isopropanol for about 2 hours for each step. Longer times will not hurt. Then one makes up isopropanol–toluene mixtures (25%, 50%, 75% v/v) conditioning the membrane in each mixture as above. Then one transfers the membrane to pure toluene and uses several changes of toluene to remove the isopropanol. (3) Degas the membrane as described previously. (4) Also, degas the solvents; here one can boil the solvent for 5–10 min or use a vacuum. See Overton's[22] review for another suggested pretreatment schedule.

Some Precautions and Observations about Membranes. (1) If the membranes are packed in an alcoholic solution, then they should be kept in an alcoholic solution (20% isopropanol) and stored in a closed container in the refrigerator. (2) When cutting a membrane to the right diameter for the osmometer chamber, avoid tears and rough edges. Be sure that the cutting blocks are clean. (3) Once a membrane has been immersed in water or solvent, keep it wet or in contact with the solvent. Do not let the membrane dry out. If the membrane dries out, cracks may occur, so discard dried out membranes. (4) Cellulose type membranes used with aqueous solutions are subject to attack by cellulases produced by various microorganisms, such as *Aspergillus niger*. They will not last indefinitely. Once you have found a good membrane, do as much as you can with it in as short a time as possible. According to Tombs and Peacocke,[1] one cellulose type membrane is good for about 50 experiments. (5) Handle membranes with care. Use forceps with rubber tips for handling them. (6) When experiments are done with proteins in denaturing solutions, such as

6 M urea or 4 M guanidinium hydrochloride, the membranes may have to be changed more frequently as these denaturants can attack membranes made of cellulose or its derivatives.

How to Use the Wescan High Speed Membrane Osmometer

There are three things that must be done to use the osmometer. First, one has to install a membrane and assemble the chamber. Second, the instrument must be calibrated. Third, the measurement of the osmotic pressure is performed.

Membrane Installation. The following steps should be followed:

1. Remove the foam insulating cover by opening the three valves and lifting off the cover.

2. Disassemble the cell CAREFULLY by first removing the four cap screws at the top of the cell assembly. Then remove the top of the cell, using thumb and forefinger to apply upward pressure while holding the lower part in its seat. Do NOT set the machined surface on the desk top; be sure that it is protected from scratching or scoring.

3. Make sure that the large Teflon ring is in place. Use a glass or plastic tip syringe (without a needle) to force solvent through the solvent drain (valve seat) hole and then through the solvent inlet (valve seat) hole until a puddle is formed in the grooves of the lower part of the cell.

4. Carefully place a wet membrane on the puddle. Start at one side, and lower the membrane gently to avoid trapping air bubbles. Press the membrane lightly and cover it with solvent.

5. Replace the upper part of the cell. Be sure that the four small Teflon O rings are seated in the grooves of the bottom part of the cell. Open the valves on the upper cell part. You may wish to rinse the upper cell part with some solvent (or buffer) before replacing it. Note that there are taper pins to align the upper part of the cell, and when it is aligned, slide it down carefully and slowly to a firm seat.

6. Replace the four screw caps and tighten the cell. Use the appropriate allen wrench and apply a small and even amount of torque successively to each screw. Make sure that the screws are as tight as possible (hand tight).

7. Carefully soak up all solvent (buffer) squeezed out from between the cell parts during the tightening process. Keep the outside of the cell dry at ALL times to prevent corrosion. Close all valves.

8. In the steps that follow, solvent (buffer) should be added by having the syringe needle tip at the bottom of the inlet tube. Do not let the solvent level go below the Swagelok fitting.

9. Fill the inlet tube with solvent. Open all valves and let solvent drain. If solvent does not drain freely, place a rubber bulb on top of the inlet tube

and slowly force solvent through the cell. Do the flushing procedure at least three times.

10. Close the SOLUTION DRAIN VALVE. Refill the inlet tube with solvent and flush the solvent chamber, using the rubber bulb if necessary. Repeat at least three times.

11. Close the SOLVENT DRAIN VALVE. Refill the inlet tube with solvent, then open the SOLUTION DRAIN VALVE. The solvent should drain freely; if it does not, use the rubber bulb until free drainage occurs. Repeat the flushing of the solution chamber at least three times.

12. Close the SOLUTION DRAIN VALVE. Refill the inlet tube with solvent. The level indicator should read at least 70 μA (if it does not, see the manual for trouble shooting).

13. Now use the SOLUTION DRAIN VALVE to lower the solvent level in the inlet tube to approximately the top of the Swagelok fitting. Close the valve.

14. At this point if there is no obvious leakage between the two cell compartments, then do the following: Zero the chart recorder using the control ON THE CHART RECORDER, or by disconnecting the recorder from the osmometer and then using a jumper wire on the recorder. Reconnect the chart recorder to the osmometer, set the voltage range on the recorder to 1 mV. Then use the recorder zero control ON THE OSMOMETER; this control potentiometer should read between 200 and 525.

15. If the OSMOMETER recorder zero control cannot zero the recorder, then it is necessary to use the flat wrench to readjust the coupling between the strain guage and the diaphragm to bring the chart recorder pen to approximately zero. Counterclockwise adjustment moves the pen upscale; clockwise adjustment moves the pen downscale. Since this is a crude adjustment, use the OSMOMETER recorder zero control to do further adjustment.

16. Fill the inlet tube with solvent. Then do the following sequence as fast as possible to prevent draining the cell: (a) Open all three valves, (b) replace the insulating cover, and (c) close the solvent and solution drain valves only.

17. Refill the inlet tube with solvent. Flush each chamber once more and refill the inlet tube. Let the osmometer stand at the desired temperature for at least two hours before going further.

18. Slowly and gently work the SOLVENT INLET VALVE between the open and partly closed positions until the air bubbles stop rising in the inlet tube. Leave this valve open.

19. Open the SOLUTION DRAIN VALVE to reduce the solvent level to approximately 60 μA on the level meter. Close the valve. After a few minutes of stabilizing, the chart recorder should give a straight line (within

±0.2% of the full scale). Usually we set the chart recorder pen in the middle of the paper (4–5 inches on a 10-inch recorder) for this test.

20. Now close the SOLVENT INLET VALVE slowly and gently, and reset the solvent level with the SOLUTION DRAIN VALVE to a level meter reading of 60 μA. After 15–45 min, the recorder should return to the same level obtained in the previous steps within these tolerances: −5% to 0% full scale.

If the membrane has been installed properly, then the instrument should perform as indicated in steps 19 and 20. When the solvent inlet valve is closed, the solvent is sealed off. In addition, the closing of the valve raises the pressure in the solvent chamber relative to the solution chamber. The amount of pressure increase depends on how the valve is closed. Ideally, it should be closed gently, not snapped. The increased pressure in the solvent chamber is dissipated by flow of solvent through the membrane into the solution chamber. This is somewhat similar to the process involved in measuring osmotic pressure. The time involved in this equilibration (usually between 15 and 45 min) will be slightly longer than the time that it takes to perform an osmotic pressure experiment. When the valve is closed, the pen will show a downscale deflection, and then it will return slowly in an exponential fashion (cf. with Fig. 2).

If the recorder returns almost immediately to the original chart reading, then there may be a large hole in the membrane or a large leak around it. Failure of the recorder to stabilize near its original reading within 60 min may be due to the membrane being impermeable to the solvent, or there may be minute leaks around the membrane. If either of these problems is encountered, change the membrane. If the instrument does not give a steady reading in step 19 and shows a progressive rise in the chart recorder pattern, the osmometer cell may not have reached thermal equilibrium. Allow the instrument to equilibrate for awhile and repeat step 19. Bubbles in the membrane can shrink and expand, and this will cause chart recorder readings to fluctuate up and down. The bubbles should be forced out by flushing with solvent before any further operation with the osmometer. Once steps 19 and 20 are performed satisfactorily, the instrument is ready for calibration.

Calibration. This is performed in the following way:

1. Set the instrument for the desired temperature. Since the Wescan osmometer uses a thermoelectric cooler, be sure that cooling water flows through it to dissipate the heat. The instrument has an automatic cutoff in the event that the cooling water is shut off.

2. Fill the inlet tube with solvent, and then open the SOLVENT INLET VALVE.

3. Carefully attach the calibration tube to the inlet tube, using a piece

of rubber tubing to connect the tubes. One must attach or remove the calibration tube carefully, since the inlet tube is held against an O-ring by a pressure fit. A longer tube, which combines the inlet and calibration functions, with calibration marks on it can prevent the possible leakage caused by fixing or removing the rubber tubing. One has to use syringe needles with longer cannulas if a longer inlet/calibration tube is used.

4. Fill the calibration tube with solvent beyond the uppermost calibration mark. Be sure that the osmometer pressure range is set to 5 cm. Then open the SOLVENT DRAIN VALVE carefully and let the liquid level come to the uppermost calibration mark. Close the valve gently and smoothly so that it is completely closed when the solvent reaches the uppermost calibration mark. There are three calibration marks, with a 5-cm spacing between successive marks.

5. Using the OSMOMETER recorder zero control, set the chart recorder to zero. Then carefully open the SOLVENT DRAIN VALVE and lower the solvent to the middle (5 cm) calibration mark. The recorder should read full scale (10 inches on a 10-inch chart recorder is 1 mV). If the recorder reads more or less than full scale, one must try to adjust the osmometer calibration control potentiometer to make the recorder read full scale. Note that the OSMOMETER RECORDER ZERO and CALIBRATION controls interact with each other, so that one should add more solvent and lower the solvent to the uppermost calibration mark. Reset the OSMOMETER RECORDER ZERO control so that the chart recorder pen reads zero. Then open the SOLVENT DRAIN VALVE and let the level of solvent drain to the middle mark (5 cm) and see whether one obtains a full-scale reading. Readjust the CALIBRATION potentiometer, if necessary. This procedure may have to be repeated 3 or 4 times before the final adjustment is made.

6. In the event that one cannot set the chart recorder properly in the calibration procedure, because there are no more turns left in the osmometer calibration or recorder zero potentiometer, then it is necessary to remove the insulating upper cover from the osmometer (all valves must be open to remove the cover; close them as soon as the cover is removed) and adjust the mechanical coupling of the strain guage to the diaphragm in the solvent chamber. This adjustment is made with the flat wrench that fits the hexagonal nut in the slot in the lower part of the osmometer cell. One must turn this nut slightly counterclockwise to move the pen upscale or clockwise to move the pen downscale. Before doing this crude adjustment, fill both chambers of the osmometer with solvent. Close the SOLVENT and SOLUTION DRAIN VALVES. Leave the SOLVENT INLET VALVE open. Fill the calibration tube with solvent and lower it to the uppermost calibration mark, using the SOLVENT DRAIN VALVE. Then do the crude

adjustment first. Second, see if you can zero the recorder. Third, see if you can lower the solvent 5 cm to the middle mark and get a full-scale (or close to full scale) reading on the chart recorder. If you cannot do this, repeat the crude adjustment. If you can do the third step successfully, then open all valves and replace the top cover as soon as possible. Close the SOLUTION and SOLVENT DRAIN VALVES. Let the osmometer equilibrate at the desired temperature for at least 2 hr. Then try the calibration again. Eventually one can get the instrument calibrated.

7. Now calibrate the instrument between the uppermost and lowest marks (5 cm). Then reset the OSMOMETER recorder zero control and repeat the calibration using the middle and lowest marks (5 cm).

8. Refill the calibration tube. Adjust the level to the uppermost mark. Set the pressure control to 10 cm. Reduce the solvent level 10 cm and note whether a full-scale deflection (or within 98–102% of full scale) is produced on the recorder pen. If a full-scale deflection is not produced, this may be due to bubbles in the cell or the calibration tube. Use the syringe cannula carefully to break up bubbles in the tube. Gently open and close the SOLVENT INLET VALVE several times to try to remove bubbles. If this does not work, then flush the cell several times with solvent.

9. Once the calibration has been done, reduce the liquid level in the inlet tube to read 60 (or your predetermined level) on the level meter. Let the recorder run for 10–20 min. If the recorder does not keep a stable tracing, then check the following possibilities: First, examine the liquid level in the inlet tube frequently. If the liquid continues to drain lower, this may indicate a possible leaky valve or a leak through the cell compartment. Check the screws for tightness, and retighten if necessary. If the liquid level stays constant, then try to get rid of the possible hidden air bubbles in the cell assembly. A temperature fluctuation or ruptured membrane can cause an unstable base line. Check the temperature of the cell body; there is a thermometer well in the top of the foam cover. If a ruptured membrane seems to be the problem (check the other possibilities first), then replace the membrane. The valves may come loose after many manipulations. See that they are screwed in tightly (hand tight). After some use (several months), the valve seats, which are made of plastic, will need to be replaced because of wear; consult the manual for how to do this.

Now you are ready to do osmotic pressure measurements. Note that step 6 is to be done only if there is difficulty in adjusting the calibration span. Usually we have done a preliminary calibration on the 5-cm pressure range with the middle and lowest marks first. If trouble is going to occur, it seems to show up at this stage first.

Measurement of Osmotic Pressure. This is done in the following manner:

1. Be sure that the chart recorder is connected properly, and that the calibration of the osmometer gives a full-scale deflection on the chart recorder for the desired pressure range (0–5 cm, 0–10 cm, etc.).

2. After the calibration step, open the SOLUTION DRAIN VALVE carefully and lower the solvent level to a predetermined reading on the level meter (the manufacturer suggests a reading of 60 μV).

3. Slowly and carefully close the SOLVENT INLET VALVE and set the solvent level, using the SOLUTION DRAIN VALVE to the predetermined reference level on the level meter.

4. Allow the instrument to stabilize. This will be indicated by a straight line plot on the chart recorder; the pen should stay at a constant position.

5. Then use the OSMOMETER recorder zero control to zero the chart recorder. This procedure sets p_0 at 0 cm.

6. Now one is ready to start the measurements. Usually we start with the lowest concentration. One should also rinse the sample chamber with the sample solution at least twice and do the measurement of osmotic pressure with the third filling. To rinse the chamber add 0.3–0.5 ml of test sample. Open the SOLUTION DRAIN VALVE until the level meter reads 60 μV (or your predetermined value). Let the solution sit in the chamber for 1–2 min. Then repeat the rinsing procedure.

7. For the measurement, add 0.3–0.5 ml of sample. Open the SOLUTION DRAIN VALVE until the level meter reads 60 or your predetermined value. Allow the recorder to stabilize. The recorder will usually go off scale as the sample is being introduced. Then, within 5–20 min, usually it will reach a constant value (cf. with Fig. 2). This time depends on the solute concentration and also on the type of membrane used.

8. Carefully readjust the level (it may rise due to solvent flow), if necessary, to the reference level set in step 6. Allow 2–3 min for restabilizing, then read the osmotic pressure on the recorder. If a 10-inch recorder is used and the pressure range is 0–5 cm, for example, then the osmotic pressure is given by

$$h = (y/10)(P_{max}) \tag{230}$$

Here h is the osmotic pressure in centimeters of solvent, y is the number of inches on the (10 inch) recorder, and P_{max} is the maximum pressure of the pressure range selected. Thus, on the 0–5 cm pressure range, 2 inches on a 10-inch recorder correspond to

$$h = (2/10)5 = 1 \text{ cm of solvent} \tag{231}$$

To convert h (cm) to π (atm), simply divide h by the number of centimeters of solvent per atmosphere (CF). Thus (see Table III)

$$\pi = h/(CF) \tag{232}$$

9. Repeat the measurement at least once. For more accurate work, one should do each measurement 3 or 4 times.

10. Start with step 6 with the next sample. After all measurements have been made, one is ready to calculate the osmotic pressure as described earlier in Tables III or IV for nonassociating solutes. For associating solutes refer to Tables VI or VII.

The Knauer High Speed Membrane Osmometer

We do not have instructions regarding the Knauer membrane osmometer. Nonetheless, there will be three similar procedures to be followed: membrane installation, calibration, and operation.

The Hewlett-Packard Membrane Osmometer

Although Hewlett-Packard high speed membrane osmometers are no longer manufactured, there are still many in use. Thus we are presenting a brief description of the operating procedures.

Preparation of the Osmometer. Make sure that the optics are aligned, the top membrane clamp is off, and the membrane and the solutions to be tested have been prepared. Then do the following:

1. Turn on the thermostat and let the cell block equilibrate to the new temperature.

2. Turn on the SERVO. Set the CONTROL to REVERSE (the solvent reservoir moves to a fully lowered position in the elevator).

3. With sufficient solvent in the bottom membrane clamp (or block), apply suction to the reservoir to clear bubbles from the system. Taking a syringe, withdraw the plunger a short distance, and then insert the needle into the hole in the center of the membrane clamp. Depress the plunger to inject a small bubble into the capillary; this bubble should not be more than ¼ inch and not less than $^1/_{16}$ inch. (If this condition is not met, apply pressure on the solvent reservoir with a rubber bulb to flush the bubble out of the system; repeat the process for introducing the bubble until you are successful.)

4. With a proper-sized bubble in the capillary below the level of the detector light, quickly transfer the already conditioned membrane to the lower clamp; be careful not to trap air bubbles in the solvent beneath the membrane. Cover with the top membrane clamp and flange; then tighten the clamp screws evenly in several stages. Replace the chamber cover and

tighten the Swagelok nut around the drain stack. Run new solvent through the glass sample stack until air bubbles are flushed from above the membrane.

5. Switch the SELECTOR to RUN, and the CONTROL to NORMAL. This causes the elevator to raise the solvent reservoir to its maximum height.

6. When the bubble in the capillary rises and intercepts the light (detector) beam, the servomechanism lowers the reservoir so that the top of the bubble is kept in the light beam. The GAIN is adjusted to minimize the response time of the servo without inducing erratic oscillations.

Calibration. Since the instrument has a direct digital readout, no calibration step is needed.

Operation. To establish the pressure reading for solvent against solvent (h_0) and for solution against solvent (h), the following procedures are suggested:

1. Add some of the most concentrated solution to the solution stack, and drain off the solvent above the membrane. This serves two purposes: conditioning the membrane, and testing whether the pressures required for the most concentrated solution are beyond the range of the instrument. Allow the instrument to stabilize. The meniscus level should be stopped at the line (reference mark) seen on the solution stack through the observation window on the block cover.

2. Now flush with solvent, and allow the instrument to stabilize. After 1 min mark the height of the reservoir (in centimeters) from the reading of the digital readout at the top of the elevator.

3. Repeat the flushing with new solvent until the height of the reservoir is reproducible to within ±0.02 cm. This should not take more than 6 rinsings. Be sure to allow the instrument to stabilize briefly between rinsings. If more than 6 rinsings are needed, this usually indicates a torn membrane or a leakage of low molecular weight material through the membrane. If the system is equilibrated, the value of the digital readout for solvent against solvent (h_0) is recorded.

4. Now one is ready to determine the osmotic pressures of the various solutions. Start with the lowest concentration. Flush the chamber twice with this solution. Then refill the stack a third time and lower the level to the reference mark. Allow the instrument to stabilize and record the digital readout reading for solution against solvent (h). Repeat this measurement one or two more times. Then start with the next solution.

5. The osmotic pressure (π) in atmospheres is given by

$$\pi = (h - h_0)/(CF) \qquad (233)$$

Here CF is the number of centimeters of solvent per atmosphere at temperature T. The data are then treated as described in Tables III and IV for nonassociating solutes, or as described in Tables VI and VII for associating solutes.

Vapor Pressure Osmometers

Currently there are two vapor pressure osmometers available. These are the Corona-Wescan Vapor Pressure Osmometer (available from Wescan Instruments, Inc.) and the Knauer Vapor Pressure Osmometer (available from Utopia Instrument Co.). Both osmometers can be used with aqueous and nonaqueous solutions. A universal probe is available that can be used with either type solution. Figure 19 is a photograph of the Corona/Wescan vapor pressure osmometer; the upper limit of molecular weight for this instrument (according to the distributor's brochure) is 50,000. Figure 20 is a photograph of the Knauer vapor pressure osmometer; the upper limit of molecular weight for this instrument (according to the manufacturer's brochure) is 25,000. The Knauer vapor pressure osmometer has two parts: the vapor pressure module, which has the thermal chamber with the probes, and the universal temperature measuring device, which has the Wheatstone bridge circuit. Both of these vapor pressure osmometers can be operated with or without a chart recorder;

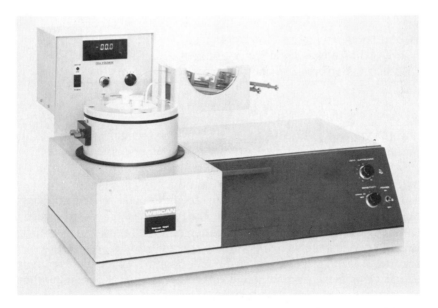

FIG. 19. Photograph of the Corona/Wescan vapor pressure osmometer. (Courtesy of Dr. David E. Burge, Wescan Instrument Co., Santa Clara, CA 95050).

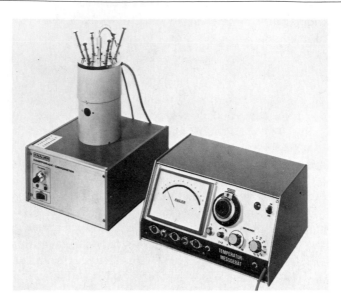

FIG. 20. Photograph of the Knauer Vapor Pressure Osmometer. (Courtesy of Mr. J. L. Armstrong, Utopia Instrument Co., Joliet, IL 60434).

however, we strongly recommend the use of a chart recorder with a vapor pressure osmometer. The Hewlett-Packard (formerly Mechrolab) vapor pressure osmometer is no longer manufactured. The Corona/Wescan and the Knauer vapor pressure osmometers are more sensitive instruments.

How To Do Vapor Pressure Osmometry

All the vapor pressure osmometers work on the same principle; this is illustrated as a schematic diagram in Fig. 21. When solvent condenses on the drop of solution (on the solution thermistor), there is a temperature change due to the heat of condensation effect. This temperature change causes the resistances of the thermistors to change and cause an imbalance in the Wheatstone bridge. Turning the slide wire resistance will rezero the null meter or the bridge circuit, and one can record the resistance change Δr or the microvolts imbalance ΔE when the steady state is obtained. A better procedure is to use a chart recorder. Although Hewlett-Packard Vapor Pressure Osmometers are no longer manufactured, there are a number of them still in use. So we will illustrate the operating procedure for vapor pressure osmometry based on our experience with this instrument. Similar procedure should apply to other vapor pressure osmometers.

Preparation of the Vapor Pressure Osmometer. The instrument should be prepared in the following manner:

1. Select the proper solvent. The total volume required is at least 100 ml beyond that needed for solution preparation (both polymer and standard benzil or other calibration solute solutions).

2. Thoroughly rinse the thermal cavity with solvent. It is quite important that the 6 needle guides be rinsed. If a log of instrument use is maintained, first rinse the guide with the last solvent used and then rinse with new solvent.

3. Construct the vapor wicks from the prepunched wick sheets, and insert a wick into the clean solvent cup. Fill the cup about half-full with solvent. Set the cup into the recessed center of the base of the thermal block, and replace the thermal cavity, being careful not to bend the two large guide prongs.

4. Depending on the type of solvent to be used, select and insert into the top of the thermal block either an aqueous or a nonaqueous probe (some vapor pressure osmometers use a universal probe, which can be used with aqueous or nonaqueous solutions). Set the temperature (an operating temperature range is usually given on the probe) using a variable temperature controller. Some models use fixed-temperature thermostats; these are preset at 37°, 50°, 67°, or 130°. Once the electrical connections have been made, turn on the thermostat or the variable temperature controller. Allow 2–4 hr for thermal stabilization before beginning the next step.

5. Switch on the NULL DETECTOR and the BRIDGE POWER. Fill two syringes with solvent only; fit each into a holder; then insert the units

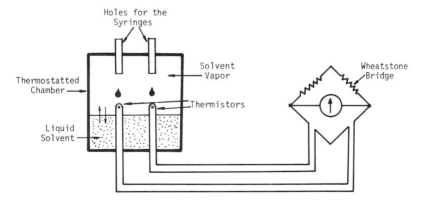

FIG. 21. Schematic diagram illustrating how a vapor pressure osmometer works. The temperature difference detected by the thermistor beads in the thermal block is measured on a Wheatstone bridge. Each thermistor is part of the Wheatstone bridge circuit.

(syringe plus holder) into the REFERENCE and SOLVENT syringe holes (numbers 5 and 6, respectively) in the thermal block. Fill each of the four other syringes with successive, different sample concentrations; fit each into a holder; then insert the units into holes 1 through 4 in succession. Wait at least 30 min before continuing; this permits a warm-up time for both the electronics and the samples in the syringes.

Operation of the Vapor Pressure Osmometer

1. The use of a recorder to monitor the experiment is strongly recommended. In setting the zero base line, for example, if the null meter and a timer were used, output voltage changes, which would prove small on a recorder, could appear large enough on the meter so that constant, unnecessary readjustment of the zero base line would be made. If a recorder is to be used, switch the function knob to RECORDER. Use the zero control on the RECORDER to zero the pen; then switch the recorder to the 1 mV input range; and then use the vapor pressure osmometer ZERO ADJUST POTENTIOMETER to set the base line.

2. A solvent drop is first put onto the reference bead by carefully rotating the reference syringe holder clockwise and lowering the unit until the needle almost touches the thermistor bead. Progress of the needle is monitored by watching through the observation window of the thermal block; under no circumstances should the needle impact the bead. Slowly rotate the knurled knob at the top of the syringe holder until a drop of solvent appears. Place the drop on the REFERENCE BEAD, then rapidly rinse the bead with 2–3 more drops. Finally, a single drop is put onto the bead; the total size (bead plus drop) should be about twice the size of the bead alone.

3. In exactly the same manner as with the reference bead, place a drop of solvent onto the sample bead using the solvent syringe (syringe number 6).

4. As soon as the sample drop is on the bead, a timer is started; at about 30 sec depress the AMPLIFIER ZERO button on the front of the instrument chasis, and center the meter needle by rotating the knob next to the button. Release the button.

5a. If a recorder is used, merely observe the curve produced. When the curve reaches a relatively constant value (usually in 2–8 min), this value associated with the solvent base line is then determined from the graph. Repeat steps 3, 4, and 5a until a reproducible base line is found.

5b. If the experiment is to be monitored with the instrument meter, the following procedure is to be used. With the drops in place on the thermistor beads, the meter needle, originally on the left side of the dial, should be swinging toward the center. With the resistance dials set at 0.00, and the needle moving no more than two small divisions (about ¹/₅ cm) per 10 sec, turn the BRIDGE ZERO DIAL until the needle is centered on the dial when the timer, which was started as soon as the second drop was put on its bead, reaches a half-integral multiple unit of time (½ min, 1 min, etc.). This will be the time period at which each reading will be taken; intervals usually vary from about 2 to 8 min. Repeat steps 3, 4, and 5b until reproducible zeros are found.

6. The resistance or voltage change (on a recorder it is a voltage change) for each concentration is measured in a manner analogous to that for determining the solvent zero base line. If the solvent is pure, the drop on the reference bead may be changed only as different concentrations are used; otherwise, replace the reference drop as each new solution drop is put on its thermistor bead.

7. The drop size is not really very critical; as long as the same approximate drop size is used, the equilibrium values will be the same. For very short times and dilute concentrations, however, the drop size effects may be pronounced.

8. If instrumental noise is too large, a zero base line may be difficult to establish. It may be accomplished by treating the solvent–solvent trials as solvent–solution trials by offsetting the BRIDGE ZERO and taking a set of readings. This set of readings is averaged to derive a bias, which is then subtracted from each subsequent solution resistance change. This procedure is automatically done when the recorder is used.

9. The instrument and solvent are calibrated using a suitable standard (usually benzil for nonaqueous applications, and KCl or mannose for aqueous applications). All resistance changes $(\Delta r)_i$ or voltage changes $(\Delta E)_i$ are divided by the appropriate concentrations, c_i, to produce a set of $(\Delta r)_i/c_i$ or $(\Delta E)_i/c_i$, which are then plotted against concentration and extrapolated to zero concentration, as shown in Fig. 22. The data used to obtain this figure are shown in Table IX. For the standard trial, this produces a calibration constant, K_{VP}, or K_{VP*}, i.e.,

$$\lim_{c \to 0} (\Delta r/c) = K_{VP*}/M \tag{234}$$

or

$$\lim_{c \to 0} (\Delta E/c) = K_{VP}/M \tag{235}$$

Here M is the molecular weight of the standard (benzil, KCl, or mannose).

10. For the experiments on unknowns in the same solvent, one can use K_{VP} to calculate M_{app} or $M_{n\ app}$ [see Eq. (43)]. These experiments are run in the same manner, except that with associating solutes, more solutions may be used than are needed with nonassociating solutes. For details on how to calculate data with nonassociating solutes, see Tables III and IV

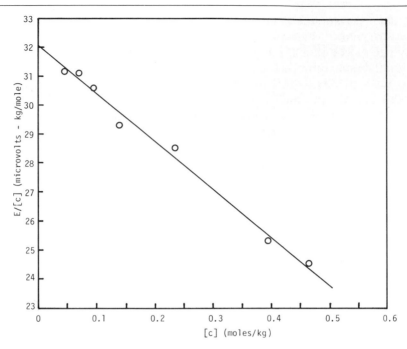

FIG. 22. Calibration plot for benzil in chloroform at 37°. The data for this plot are shown in Table IX. The slope of this plot depends on the temperature and the solute–solvent combination. (By courtesy of J. E. Flannery.)

and the discussion accompanying them. For details on how to treat the data with associating solutes, consult Tables VI and VII and the discussion accompanying them.

An Illustration–Determination of the Instrument Constant (K_{vp}) for a Vapor Pressure Osmometer

As we mentioned earlier, it is simpler to calibrate the vapor pressure osmometer by performing experiments at the desired temperature with the chosen solvent using as a standard a nonvolatile solute of known molecular weight, such as benzil or benzoic acid for nonaqueous solutions, of KCl or mannose for aqueous solutions. The data listed in Table IX were obtained from measurements of benzil solutions in chloroform at 37°. Recrystallized benzil was used, and the solutions were made up by weight. Vapor pressure osmometry experiments were performed on the series of standard solutions, and values of the microvolts imbalance, E, at various solute concentrations, $[c]$, in moles per kilogram were collected.

Replicate determinations were done (3 or more) at each concentration, and the average values of E were determined. These average values of E were divided by the corresponding values of $[c]$, and a table of $E/[c]$ and $[c]$ values (see Table IX) was constructed. These values of $E/[c]$ were plotted against the appropriate values of $[c]$; Fig. 22 shows this plot. Here seven solutions with different concentrations were used; ordinarily, one would use at least four solutions. The experimental points in Fig. 22 suggested that the data could be fit by a straight line; hence, linear regression analysis was used. The intercept of this plot [see Eq. (43)] is K_{vp}. This same procedure would be used with an aqueous solution. The slope of the plot shown in Fig. 22 is negative. The slope of these calibration plots will depend on the temperature and the solute–solvent combination. It should be emphasized that the value of K_{vp} reported here is peculiar to our particular instrument (a Hewlett-Packard Model 302B vapor pressure osmometer). The calibration constant is a working constant and must be determined individually for each vapor pressure osmometer. In addition, owing to aging of the thermistors and changes in solvent batches, K_{vp} changes with time, and must be redetermined periodically.

Although the concentration units used here were moles per kilogram, it is often more convenient to use concentrations in weight per volume

TABLE IX

CALIBRATION OF THE VAPOR PRESSURE OSMOMETER AT 37° WITH BENZIL
IN ETHANOL-FREE, DRY CHLOROFORM[a,b]

E (μV)	$[c]$ (mole/kg)	$E/[c]$ (μV-kg/moles)
1.472	0.0472	31.186
2.247	0.0723	31.079
2.888	0.0944	30.593
4.120	0.1406	29.303
6.655	0.2339	28.452
10.100	0.4001	25.244
11.400	0.4646	24.537

[a] From linear regression, one obtains:

7 points, correlation coefficient $= -0.9958$
slope $= -16.4534 \ \mu V \cdot kg^2 \cdot mol^{-2}$
intercept $= K_{vp} = 32.043 \ \mu V \cdot kg \cdot mol^{-1}$

Note that $E/[c] = K_{vp}(1 + BM^2[c])$, for $[c]$ in moles per kilogram. The slope of this plot depends on the temperature and the solute–solvent combination. Figure 22 shows a calibration plot based on these data.

[b] This was done on a Hewlett-Packard Model 302 B vapor pressure osmometer.

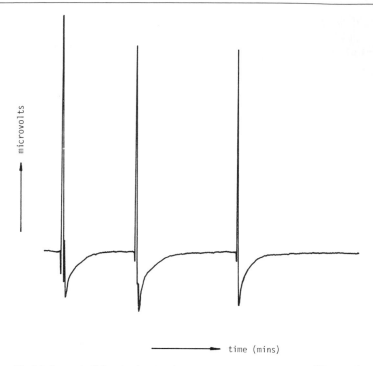

FIG. 23. Attainment of the steady state in vapor pressure osmometry. The results of three replicate experiments are shown. The spikes occur when a new sample is introduced, then the chart recorder voltage decreases and rises again in an exponential manner to a constant value at the steady state.

(grams/liter, etc.) units. The calibration plot in these concentration units is done in the same manner, and one can plot E/c vs c to get K_{vp}/M from the intercept [see Eq. (43)]. Alternatively, one can plot $E/[P]$ vs c to get K_{vp} directly, since $[P] = c/M$ for c in grams/liter. Figure 23 is a redrawing of some chart recorder traces for some vapor pressure osmometry experiments. Note how the micovolts imbalance, E, levels off with time. This is the steady state, and the value of E, under steady-state conditions, is the one used in the calculation. Three runs are shown here.

Acknowledgment

This work was supported in part by a grant (A485) from the Robert A Welch Foundation. Many thanks are due to Miss Carla Wood for her help in searching the literature and to Mrs. Catherine Mieth and Mrs. Lisa Brundrett for their excellent typing of the manuscript. We thank Professor C. N. Pace for his interest and comments on this work.

[6] Protein Concentration Measurements: The Dry Weight

By D. W. KUPKE and T. E. DORRIER

The magnitude of any activity or property assigned to a protein is based ultimately on a unit of mass, determined or assumed. For some parameters in common use the masses underlying their values depend on measurements or assumptions outlined years ago (e.g., protein partial specific volumes calculated from amino acid compositions). Hence, the value derived for such a parameter, while independent of a direct mass determination, may in fact not be adequate for the purpose at hand. In other cases, the concentration assigned to the protein in a solution is somewhat arbitrary; on occasions, the consequent error in the value of a parameter dependent on this assignment is sufficiently magnified in the succeeding relationships that the number of subunits of an enzyme may be in error by a whole number. Whenever possible, all concentrations assigned in the study of a protein should be related to a consistently defined dry mass of the same material actually being used in the investigation. As an example, the β function (which is related to the shape of a protein) depends on 4 different kinds of measurements (sedimentation velocity, viscosity, density, and an independent molecular weight determination), each involving the protein concentration. The usefulness of the β function would be enhanced, even for the globular proteins, if all concentrations in these determinations were related to a defined dry mass from the same stock solution.

Concentrations are usually evaluated via rapid secondary methods for convenience or necessity. The coefficients relating these values (light absorption or refraction, chemical procedures, etc.) to a unit of mass are frequently obtained from data published by other investigators on other protein preparations. This may be justified for some purposes; the determination of the concentration coefficient on the basis of an independent dry weight measurement is a tedious delay in the accomplishment of a more gainful result. Proteins of the same name prepared elsewhere, however, may differ appreciably in the magnitude of a property per unit of its dry mass. Hence, the practice of applying a concentration coefficient from one source and interrelating the results so derived with a property based on a concentration defined at yet another locale may lead different investigative groups to conflicting conclusions. On the other hand, if all laboratory operations are related to a locally defined dry mass of suitable precision, the internal consistency of the parameters determined for the protein can be quite good. The overall conclusions of the investigation may agree with those of another laboratory which used a different definition of the dry mass and/or other methods of preparation. The values of the mass-dependent properties, however, (absorptivity, refractive increment, partial and isopotential specific volumes, etc.) may remain distinguishable between the two laboratories even if the analytical precisions in both locales are demonstrably the same. The range of values in the literature for these fundamental properties of the most common proteins is considerably greater than can be accounted for by analytical imprecision. The history of a preparation and the definition of its dry mass appear to be major factors contributing to the larger than expected differences in the published values.

It is not the intent here to explore or describe methods for estimating the concentration of proteins in solution. These are secondary methods and each investigator usually develops the techniques and precisions necessary for achieving the goals of the investigation. The underlying question is whether the concentration coefficient which is being applied to the material in solution is an adequate reflection of the amount of a realizable mass being assigned as the protein component. The emphasis to be given here is on the precision of an operationally defined dry weight rather than on accuracy in terms of an absolute value. Indeed, the latter, even as an abstraction, is hardly definable.[1]

Along with refinements in purification and the trend toward standardization of protein preparations,[2] some trend toward standardization of the definition of the protein dry weight ought to follow. No such trend is visible and one seems long overdue. There is no unanimity, however, on how to establish the dry weight of proteins. "What method will enable one to get a completely dry protein without concomitant oxidation" is the rhetorical question posed by Professor M. J. Hunter.[3] Nonetheless, a beginning is effected by enhancing the general level of precision for acquiring a given kind of dry weight. This is a more modest goal and one to which we draw attention in the following protocol. This protocol over the past 9 years has led to a precision on a routine basis of 1 to 2 parts in 10^3 in terms of the unsmoothed data. The method follows closely one described by Hunter[4] in her careful and credible comparison of protein dry weight methodology.

A Dry-Weight Protocol

For introduction, we point out that via the commonly taught method of yesteryear, the best precision attained in our laboratory for the dry weight

[1] H. Eyring, *Anal. Chem.* **20,** 96 (1948).
[2] T. Peters, Jr., *Clin. Chem.* **14,** 1147 (1968); B. T. Doumas, *Clin. Chem.* **21,** 1159 (1975).
[3] M. J. Hunter, personal communication, 1976.
[4] M. J. Hunter, *J. Phys. Chem.* **70,** 3285 (1966)

of a simple protein was 1 part in 90, the variation being the difference between the extremes of the observed values from 5 or more samples handled concurrently. This level of uncertainty could not be reduced by increasing the technical competence. Approximately, this method consists of pipetting 5–10 ml of a 1–3% protein solution into standard 25–50 ml preweighed, clean, and heat-dried weighing bottles; weighing to 0.1 mg; heating at ~105° in an air oven; cooling in a desiccator and reweighing. The heating and weighing operations are repeated until constant weights, ±0.1 mg, are achieved. In theory, the precision as defined above at the 0.1-mg weighing limitation should approach about 1 part in 10^3.

The procedure described here to obtain better precision on this basis is one which can be practiced in almost any well-equipped laboratory; no unusual equipment or specialized expertise are required. It is assumed that a small vacuum oven and a well-maintained analytical balance, sensitive at least to 10 μg, are available. Balances with maximum loads of the order of 20 g and graduated to 5 μg are very suitable. Protection against drafts and unusual vibrations is required, and the temperature and relative humidity of the room should not change rapidly during a series of weighings.

In order to realize a precision in which the maximum difference between replicate samples is 1 to 2 parts in 10^3, each sample should contain about 10 mg of the protein. Thus, a dry weight determination consisting of quadruplicate samples (which we have come to regard as an adequate number) entails the expenditure of 30–50 mg of the protein. If true microbalances and the expertise for their correct operation are available, the amount of protein may be reduced appropriately. Details of a method for determining dry weights of proteins with a vacuum microbalance on samples containing 1–10 mg has been described.[5]

An important consideration in achieving good precision involves the use of weighing vessels that are as small as possible. This reduces substantially the uncertainties resulting from the variable adherence of moisture and dust particles. For this protocol, glass weighing vessels, no larger than 600 μl in total volume, have been found adequate; included are snug though easily removed glass caps overlapping the top half of the outside walls of the vessels. This type of weighing bottle is available commercially. The caps and vessels are matched as well as possible, then ground further by hand turning until the caps are well seated yet not sticky. For convenience in handling these small vessels during the drying and cooling operations, circular block trays of aluminum are useful. Aluminum blocks, about 3 cm thick, are milled to a diameter a little smaller than the inside diameter of medium-sized vacuum desiccators (~15 cm). Twelve holes are bored around the perimeter of the circular block of

[5] R. Goodrich and F. J. Reithel, *Anal. Biochem.* **34**, 538 (1970).

sufficient depth and width to hold the caps when not in use; 12 narrower holes are bored radially toward the axis in order to hold the weighing vessels upright when capped, but shallow enough to allow space under the caps for inserting curved forceps around the vessels. A center hole is tapped into the tray so that a suitable carrying handle with threaded end can be inserted and removed conveniently (in some ovens and vacuum desiccators a carrying handle of suitable length will not fit inside). During the heating period, the tray rests on an insulated pad rather than on the metal shelves of the oven. In our practice, wetted asbestos tape is molded around the bottom and sides of the tray; this material, when baked, is very hard and does not shower particles during the vacuum drying. The inside of the oven and its gaskets are thoroughly cleaned before each use, and the vacuum is tested with an independent pressure gauge (<50 μm mercury). The vessels are handled with Teflon-wrapped, curved forceps whose tips are maintained clean and lint free at all times. A collar of glass surrounds each vessel when in the oven and the desiccator so that the caps can rest on the collars above the vessels in order to prevent dust particles from falling in. These collars ($\sim$5 mm high) are nicked along the bottom edge to permit gas exchange.

The simplest dry-weight determinations are performed on deionized proteins in water. This is accomplished by passing the protein solution through a column of a water-washed, mixed-bed ion exchanger such that the pH of the effluent containing the protein remains at approximately the known isoionic point. This solution may be concentrated, as in a Diaflow ultrafilter equipped with a washed PM-10 membrane. For some purposes, solvents other than pure water are required. If the solvent medium requires the addition of small-molecule solutes to water, these solutes should be either completely nonvolatile or completely volatile; this is determined beforehand by performing solvent dry-weight determinations. Such solutes can then be added in known amounts to a known amount of water on the analytical balance. Known amounts of this medium are then added to a weight of the solute-free protein preparation in order to obtain a stock solution of 3–5% protein (complete solubility is required). Evaporation must be prevented during these weighings. The stock solution may be kept thereafter in a screw-cap vial containing a silicone plug, with narrow axial hole, which fits tightly into the neck below the screw cap. The axial hole is drilled so that a narrow-gauge syringe needle can be inserted for the withdrawing of samples to be used in the determination of the secondary concentration coefficients, densities, and other analyses in concert with the dry-weight measurement.

A single dry-weight experiment involves 6 weighing vessels, 4 containing the protein solution and 2 for the solvent water. Hence, one tray can accommodate 2 experiments where water is the solvent. (More than 2 solvent vessels may be indicated with solvents other than water; in this case, empty or dummy vessels can be used to set a confidence level on the weight changes of the glass itself. We have consistently found no demonstrable change in the anhydrous weight of the glass, whether empty or filled initially with deionized water, within the level of precision exercised by this protocol. We routinely filter the deionized water, however, through a prewashed 0.22 μm Millipore filter.) The scrubbed weighing vessels are preheated at 105° for 24 hr *in vacuo* and are then allowed to cool in a desiccator under vacuum. Each vessel is capped and taken out separately for the weighings during which time the desiccator is reevacuated. Each empty vessel is weighed until a constant weight is obtained; these weights are extrapolated to the time when the vacuum seal was broken for estimating the anhydrous weight. About 250 μl of the stock solution is inserted into the vessel with a gas-tight syringe, and the vessel plus solution is weighed with cap on. The difference between this weight and the constant weight of the empty vessel is the weight of the solution uncorrected for vacuum conditions. The completed tray with weighed samples is placed in the vacuum oven with caps raised on the collars and dry nitrogen gas is passed into the chamber at 45° until the liquid has virtually evaporated (~24 hr or more). [Alternatively, the aluminum tray with weighed samples may be partially immersed in liquid nitrogen to freeze the solutions.[4] The vacuum-dried protein after the final heating *in vacuo* is fluffy and sometimes is completely soluble, but special precautions may be required to avoid slight losses during the drying under vacuum.] The nearly dry samples are then heated under vacuum at ~105°. In our experience, 3–4 days of heating yield weights which are virtually at the constant weight. A second and third heating of ~2 days each between weighings is usually practiced to prove the attainment of a constant weight (± 10 μg). After each vacuum heating, the tray, with vessels capped, is cooled under vacuum in a very clean desiccator at the temperature of the analytical balance. The time is noted when the vacuum seal is broken for removing a sample. The latter is weighed versus time, rapidly at first (3–5 min intervals), until the weight remains unchanged (30–60 min). These weights are then extrapolated to the value at the time the vacuum seal of the desiccator was broken. In this way the variable pick up of moisture by a sample while on the analytical balance is accounted for. The solvent vessels have been found to gain only 10 μg or less at relative humidities of 50–60%, whereas those containing the protein often gain 50 μg or more per 10 mg of protein. With this kind of care the observed precision of the entire dry-weight operation, by this protocol, can closely approach the limit dictated by the analytical balance itself.

For example, with a balance reading to 5 μg, our data show that the maximum differences in the weight fraction among 4 or more samples of a set correspond to a range in net dry weight of between 10 and 16 μg, or ±5 to 8 μg about the mean value; these data represent the sum of all experiments on different preparations of ribonuclease and of serum albumin over a continuous 6-month period by 2 operators working nearly independently.

A small air correction is made for the difference in density between the solution and the dried material. Although this correction is usually no more than 1 part in 2000 in the weight fraction of proteins, it is good practice to routinely include the effect of buoyancy. For this purpose, it is sufficient to assume an air density of 1.2 mg/ml for the buoyant force exerted on the material in the weighing vessels. The approximate densities of the solution and solid are related to those of the counterweights of the analytical balance. Details for this correction are described by Corwin,[6] and tables for the case of steel counter weights with substitution-type balances are given by Macurdy.[7] For the dry protein, the reciprocal of its approximate specific volume may be used for the density (usually between 1.33 and 1.43 g/ml). If the density of the predried solution is not known, it is sufficient to increase the density of water (at the temperature in the balance chamber) by 0.003 g/ml per 1% of protein. If pure water is not the solvent, the density increment to be added to the density of water plus protein owing to the presence of salts or other added solutes can sometimes be obtained from existing tables—or the approximate density may be determined directly via a simple procedure described presently in the succeeding section of this article.

Finally, it should be emphasized that standardized techniques must be employed in the weighings in order to achieve the maximum precision with a given analytical balance.[6,7] An experienced operator is also concerned with maintaining the performance of the instrument while carrying out the weighings. The balance is indeed a sensitive instrument; the mass of only 1 μl of an aqueous solution can be weighed to an accuracy of 1 part in 10^2 or 10^3, depending on the type of instrument. Routine checks with carefully maintained standard weights and periodic servicing by trained personnel are highly recommended.

Correlation with Secondary Methods

Weighed aliquots of the same stock solution as that used for the dry-weight analysis should be taken to establish the concentration coefficient

[6] A. H. Corwin, in "Physical Methods of Organic Chemistry" (A. Weissberger, ed.), 3rd ed., Vol. I, Part I, Chapter III, p. 71. Wiley (Interscience), New York, 1959.
[7] L. B. Macurdy, in "Treatise on Analytical Chemistry" (I. M. Kolthoff, P. J. Elving, and E. B. Sandell, eds.). Vol. 7, Part I, p. 4247. Wiley (Interscience), New York, 1967.

by the other methods that are to be used in the routine assays of a protein. The ultraviolet absorbancy, which is often utilized in assaying for concentration, provides such an example. Usually the stock solution must be diluted substantially so that the most accurate region of the optical density scale for a particular instrument is utilized; the concentration of solutions of simple proteins often must be reduced to ≤ 1 mg/g of solution. This dilution is accomplished accurately by weighing the diluent (water or other suitable medium) first and then adding the smaller volume of the stock solution through a gas-tight syringe. Evaporation error attending the second addition can be reduced by prior fitting of a drilled plug (e.g., silicone stoppers) into the mouth of the weighing container. The drill hole should be wide enough to easily accommodate the syringe needle ($\sim$27 gauge); the latter must be wiped dry before insertion, and it should not be immersed into the diluent when the syringe is discharged. Screw-cap vials not much larger than the final volume are satisfactory containers; the screw cap is tightened for each weighing. Several samples are prepared and the weight fraction, W_d, of the protein in each diluted mixture is calculated from the dry weight of the stock solution. Thus,

$$W_d = Wm/(m + m') \tag{1}$$

where W is the known weight fraction of protein in the stock solution, m is the weight of the stock solution and m' the weight of the diluent. The correction for air buoyancy is not significant here unless the density of the diluent is grossly different from that of the stock solution.

Concentrations for biochemical studies are most often given in terms of a weight of protein per unit of volume, and the absorptivity is usually stated in terms of the net absorbance relative to such dimensions. Hence, the number of grams of dry protein per gram of the diluted solution, W_d, is converted to grams per milliliter, c_d, according to the general relation $c = \rho W$, where ρ is the density of the solution. The approximate density ($\sim \pm 10^{-4}$ g/ml for this purpose) of the diluted solution should be determined or calculated because c_d may differ significantly from W_d, depending on the density of the diluent. A satisfactory procedure for determining this density (if a rapid densimeter is not available) is by the use of constriction pipettes (e.g., the Carlsberg type). The ratio of the net weights at the temperature of the analytical balance chamber for a given pipette ($\sim$2 ml) when filled with the diluted protein solution and then with water (after many rinsings) provides an adequate value for the specific gravity ($= \rho / \rho^0$); the density of water, ρ^0, at the observed temperature is found from tables[8] ($\rho^0 = 0.9982$ g/ml at 20°, and this value decreases from $\sim$0.0002 to

[8] G. S. Kell, *J. Chem. Eng. Data* **12;** 66 (1967). Cf. H. Wagenbreth and W. Blanke, *Phys. Tech. Bundestagen Mitteilungen* **6,** 412 (1971). Density units in the latter reference are in kg/cm³ (or g/cm³), which differs in the fifth decimal place from density units in g/ml; i.e., 1.000027 cm³ = 1 ml.

0.0003 g/ml per degree of temperature rise from 20° to 30°). Alternatively, if the density of the stock solution and that of the diluent are sufficiently well known (to $\sim 10^{-4}$ g/ml), the concentration, c_d, of the diluted solution can be calculated. In this case, the concentration, c, of the stock solution (given by ρW) is divided by a volume dilution factor, V_d, in which

$$V_d = (V + V')/V = (m/\rho + m'/\rho')/m/\rho = 1 + m'\rho/m\rho' \qquad (2)$$

where V = volume (milliliters) and primes refer to the diluent. For this purpose the volume change of mixing, ΔV, is of negligible consequence as is assumed in the relation. Thus,

$$c_d = c/V_d \qquad (3)$$

For simple proteins in water (or dilute aqueous solvents), a density increment of 0.0003 g/ml can be added empirically to the density of water for each 1 mg of the protein per gram of stock solution in order to obtain ρ. If the diluent is a salt solution or common buffer instead of water, ρ', can sometimes be obtained from existing density-composition tables.

The absorbance coefficient, $\epsilon_{(\lambda,T)}$, at a specified wavelength (λ) and temperature T (1 cm pathlength) is often defined as the net absorbance of a solution which contains 1 mg of protein per milliliter of solution. Thus, in terms of the concentration c_d of the diluted solution

$$\epsilon_{(\lambda,T)} = A/10^3 c_d \qquad (4)$$

where A is the observed net absorbance for a solution of concentration c_d at the wavelength and temperature specified. For aqueous systems in the room-temperature range, the volume increases between 2 and 3 parts in 10^4 per degree of rise in T.[8] Such a correction may be required if the temperature difference between the analytical balance chamber and the spectrophotometer is large. For best precision, the temperature of the sample must be controlled during measurements by secondary methods that are temperature dependent, such as light absorption and refraction.

Acknowledgments

We gratefully acknowledge the helpfulness of Drs. M. J. Hunter and Theodore Peters, Jr. during the preparation of this article. The previously unpublished procedures described herein were developed with the aid of Grants GB-27331 and BMS-75-01599 from the U. S. National Science Foundation.

[7] Molecular Weight Measurements by Sedimentation Equilibrium: Some Common Pitfalls and How to Avoid Them[1]

By Kirk C. Aune

The biochemical literature abounds with studies of macromolecules wherein a report of the molecular weight of the entities in a system is given. Many techniques have been utilized to obtain the data throughout the history of biochemical research, but probably the most prevalent method since its introduction has been analytical ultracentrifugation.[1a] The reasons for this are many and varied, but most can be related through the single observation that the information content in a sedimentation experiment is exceedingly high.

The science of the use of analytical ultracentrifugation, both theoretical and experimental, has been extensively developed and recorded in the literature over the years. This development, although having received substantial initial attention in the 1920s and 1930s through such workers as Svedberg, Tiselius, Faxéx, Lamm, and others, has continued to the present time. There are many, more than adequate, treatise and review reference sources that follow this development, even up to the present.[2-7] This list is not to be considered complete by any means, for there are many individual offerings that have made significant advances in specific areas of ultracentrifugation. Some of these will be addressed when encountered in the text of the chapter.

Because of the wealth of information already available to the biochemist who may want to use the ultracentrifuge, this chapter needs to provide a somewhat different perspective in order to be anything less than redundant. The particular goal of this contribution is to discuss the proce-

[1] This research was supported in part by U.S. Public Health Service Research Grant GM22244 and Career Development Award GM00071 from the National Institute of General Medical Sciences and by the Robert A. Welch Foundation Grant Q-592.

[1a] T. Svedberg and J. B. Nichols, *J. Am. Chem. Soc.* **45**, 2190 (1923).

[2] T. Svedberg and K. O. Pedersen, "The Ultracentrifuge." Oxford Univ. Press, London and New York, 1940.

[3] H. K. Schachman, "Ultracentrifugation in Biochemistry." Academic Press, New York, 1959.

[4] H. Fujita, "Mathematical Theory of Sedimentation Analysis." Academic Press, New York, 1962.

[5] J. W. Williams, "Ultracentrifugal Analysis in Theory and Experiment." Academic Press, New York, 1963.

[6] D. C. Teller, this series Vol. 27, p. 346.

[7] K. E. Van Holde, *in* "The Proteins," 3rd Ed. (H. Neurath and R. L. Hill, eds.), Vol. 1, p. 225. Academic Press, New York, 1975.

dures one can and should readily perform in order to extract a minimal conclusion from sedimentation equilibrium experiments.

There is a substantial difference in the manner of data analysis carried out by a researcher who is, for example, interested in an associating system and a researcher who wants an estimate of the molecular weight of his favorite enzyme. The type of analysis used in the study of associating systems requires somewhat more extensive computations to yield meaningful answers. Moreover, that system is analyzed in a manner where heterogeneity is expected. In the second type of study, it has been customary to observe a linear plot of the logarithm of concentration versus the square of the radial position and relate the obtained slope to a molecular weight. It will be shown here that since rapid calculations can now be performed through the use of calculators, programmable calculators, and computers, which are readily available, more information can be obtained regarding a system with a smaller probability of making an erroneous assessment.

Fundamental Equations

The condition of sedimentation equilibrium leads to a state where the total chemical potential is invariant with radial position. This is a statement of the thermodynamics of a system that has been exposed to a centrifugal force field for a sufficient period of time so that net transport of mass has ceased.

The ramifications of the above thermodynamic statement are seen only when a physical model of the system is developed. The purpose of this chapter is to highlight procedures for obtaining as much information as possible regarding a physical picture from data collected on a system that has been brought to the condition of sedimentation equilibrium. It will be necessary to have the equations of state in order to discuss the data. Hence, the development of data description commences with the general equations of state.

Nonideal Multicomponent System

Considering a system of q components exposed to a centrifugal force field due to an angular velocity, ω, equations can be written, whereby the molalities, m_k, of all species i are related to radial positions, r, within the field and thermodynamic parameters of each component. One such equation[8] is

$$\sum_{j=2}^{q} (\partial \mu_i / \partial m_j)_{T,P,m_{k \neq j}} \, dm_j/dr = M_i(1 - \bar{v}_i \rho)\omega^2 r \tag{1}$$

[8] C. Tanford, "Physical Chemistry of Macromolecules," p. 257. Wiley, New York, 1961.

where μ_i, M_i, and $\bar{v}_i$ are the chemical potential, molecular weight, and partial specific volume of component i, respectively. The quantities, T, P and ρ are the temperature in degrees Kelvin, pressure and density, respectively. The partial derivative, $(\partial\mu_i/\partial m_j)_{T,P,m_{k\neq j}}$, is taken at constant T, P, and molality of all components k except component j.

By utilizing the molality scale where one component is taken as principal solvent with its molality invariant by definition, there remain $q - 1$ equations of the above type that provide the complete description of the system.

Now depending upon whether the method of observation can discriminate between species j and whether there are interactions between species j (all j), the complexity of Eq. (1) may be reduced. The problem of interactions between species j has been dealt with in considerable detail by Casassa and Eisenberg.[9]

Operationally, it is useful to define and discriminate interactions. If a system is considered to contain q components, it is generally convenient to subcategorize. Principal solvent is taken as component 1, molecules of "central interest" as component 2, and all other components are defined with a specific subscript. This will be important, for it is simpler to consider "preferential interactions" between component 2 and other components in the system to be a separate class of interactions, as did Casassa and Eisenberg[9] and Hade and Tanford.[10] All other interactions become those interactions which affect a physical picture of the system, i.e., self-association or heterologous association of molecules of "central interest." These latter interactions have been dealt with as an entirely separate class in great detail.[6,11-15]

The homogeneous three-component system is tractable. The two equations of interest become:

$$(\partial\mu_2/\partial m_2)_{T,P,m_3}\, dm_2/dr = M_2(1 - \bar{v}_2\rho)\omega^2 r - (\partial\mu_2/\partial m_3)_{T,P,m_2}\, dm_3/dr \quad (2)$$

and

$$(\partial\mu_3/\partial m_3)_{T,P,m_2}\, dm_3/dr = M_3(1 - \bar{v}_3\rho)\omega^2 r - (\partial\mu_3/\partial m_2)_{T,P,m_3}\, dm_2/dr \quad (3)$$

Assuming that redistribution of component 3 is due primarily to the me-

[9] E. F. Casassa and H. Eisenberg, *Adv. Protein Chem.* **19**, 287 (1964).
[10] E. P. K. Hade and C. Tanford, *J. Am. Chem. Soc.* **89**, 5034 (1967).
[11] E. T. Adams, Jr, "Fractions," No. 3. Spinco Division, Beckman Instruments, Inc., Palo Alto, California, 1967.
[12] D. C. Teller, T. A. Horbett, E. G. Richards, and H. K. Schachman, *Ann. N. Y. Acad. Sci.* **164**, 66 (1969).
[13] D. E. Roark and D. A. Yphantis, *Ann. N. Y. Acad. Sci.* **164**, 245 (1969).
[14] E. T. Adams, Jr., *Ann. N. Y. Acad. Sci.* **164**, 226, (1969).
[15] K. E. Van Holde, G. P. Rossetti, and R. D. Dyson, *Ann. N. Y. Acad. Sci.* **164**, 279 (1969).

chanical force applied to component 3, $(\partial\mu_3/\partial m_2)_{T,P,m_3} \cdot dm_2/dr \ll M_3(1 - \bar{v}_3\rho)\omega^2 r$, the relationship describing the redistribution of the molecule of "central interest" becomes

$$\frac{d \ln m_2}{dr^2} = \frac{M_2(1 - \bar{v}_2\rho)\omega^2}{2RT\{1 + [(\partial \ln \gamma_2)\partial m_2]_{T,P,m_3}\}}\left[1 - \frac{M_3(1 - \bar{v}_3\rho)}{M_2(1 - \bar{v}_2\rho)}\left(\frac{\partial\mu_2}{\partial\mu_3} \right)_{T,P,m_2}\right] \quad (4)$$

In this form the concentration of component 2 is seen to vary exponentially with the independent variable, r^2, as well as with molecular properties. The nonideality is present in the form of $(\partial \ln \gamma_2/\partial m_2)_{T,P,m_3}$ while the preferential solvation effect is seen to be due to nonzero values of $(\partial\mu_2/\partial\mu_3)_{T,P,m_2}$.

Homogeneous Ideal Two-Component System

Generally, for dilute salt solutions $(\partial\mu_2/\partial\mu_3)_{T,P,m_2}$ is very small. Hence, proteins in dilute salt solutions can usually be regarded as two-component systems. Therefore, a macromolecule behaving ideally will be found to be redistributed according to the equation:

$$(d \ln m_2)/dr^2 = [M_2(1 - \bar{v}_2\rho)\omega^2]/2RT \quad (5)$$

This equation requires that a plot of $\ln m_2$ versus r^2 will be linear with the slope equal to the right-hand side. Conditions can be arranged such that this is the usual case for data analysis. The molalities employed in experiments are such that

$$d \ln m_2 \simeq d \ln C_2 \quad (6)$$

Hence, the more readily determined concentration (unit of grams per volume), C_2, or a quantity proportional to C_2, is acceptable for data analysis.

Heterogeneous Ideal "Two-Component" System

When the molecule of "central interest" becomes a population of molecules of different molecular weight, the system containing these can no longer be regarded rigorously as a two-component system. However, such a system may be simplified. Note that for the multicomponent system [Eq. (1)], there is a term involving the concentration gradient of each component. If there is *no interaction* between any of the components of "central interest," each component contributes a term to the concentration gradient. If component 2 is now considered to be comprised of several species i, each will contribute a term:

$$(d \ln C_i)/dr = [M_i(1 - \bar{v}_i\rho)\omega^2]/2RT \qquad (7)$$

when species i is considered to be ideal.

If the measure of the concentration gradient weighs each component equally (as is generally true for refractometric techniques involving similar chemical species), the total apparent concentration distribution and gradient are just the sums:

$$C = \sum_{i=1}^{n} C_i \qquad (8)$$

and

$$(d \ln C)/dr^2 = 1/C \sum_{i=1}^{n} dC_i/dr^2 \qquad (9)$$

for n components of central interest. The gradient $d \ln C/dr^2$ then becomes:

$$(d \ln C)/dr^2 = [M_w(r) \cdot (1 - \bar{v}\rho)\omega^2]/2RT \qquad (10)$$

assuming $\bar{v}$ is the same for all the components comprising the heterogeneous system. The quantity $M_w(r)$, is the weight average molecular weight at position r where

$$M_w(r) = \sum_{i=1}^{n} M_i C_i(r)/ \sum_{i=1}^{n} C_i(r) \qquad (11)$$

It is seen for a heterogeneous system that a plot of $\ln C$ vs r^2 will not necessarily be linear. This will be discussed more fully in a later section.

If there are significant interactions between the macromolecules in the heterogeneous system, the terms involving the partial derivatives, $(\partial\mu_i/\partial\mu_j)_{T,P,mi}$ from the many components that are interacting, are necessarily nonzero. However, contrary to the case where preferential solvation is involved (small molecule effect), the partial derivative will be a function of position, because the concentrations of components i and j are varying significantly. The complexity is easily reduced by considering an additional component to be present in the system for each interacting pair of molecules. The constraints imposed on the system due to the interaction between molecules i and j, for example, then become:

$$(d \ln C_{ij})/dr^2 = [M_{ij}(1 - \bar{v}\rho)\omega^2]/2RT \qquad (12)$$

and, because chemical equilibrium also exists,

$$C_{ij} = K'C_i \cdot C_j \qquad (13)$$

where K' is the thermodynamic equilibrium constant of association between components i and j.

The ramifications of Eqs. (12) and (13) will be seen in a later section.

Types of Sedimentation Equilibrium Methods and Data

Schlieren Optics

The schlieren optical system in the ultracentrifuge yields directly a quantity proportional to the radial concentration gradient, dC/dr. If this particular optical system is utilized, it is generally applied in conjunction with the Archibald *approach* to equilibrium technique.[16] This technique had considerable popularity prior to the advent of the use of Rayleigh interference optics and the high speed sedimentation equilibrium technique outlined by Yphantis.[17] This was because the Archibald technique does not require a large amount of time, as was required by the contemporary equilibrium methods.

In the case of either the Archibald technique, where only two positions in the cell contained information or conventional sedimentation equilibrium, all the data analysis techniques have some relation to the fundamental sedimentation equilibrium Equation (1). Equation (7) can be rewritten in the following form when considering a single ideal solute component 2:

$$(RT/r)(dC_2/dr) = M_2(1 - \bar{v}_2\rho)\omega^2 C_2 \qquad (14)$$

It is seen in Eq. (14) that if the absolute concentration of component 2 can be established at any radial position, and if dC_2/dr is the direct raw measure, the quantity $(1/r) \cdot dC_2/dr$ is linearly related to the concentration C_2, and yields a slope that is proportional to the molecular weight through known parameters. The absolute concentration, C_2, can be readily established by numerical integration and conservation of mass. Therefore, in theory, such an analysis would provide a reliable measure of M_2. A complete discussion of these methods can be found in several sources.[18-20]

In theory, such a measurement is quite acceptable. In practice, more frequently than not, considerable error can be involved in such a result. First, an accurate measure of dC/dr from the schlieren optical system is difficult to obtain. Second, in order to have a measurable dC/dr, concentrations of component 2 are required that are sufficiently high to involve substantial nonideality effects in many cases. Third, because dC/dr is measured and because relatively high concentrations are utilized, there is very little variation in C_2 from the meniscus to the base of the cell. Consequently, the molecular weight (with nonideality and impreciseness) is a weight average at essentially a single concentration. The impreciseness

[16] W. J. Archibald, *J. Phys. Chem.* **51,** 1204 (1947).
[17] D. A. Yphantis, *Biochemistry* **3,** 297 (1964).
[18] K. E. Van Holde and R. L. Baldwin, *J. Phys. Chem.* **62,** 734, (1958).
[19] C. Tanford, "Physical Chemistry of Macromolecules," p. 259. Wiley, New York, 1961.
[20] H. K. Schachman, *Biochemistry* **2,** 887 (1963).

and nonideality are, therefore, eliminated only by an extrapolation of the experimental quantity M_2 (app), to infinite dilution. It is clear that since several experiments are required to obtain only the single result M_2, the information yield is low.[21] The more ubiquitous use of Rayleigh interference optics with the "high speed" method of Yphantis[17] has allowed ultracentrifugation to explore more areas of macromolecular analysis than simply providing a report of molecular weight results. It is for that reason that only the interference and absorbance optics data will be dealt with in this chapter.

Rayleigh Interference Optics

The use of Rayleigh interference optics for sedimentation equilibrium studies provides a considerable improvement in precision over the schlieren optical method. Because of the increased precision and because reliable optical alignment techniques are available with the adjustable optics,[22–24] the Rayleigh interference technique should be the preferred optical technique.

The technique is a differential technique, in that the refractive index of a solution containing redistributed material is compared to the refractive index of solvent as a function of radial position. The principal "catch" in the technique is that, generally, an absolute measure of the refractive index is not directly obtainable from the measurement. Certain *experimental* procedures have been developed to obtain an absolute measurement of the refractive index at some position. Alternatively, some *numerical procedures* have been developed for the same purpose. Both of these are discussed below.

The concentration of solute or solutes (mass/volume) is related to refractive index by the equation

$$n_s = n_0 + (dn/dc) \cdot C + \cdots \tag{15}$$

where n_s is the refractive index of a solution containing solute at a concentration C, and n_0 is the refractive index of the solvent. The refractive index is generally linearly dependent on solute concentrations employed in sedimentation studies. Hence, dn/dC is a constant and higher-order terms

[21] Although it is true that the second virial coefficient can be obtained as an additional piece of information, a heterogeneous system, whether involving chemical interacting species or not, is extremely difficult to quantitate by this procedure.

[22] R. D. Dyson, *Anal. Biochem.* **33,** 193 (1970).

[23] E. G. Richards, D. C. Teller, and H. K. Schachman, *Anal. Biochem.* **41,** 189 (1971).

[24] E. G. Richards, D. C. Teller, V. D. Hoagland, Jr., R. H. Haschemeyer, and H. K. Schachman, *Anal. Biochem.* **41,** 215 (1971).

in concentration can be neglected. The refractive index difference, $(n_s - n_o)$ is simply $(dn/dC) \cdot C$ and the interference fringe shifts are directly proportional to concentration differences because the interference fringe shifts are directly proportional to refractive index differences. The relationship between interference fringe patterns and concentration is, therefore,

$$f = kC \tag{16}$$
$$f_a = kC_a \tag{17}$$
$$y = f - f_a = k(C - C_a) \tag{18}$$

where k is an optical constant and f is the *absolute fringe displacement* for an *absolute concentration, C*. The subscript, a, represents a position at which the raw measurement of the relative fringe displacement, y, is taken. Generally, the displacement, y, is measured relative to the displacement at or near the meniscus.

The optical constant, k, is of no consequence, except when it is desirable to establish absolute concentrations, such as in the evaluation of equilibrium constants. Whether f_a, the absolute fringe displacement at position a, is determined from numerical procedures or from experimental procedures, it is required for complete data analysis.

The interference fringe pattern is generally obtained in permanent form by recording the image on photographic plates with Kodak spectroscopic emulsion 11-G. The plate is mounted on a suitable microcomparator, and fringe displacements are measured. Teller has outlined a procedure that should be adopted as the minimal requirement for plate reading.[6] The fringe displacement relative to the meniscus position, in such an analysis, should have a standard deviation on the order of the 5–10 μm or less when sufficient care is taken. A data set, therefore, consists of fringe displacements, y_i, relative to the meniscus, which are a function of radial position r_i. The information extracted from that raw data regarding the system then depends on the data processing to follow.

If Eq. (7) is written in the form where the fringe displacements are represented as data, the relationships

$$1/(y + f_a) \cdot dy/dr = [M_2(1 - \bar{v}_2\rho)\omega^2 r]/RT \tag{19}$$

and

$$d \ln y/dr^2 = [M_2(1 - \bar{v}_2\rho)\omega^2]/2RT \cdot (1 + f_a/y) \tag{20}$$

are obtained for pure ideal solute.

The utility of Eq. (20) is quickly seen, for if experimental conditions can be arranged where f_a is very small compared to the fringe displacements measured, the derivative, $d\ln y/dr^2$, becomes invariant with y:

$$d \ln y/dr^2 \simeq [M_2(1 - \bar{v}_2\rho)\omega^2]/2RT \tag{21}$$

This observation was put forth by Yphantis[17] and led to the development of the Yphantis, or high speed sedimentation, technique for clearing the meniscus, whereby data may be plotted in a linear relationship to obtain the quantity, M_2. That technique, probably the most ubiquitous one, represents one of the experimental methods of disposing of the problem presented by the quantity, f_a.

If the meniscus displacement, f_a is not negligible, one can also establish the quantity by adopting the technique of LaBar,[25] where the fringe pattern is recorded as a function of time during the approach to equilibrium. In that procedure, specific fringes are followed to establish directly the decrease in fringe displacement at the meniscus relative to the initial fringe displacement, f_0.

Other equally satisfactory techniques have been presented for the establishment of f_a at the more moderate speeds where f_a is quite significant.[18,26-29]

When a system is heterodisperse, the experimental condition of high speed becomes less desirable. This is because although the high molecular weight material is cleared from the meniscus, material of low molecular weight might complicate the situation owing to a significant concentration at the meniscus. Moreover, the high speed technique tends to impose a limit on the amount of data yield, for all of the information regarding mass redistribution and system heterogeneity is near the bottom of the cell.

If the low speed techniques are employed, the concentration redistribution is quite shallow providing large amounts of data, but only over a narrow concentration range.

A compromise condition is where moderate centrifugation speeds are utilized to generate a situation where the meniscus concentration can not be considered to be clear. However, sufficient redistribution occurs whereby the concentration range varies 50-fold with substantial amounts of data.

Absorbance Optics

The UV absorbance optics of the model E ultracentrifuge proved to be satisfactory for detecting ultraviolet-absorbing species during sedimentation. The nonuniformity of the photographic recording and the problems

[25] F. E. LaBar, *Proc. Natl. Acad. Sci.*, U.S.A. **54**, 31 (1965).
[26] H. K. Schachman, *Biochemistry* **2**, 887 (1963).
[27] P. A. Charlwood, *Biopolysomes* **5**, 663 (1967).
[28] S. Hanlon, K. Lamers, G. Lauterback, R. Johnson, and H. K. Schachman, *Arch. Biochem. Biophys.* **99**, 157 (1962).
[29] H. K. Schachman, and S. J. Edelstein, *Biochemistry* **5**, 2681 (1966).

associated with ultimate translation of the photograph employing a densitometer kept the absorbance optics from being usable for analysis of a system at sedimentation equilibrium. The pioneering work of Schachman and co-workers[28,29] introduced the method of direct-scanning absorption optics.

The scanning optics, in the form developed, became suitable for sedimentation equilibrium analysis. The electronic nature of the output and ability to sample several cells during the same run has promoted considerable activity in the direction of computerization of the data collection process.[30-32] The data from the latter process approaches the precision of the interference optics data. In theory, the scanning absorbance system can yield more information about a system containing species with different chromophores. This can be done by collecting data at several wavelengths in order to observe each species independently. If the absorption spectra of the chromophores are not distinctly isolated from each other, a more difficult task is encountered, but it need not be insurmountable. This will be discussed in a later section.

The data collected from the scanner, whether it be in digital or analog form traced on the Beckman supplied strip chart recorder, is proportional to the absorbance at each position radial to a reference hole. Considerable caution should be exercised with regard to the use of the absorbance information directly. If the slit width at the monochromator is 2 mm, the band width is sufficiently wide so that stray light reduces the absorbance. Generally, this is of no concern for the Beer–Lembert law is still obeyed and the output is, therefore, proportional to concentration. The recorder output yields true absorbances for most protein solutions at 280 nm when a monochromator slit width of 0.5 mm is used, although the more narrow slit width sacrifices energy and, thus, introduces more noise in data. The higher noise level is generally not satisfactory for routine use.

Another observation made in this laboratory is that, in the double-beam mode, the air–air base line in the reference hole is not necessarily coincident with the air–air base line or solvent–solvent base line where the light has passed through the sapphire windows. Moreover, the base lines are observed to be speed dependent and cell dependent. Thus, a proper base line may not necessarily be obtained by overspeeding to clear the meniscus or by subtracting the base line obtained with a cell containing solvent in both sectors during the same run.

Because of the aforementioned uncertainties, this laboratory treats the scanner data in much the same way as the interference data is treated. The

[30] S. P. Spragg and R. F. Goodman, *Ann. N. Y. Acad. Sci.* **164**, 294, (1969).
[31] R. H. Crepeau, S. J. Edelstein, and M. J. Rehmar, *Anal. Biochem.* **50**, 213 (1972).
[32] A. H. Pekar, R. E. Weller, R. A. Byers, and B. H. Frank. *Anal. Biochem.* **42**, 516 (1971).

absorbance near the meniscus is subtracted from the absorbances radial to the meniscus. These values are retained in a data set of displacements, y_i as a function of radial position. Equations (16)–(21) then become applicable to scanner data where f_a and f are proportional to the absolute absorbance at the meniscus and absolute absorbances at the other radial positions.

Methods of Data Analysis

Rayleigh Interference

The low speed sedimentation equilibrium technique has been discussed quite thoroughly by Van Holde.[33] The principal feature of the low speed method is that much higher concentrations of material are employed and more shallow concentration gradients result. Van Holde[33] pointed out that the basic equation used for analysis of low speed data is of the form:

$$(f_b - f_a)/f_o = [M(1 - \bar{v}\rho)\omega^2]/2RT \cdot (r_b^2 - r_a^2) \tag{22}$$

The quantity, $f_b - f_a$, is measured directly from the interference plate. The quantity, f_o, is determined from a separate synthetic boundary experiment. The quantity, M, becomes the single result from the experiment.

The Yphantis or high speed technique[17] or somewhat more moderate speed technique[25,34,35] provides data that span a wider concentration range. (Perhaps the most serious drawback to these techniques is that the wide concentration range is obtained in a narrow radial distance range.) Nevertheless, the form of the data is such that the integrated version of Eq. (19) can be utilized to considerable advantage:

$$\ln(y + f_a) = \ln f_a + [M_2(1 - \bar{v}_2\rho)\omega^2(r^2 - r_a^2)]/2RT \tag{23}$$

With $f_a \ll y$, a plot of $\ln y$ versus r^2 will provide a measure of homogeneity. If the plot is linear, it is *suggestive of a homogeneous system* containing a species with molecular weight, M_2. Because of this latter attribute, the plot has become the most used as well as abused representation of sedimentation equilibrium data.

Although there are certain objections to the use of the plot indicated in Eq. (23) from a statistics point of view (to be discussed), the principal objection is based on the problems that can arise owing to heterogeneity and nonideality.

[33] K. E. Van Holde, *"Fractions"* No. 1, Spinco Division, Beckman Instruments, Inc., Palo Alto, California, 1967.
[34] J. E. Godfrey and W. F. Harrington, *Biochemistry* **9**, 886 (1970).
[35] E. G. Kar and K. C. Aune, *Anal. Biochem.* **62**, 1 (1974).

Nonideality expresses itself by affecting the value of the observed molecular weight. The *first* virial coefficient is, in fact, the reciprocal of the molecular weight. Thus, if the *second* virial coefficient is significant, the *apparent first* virial coefficient is

$$1/M_2 \text{ (app)} = 1/M_2 + BC_2 \tag{24}$$

where B is the second virial coefficient.

In order to illustrate the effect of the second virial coefficient on the collected data, the data in the table were computed. A typical low speed sedimentation equilibrium experiment might have a concentration variation of 1–4 mg/ml from meniscus to the bottom of the cell. The data illustrate an example where the virial coefficient has a value of 4.44×10^{-7} ml/mg in order to cause a 10% decrease in the observed molecular weight, which ideally would be 100,000.

The apparent molecular weight, the actual result obtained from the experiment, is computed to be 90,000. The concentration dependence would not express itself in the typical data analysis. The virial coefficient would only become evident when the sample is analyzed at several other concentrations.

Under conditions of high speed equilibrium, this same sample would also tend to mask nonideality effects if a conventional plot of ln C versus the square of the radial position is utilized. However, because the initial concentration is much lower (0.1–0.2 mg/ml), the effect of the virial coefficient is minimal. A plot of "typical" high speed equilibrium data (error free) is illustrated in Fig. 1. The slope of the line appears invariant with concentration and, therefore, suggests that the system contains a pure

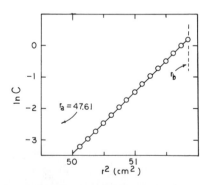

FIG. 1. Dependence of the natural logarithm of concentration (g/liter) on the square of the radial distance. Synthetic data for homogeneous species with molecular weight $M = 100,000$; second virial coefficient, $B = 4.44 \times 10^{-7}$ ml mg^{-1}, $[M(1 - \bar{v}\rho)\omega^2]/2RT = 2.0$. $r_a = 6.9$ cm, and $C_a = 2.6805 \times 10^{-4}$ g/liter.

species. Although the apparent molecular weight at the base of the cell is 94,900 ($C = 1.2$ mg/ml) and 100,000 at the meniscus ($C = 2.68 \times 10^{-4}$ mg/ml), the least-square plot utilizing Eq. (23) yields a molecular weight of $98,663 \pm 186$. The logarithmic plot masks the virial coefficient effect (almost a 5% variation in apparent molecular weight) and provides a number that has a low standard deviation, which leads to false confidence in the result. This is not necessarily serious, for the result is within 2% of the "ideal" molecular weight and, therefore, well within the expected error of typical sedimentation equilibrium experiments. However, the example serves to introduce some of the problems associated with logarithmic plots in molecular weight analysis.

Although the above result suggests that the high speed method may typically provide an ideal molecular weight, there may be instances where BM_2 is considerably greater than 4.4×10^{-2} ml mg^{-1}. Munk and Cox[36] have shown that proteins in guanidine hydrochloride may exhibit considerable nonideality in low speed as well as in high speed methodology.

If a system is heterogenous with respect to solvent, preferential solvation can occur. Because the concentration gradient of the solvent is very shallow compared to the concentration gradient for the macromolecule and because $(\partial\mu_2/\partial\mu_3)_{T,P,m_2}$ can be expected to be independent of the concentration of macromolecule, the preferential solvation effect is seen to be a constant factor in Eq. (4). Hence, a linear plot of $\ln C$ vs. r^2 will be obtained for homogeneous macromolecules in multicomponent solvents. If the preferential solvation effects are ignored, the apparent molecular weight obtained from the slope of the line may simply be incorrect. This was illustrated in a study of glyceraldehyde-3-phosphate dehydrogenase[37] (GPDH) where linear plots for $\ln C$ vs r^2 were obtained in both 0.1 M potassium chloride and 1.30 M potassium phosphate solvents. The molecular weight of GPDH was computed to be 137,600 in the former solvent and 86,400 in the latter solvent. Although the protein is tetrameric, no observable dissociation was actually occurring, and the apparent decrease in molecular weight could be accounted for solely by preferential hydration of GPDH in phosphate buffer.

Two other solvent systems, guanidine hydrochloride and sodium dodecyl sulfate (SDS), are frequently used to obtain the molecular weight of the ultimate subunits of proteins. Preferential interactions with proteins in these solvents can greatly influence the reported molecular weight and, therefore, if erroneously assessed, alter the definition of the stoichiometry of polymeric proteins.

Lee and Timasheff have presented considerable careful data on 12

[36] P. Munk and D. J. Cox, *Biochemistry* **11**, 687 (1972).
[37] K. C. Aune and S. N. Timasheff, *Biochemistry* **9**, 1481 (1970).

proteins showing that the effects of preferential solvation in guanidine hydrochloride provide no simple general rule, such as decreasing the partial specific volume by 0.020 ml/g for all proteins.[38] They do provide a method for estimating the partial specific volume based on amino acid composition in lieu of densimetric measurements of $\bar{v}$. Such a method should be less prone to make an erroneous assignment that could lead to a 20% error in the molecular weight.

Tanford and co-workers have shown that if the proper measure of bound SDS is performed and accounted for in the data analysis, molecular weights of proteins may be obtained by sedimentation equilibrium in SDS.[39] Since proteins bind preferentially, 1.4 g of SDS per gram compared to up to 0.15 g of guanidine hydrochloride per gram, it is imperative that the multicomponent effects be carefully accounted for in the former solvent.

If a system is heterogeneous with respect to macromolecular component whether interacting or not, the heterogeneity is quite likely to be masked in a $\ln C$ versus r^2 plot. This problem has been pointed out by Munk and Cox[36] and Dyson and Isenberg.[40] Both papers illustrated the seemingly linear $\ln C$ vs r^2 plots for quite heterogeneous systems. When these plots express significant curvature, the temptation exists to take limiting slopes at low and high concentrations to yield the molecular weights of the small and large species, respectively, in the system. In view of the inability of $\ln C$ vs r^2 plots to reveal *directly* a heterogeneous system[36] where two proteins are present in equal numbers with molecular weights differing by a factor of 2, it would seem that an erroneous assignment of molecular weights may be made by using such a procedure.

The plot of the logarithm of fringe displacement versus the square of the radial position for a synthetic set of data is illustrated in Fig. 2. The heterogeneity is expressed by the significant curvature and represents two noninteracting species at equal initial concentrations. One procedure frequently observed in the literature is to take limiting slopes as representative of the molecular weights of each of the species in the system. The values of 31,000 and 39,800 were obtained by such a procedure. It has been recognized that this procedure is somewhat falacious, because the slopes represent weight average molecular weights. It may be suggested, that since the distributions are exponential, the molecular weight near the bottom of the cell reflects the presence of *both* species. The molecular weight at the lower fringe displacements, closer to the meniscus, will

[38] J. C. Lee and S. N. Timasheff, *Biochemistry* **13**, 257 (1974).
[39] C. Tanford, Y. Nozaki, J. A. Reynolds, and S. Makino, *Biochemistry* **13**, 2369 (1974).
[40] R. D. Dyson and I. Isenberg, *Biochemistry* **10**, 3233 (1971).

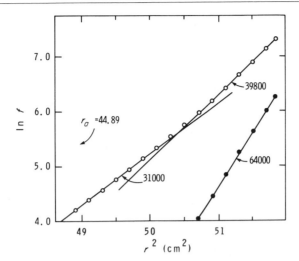

FIG. 2. Dependence of the natural logarithm of fringe displacement on the square of the radial distance. Synthetic data for heterogeneous noninteracting system of two species with molecular weights of 25,000 and 50,000 at equal weight concentrations; $[(1-\bar{v}\rho)\omega^2]/2RT) = 3 \times 10^{-5}$, $r_a = 6.7$cm, $f_{1a} = 2.72$, and $f_{2a} = 0.0297$.

reflect, primarily, the low molecular weight species (1). Therefore, only the lower slope should be valid.

Accepting that premise, the fringe displacement due to the larger component (2) could be obtained by the proper subtraction of curves:

$$\ln f_2(r) = \ln[\exp\{\ln\ f(r)\} - f_1(r)] \tag{25}$$

where

$$f_1(r) = \exp\{\text{intercept (1)} + \text{slope (1)} \cdot r^2\} \tag{26}$$

The line through the solid circles (Fig. 2) is generated from the data using Eq. (25) and suggests that the molecular weight of species (2) is 64,000 rather than 39,800. It remains for the reader to choose between the two procedures in view of the fact that the actual molecular weights used to generate the data are 25,000 and 50,000.

The *selected example* points out that: (a) the molecular weight of the species (2) is greater than immediately revealed and that the characterization of the system requires further analysis; (b) but also it cannot be generally assumed that the lowest slope is indicative of the true molecular weight of the smallest species, for a more rigorous analysis provided a description of the system that was equally falacious.

The illustrated example was a system containing species at equal weight concentrations differing in molecular weight by a factor of 2. Although the example revealed heterogeneity, the data set did not success-fully lend itself to the simple procedure outlined above for resolving the molecular weights in the system. Clearly, considerable caution should be exercised when employing Eqs. (25) and (26). It would seem necessary that the high molecular weight material be larger than a factor of 2, and be present as a genuine contaminant rather than accounting for 50% of the material. If slightly higher speeds are employed, a better separation of small and large species would be achieved. Therefore, if such a method is employed, conditions should be adjusted to obtain optimal data for the species of lowest molecular weight. Since the molecular weight of the smaller species is generally unknown, there is a certain amount of circular-ity in the approach. Hence, several run speeds must be employed before this type of characterization can be used cautiously, with any confidence.

Scanning Absorbance Optics

The problems associated with studying homogeneous macromolecules using interference optics are generally encountered and dealt with in a similar fashion using the scanner optics. This is due to the uncertainty in the base line as already discussed, as well as to the inherent increased imprecision of the commercially available scanner optics recording system.

If the macromolecular system is heterogeneous, particular caution must be exercised when interpreting the data. The distribution of mass in the cell is not analyzed on a total mass/volume basis as in general with the interference optical system. To a first approximation, the species that exhibits the largest contributing absorptivity–concentration product, $(E \cdot C)$ at a particular wavelength and radial position, will dominate the characteristics of a $\ln A$ vs r^2 plot. The absorption at any radial position is

$$A(r) = \sum_{i=1}^{n} E_i \cdot C_i(r) \tag{27}$$

The *observed* absorption will depend upon the slit width at the monochromator and the nature of the individual chromophores. If the absorptivities of the n species are identical at the wavelength of observa-tion, and if the stray light transmission characteristics are similar for each of the chromophores, the plot of $\ln A$ vs r^2 will contain no additional information over that obtained with interference optics. Otherwise, the various E_i become weighting factors in the $\ln A$ term. It is readily seen that, in the event the absorption spectra of the redistributed species in the system are significantly different, it may be possible to scan for individual concentration distribution of each species in the system by proper selec-tion of wavelengths.

Aspects That Could be Considered

Although a plot of the logarithm of some quantity proportional to concentration versus the square of the radial position provides a graphical image of the raw data obtained in a sedimentation equilibrium experiment, considerably more information can be made available.

With careful data analysis the presence of heterogeneity can be more readily detected. Perhaps, the best means of *detecting* the heterogeneity is to compute molecular weight averages as a function of macromolecular concentration. The most readily computable quantities are M_n, M_w, and M_z, the number, weight, and second moment, or Z average, molecular weights. The quantities M_n and M_w are derived from numerical integration and differentiation of the primary data, and can be obtained with reasonable accuracy manually, or by using the simple programmable calculators.[41,42] The quantity, M_z, is related to the second derivative of the primary data and requires somewhat more smoothing to yield useful information. This generally involves more extensive computer analysis. Such methods have been discussed.[12,13,35] The two methods of visual representation are illustrated in Fig. 3.

In addition to providing a more visual representation of the data than ln C vs r^2 plots, the molecular weight averages can be directly processed to yield the number and molecular weights of the species in the system.

One method for dealing with these molecular weight averages was discussed by Sophianopoulos and Van Holde[43] and further expanded by Roark and Yphantis.[13] These workers showed that a molecular weight average, M_k is related to the molecular weight average of next lower moment, M_{k-1}, for any system containing two species of molecular weights M_1 and M_2 in a system:

$$M_k = (M_1 + M_2) - (M_1 M_2)(1/M_{k-1}) \tag{28}$$

The linearization technique suggested in Eq. (28) represents a far superior method for obtaining the molecular weights of the two species in the system compared to the method of taking limiting slopes at the low and high concentration ends of ln C vs r^2 plots.

In addition, Roark and Yphantis presented relationships that could yield molecular weights in three species systems and also presented procedures for circumventing the difficulties associated with nonideality.[13] These techniques require much more highly refined data than are required for the "two species" plot.

[41] P. D. Jeffrey and M. J. Pont, *Biochemistry* **8**, 4599 (1969).
[42] K. C. Aune and S. N. Timasheff, *Biochemistry* **10**, 1609 (1971).
[43] A. J. Sophianopoulos and K. E. Van Holde, *J. Biol. Chem.* **243**, 1804, (1964).

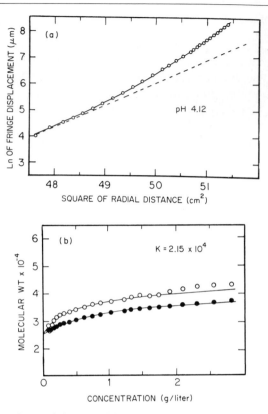

Fig. 3. (a) Dependence of the natural logarithm of the fringe displacement (in micrometers) on the square of the radial distance. Conditions: 24,000 rpm, 25°, 0.178 M NaCl, 0.01 M acetate buffer, pH 4.12, α-chymotrypsin concentration 0.76 g/liter. The dashed line corresponds to monomer. (b) Dependence of the molecular weight averages on α-chymotrypsin concentration in grams per liter calculated from data in Fig. 3a. Filled circles represent point number-average molecular weights, and open circles represent point weight-average molecular weights. From K. C. Aune and S. N. Timasheff, *Biochemistry* **10**, 1609 (1971).

The molecular weight averages may also be utilized to yield equilibrium constants in associating systems, because there is a defined relationship between the various moment averages and an interaction constant for a specific model. Hoagland and Teller studied the dissociation of glyceraldehyde-3-phosphate dehydrogenase presenting the formulations relating the dissociation constant to the various molecular weight averages.[44] Similarly, Aune and Timasheff studied the dimerization of α-chymotrypsin evaluating the association constant from the number and weight average molecular weights[42] (see Fig. 3).

[44] V. D. Hoagland, Jr. and D. C. Teller, *Biochemistry* **8**, 592 (1969).

Analyses of associating systems have also been performed utilizing the approach outlined by Steiner where the concentration of the monomeric species as a function of radial position becomes the derived quantity once the monomeric molecular weight is established.[45] The derived quantities lead to a definition of the system in terms of association constants for a particular model. Data analysis using this methodology has been outlined extensively by Adams.[11]

Direct functional analysis of the concentration distribution can also be employed quite successfully. Howlett and Nichol describe a procedure involving the solution of a set of simultaneous equations, whereby, concentrations and molecular weights are obtained directly from the raw data.[46]

Haschemeyer and Bowers also approached the problem of direct functional analysis of the concentration distribution.[47] Simultaneous equations formed from the data collected at several speeds are constrained to form a consistent set of parameters including the molecular weights and meniscus concentrations of the species in the system.

Rohde et al.[48] and Aune and Rohde[49] have described a procedure for direct functional analysis that is somewhat less ambitious. That is, the molecular weights of the species in the system are considered to be known from independent experimentation on homogeneous systems. The concentration distribution is then analyzed by fitting the resultant linear equations with an iterative search technique that constrains the meniscus concentrations, $C_i(a)$ to quantities ≥ 0:

$$C = \Sigma C_i(a) \exp\{[M_i(1 - \bar{v}_i\rho)\omega^2]/2RT(r^2 - r_a^2)\} \qquad (29)$$

That method also completely defines the composition of the system within the limits of experimental error.

An illustration of this type of direct functional fit data is seen in Fig. 4. The data are seen to satisfy the model where the two proteins in the system are forming a 1 : 1 complex. The parameters obtained from the fit define the composition of the system. Since the composition is known at all positions in the cell, the equilibrium constant is

$$K' = C_i(a) \cdot C_j(a)/C_{ij}(a) \qquad (30)$$

Now the fitting procedures yield parameters proportional to the meniscus concentrations. The constant, k, in Eqs. (16)–(18) must be ascer-

[45] R. F. Steiner, Arch. Biochem. Biophys. 39, 333 (1952).
[46] G. J. Howlett and L. W. Nichol, J. Biol. Chem. 248, 619 (1973).
[47] R. H. Haschemeyer and W. F. Bowers, Biochemistry 9, 435, (1970).
[48] M. F. Rohde, S. O'Brien, S. Cooper, and K. C. Aune, Biochemistry 14, 1079 (1975).
[49] K. C. Aune and M. F. Rohde, Anal. Biochem. 79, 110 (1977).

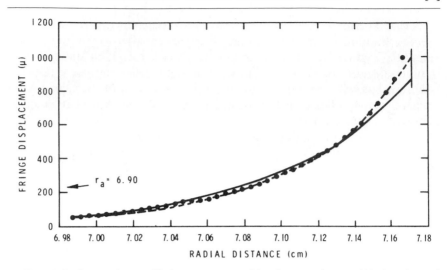

FIG. 4. Sedimentation equilibrium pattern resulting from a mixture of *Escherichia coli* ribosomal proteins S3 and S5 in TMK buffer (0.35 *M* KCl, 0.03 *M* Tris, 0.02 *M* MgCl$_2$, pH 7.4). Rotor speed 28,000 rpm; temperature 5°; column height 2.7 mm; initial protein concentration 0.126 mg/ml of S3 and 0.125 mg/ml of S5. ●, Experimental points; - - -, the theoretical fit for the three species S3, S5, and S3–S5 dimer; ——, the best obtainable fit for the two species S3 and S5. From M. F. Rohde, S. O'Brien, S. Cooper, and K. C. Aune, *Biochemistry* **14**, 1079 (1975).

tained from a known solution to obtain the appropriate association or dissociation constants where a change in the number of moles is involved. Thus, if fringe displacements are measured, the equilibrium constant in terms of molar concentrations

$$K = (M_iM_j)/(M_i + M_j) \{f_{ij}(a)/[f_i(a) f_j(a)k]\} \tag{31}$$

is computed for the above interacting systems. The extent of heterogeneity can be assessed in the case of simple mixed systems.

The direct functional analysis has the limitation of dealing with probably no more than four species in a system. In the event a system is extremely heterogeneous, the numerical techniques of Scholte may be utilized to define the mass distribution.[50]

The discussion has progressed from a rather simple discussion of the ln *C* vs *r*^{2} plots to more complex curve-fitting involving, perhaps, multiple exponential terms. Clearly, computer analysis is suggested. What about the researcher who is interested only in the molecular weight of his favorite enzyme? Must he also be deeply involved in numerical analysis of

[50] T. G. Scholte, *Ann. N. Y. Acad. Sci.* **164**, 156 (1969).

his data? Perhaps not, but there are a few simple considerations he might make. First, one aspect that is clearly emphasized in the direct functional analysis approach is the error in the data and how it affects the final result. In careful plate-reading procedures where several fringes are measured, the uncertainty of fringe displacement is relatively uniform from low displacements to high displacements. Hence, the error in direct functional fitting is equally weighted throughout the concentration range. However, if ln C versus r^2 plots are utilized, a simple least-square analysis is performed on data arbitrarily cut off at 100 μm displacement. The reason there is an arbitrary data cutoff at 100 μm is that generally, on a logarithm scale, the ordinate has considerable fluctuation. Including such data would influence the computed slope in an inordinately large fashion. A far better procedure is to not disregard any data in the fitting procedure and employ (a) a single species nonlinear curve fitting procedure to obtain the molecular weight; or (b) a weighted least-square fitting routine which will provide a molecular weight that very clearly approximates the result obtained in (a).

The error in the logarithm of concentration is related directly to the error in the concentration and inversely to the concentration, $\delta \ln C = \delta C/C$. Hence, if all data have equal weight in direct function fitting, the ln $C(r)$ in a least-square determination versus r^2 should be weighted by the concentration $C(r)$. If the data in Fig. 1 are treated by the outlined weighted least-square procedure, the apparent molecular weight is computed to be 98,067, or 0.6% lower than the simple least-square result. The data in Fig. 1 are nonideal data. Therefore weighting the fit more at higher concentrations will necessarily reveal more nonideality through a lower molecular weight. The most important point is not that the computed molecular weight may be higher or lower, but rather that the fit is an unbiased fit assuming that the data are representative of a pure ideal species at equilibrium. Ignoring data below 100 or 200 μm displacement biases in favor of high concentrations, whereas, inclusion of the low concentration data in an unweighted fit biases towards low concentration data.

One last consideration to be made involves an accounting of material in the system. A numerical integration of the fringe displacement can be readily performed as discussed elsewhere[6,35]:

$$\int_{r_a}^{r_b} f \, dr^2 = f_0(r_b^2 - r_a^2) \tag{32}$$

The quantity, f_0, obtained from that numerical integration should: (a) agree with the known initial concentration placed in the cell; (b) agree with the value obtained from an analytical integration.

If item (a) is satisfied above, then no major precipitation of material has occurred. It is certainly desirable to know that the obtained molecular weight is applicable to the entire sample under inspection.

The analytical integration is that which can be performed when the derived parameters describing the system are available. For example, if the system is declared homogeneous, with molecular weight, M, from a least-square calculation of a ln C versus r^2 plot, then

$$f_0 = f_a \cdot \{\exp (2H) - 1\}/2H \tag{33}$$

where

$$H = M(1 - \bar{v}\rho)\omega^2(r_b^2 - r_a^2)/4RT \tag{34}$$

and f_a is obtained from the intercept. The preceding calculation can confirm or deny the accounting of all the material in the system as being a single species.

If the system is heterogeneous, the amount of the individual contributing species can be assessed in a similar fashion where there are exponential terms for each species.[49]

Summary

Several journals, because of the demand for publishing space, have tended to discourage the use of figures that contain little information that can be effectively transmitted in the text of an article. It has been suggested that plots of ln C vs r^2 fit into that category. The present writing would support that position; for computed molecular weights from least-square calculations reported with the appropriate error associated in the fit will convey an equal amount of information as a statement of linearity in the text and a figure supporting the contention.

In addition to performing a least-square analysis of ln C versus r_2 plots,

EFFECT OF THE SECOND VIRAL COEFFICIENT ON OBSERVED MOLECULAR WEIGHT[a]

M_2 (app)	c_2 (mg/ml)
100,000	0.0
99,960	0.01
99,120	0.2
95,700	1.0
90,000	2.5
84,900	4.0

[a] $B = 4.44 \times 10^{-7}$ ml/mg.

it is incumbent upon authors to analyze their data for complicating factors such as heterogeneity and nonideality. Some of the examples discussed illustrate how erroneous the conclusions derived can be in the absence of the more rigorous analysis. Data analysis, in any event, should not end with a ln C vs r^2 plot. The system should be subjected to composition accounting by numerical integration of data, and to analysis for the point average molecular weight distributions. These considerations are well within the scope of all ultracentrifuge users and should be employed as a minimum during data analysis.

[8] Continuous Laser Optics in the Ultracentrifuge

By ROBLEY C. WILLIAMS, JR.

The use of lasers as light sources in interferometry is now quite widespread. A small helium–neon laser (1–10 mW output power) makes a satisfactory and inexpensive light source for the Rayleigh interference optical system of the ultracentrifuge.[1,2] No commericially built installation is available at present, but a number of investigators have added lasers to their instruments. The system described here is that used routinely by the author during the last 5 years.

Apparatus

Installation

A schematic drawing of a straightforward laser mounting is shown in Fig. 1. The laser is simply fastened horizontally to the side of the inner barricade of the ultracentrifuge on a platform that provides for leveling and for about 1 cm of adjustment in the horizontal plane. The light beam from the laser is directed through a 2 cm hole in the barricade and into a spatial filter, which consists essentially of a microscope objective lens with a pinhole in its focal plane. The spatial filter provides a divergent beam of light from an effective point source, and it greatly improves the quality of the final image by blocking noncollimated light present in the output beam of the laser. The divergent cone of light from the spatial filter is reflected upward into the lower (collimating) lens of the rotor chamber by a prism or front surface mirror. This element is made to be rotationally adjustable about a vertical axis and to slide left–right along the same

[1] R. C. Williams, Jr., *Anal. Biochem.* **48**, 164 (1972).
[2] J. A. Lewis and J. W. Lyttleton, *Anal. Biochem.* **56**, 52 (1973).

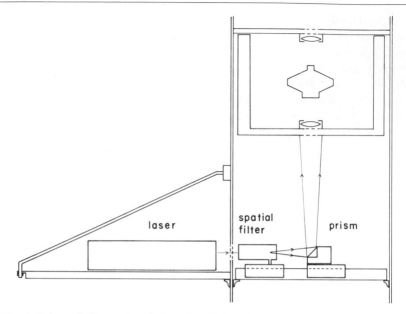

FIG. 1. Schematic illustration of a laser installation. The laser, spatial filter, and prism are shown in their approximate positions relative to the rotor of the ultracentrifuge. The view is from the front of the instrument. A detailed drawing is given by R. C. Williams, Jr., *Anal. Biochem.* **48,** 164 (1972).

optical track that holds the spatial filter assembly. Two alternatives to this geometry should be noted. First, to conserve space, the laser can be mounted vertically on the side of the barricade and its beam directed into the spatial filter by means of a mirror or prism placed near the hole in the barricade.[3] Second, a mechanically more stable but more expensive assembly can be obtained by installing a beam-expanding telescope (spatial filter and collimating lens combined) directly on the laser and replacing the lower chamber lens by an optical flat.

Continuous and Modulated Illumination

One of the advantages of lasers is that they can be turned on and off rapidly. Switched helium–neon lasers are available that can be turned on and off in times less than 1 μsec.[4] Electrooptical and acoustooptical modulators are also available and can be placed in the beam of a continuous laser to accomplish rapid switching.[2] By the use of either switching

[3] Dr. Marc S. Lewis, personal communication.
[4] The Model 607M laser manufactured by Liconix, Inc., Mountain View, California, is an example.

system, the light can be synchronized with the turning of the rotor. Synchronization allows measurements to be made on each of the cells in a multicell rotor. In addition, fine synchronization allows one to illuminate the cell only when its double slits are centered over the stationary double slits in the lens mask.[5] A striking increase in fringe contrast then results from the elimination of the unwanted light that is ordinarily passed to the photographic plate when only one of the cell's slits is in line with a slit in the stationary mask.

Simple timing circuits can easily be devised to trigger the laser or modulator in synchrony with the rotor. The necessary timing information can be obtained from the scanner-multiplexer circuit, if one is present on the ultracentrifuge, or from detection of an auxiliary light beam passed through the lenses of the absorption optical system to measure the time of passage of the counterweight.

A Specific System

The system now in use in our laboratory is described for purposes of illustration. The reader who builds his own should have little difficulty improving upon it. We employ the optical configuration shown in Fig. 1. The laser is a Spectra-Physics Model 126,[6] which has 3 mW output and can be modulated by voltage levels of 0.2 V (on) and 5 V (off). The spatial filter (Oriel Optics Co., Model 1522) has a 20× objective lens and a 10 μm pinhole. This combination provides full illumination of the collimating lens. The 90° prism (Oriel Optics Co., Model A-72-143-00) has 2.5-cm square faces flat to 0.1 wave. It produces no noticeable distortion of the final image. Both the spatial filter and the prism are mounted on a section of triangular cast iron optical rail (Klinger Scientific Apparatus Corp., Model 02-2053) atop convenient carriers. The optical alignment of this system has been described in detail.[1] Essentially, one positions the spatial filter and prism to provide a collimated beam parallel to the axis of rotation of the rotor and then utilizes one of the standard sets of procedures[7-10] for the rest of the alignment. The only modification to these procedures necessitated by the use of the laser is the substitution of a flat-sided flask (a small "Falcon Flask" is suitable) filled with dilute milk

[5] C. H. Paul and D. A. Yphantis, *Anal. Biochem.* **48**, 588 (1972).

[6] This laser is no longer manufactured (see footnote 4). A suitable nonmodulated laser is the Spectra-Physics Model 135.

[7] L. Gropper, *Anal. Biochem.* **7**, 401 (1964).

[8] E. G. Richards, D. C. Teller, and H. K. Schachman, *Anal. Biochem.* **41**, 189 (1971).

[9] E. G. Richards, D. C. Teller, V. D. Hoaglund, Jr., R. H. Haschemeyer, and H. K. Schachman, *Anal. Biochem.* **41**, 215 (1971).

[10] A. W. Rees, E. A. Lewis, and M. S. DeBuysere, *Anal. Biochem.* **62**, 19 (1974).

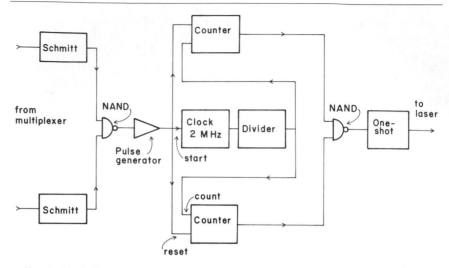

FIG. 2. Block diagram of a circuit for synchronization of laser with rotor. Signals from the Beckman multiplexer are converted to TTL levels by two Schmitt triggers. Once per revolution of the rotor, both signals will be high simultaneously. This condition is detected by the first NAND gate and is used to generate a brief pulse. The pulse simultaneously resets the two counters and starts the clock. The counters then count clock pulses until a preset number is reached for each, whereupon each triggers a one-shot multivibrator that turns on the laser. One counter is used to illuminate the cell and the other to illuminate the counterweight, so that a composite record of the images of both is present on the film. Control of the delay between the resetting of the clock and the starting of the laser is obtained by adjusting the modulus of the divider (coarse) and by setting the counters (fine). Control of the time of illumination is obtained by adjustment of the external resistance and capacitance of the one-shot multivibrator. The primary requirements for a timing circuit are that delay times from zero to about 10 msec be provided, with a precision of 0.1% or better, and that a range of illumination times from about 5 to 300 μsec be available.

for ground glass in those steps that require a diffuser. This substitution reduces the "speckle" (an annoying result of the coherence of the laser light) which otherwise renders some operations (e.g., focusing of the camera lens) difficult.

The laser is modulated by the circuit shown schematically in Fig. 2. The Schmitt triggers of the scanner-multiplexer are used as the synchronization time signal. A detailed description of another similar circuit is given by Paul and Yphantis.[11]

Red-sensitive photographic films and plates provide exposure times, with and without modulation, in the range of 1–30 sec, when stationary double slits of width 0.5 mm are employed. Kodak type SO-410 film, which has a dimensionally stable "Estar" base, Kodak type IV-F plates, and

[11] C. H. Paul and D. A. Yphantis, *Anal. Biochem.* **48**, 605 (1972).

Kodak type 3414 film all give excellent results. Kodak type I-N plates, while sensitive, have been found to have a grain too coarse to be compatible with work of high precision.

Discussion

Advantages of the Helium–Neon Laser over the Mercury Arc

The interference optical system of the ultracentrifuge requires that the light passing through the solution in the cell be well collimated. Because the mercury arc must be used with a slit of finite extent (approximately 2 mm in the radial direction), the collimation of its light is necessarily imperfect. The laser, on the other hand, emits a nearly parallel beam that can be focused to an extremely fine spot, which is a good approximation to a point source. The laser thus provides better collimation of the light passing through the ultracentrifuge cell. In addition, the light from the laser is much more nearly monochromatic than the light from the mercury arc. The contrast of the fringes is governed both by the spectral bandwidth of the illuminating light and by the collimation of the light passing through the cell. Both factors are improved when a laser is used as the light source, and the contrast of the resulting fringes is thus greatly improved over the contrast of the fringes obtained with the mercury arc. If the laser is also modulated, as discussed above, the improvement in contrast is still more marked.

Some typical fringe patterns are shown in Fig. 3. The higher contrast obtained with the laser is obvious. In reading these patterns manually, we find that the standard deviation of reading the vertical displacement of a fringe is about two-thirds of the standard deviation obtained with patterns produced by the mercury arc. This gain in precision is small, but it is significant in sedimentation equilibrium work. The relatively high contrast of the laser fringes will doubtless lend itself to use with automatic image-scanning equipment,[12,13] where the improvement in precision is likely to be much greater.

In the long run, a laser light source can be cheaper to operate than is the mercury arc. A 3 mW, continuous-wave laser costs about 350 dollars and lasts approximately 10,000 hr. The other optical elements in the system described above cost approximately 800 dollars. By contrast, the AH-6 lamps generally employed cost 30 dollars each, and last about 150 hrs. The monetary break-even point is thus reached at about 5000 hr of operation.

[12] D. J. DeRosier, P. Munk, and D. J. Cox, Anal. Biochem. **50**, 139 (1972).
[13] R. M. Carlisle, J. I. H. Patterson, and D. E. Roark, Anal. Biochem. **61**, 248 (1974).

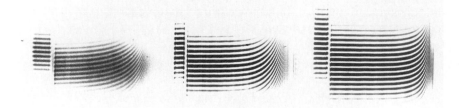

FIG. 3. Rayleigh interference fringes obtained with three different light sources. *Left:* mercury arc lamp (Kodak type II-G plate); *middle:* laser, unmodulated (Kodak type SO-410 film); *right:* laser, modulated (Kodak type SO-410 film). Each of the three patterns represents a solution column approximately 2.5-mm long, at sedimentation equilibrium. Although three different experiments are represented, the slit width is the same for each. The patterns were photographically reproduced under identical conditions of contrast. The vertical striations of intensity ("stripes") which can be seen in both of the laser-produced images have their origin in stress effects in the windows of the ultracentrifuge cell (Williams, text footnote 1). Their intensity and severity can be reduced by the use of PVC window liners as recommended by A. T. Ansevin, D. E. Roark, and D. A. Yphantis [*Anal. Biochem.* **34,** 237 (1970)] and by the adjustment of the plane of polarization of the light so that the electric vector is parallel to the radius of the rotor. The striations have been found to have no discernible effect upon the final results obtained (Williams, text footnote 1).

Besides these general advantages, the laser system is useful in special situations. Because of the coherence of the laser light, work can be done with large (many centimeters) differences in total optical path between solvent and solution. The mercury arc is limited to differences in optical path of about 0.004 cm. Solutes that absorb strongly in the green, such as hemoglobin[14] and rhodopsin,[15] can be examined at relatively high concentrations.

A Disadvantage and a Pitfall

A disadvantage of the laser relative to the mercury arc is that a theoretically well grounded and operationally simple schlieren optical system is difficult to construct. The difficulty stems from the necessity of locating an effective slit source at the focal plane of the chamber collimating lens. A slit and (for good illumination) a cylinder lens must be placed in approximately the same location as the spatial filter, necessitating repositioning of that element of the optical system each time a change from the interference to the schlieren mode is carried out. If great accuracy and precision are not required, however, a useful "rough and ready" schlieren

[14] R. C. Williams, Jr., *Proc. Natl. Acad. Sci. U.S.A.* **70,** 1506 (1973).
[15] Dr. Marc S. Lewis, personal communication.

image can easily be obtained by introducing a cylindrical lens of about 10 cm focal length into the beam in front of the spatial filter.[4]

An unexpected pitfall in the use of the laser light source stems from its coherence. The bending of light which occurs within the cell as a result of the presence of a gradient of refractive index there (the "Wiener skewing"[16,17]) is always potentially a source of systematic error in interferometric measurements in the ultracentrifuge. However, with the extended and relatively incoherent light source provided by the mercury arc, the interference fringes become blurred and indistinct as values of the gradient are approached which would produce significant errors from this source.[17] With the laser it is quite possible to measure perfectly distinct fringes well into the region in which Wiener skewing is a major cause of error. Molecular weights lower than the true value are observed in this region. The problem is particularly severe with cells of 30-mm optical path. Correction for the effect of Wiener skewing is possible in principle but very consuming of computer time in practice.[18] The best policy for the present would seem to be to avoid steep gradients, preferably by calculating in advance from Rilbe's relations[17] a maximum acceptable error in the observed molecular weight.

[16] H. Svensson, *Opt. Acta* **1**, 25 (1954).
[17] H. Svensson, *Opt. Acta* **3**, 164 (1956).
[18] Unpublished computations of the author.

Section II

Interactions

[9] Molecular Transport of Reversibly Reacting Systems: Asymptotic Boundary Profiles in Sedimentation, Electrophoresis, and Chromatography

By Lilo M. Gilbert and Geoffrey A. Gilbert

Protein–protein interactions in solution[1] are easily detected by the commonly used methods of molecular transport such as electrophoresis,[2] sedimentation[3] and chromatography.[4,5] In contrast, it is very difficult indeed to determine unambiguously the stoichiometry and thermodynamic parameters of any such interactions. Apart from experimental difficulties, it is inherent in the nature of the problem that awkward correlations between parameters can be expected to arise. The present article describes some preliminary steps that can be helpful to anyone attempting to proceed from the initial recognition of the presence of interactions toward an understanding of their nature and magnitude. For this purpose, we show how to simulate boundary patterns for reversibly reacting systems, which can be used for comparison with actual transport experiments.

When calculating so-called "reaction"[2] boundaries, it leads to a great simplification if one ignores diffusion and assumes instantaneous reequilibration, and this we shall do while recommending that more powerful and comprehensive treatments of the kind originated by Cann and Goad,[6,7] and under active development,[8,9] should then be considered. By ignoring diffusion, one is effectively viewing a boundary in its "asymptotic" state, to which it would tend with time. However, as diffusion does not, in general, introduce totally new features into a boundary, but merely damps out and softens what is predicted by a diffusion-free treatment, the essential features of a boundary can still be expected to be preserved.

Before we proceed to our main discussion of boundary forms, a point of some interest should be mentioned. Moving-boundary methods are

[1] L. W. Nichol, J. L. Bethune, G. Kegeles, and E. L. Hess, *in* "The Proteins," 2nd ed. (H. Neurath, ed.), Vol. II, Chap. 9. Academic Press, New York, 1964.
[2] L. G. Longsworth, *in* "Electrophoresis Theory, Methods and Applications" (M. Bier, ed.), Chap. 3 and 4. Academic Press, New York, 1959.
[3] H. Fujita, "Foundations of Ultracentrifugal Analysis." Wiley, New York, 1975.
[4] D. J. Winzor and H. A. Scheraga, *Biochemistry* **2**, 1263 (1963).
[5] G. K. Ackers, *in* "The Proteins," 3rd ed. (H. Neurath and R. L. Hill, eds.), Vol. 1, Chap. 1, Academic Press, New York, 1975.
[6] J. R. Cann, "Interacting Macromolecules." Academic Press, New York, 1970.
[7] J. R. Cann and W. B. Goad, see this series Vol. 27 [12].
[8] J. R. Cann and N. D. Hinman, *Biochemistry* **15**, 4614 (1976).
[9] J.-M. Claverie, *Biopolymers* **15**, 843 (1976).

accepted as complementary to "equilibrium" methods (light scattering, osmotic pressure, equilibrium centrifugation, etc.), but it is not always realized that they themselves yield true equilibrium information if arranged properly. Thus if a boundary is formed between solution and solvent, the boundary centroid[2,10] (or the "equivalent sharp boundary"[2]) moves with a velocity characteristic of the material at equilibrium in the *plateau* region of the solution independently of any material undergoing reequilibration within the *boundary*. Only for velocity centrifugation need this not be precisely true because of the progressive but slight dilution of the plateau region due to the effect of migration in a nonhomogeneous field and sector-shaped cell.

We next proceed to the flux equations from which all calculations stem.

Flux Equations

The equations governing the velocity of a single *stable* differential boundary can be written down immediately. In a previous article by us,[11] the motion of a differential boundary formed between two solutions containing a solitary substance A at concentrations $\bar{m}_A$ and $\bar{m}_A + d\bar{m}_A$, respectively, was discussed in detail. Here we consider the effect of the presence of a second substance B on such a boundary.[12–18] If B affects the velocity of A, it will necessarily also affect the velocity of the differential boundary of A. Simultaneously, of course, there must be a reciprocal effect of A on the velocity of B. Let the constituent velocities[2,19] of A and B (in the sense that Tiselius[20] used the term) be $\bar{v}_A$ and $\bar{v}_B$, respectively, and let the velocity of the differential boundary be v. Suppose that the boundary is stable with time, and that it lies at zero time at the junction of the plateau regions α and β, shown in Fig. 1. In unit time the boundary will have moved a distance v from its initial position, extending the α region at the expense of the β region. Conservation of mass requires that the shaded rectangle of area $v\, d\bar{m}_A$ should correspond to the difference in

[10] R. J. Goldberg, *J. Phys. Chem.* **57**, 194 (1953).
[11] L. M. Gilbert and G. A. Gilbert, see this series Vol. 27 [11].
[12] V. P. Dole, *J. Amer. Chem. Soc.* **67**, 1119 (1945).
[13] J. P. Johnston and A. G. Ogston, *Trans. Faraday Soc.* **42**, 789 (1946).
[14] B. Davison, Discuss. Faraday Soc. **7**, p. 45 (1949).
[15] G. A. Gilbert and R. C. L. Jenkins, *Nature (London)* **199**, 688 (1963).
[16] R. C. L. Jenkins, *J. Phys. Chem.* **69**, 3785 (1965).
[17] L. W. Nichol and A. G. Ogston, *J. Phys. Chem.* **69**, 1754 (1965).
[18] L. M. Gilbert and G. A. Gilbert, *Biochem. J.* **97**, 7C (1965).
[19] L. W. Nichol and A. G. Ogston, *Proc. R. Soc. London, Ser. B* **163**, 343 (1965).
[20] A. Tiselius, *Nova Acta Regiae Soc. Sci. Upsal.* **7**, [4], 1 (1930).

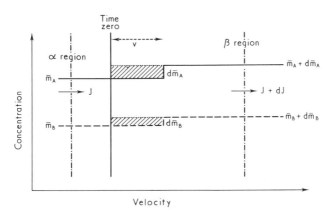

FIG. 1. Diagrammatic representation of a differential boundary for an interacting mixture of substances A and B undergoing molecular transport.

flux of A into and out of the region bounded by two arbitrarily placed planes situated well into the plateau regions. Let the flux of A in the region α be J_A and in the region β be $J_A + dJ_A$. Then it follows that

$$dJ_A = v \, d\overline{m}_A \tag{1}$$

Since the mass of B is also conserved, a similar equation holds simultaneously for B. Therefore

$$dJ_B = v \, d\overline{m}_b \tag{2}$$

(and further similar equations would hold for any other independent components in the solution[12,14]). The contribution of B to the boundary is shown by a dashed line in Fig. 1.

Equations (1) and (2) are deceptively simple. At first sight $d\overline{m}_A$ and $d\overline{m}_B$ are arbitrary increments in $\overline{m}_A$ and $\overline{m}_B$, respectively, but in fact their ratio, R say, is fixed by the ratio of dJ_A to dJ_B, as can be seen by combining Eqs. (1) and (2) to give Eqs. (3a,b) and (4).

$$v = dJ_A/d\overline{m}_A = dJ_B/d\overline{m}_B \tag{3a,b}$$

i.e.,

$$R = d\overline{m}_A/d\overline{m}_B = dJ_A/dJ_B \tag{4}$$

In order to evaluate the ratio dJ_A/dJ_B, we have to consider the coupling of A and B. The flux J_A of A is the product of the constituent concentration $\overline{m}_A$ and the constituent velocity $\overline{v}_A$ as defined in Eq. (5).

$$J_A = \overline{m}_A \, \overline{v}_A \tag{5}$$

Similarly

$$J_B = \bar{m}_B \, \bar{v}_B \tag{6}$$

The interaction of B with A, which influences $\bar{v}_A$ and hence J_A, is taken account of by writing for the increment in J_A across the differential boundary the total differential equation,

$$dJ_A = (\partial J_A/\partial m_A)_{m_B} \, d\bar{m}_A + (\partial J_A/\partial \bar{m}_B)_{\bar{m}_A} \, dm_B \tag{7}$$

Similarly, for B

$$dJ_B = (\partial J_B/\partial \bar{m}_A)_{\bar{m}_B} \, d\bar{m}_A + (\partial J_B/\partial \bar{m}_B)_{\bar{m}_A} \, d\bar{m}_B \tag{8}$$

Then, by combining Eqs. (4), (7), and (8), we obtain a quadratic equation Eq. (9)[21, 22] in R

$$R = \frac{(\partial J_A/\partial \bar{m}_A)_{\bar{m}_B} \, R + (\partial J_A/\partial \bar{m}_B)_{\bar{m}_A}}{(\partial J_B/\partial \bar{m}_A)_{m_B} \, R + (\partial J_B/\partial m_B)_{\bar{m}_A}} \tag{9}$$

with roots R_1 and R_2 given by

$$R = \frac{\left(\dfrac{\partial J_A}{\partial \bar{m}_A}\right)_{\bar{m}_B} - \left(\dfrac{\partial J_B}{\partial \bar{m}_B}\right)_{\bar{m}_A} \pm \sqrt{\left\{\left(\dfrac{\partial J_A}{\partial \bar{m}_A}\right)_{\bar{m}_B} - \left(\dfrac{\partial J_B}{\partial \bar{m}_B}\right)_{\bar{m}_A}\right\}^2 + 4\left(\dfrac{\partial J_B}{\partial \bar{m}_A}\right)_{m_B}\left(\dfrac{\partial J_A}{\partial \bar{m}_B}\right)_{\bar{m}_A}}}{2(\partial J_B/\partial \bar{m}_A)_{\bar{m}_B}} \tag{10}$$

The underlying reason for a quadratic becomes apparent when we calculate v from Eqs. (3a) and (7) or Eqs. (3b) and (8). Thus we find from Eqs. (3a) and (7), for example,

$$v = dJ_A/d\bar{m}_A = (\partial J_A/\partial \bar{m}_A)_{\bar{m}_B} + (\partial J_A/\partial \bar{m}_B)_{\bar{m}_A} \, 1/R \tag{11}$$

and we realize that the two values of R give rise to two values of v, v_1, and v_2, say, corresponding to *two* stable differential boundaries between the plateau regions α and β, each with a precise value of $d\bar{m}_A/d\bar{m}_B$ determined by Eq. (10). What this means in practice is that, if the plateau levels are set entirely arbitrarily, the initial single boundary splits into two separate stable boundaries (or in the general case into as many boundaries as there are independent components[12,14]). The values of the ratios R_1 and R_2 for the distribution of material between the two boundaries are governed by Eq. (10), and the absolute quantities $(d\bar{m}_A)_1$, $(d\bar{m}_B)_1$, and $(d\bar{m}_A)_2$, $(d\bar{m}_B)_2$ have always to add up to the total differences, $(d\bar{m}_A)_T$, $(d\bar{m}_B)_T$ say, set

[21] A. C. Offord and J. Weiss, *Nature (London)* **155**, 725 (1945).
[22] L. M. Gilbert and G. A. Gilbert, *in* "The Regulation of Enzyme Activity and Allosteric Interactions" (E. Kvamme and A. Phil, eds.), p. 73. Academic Press, New York, 1968.

initially between the plateau regions. Thus

$$(d\bar{m}_A)_1 + (d\bar{m}_A)_2 = (d\bar{m}_A)_T \tag{12}$$

and

$$(d\bar{m}_B)_1 + (d\bar{m}_B)_2 = (d\bar{m}_B)_T \tag{13}$$

From Eq. (12) and the definition of R [Eq. (4)] it follows that

$$R_1(d\bar{m}_B)_1 + R_2(d\bar{m}_B)_2 = (d\bar{m}_A)_T \tag{14}$$

and therefore from Eqs. (13) and (14) that

$$(d\bar{m}_B)_1 = [(d\bar{m}_A)_T - R_2(d\bar{m}_B)_T]/(R_1 - R_2) \tag{15}$$

The corresponding increments $(d\bar{m}_B)_2$, $(d\bar{m}_A)_1$, and $(d\bar{m}_A)_2$ can be calculated from Eqs. (12), (13), and (15).

It will be obvious that in the lower limit of there being no interaction at all between A and B, two easily identified boundaries would be found with velocities v_A and v_B, which would be those to be expected for difference boundaries of each component present on its own. Otherwise the values of v_1 and v_2 depend, according to Eqs. (10) and (11), upon the values of the partial derivatives $(\partial J_A/\partial \bar{m}_B)_{\bar{m}_A}$, $(\partial J_B/\partial \bar{m}_A)_{\bar{m}_B}$, etc. If we are given these, we can set up the boundary system in a very interesting way. We can choose $(d\bar{m}_A)_T$ and $(d\bar{m}_B)_T$ so that their ratio exactly satisfies one or other root of Eq. (10).[18,22] By this artifice we can ensure that the entire change in $\bar{m}_A$ and $\bar{m}_B$ occurs at one boundary and that the other boundary is completely suppressed. Let us assume that we have satisfied the root R_1 in this way. Then we will see only a single stable differential boundary moving with velocity v_1. This opens up a way of constructing a *finite* boundary by numerical integration. To do this we set up at an arbitrary but sufficient distance in the plateau region another quite independent and separate differential boundary with $(\bar{m}_A + d\bar{m}_A)$, and $(\bar{m}_B + d\bar{m}_B)$ instead of $\bar{m}_A$ and $\bar{m}_B$ as starting concentrations for A and B, again choosing a value for the ratio $(d\bar{m}_A)_T/(d\bar{m}_B)_T$ which satisfies the root R_1 of Eq. (10) and suppresses the complementary boundary. Successive repetition of this process gives us a set of stable differential boundaries, with each boundary separated by an arbitrary length of plateau from its neighbor. If we decide to decrease these lengths to zero, we achieve a continuous finite boundary as the individual differential boundaries merge. Only if a boundary tends to be "involute," curving back over itself, do we have to do a further calculation to find the position of the resulting hypersharp boundary, as we explain later.

We have stressed this physical approach to boundaries for two reasons. First, it is not too unlike what happens in actuality during the development of a boundary from an initial finite sharp boundary, and second, if a boundary profile has to be calculated, numerical methods, ultimately involving the summation of a series of difference boundaries, have to be used in all but the most elementary cases,[23] since analytical solutions have in most cases not been discovered and are not likely to be.

By integrating Eq. (4) for appropriate model systems, we are able to demonstrate the range of types of boundary that are theoretically possible and might be found experimentally. Since, if effective use is to be made of this method for interpreting actual systems, readers will need access to a programmable calculator or computer with graphical output and in that case will be able to generate boundary patterns for themselves, only a few representative patterns will be displayed here. For these the necessary algebra will be given in some detail. We will use as example[18] the sequential reactions defined by the chemical and mass-action equations [Eqs. (16) to (19)].

$$A + B = C_1 \tag{16}$$
$$c_1 = k_1\, a\, b \tag{17}$$
$$B + C_1 = C_2 \tag{18}$$
$$c_2 = k_2\, b\, c_1 = k_1\, k_2\, a\, b^2 \tag{19}$$

where a, b, c_1, and c_2 are molar concentrations, and k_1 and k_2 are interaction constants.

The species A, B, C_1, and C_2 will be assumed to migrate with characteristic velocities v_A, v_B, v_{C_1}, and v_{C_2}, respectively, which, where necessary, will be treated as functions of solution composition. Concentrations will be measured either in units of molar concentration, e.g., c_1 moles/liter, or of weight concentration, e.g., w_{C_1} g/dl. Molecular weights will be denoted by M_{C_1} etc. Clearly, even for this limited system, a large number of permutations are possible; for instance, one could consider cases where the complexes move either slower or faster than their components, or at intermediate velocities. Likewise, all the species could move toward solution, or alternatively toward solvent, as in the trailing and leading boundaries, respectively, in free-boundary electrophoresis and chromatography.

We shall deal first with an electrophoresis example for which it is reasonable to postulate a sequence of velocities of the kind

$$v_A < v_{C_1} < v_{C_2} < v_B$$

and in this first instance each velocity will be treated as independent of concentration. We have to demonstrate the effect of changing the plateau concentrations, and the ratios of the concentrations, of the reactants on

[23] G. A. Gilbert and R. C. L. Jenkins, *Proc. R. Soc. London, Ser. A* **253**, 420 (1959).

the profiles of the conjugate (leading and trailing) boundaries. Besides concentration profiles we have also to calculate the derivatives of concentration with respect to distance, since these are the analogs of conventional schlieren patterns.

As an example of an actual experimental study we draw the readers' attention to the pioneer work of Pepe and Singer[24] on soluble monovalent antigen–bivalent antibody complexes. (It was, in fact, to the electrophoretic results of those experiments that this type of calculation was probably first applied.[18,25])

We begin the calculation by finding the values of the velocities v_1 and v_2, but before doing so we would like to note that all computations become simpler if the free concentrations a and b are used as prime variables rather than the constituent concentrations $\bar{m}_A$ and $\bar{m}_B$. (This approach is particularly advantageous when there are more than two independent components.) The appropriate equations will therefore be developed in terms of a and b. It is also helpful, particularly as regards flexibility in changing from one model system to another, if the algebra is not refined, even though this means some clumsiness in the expressions to be programmed.

I. Special Case: Velocities of Species Independent of Concentration

Basic Equations

Constituent concentrations (moles per liter)

$$\bar{m}_A = a + c_1 + c_2 \tag{20}$$
$$\bar{m}_B = b + c_1 + 2c_2 \tag{21}$$

Flux of A and B, from Eqs. (5) and (6)

$$J_A = m_A\, v_A = a\, v_A + c_1\, v_{C_1} + c_2\, v_{C_2} \tag{22}$$
$$J_B = \bar{m}_B\, \bar{v}_B = b\, v_B + c_1\, v_{C_1} + 2c_2\, v_{C_2} \tag{23}$$

Complete differentials in terms of a and b

$$dc_1 = d(k_1 ab) = k_1\,(a\, db + b\, da) \tag{24}$$
$$dc_2 = d(k_1 k_2 ab^2) = k_1 k_2\,(2\,ab\, db + b^2\, da) \tag{25}$$
$$d\bar{m}_A = da + dc_1 + dc_2 \tag{26}$$
$$d\bar{m}_B = db + dc_1 + 2\,dc_2 \tag{27}$$
$$dJ_A = d(\bar{m}_A\, \bar{v}_A) = v_A\, da + v_{C_1}\, dc_1 + v_{C_2}\, dc_2 \tag{28}$$
$$dJ_B = d(\bar{m}_B\, \bar{v}_B) = v_B\, db + v_{C_1}\, dc_1 + 2\,v_{C_2}\, dc_2 \tag{29}$$

[24] F. A. Pepe and S. J. Singer, *J. Am. Chem. Soc.* **81**, 3878 (1959).
[25] S. J. Singer, F. A. Pepe, and D. Ilten, *J. Am. Chem. Soc.* **81**, 3887 (1959).

Velocities v_1 and v_2 of the Differential Boundaries

Our nomenclature will be such that $v_2 > v_1$ with boundary 1 lying adjacent to region α, boundary 2 adjacent to region β. The plateau between boundaries 1 and 2 will be termed γ. The conservation equations (3a,b) have now to be expressed as functions of a and b.

In place of R we shall use the analogous ratio $da:db$ for which we shall adopt the symbol μ.[23]

$$\mu = da/db \tag{30}$$

The conservation of mass Eq. (3a,b) for the boundaries can be rewritten as

$$v = (dJ_A/db)/(d\bar{m}_A/db) = (dJ_B/db)/(d\bar{m}_B/db) \tag{31a,b}$$

Then with the help of Eqs. (24)–(29) the differentials can be expressed in terms of da and db, and finally μ, to give as end result the simultaneous Eqs. (32a,b) in μ.

$$v = \frac{\mu(v_A + v_{C_1}c_1/a + v_{C_2}c_2/a) + (v_{C_1}c_1/b + 2v_{C_2}c_2/b)}{\mu(1 + c_1/a + c_2/a) + (c_1/b + 2c_2/b)} \tag{32a,b}$$

$$v = \frac{\mu(v_{C_1}c_1/a + 2v_{C_2}c_2/a) + (v_B + v_{C_1}c_1/b + 4v_{C_2}c_2/b)}{\mu(c_1/a + 2c_2/a) + (1 + c_1/b + 4c_2/b)}$$

In order to program these equations to solve for μ, it is convenient to designate subscripted symbols (Z) for the coefficients of μ and for the constant terms. Equations (32a,b) condense to

$$v = (Z_1\mu + Z_2)/(Z_3\mu + Z_4) = (Z_5\mu + Z_6)/(Z_7\mu + Z_8) \tag{33a,b}$$

and hence lead to a quadratic in μ, the roots μ_1 and μ_2 of which are given by the equation

$$\mu = \frac{-(Z_1Z_8 + Z_2Z_7 - Z_4Z_5 - Z_3Z_6) \pm [(Z_1Z_8 + Z_2Z_7 - Z_4Z_5 - Z_3Z_6)^2 - 4(Z_1Z_7 - Z_3Z_5)(Z_2Z_8 - Z_4Z_6)]^{1/2}}{2(Z_1Z_7 - Z_3Z_5)} \tag{34}$$

These two values of μ apply, respectively, to the two differential boundaries of velocity v_1 and v_2. From Eqs. (26) and (27) it follows immediately that each μ is related to each R by the equation

$$R = d\bar{m}_A/d\bar{m}_B = (Z_3\mu + Z_4)/(Z_7\mu + Z_8) \tag{35}$$

the Z's being defined by the correspondence between Eqs. (32a,b) and (33a,b).

Tracing out a boundary formed initially between solvent and solution of plateau concentrations $\bar{m}_A{}^0$, $\bar{m}_B{}^0$ requires (1) selection of the appropriate root μ_1 or μ_2, (2) calculation of the breakaway point v^0 of the boundary from the plateau level, and (3) calculation of the change in $\bar{m}_A$, $(da = \mu \, db)$, for a given change $d\bar{m}_B$ in $\bar{m}_B$, the integration being effected numerically and carried on until only one component is left.

To decide which root of Eq. (34) is relevant in any given situation, we note that when we are concerned with a moving-boundary experiment in which the β region is a permanent plateau region [of concentration $(\bar{m}_A^0)_\beta$, $(\bar{m}_B^0)_\beta$] then the parameters to be used will be v_2 and μ_2, and the boundary will correspond to a descending boundary in electrophoresis, a trailing boundary in chromatography, or a normal sedimentation velocity boundary in ultracentrifugation. On the other hand, integration based on v_1 and μ_1 would be appropriate for an ascending boundary in electrophoresis or a leading boundary in chromatography, the α region then containing the solution plateau $(\bar{m}_A^0)_\alpha$, $(\bar{m}_B^0)_\alpha$.

Sometimes both roots are required as, for instance, in a more general type of experiment in which an initial boundary is set up between two solutions of concentration $(\bar{m}_A^0)_\alpha$, $(\bar{m}_B^0)_\alpha$ and $(\bar{m}_A^0)_\beta$, $(\bar{m}_B^0)_\beta$, respectively. Two finite difference boundaries will then be generated, and integration must be carried out, using both sets of parameters, from the two plateau regions α and β toward the common region γ which will link the two boundaries. This will be more easily appreciated after a numerical example has been worked through.

We now set up a model for the simple electrophoresis case for which we choose arbitrary but reasonable values of the various constants.

Numerical Example: Ascending and Descending Boundaries in Electrophoresis

Parameters. Mobilities $\times 10^5$: $v_A = 1$, $v_B = 4$, $v_{C_1} = 2$, $v_{C_2} = 3$ cm^2 V^{-1} s^{-1}

Equilibrium constants: $k_1 = 4$ $k_2 = 1 \times 10^5$ liters/mol
Molecular weights: $M_A = 150,000$, $M_B = 100,000$
Total concentration: $\bar{w}^0 = 2$ g/dl
Ratio of concentrations: $\bar{w}_A^0 : w_B^0 = 3:2$

Computation. For the given plateau *constituent* concentrations by weight ($\bar{w}_A^0 = 1.2$ g/dl, $\bar{w}_B^0 = 0.8$ g/dl), the corresponding molar concentrations of all the *species* must be calculated. The constituent molar concentrations are given by

$$m_A^0 = 10 \, \bar{w}_A^0/M_A = 0.8 \times 10^{-4} \text{ moles/liter}$$
$$\bar{m}_B^0 = 10 \, \bar{w}_B^0/M_B = 0.8 \times 10^{-4} \text{ moles/liter}$$

The simplest route for finding the concentrations of the species is via c_1.

Although trivial, carrying through the algebra can be time-consuming, and so this is sketched below.

Calculation of c_1. Eliminate b between Eqs. (17) and (19) to obtain

$$a = (k_2 c_1^2)/(k_1 c_2) \tag{36}$$

and from Eq. (21) to obtain

$$c_2 = (\overline{m}_B - c_1)/[(2 + 1)/k_2 c_1] \tag{37}$$

Eliminate a between Eqs. (20) and (36) to obtain

$$\overline{m}_A = (k_2 c_1^2)/(k_1 c_2) + c_1 + c_2 \tag{38}$$

Eliminate c_2 between Eqs. (37) and (38). The result is a cubic equation in c_1 which reads

$$c_1^3(4k_1 k_2^2 - k_1^2 k_2) + c_1^2(4k_1 k_2$$
$$- k_1^2 + 2k_1^2 k_2 \overline{m}_A) + c_1(k_1 + k_1^2 \overline{m}_B + k_1^2 k_2 \overline{m}_B^2$$
$$- 2k_1^2 k_2 \overline{m}_A \overline{m}_B + k_1^2 \overline{m}_A) - k_1^2 \overline{m}_A \overline{m}_B = 0 \quad (39)$$

Equation (39) has been multiplied throughout by k_1 ($k_1 \neq 0$) to give a "well behaved" equation, which is dimensionless and ready for numerical solution. The root with physical meaning is the root for which a and b are both positive. For the given example, solution of Eq. (39) leads to the following values for the molar concentrations (moles per liter) of the species: $a = 0.276 \times 10^{-4}$; $b = 0.140 \times 10^{-4}$; $c_1 = 0.388 \times 10^{-4}$; $c_2 = 0.136 \times 10^{-4}$.

The values of Z can now be calculated and substituted in Eq. (34). The values of μ found are $\mu_1 = -4.431$ and $\mu_2 = -0.7122$, and the corresponding values of v_1 and v_2 from Eqs. (32a,b) are $v_1 = 1.485 \times 10^{-5}$ and $v_2 = 2.871 \times 10^{-5}$ cm^2 V^{-1} s^{-1} ($v_1 < v_2$). The values of R are $R_1 = 2.765$ and $R_2 = 0.4444$.

As stated earlier, v_2 determines the breakaway point from the plateau of the descending boundary, and v_1 the breakaway point of the ascending boundary. As some help toward understanding the role of v_1 and v_2, these two velocities are shown in Fig. 2 as a function of k_1 ($k_1 = 4k_2$) from relatively low values of k_1 where $v_1 \simeq v_A$, $v_2 \simeq v_B$ to very high values of k_1.

The calculation of the complete boundary sequence is described next.

Numerical Integration

Descending Boundary. Starting from plateau levels $\overline{m}_A^0$, $\overline{m}_B^0$, $\overline{m}_B$ is decreased by subtracting a suitably small increment in b, and then the new corresponding value of the concentration of a is found by applying Eq. (30) in the form

$$(da/db)_2 = \mu_2$$

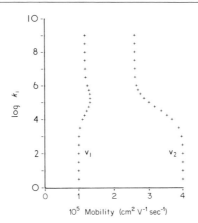

FIG. 2. Mobilities v_1 and v_2 for differential boundaries generated by the electrophoresis of a reversibly interacting system $A + B = C_1$, $B + C_1 = C_2$: $k_1 = c_1/ab$, $k_2 = c_2/bc_1$, $k_2 = k_1/4$ liters/mole; $v_A = 4 \times 10^{-5}$, $v_B = 1 \times 10^{-5}$, $v_{C_1} = 2 \times 10^{-5}$, $v_{C_2} = 3 \times 10^{-5}$ cm²V⁻¹sec⁻¹. Concentration of $A = \frac{1}{3}$ g/dl, concentration of $B = 1/6$ g/dl. Molecular weight of $A = 150,000$; molecular weight of $B = 100,000$.

Since Eq. (30) applies strictly only for infinitesimal changes, a suitable numerical integration procedure (for instance by Runge-Kutta) has to be used to allow for a finite step length. Instructions are usually readily available at any computer center. The step size must be chosen small enough for there to be an insignificant change in values on halving the step length. Some computers may be found to have insufficient word length, and these require the use of double precision.

For each new value of $\overline{m}_A$, $\overline{m}_B$ a new value of μ_2 is calculated by Eq. (34) and then a value for v_2 by Eq. (33). One continues through the boundary, plotting $\overline{m}_A$ and $\overline{m}_B$ against v_2, until b has reached zero (or the lowest practical value for calculation). The residue of a when b is zero is the concentration of A which emerges as pure component on the solvent side of the boundary. The variation of the actual species through the boundaries is shown in Fig. 3.

A close approximation to the concentration gradient at each point in the boundary can be obtained by dividing the change in $\overline{w}_A + \overline{w}_B$ during a step by the change in v_2 ($\Delta\overline{w}/\Delta v_2$), where $\overline{w}_A$ and $\overline{w}_B$ are the weight concentrations in g/dl calculated by Eqs. (40) and (41).

$$\overline{w}_A = \overline{m}_A \, (M_A/10) \tag{40}$$
$$\overline{w}_B = \overline{m}_B \, (M_B/10) \tag{41}$$

The total concentration will be defined by

$$\overline{w} = \overline{w}_A + \overline{w}_B \tag{42}$$

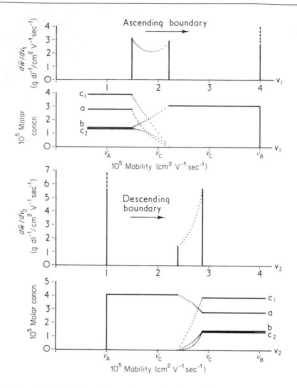

FIG. 3. "Diffusion-free" electrophoresis reaction boundaries (ascending and descending) for a model reversible reaction $A + B = C_1$, $B + C_1 = C_2$: $k_1 = c_1/ab = 100,000$, $k_2 = c_2/bc_1 = k_1/4$ liters/mole; $v_A = 1 \times 10^{-5}$, $v_B = 4 \times 10^{-5}$, $v_{C_1} = 2 \times 10^{-5}$, $v_{C_2} = 3 \times 10^{-5}$ cm²V⁻¹sec⁻¹. Molecular weight of $A = 150,000$; molecular weight of $B = 100,000$. $\overline{w}^0 = \overline{w}_A^0 + \overline{w}_B^0 = 2$ g/dl, $\overline{w}_A^0 : \overline{w}_B^0 = 3:2$.

This boundary too is shown in Fig. 3. The variety of patterns attainable is readily visualized by plotting boundaries produced for a range of parameters.

In Fig. 4 we show for both ascending and descending boundaries the result of altering the ratio of the concentration of A to B, while keeping the total weight concentration constant.

Finite Difference Boundaries in Electrophoresis

A complete exploration of a system might profitably include a study of boundaries generated from an initial concentration boundary formed by the juxtaposition of solutions of concentration $(\overline{m}_A^0)_\alpha$ $(\overline{m}_B^0)_\alpha$ and $(\overline{m}_A^0)_\beta$ $(\overline{m}_B^0)_\beta$, provided that means are available for stabilizing the boundaries against convection (as, for instance, in gel electrophoresis). Instead of one

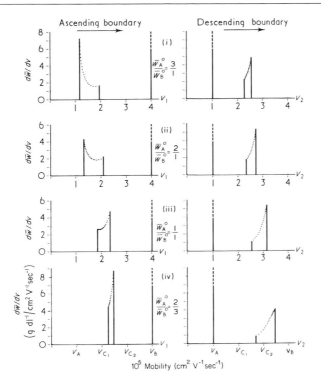

FIG. 4. "Diffusion-free" electrophoresis reaction boundaries (ascending and descending) for a model reversible reaction $A + B = C_1$, $B + C_1 = C_2$: $k_1 = c_1/ab = 100,000$, $k_2 = c_2/bc_1 = k_1/4$ liters/mole; $v_A = 1 \times 10^{-5}$; $v_B = 4 \times 10^{-5}$, $v_{C_1} = 2 \times 10^{-5}$, $v_{C_2} = 3 \times 10^{-5}$ cm^{2}V^{-1} sec^{-1}. Molecular weight of $A = 150,000$; molecular weight of $B = 100,000$. $\bar{w}^0 = \bar{w}_A^0 + \bar{w}_B^0 = 2$ g/dl, (i) $\bar{w}_A^0 : \bar{w}_B^0 = 3 : 1$, (ii) $\bar{w}_A^0 : \bar{w}_B^0 = 2 : 1$, (iii) $\bar{w}_A^0 : \bar{w}_B^0 = 1 : 1$, (iv) $\bar{w}_A^0 : \bar{w}_B^0 = 2 : 3$.

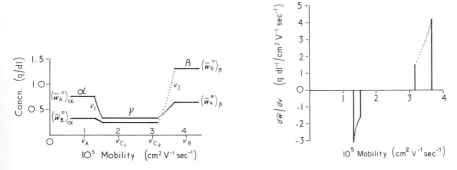

FIG. 5. "Diffusion-free" finite difference boundary in electrophoresis for the reversible reaction $A + B = C_1$, $B + C_1 = C_2$. See text.

reaction boundary and one pure component, two reaction boundaries will develop. An example is shown in Fig. 5.

II. General Case: Velocities Dependent upon Concentration

Sedimentation

In the electrophoresis case just described some simplicity was achieved by treating the velocities of all species as constants. However, since an analytical solution is avoided by the use of numerical methods, there is nothing in principle beyond the handicap of more complicated algebra, to prevent the introduction of concentration dependent terms for the velocities (or for the equilibrium constants). In sedimentation experiments the effect of the concentration dependence of velocities can be so dominant that its neglect is not wise. We therefore extend our treatment to include concentration dependent velocities in sedimentation, and we show how to calculate any consequent hypersharp region. For the sake of illustration we shall assume that the velocities of all species are modified by the multiplying term $(1 - g\,\bar{w})$ where $\bar{w}$ is the total concentration of material present. Any other functional dependence could be treated by analogous methods. We shall retain the same model system and shall assume that the sedimentation velocities of the species are given by

$$v_A = (v_A)_0\,(1 - g\,\bar{w}) \tag{43}$$
$$v_B = (v_B)_0\,(1 - g\,\bar{w}) \tag{44}$$
$$v_{C_1} = (v_{C_1})_0\,(1 - g\,\bar{w}) \tag{45}$$
$$v_{C_2} = (v_{C_2})_0\,(1 - g\,\bar{w}) \tag{46}$$

Consequent upon the presence of the term $g\,\bar{w}$, the total differentials of the fluxes in the conservation Eqs. (1 and 2) are considerably extended. For instance, for the flux of A

$$dJ_A = d(av_A + c_1 v_{C_1} + c_2 v_{C_2})$$
$$= \{(v_A)_0 da + (v_{C_1})_0 dc_1 + (v_{C_2})_0 dc_2\}(1 - g\,\bar{w})$$
$$+ a\,dv_A + c_1\,dv_{C_1} + c_2\,dv_{C_2} \tag{47}$$

where

$$dv_A = -g(v_A)_0 d\bar{w}$$
$$= -g(v_A)_0 \left\{ \frac{M_A}{10}\,d\bar{m}_A + \frac{M_B}{10}\,d\bar{m}_B \right\} \tag{48}$$

with similar expressions for dv_{C_1} and dv_{C_2}. A parallel relationship holds for dJ_B. The terms required for solving Eqs. (32a,b), e.g., $d\bar{m}_A$ (Eq. 26) and $d\bar{m}_B$ (Eq. 27) have been given previously, and the way is open to calculate μ_2 for the sedimentation boundary once we have chosen numerical values for the new parameters in our model. We shall assume the following values for the sedimentation coefficients (in Svedberg units) and for the concentration dependence coefficient g.

$$(v_A)_0 = 8.000 \text{ S} \qquad (v_B)_0 = 6.105 \text{ S}$$
$$(v_{C_1})_0 = 11.246 \text{ S} \qquad (v_{C_2})_0 = 14.074 \text{ S}$$
$$g = 0.07 \text{ dl/g}$$

The molecular weights and interaction constants are the same as in the example for electrophoresis. The distance axis of the boundary pattern will be normalized as before by relating it to unit field and unit time.

The situation differs considerably from that holding in electrophoresis, for one cannot know immediately whether A or B will emerge as the pure component behind the reaction boundary, even though B is the more slowly sedimenting species. There is even the possibility that neither species will, and that both A and B will terminate at the lower edge of the reaction boundary. Further, except for dilute solutions, hypersharpness is likely to occur. If we choose a total concentration of $\overline{w}^0 = 1.25$ g/dl and a ratio of $\overline{w}_A{}^0 : \overline{w}_B{}^0 = 4 : 3$, i.e., $\overline{w}_A{}^0 = 0.714$ g/dl and $\overline{w}_B{}^0 = 0.536$ g/dl, we can illustrate this possibility.

An initial trial integration shows that for these plateau values it is B that emerges from the reaction boundary and that a is therefore the appropriate variable to reduce by steps to zero. When this is done, the integration procedure generates the involuted curves for the boundary profile shown in Fig. 6. The physical interpretation of this purely mathematical profile is that the higher concentration region of the boundary

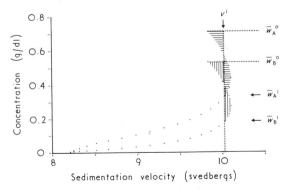

FIG. 6. Hypersharp "diffusion-free" sedimentation reaction boundary for the model reversible reaction $A + B = C_1$, $B + C_1 = C_2$: $M_A = 150,000$, $M_B = 100,000$; concentration dependence $g = 0.07$ dl/g, $(s_A)_0 = 8.000$, $(s_B)_0 = 6.105$, $(s_{C_1})_0 = 11.246$, $(s_{C_2})_0 = 14.074$ S. $k_1 = c_1/ab = 100,000$, $k_2 = c_2/bc_1 = k_1/4$ liters/mole; $\overline{w}^0 = \overline{w}_A{}^0 + \overline{w}_B{}^0 = 1.25$ g/dl, $\overline{w}_A{}^0 : \overline{w}_B{}^0 = 4 : 3$.

adjacent to the plateau is hypersharp ("self-sharpening"), whereas the lower concentration region is broad. To find the position v' where the break in the profile occurs, and also the magnitude of the sudden fall there in the concentrations of A and B (say to $\overline{m}'_A$ and $\overline{m}'_B$, respectively), one must iterate from initial guessed values to establish the values of $\overline{m}'_A$ and $\overline{m}'_B$ (or $\overline{w}'_A$, $\overline{w}'_B$ equivalently) which cause the centroids of the boundaries of A and B in the hypersharp region to coincide. A rough approximation, sufficiently good for many purposes, can be achieved by equalizing the shaded areas in Fig. 6 by eye. To obtain an exact solution it is necessary to refine this first estimate in the following way.

Calculation of Hypersharp Reaction Boundaries

We are required to determine the exact size and position of the step in the boundary in Fig. 6. (The situation in Fig. 6 is very likely to be encountered also in chromatography, particularly in "difference" chromatography, for which one example was evaluated earlier,[22] but without any detailed description.)

The criteria to be satisfied at the breakpoint where the hypersharp boundary region joins the broad region are the following. (1) The value of v'_A of the velocity of the sharp front of A has to coincide with the corresponding value v'_B for B. (2) The value v'_2 of the velocity of the boundary profile, at the junction of the broad region with the step, has to be identical to the values v'_A and v'_B.

We emphasize that v'_2 is the velocity of the boundary profile at the concentration level $\overline{m}'_A$, $\overline{m}'_B$ and that it is calculated as before by first finding μ_2 from Eq. (34) and then v_2 from Eq. (33).

The velocities v'_A and v'_B are found by applying mass conservation principles, but now to *finite* difference boundaries. With this proviso, the treatment that produced Eq. (1) leads to Eq. (49).

$$v'_A (\overline{m}_A{}^0 - \overline{m}'_A) = \overline{m}_A{}^0 \overline{v}_A{}^0 - \overline{m}'_A \overline{v}'_A \tag{49}$$

where $\overline{v}'_A$ is the constituent velocity of A for the concentration $\overline{m}'_A$ of A at the bottom of the hypersharp step. Similarly, for B

$$v'_B (\overline{m}_B{}^0 - \overline{m}'_B) = \overline{m}_B{}^0 \overline{v}_B{}^0 - \overline{m}'_B \overline{v}'_B \tag{50}$$

A rough estimate by eye in Fig. 6 gave the values 0.31 and 0.16 g/dl, respectively, for $\overline{w}'_A$ and $\overline{w}'_B$ (i.e., $\overline{m}_A = 2.1 \times 10^{-5}$ moles/liter $m_B = 1.6 \times 10^{-5}$ moles/liter). Application of Eqs. (49) and (50) with these concentrations inserted gave values of v'_A and v'_B of 10.06 S and 10.01 S, whereas v_2 for the top of the *broad* boundary region for the same concentrations was found to be 9.98 S. The first estimate was therefore not too bad. To proceed further a criterion for imbalance is needed and is provided by the function $\phi(v')$ where

$$\phi(v') = (v'_A - v'_2)^2 + (v'_B - v'_2)^2 \tag{51}$$

which is zero at the correct value of $\overline{m}'_A$, $\overline{m}'_B$ when v'_A, v'_B and v'_2 coincide. $\overline{m}'_A$ and $\overline{m}'_B$ are therefore varied (by varying a' and b') either by hand, or by a machine-controlled process, until $\phi(v')$ is less than a given small quantity. In the present case $\phi(v')$ was reduced to $<10^{-7}$ with v'_2, v'_A, and v'_B having the value 10.025 in common after $\overline{w}'_A$ and $\overline{w}'_B$ had been adjusted to be 0.329 and 0.182 g/dl, respectively. Thus the hypersharp step occurs at $v' = 10.025$ and consists of a total drop in concentration of $\overline{w} = 0.739$ g/dl.

The rest of the broad part of the boundary has now to be recalculated, starting from the newly calculated point $\overline{w}'_A$, $\overline{w}'_B$, by reducing a in steps to zero. A residue of pure B emerges from the boundary at a concentration of 0.007 g/dl. The new broad part of the boundary is, to within the limits of resolution of the drawing in Fig. 6, superimposable on the original.

Summary and Conclusions

We have dealt in this article only with reaction boundaries generated in electrophoresis and sedimentation experiments, but the same approach is equally applicable to chromatography experiments.[5,22] Our treatment is most appropriate for an initial exploration of a system, since it ignores important effects that include diffusion, finite time of reaction, inhomogeneous fields, nonrectangular geometry of cells, and nonideality. The theoretical concepts have been kept simple, being merely the application of mass-conservation equations to differential boundaries, or to finite difference boundaries when hypersharp regions occur, and advice has been given on how to minimize the potentially tedious algebra. Descriptions of much more ambitious procedures, best applied after a preliminary survey by the above methods, can be found in recent papers by Cann and Kegeles,[26] Payens and Nijhuis,[27,28] and Schmidt and Payens,[29]

[26] J. R. Cann and G. Kegeles, *Biochemistry* **13**, 1868 (1974).
[27] T. A. J. Payens and H. Nijhuis, *Biochim. Biophys. Acta* **336**, 201 (1974).
[28] H. Nijhuis and T. A. J. Payens, *Biochim. Biophys. Acta* **336**, 213 (1974).
[29] D. G. Schmidt and T. A. J. Payens, *in* "Surface and Colloid Science" (E. Matijevic, ed.), Vol. 9, Chap. 3. Wiley, New York, 1976.

who have applied their methods for simulating boundary patterns to the study of hemocyanin, ligand complexes, and α-casein, β-casein complexes, respectively.

Acknowledgment

Support was given by the Science Research Council of Great Britain during the writing of this article.

[10] Calculation of Simulated Sedimentation Velocity Profiles for Self-Associating Solutes

By DAVID J. COX

Chemical equilibria involving macromolecules are usually examined by such equilibrium techniques as light scattering, equilibrium sedimentation, and osmotic pressure. These measurements define the dependence of various molecular weight averages on the solute concentration. The stoichiometry and equilibrium constants for the chemical equilibrium are obtained from the concentration dependence of the average molecular weights by appropriate curve-fitting procedures.

The treatment of macromolecular interactions by equilibrium methods makes heavy demands on the quality of the data and is sensitive to vexing artifacts, particularly if macromolecular contaminants are present that do not participate in the chemical equilibrium. The results are sometimes ambiguous, in that the data can be made to fit more than one reaction scheme.

The behavior of macromolecular solutes during transport experiments—sedimentation, chromatography, or electrophoresis—is affected in striking and characteristic ways by chemical interactions among the solute molecules.[1,2] The shape of the boundary generated in a transport process by a chemically reacting solute contains a great deal of information about the solute and the reactions in which it participates. In principle, transport experiments can be a useful complement to equilibrium techniques in defining the nature of a chemical reaction involving macromolecules. The practical problem lies in finding a suitable way to extract the relevant information from boundary profiles.

No methods are presently available for inferring directly the nature of a solute and its interactions from the shape of the boundary it produces in

[1] L. W. Nichol, J. L. Bethune, G. Kegeles, and E. L. Hess, *in* "The Proteins," 2nd ed. (H. Neurath, ed.), Vol. 2, p. 305. Academic Press, New York, 1964.
[2] J. R. Cann, "Interacting Macromolecules." Academic Press, New York, 1970.

a transport experiment. However, it is possible to proceed indirectly by asking a related question: if one were to encounter an interacting system with certain specified properties, how would it behave in a transport experiment? A real interacting system can then be approached by predicting the behavior of a series of plausible models and identifying those models that can be made to produce the experimentally observed boundary profile. This approach has the defect that a particular successful model may not be unique; several models, only one of which is correct, may generate acceptable profiles. On the other hand, the comparison of predicted and experimental transport profiles will always limit the range of acceptable reaction models. The method can be particularly useful in choosing among several reaction schemes that may be consistent with a set of data drawn from equilibrium techniques.

There has been some recent progress in developing analytical techniques for predicting the transport behavior of interacting solutes,[3] but the current method of choice is computer simulation using finite difference models of the transport process. Several simulation procedures are now available that are capable of predicting in detail the boundary profiles that will be given by a wide variety of interacting systems, and there is a substantial literature describing the methods and displaying representative calculated profiles.

The reacting solutes that have attracted the greatest attention are the rapidly equilibrating self-associating systems that were first treated by Gilbert, using an analytical procedure that neglects the effect of diffusion on the boundary shape.[4,5] This procedure is extremely useful in showing the qualitative features of reaction boundaries, but it is obviously not suitable for detailed predictions of boundary shape. The technique of Bethune and Kegeles[6,7] was the first to incorporate the effect of diffusion, exploiting an analogy between countercurrent distribution and other transport processes, particularly velocity sedimentation. More recently, three different methods have been developed that are designed to give quantitatively accurate calculated boundary profiles for interacting systems.[8-12] Each of these methods has specific advantages and limitations

[3] H. Schönert, *Biophys. Chem.* **3**, 161 (1975).
[4] G. A. Gilbert, *Discuss. Faraday Soc.* **20**, 68 (1955).
[5] G. A. Gilbert, *Proc. R. Soc. London, Ser. A* **250**, 377 (1959).
[6] J. L. Bethune and G. Kegeles, *J. Phys. Chem.* **65**, 1761 (1961).
[7] J. L. Bethune, *J. Phys. Chem.* **74**, 3837 (1970).
[8] J. R. Cann and W. B. Goad, *J. Biol. Chem.* **240**, 148 (1965).
[9] W. B. Goad and J. R. Cann, *Ann. N. Y. Acad. Sci.* **164**, 172 (1969).
[10] M. Dishon, G. H. Weiss, and D. A. Yphantis, *Biopolymers* **4**, 449 (1966).
[11] D. J. Cox, *Arch. Biochem. Biophys.* **119**, 230 (1967).
[12] D. J. Cox, *Arch. Biochem. Biophys.* **129**, 106 (1969).

for dealing with particular kinds of problems, but the techniques are inter-changeable for many purposes.

This article will consider in detail one simulation procedure that is particularly suitable for rapidly equilibrating self-associations of the kind first treated by Gilbert. The technique will be described here as it is applied to velocity sedimentation in the ultracentrifuge. The article is intended to make this specific technique more generally available. Moreover, since all computational methods of this kind have a number of features in common, a detailed consideration of one scheme may be of some further use as an introduction to simulation of transport processes in general.

Self-Association Equilibria

A self-association reaction involves chemical equilibria between a monomeric solute and one or more aggregates of subunits: $jA \rightleftarrows A_j$. The concentrations of the monomer and any individual aggregate are related by an association constant that can be expressed on a weight scale:

$$K_j = C_j/C_1^{\,j} \qquad C_j = K_j C_1^{\,j} \tag{1}$$

At chemical equilibrium the total solute concentration is:

$$C_T = \Sigma_j K_j C_1^{\,j} \qquad K_1 \equiv 1 \tag{2}$$

where the summation is over all the species present, including the monomer.

During a transport experiment, the solute concentration at each point in the moving boundary changes with time and, if the solute undergoes self-association, the chemical equilibrium is perturbed. The transport be-havior of the solute depends strongly on how rapidly the solute relaxes to the new equilibrium appropriate for the altered total solute concentration. If reequilibration occurs much more rapidly than the local changes in solute concentration resulting from the transport process, the solute will be very close to local chemical equilibrium everywhere in the system throughout the experiment. For velocity sedimentation or for gel permea-tion chromatography, the chemical reactions leading to reequilibration do not need to be extraordinarily rapid to produce this situation. An associa-tion reaction that relaxes toward equilibrium with a half-time of 30 sec or less is indistinguishable from an infinitely rapid reaction.[13,14] In this case, the distribution of the solute among monomer and the various aggregates

[13] J. K. Zimmerman, *Biochemistry* **13**, 384 (1974).
[14] J. R. Cann and G. Kegeles, *Biochemistry* **13**, 1868 (1974).

at a particular point in the moving boundary will be unambiguously determined by the total solute concentration at that point.

The existence of local chemical equilibrium everywhere in the system means that, in simulating the transport of a self-associating solute that reequilibrates rapidly, the solute can be treated as a single component, however complex the association equilibrium may be.[15] The self-association of the solute is expressed in the concentration dependence of the transport coefficients—for sedimentation in the ultracentrifuge, the sedimentation coefficient, and the diffusion coefficient. The local sedimentation coefficient, for example, depends on the relative amounts of monomer and aggregates present, and the population of solute species depends in turn on the total concentration. Since the solute concentration varies across the boundary, the local sedimentation coefficient also varies. Different parts of the boundary migrate at different rates, and the shape of the solute profile is distorted. It is this distortion that produces the distinctive boundary shapes characteristic of different self-associating solutes.

Any general purpose program that can deal with concentration dependent sedimentation and diffusion is thus capable of treating associating solutes provided that the association reactions reequilibrate reasonably rapidly. Such a program can be adapted to this particular use simply by supplying it with a way of finding the local average sedimentation and diffusion coefficients appropriate for any total solute concentration it may encounter.

General Considerations

Notation. The notation used in the simulation procedure is displayed in Fig. 1. The centrifuge cell is divided into an array of n boxes by placing $n + 1$ boundaries at various distances r from the axis of rotation. The meniscus is r_1 and the bottom of the solution column is r_{n+1}. The boxes are normally made uniform in size at the beginning of the simulation, but, as indicated in the figure, they generally do not remain so as the computation proceeds. The distribution of the solute in the cell at any time is given by a list of values of the mean weight concentration $\bar{C}$ in each box.

The fundamental arrays that are manipulated by the simulation program are the boundary positions, r, and the concentrations in the boxes, $\bar{C}$. Several other arrays derived from these are used during the calculation. The mean concentration in box i is assigned to the midpoint of the box, $\bar{r}_i$:

$$\bar{r}_i = (r_{i+1} + r_i)/2 \tag{3}$$

[15] H. Fujita, "Mathematical Theory of Sedimentation Analysis." Academic Press, New York, 1962.

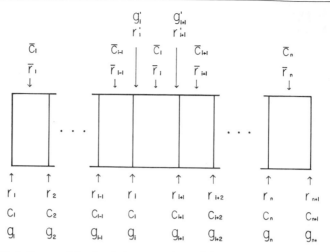

FIG. 1. Notation used in simulation procedure. C is weight concentration, r is radial position, and g is concentration gradient. Unprimed symbols without bars refer to boundaries between boxes; barred quantities refer to midpoints of boxes; primed symbols refer to positions equidistant from midpoints of adjacent boxes, which do not necessarily coincide with boundary positions.

The concentration C_i at boundary i is obtained by linear interpolation between the boxes on either side:

$$C_i = \bar{C}_{i-1} + (\bar{C}_i - \bar{C}_{i-1})[(r_i - \bar{r}_{i-1})/(\bar{r}_i - \bar{r}_{i-1})] \tag{4}$$

The model deals only with the infinite cell case, and so the concentrations at the ends of the array are set equal to the concentrations in the first and last boxes: $C_1 = \bar{C}_1$ and $C_{n+1} = \bar{C}_n$.

The concentration gradient between two boxes, g_i', is:

$$g_i' = (\bar{C}_i - \bar{C}_{i-1})/(\bar{r}_i - \bar{r}_{i-1}) \tag{5}$$

The gradient is assigned to a position equidistant from the midpoints of the two boxes:

$$r_i' = (\bar{r}_{i-1} + \bar{r}_i)/2 \tag{6}$$

Since adjacent boxes are not necessarily equal in size, r_i' need not coincide with r_i. Whenever the gradient at boundary i is needed, it is found by linear interpolation:

$$g_i = g_i' + (g_{i+1}' - g_i')\ [(r_i - r_i')/(r_{i+1}' - r_i')] \tag{7}$$

General Structure of the Calculation. A simulation is begun by selecting an array of boxes appropriate for the transport process to be described.

For sedimentation in the ultracentrifuge, the total length of the solution column is usually taken as 1 cm, and the solution column is divided into 50, 100, or 200 boxes, depending on the application. The initial interval between the boundaries, Δr, is thus 0.005–0.020 centimeters. In general, the more finely divided the space grid, the more precise the results will be. On the other hand, the number of computing operations required for the calculation is very strongly affected by the box interval, and so a compromise must be made between high precision and excessive computer time. This matter will be discussed in more detail below (see section on diffusion).

Once a suitable array of boundaries and boxes has been selected, the initial distribution of the solute is described by assigning a value to $\bar{C}$ for each box. The simultaneous sedimentation and diffusion of the solute is simulated by describing alternate brief rounds of diffusion without sedimentation and sedimentation without diffusion. The techniques used for this purpose are described in detail in following sections of the article.

The repeated application of sedimentation and diffusion operations to the concentration array results in the accumulation of round-off error. To keep the round-off within tolerable limits, a computer word size of at least forty bits is needed. For many operating systems, this will require writing most of the variables in double precision.

The alternating application of separate routines for diffusion and sedimentation is a convenient procedure, but any scheme of this kind has one significant limitation: it does not describe correctly the behavior of concentration gradients at the ends of the solution column.[16] Such a model is analogous to an infinite cell solution of the differential equation for the ultracentrifuge. Thus, a system with a uniform initial solute concentration throughout the cell, in which a boundary forms at the meniscus early in the run, cannot be treated. The initial solute distribution given to the computer must be one in which the solute boundary is far enough removed from the top of the array that the concentration at the meniscus will remain close to zero throughout the simulation. For a 100-box array with the meniscus at r_1 and a uniform box size Δr, the following initial distribution is usually suitable:

$$
\begin{aligned}
r_i &= r_{i-1} + \Delta r & 1 < i \leq 100 \\
\bar{C}_i &= 0 & 1 \leq i \leq 10 \\
\bar{C}_i &= C_0 & 10 < i \leq 100
\end{aligned}
\tag{8}
$$

Alternatively, concentration measurements from an experimental solute profile can be supplied to the computer. If a synthetic boundary cell has

[16] D. J. Cox, *Arch. Biochem. Biophys.* **112**, 259 (1965).

been used in the experiment to be simulated, the first photograph after the centrifuge rotor reaches full speed can be used. For experiments using a standard cell, it is necessary to wait until the entire experimental boundary profile has left the meniscus before taking the photograph that will be used to provide the initial solute distribution for the simulation.

The development of the solute gradient at the bottom of the solution column is also described incorrectly by the simulation procedure. Therefore, it is necessary to terminate the simulation before any part of the sedimenting boundary reaches the bottom of the array; that is, the run must continue only so long as a plateau region persists.

Average Transport Coefficients

Definition of $\bar{S}$ and $\bar{D}$. The behavior of a self-associating solute in a velocity sedimentation experiment is simulated by describing many successive, alternate rounds of sedimentation and diffusion. At the beginning of each round, the solute concentration at each boundary in the array is calculated using Eq. (4), and the average sedimentation or diffusion coefficient corresponding to each of these solute concentrations is computed.

For a solute that sediments ideally, the appropriate average sedimentation coefficient is the weight average:

$$\bar{S} = \frac{\Sigma_j S_j C_j}{\Sigma_j C_j} = \frac{\Sigma_j S_j K_j C_1{}^j}{\Sigma_j K_j C_1{}^j} \tag{9}$$

where S_j is the sedimentation coefficient of the j-mer and the summation is over all species present.[15] The use of the weight average implies that each subunit turns over rapidly as it migrates, appearing in turn as monomer and as a member of each aggregate. The fraction of time that a subunit spends as j-mer and thus migrating with the j-mer sedimentation coefficient is equal to the weight fraction of species j in the equilibrium mixture.

Real solutes do not sediment ideally; the local sedimentation coefficient of each species can be expected to show a hydrodynamic dependence on the local concentration of all species present. The hydrodynamic concentration dependence is superimposed on the variation of $\bar{S}$ resulting from the association reaction. It is possible to write quite elaborate expressions to describe the effect, but there is usually no good theoretical or experimental basis for preferring any particular description. For most purposes, a relatively simple expression is used:

$$\bar{S} = \frac{\Sigma_j S_j K_j C_1{}^j}{\Sigma_j K_j C_1{}^j} (1 - kC_T) \tag{10}$$

The proper average diffusion coefficient[12,17] for an associating solute is a little less obvious. The total mass flow due to diffusion at a particular point in the moving boundary is the sum of the flows of all the solute species present:

$$J = \Sigma_j D_j (dC_j/dr) \tag{11}$$

We need to define an average diffusion coefficient that will relate the total flow to the total solute concentration gradient:

$$\overline{D}(dC_T/dr) = \overline{D}\Sigma_j(dC_j/dr) = \Sigma_j D_j(dC_j/dr) \tag{12}$$

$$\overline{D} = \frac{\Sigma_j D_j(dC_j/dr)}{\Sigma_j(dC_j/dr)} \tag{13}$$

The concentration gradient of species j is directly related to the monomer gradient; since $C_j = K_j C_1^j$ [Eq. (1)],

$$dC_j/dr = (d/dr)K_j C_1^j \tag{14}$$

Provided that $(dK_j/dr) = 0$,

$$dC_j/dr = jK_j C_1^{j-1}(dC_1/dr) \tag{15}$$

The proviso seems trivial at first sight, but in fact it excludes certain physically significant cases. For the ultracentrifuge, the most important exclusion is the case in which the association equilibrium is pressure dependent.[18-20]

When Eq. (15) is inserted into the summations in Eq. (13), the monomer gradient appears in every term in both the numerator and the denominator and thus cancels:

$$\overline{D} = \frac{\Sigma_j j D_j K_j C^{j-1}}{\Sigma_j j K_j C^{j-1}} \tag{16}$$

Multiplying through both summations by C_1 and replacing $K_j C_1^j$ by C_j [Eq. (1)] yields an expression for $\overline{D}$ in terms of the concentrations of the various solute species:

$$\overline{D} = \frac{\Sigma_j j D_j C_j}{\Sigma j C_j} \tag{17}$$

Calculation of $\overline{S}$ and $\overline{D}$. Equations (2), (9), and (16) provide the working relations needed to find the average sedimentation and diffusion coeffi-

[17] R. F. Steiner, *Arch. Biochem. Biophys.* **49,** 400 (1954).
[18] L. F. TenEyck and W. Kauzmann, *Proc. Natl. Acad. Sci. U.S.A.* **58,** 888 (1967).
[19] R. Josephs and W. F. Harrington, *Proc. Natl. Acad. Sci. U.S.A.* **58,** 1587 (1967).
[20] W. F. Harrington and G. Kegeles, this series Vol. 27, p. 306.

cient for any solute concentration that may arise during a simulation. Using equilibrium constants supplied to the program, Eq. (2) will produce the monomer concentration that corresponds to the total solute concentration. The value of C_1 is then inserted into Eqs. (9) and (16) to give $\bar{S}$ and $\bar{D}$.

In a few simple cases, obtaining C_1 for a given C_T is a straightforward matter. For example, if the association reaction is a dimerization, Eq. (2) is a quadratic

$$C_T = C_1 + K_2 C_1^2 \tag{18}$$

which can be solved directly. In most cases, however, C_1 must be found by a trial-and-error procedure. Since C_1 must be greater than zero and less than C_T, a simple binary search beginning at $C_T/2$ serves the purpose; a flow chart for a suitable procedure is shown in Fig. 2. The residual error in C_T is normally set at $10^{-4}\, C_T$, and the search converges to this limit reliably and reasonably rapidly.

Once C_1 has been found, Eqs. (9) and (16) are used to calculate $\bar{S}$ and $\bar{D}$. For this computation, the program must have available the sedimentation and diffusion coefficients for the monomer and for the various aggre-

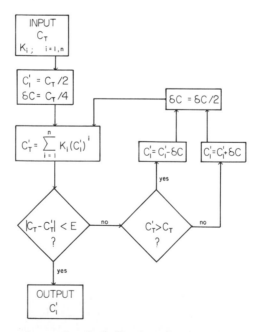

FIG. 2. Binary search procedure for finding monomer concentration (C_1) corresponding to given total concentration (C_T) for self-associating system containing one or more aggregated species in equilibrium with monomer. Primed quantities are trial values; E is maximum acceptable residual.

gates and a value for the hydrodynamic concentration dependence of the sedimentation coefficient. These values are supplied to the program at the beginning of the simulation, and their selection is somewhat arbitrary. The hydrodynamic concentration dependence varies experimentally from one system to another, depending on the nature of the solute and the solvent. A reasonable initial estimate for globular proteins carrying a small net charge in a solvent of moderate ionic strength would be $k = 0.01$ liter/g. The transport coefficients can be obtained in several ways. One general procedure is to specify the molecular weight of the monomer (M_1), the frictional ratio of each species (f/fo), the partial specific volume ($\bar{v}$), the temperature (T), and the density (ρ) and viscosity (η) of the solvent. The program is then asked to calculate S_j and D_j for each species[21]:

$$S_j = \frac{jM_1(1 - \bar{v}\rho)}{Nf_j} \qquad (19)$$

$$D_j = \frac{RT}{Nf_j} \qquad (20)$$

$$f_j = 6\pi\eta R_j(f/fo)_j \qquad (21)$$

$$R_j = \left(\frac{3jM_1\bar{v}}{4\pi N}\right)^{1/3} \qquad (22)$$

where R is the gas constant and N is Avogadro's number.

Alternatively, the program may be given sedimentation and diffusion coefficients for the monomer. The transport coefficients for other species are then given by

$$S_j = S_1(j)^{2/3} \qquad (23)$$
$$D_j = D_1(j)^{-1/3} \qquad (24)$$

These expressions have the effect of specifying equal frictional ratios for all species.

The rather considerable freedom in adjusting the hydrodynamic concentration dependence and the various transport coefficients is not a problem if one is interested only in describing qualitatively the transport behavior of a particular class of associating systems. For such a purpose, merely plausible estimates of k and of the frictional ratios are adequate. However, if one intends to fit simulated data to a real experiment, then the selection of appropriate values becomes a quite serious problem. It will generally be necessary to carry out a number of simulations, sampling the entire range of behavior for which some experimental precedent can be found.

[21] H. K. Schachman, "Ultracentrifugation in Biochemistry." Academic Press, New York, 1959.

Tabulation of $\bar{S}$ and $\bar{D}$. It will be evident from the foregoing discussion that the calculation of the value of $\bar{S}$ or $\bar{D}$ corresponding to a particular total solute concentration can be a somewhat clumsy and time-consuming procedure. A transport coefficient at each boundary in the array must be found before each of the many rounds of simulated sedimentation or diffusion. A complete simulation requires several thousand transport coefficients, and the computer time would be prohibitive if each of these were computed separately using Eqs. (2), (9), and (16) and the search procedure of Fig. 2. This problem is dealt with by constructing a table containing $\bar{S}$ and $\bar{D}$ at each of a few hundred solute concentrations before the simulation begins. Then, as the simulation proceeds, the needed transport coefficients are extracted from the table by an efficient interpolation procedure.

The table should cover a range of concentrations from zero to slightly above the initial plateau concentration. Values of $\bar{S}$ and $\bar{D}$ can be calculated for any selected solute concentration in this range using Eqs. (2), (9) or (10), and (16). A list of concentrations with equal increments between successive entries is sometimes suitable. However, the dependence of $\bar{S}$ and $\bar{D}$ on the solute concentration is often highly nonlinear, and a table constructed using equal increments of concentration may then contain large gaps between successive values of $\bar{S}$ and $\bar{D}$. To avoid interpolation errors that may arise in such a case, the usual procedure is first to outline the table by finding $\bar{S}$ and $\bar{D}$ for a limited number of entries—usually one hundred—separated by equal intervals in concentration. The maximum and minimum values of $\bar{S}$ are found in the preliminary table and 1% of the difference between these limits is taken as the largest acceptable difference between successive entries in the table. The table is inspected for unacceptably large gaps. When the difference between the $\bar{S}$ entries for two successive concentrations is found to be too large, additional values of $\bar{S}$ and $\bar{D}$ are calculated for a concentration midway between the two. These values are inserted into the table and succeeding entries are reindexed. The process is continued until all large gaps in $\bar{S}$ have been eliminated.

One may sometimes want to simulate ideally sedimenting solutes consisting of a monomer in equilibrium with a single aggregate. Such a simulation will be useless in an attempt to mimic a real system, since real solutes never sediment ideally. However, the qualitative features of such systems may be of interest, and so such simulations are often done.

The assembly of a table with entries at equal intervals in $\bar{S}$ can be done in a quite convenient way for the special case of an ideal monomer-j-mer system. If k is zero in Eq. (10), the minimum value of $\bar{S}$ is the monomer sedimentation coefficient. The maximum value of $\bar{S}$ must be that corresponding to the maximum value of C_T. If the desired number of entries in the table is m, then the number of intervals between successive entries is $m-1$. The difference between $\bar{S}_{max}$ and S_1 is divided into $m-1$ equal increments, and an array of values of $\bar{S}$ is assembled:

$$\Delta\bar{S} = (\bar{S}_{max} - S_1)/(m - 1) \tag{25}$$
$$\bar{S}_1 = S_1$$
$$\bar{S}_2 = \bar{S}_1 + \Delta\bar{S}$$
$$\cdots$$
$$\bar{S}_n = \bar{S}_{n-1} + \Delta\bar{S} \tag{26}$$
$$\cdots$$

The *monomer* concentration corresponding to each of the values of $\bar{S}$ is easily found. Since

$$\bar{S} = \frac{S_1 C_1 + S_j C_j}{C_1 + C_j} = \frac{S_1 C_1 + S_j K_j C_1^{\,j}}{C_1 + K_j C_1^{\,j}} \tag{27}$$
$$= \frac{S_1 + S_j K_j C_1^{\,i-1}}{1 + K_j C_1^{\,j-1}} \tag{28}$$
$$\bar{S} - S_1 = K_j C_1^{\,j-1}(S_j - \bar{S}) \tag{29}$$

It follows that

$$C_1 = \left(\frac{\bar{S} - S_1}{K_j(S_j - \bar{S})}\right)^{1/(j-1)} \tag{30}$$

Once C_1 is known for a particular value of $\bar{S}$, the corresponding value of C_T is calculated using Eq. (2).

It is sometimes useful to have available a no-diffusion treatment of the type developed by Gilbert for comparison with a complete simulation of the same system. If a Gilbert-type solution is wanted for an ideally sedimenting solute, it can be constructed conveniently during the assembly of the tables. In addition to the weight-average sedimentation coefficient, an average of the following kind is also computed for each solute concentration:

$$\bar{S}_g = \frac{\Sigma_j j S_j C_j}{\Sigma_j j C_j} = \frac{\Sigma_j j S_j K_j C_1^{\,j}}{\Sigma_j J K_j C_1^{\,j}} \tag{31}$$

Gilbert and Gilbert have shown that a plot of $dC_T/d\bar{S}_g$ against $\bar{S}_g$ is equivalent to the gradient profile in the absence of diffusion.[22] Once values of $\bar{S}_g$ have been tabulated for a series of total solute concentrations, the derivatives can be approximated using the differences between successive entries in the table.

[22] G. A. Gilbert, *in* "Ultracentrifugal Analysis in Theory and Experiment" (J. W. Williams, ed.), p. 73. Academic Press, New York, 1963.

During the subsequent course of the simulation, values of $\bar{S}$ and $\bar{D}$ are extracted from the table by linear interpolation. If, for example, the local sedimentation coefficient at boundary i is needed, the concentration at the boundary C_i is found using Eq. (4). Then C_i is compared with the entries in the table, beginning with the entry for $C_T = 0$ and proceeding upward in concentration. If $C_{T,n}$ is the first tabulated value larger than C_i, then C_i must be between $C_{T,n}$ and $C_{T,n-1}$, and the sedimentation coefficient at boundary i is:

$$\bar{S}_i = \bar{S}_{n-1} + (\bar{S}_n - \bar{S}_{n-1}) \frac{(C_i - C_{T,n-1})}{(C_{T,n} - C_{T,n-1})} \tag{32}$$

Values of $\bar{D}_i$ are found in the same way.

The normal procedure is to obtain values of $\bar{S}_i$ or $\bar{D}_i$ at every boundary in the array in a single continuous operation. The table search can be made more efficient by taking advantage of the fact that the solute concentration always increases continuously from the top to the bottom of the array; that is, $C_{i+1} \geq C_i$ for any two adjacent boundaries at all times during the run. If C_i has been found to be between $C_{T,n-1}$ and $C_{T,n}$, then C_{i+1} must be somewhere above $C_{T,n-1}$. In order to find the interval in the table that contains C_{i+1}, it is not necessary to return to the beginning of the table. It is, however, useful to test whether C_{i+1} is in fact greater than $C_{T,n-1}$. If it is not, the simulation has produced an inverse gradient, and the program probably contains an error. The error is reported and the simulation is terminated. Most of the common programming errors that are not detected by the compiler do produce concentration inversions, and so will be trapped by this simple test. It should be noted that the trap will usually detect only substantial inversions and will ignore small ones. This is desirable, since a certain degree of round-off error occurs during the normal operation of the program. Round-off will produce inversions in the plateau region that are inevitable and not troublesome if they are small—as they will be if the computer word size is adequate. A direct test requiring that $C_{i+1} \geq C_i$ in every case would unnecessarily terminate every simulation attempted.

Simulation of Diffusion

Simulation Procedure. The general scheme used to simulate diffusion is derived from that developed by Vink to describe concentration-independent diffusion in a system of constant cross section.[23,24] The model must be modified to deal with the concentration dependence of $\bar{D}$, the

[23] H. Vink, *Acta Chem. Scand.* **18**, 409 (1964).
[24] D. J. Cox, *Arch. Biochem. Biophys.* **112**, 249 (1965).

sector shape of the cell, and the nonuniform array of box sizes. The model calculates the flow of solute across each boundary in the array during a short time interval. Once the flow across each boundary has been found, the accumulation (or depletion) of solute in each box is calculated. Dividing by the volume of the box gives change in concentration during the time interval. A new array of concentrations is calculated, and the calculation is repeated.

The program begins by transferring arrays of $\bar{C}$ and r to the diffusion routine. The concentration C_i is calculated for each *boundary* using Eq. (4), and the corresponding value of the average diffusion coefficient $\bar{D}_i$ is extracted from the table assembled at the beginning of this program. The gradient g_i at each boundary is found using Eqs. (5) and (7). The mass of solute that flows across boundary i during the time interval at Δt_D is equal to $\bar{D}_i \Delta t_D g_i A_i$, where A_i is the area of the boundary. The area is $b \theta r_i$, where b is the vertical thickness of the centrifuge cell and θ is the sector angle. Since the geometric factors are constant and will cancel later in the calculation, it is convenient to define a flow parameter for each boundary:

$$f_i = \bar{D}_i \Delta t_D g_i r_i \qquad (33)$$

Since no solute is to cross the ends of the solution column, f_1 and f_{n+1} are set equal to zero.

The flow parameter f_i is proportional to the flow *upward* in the array, across boundary i from box i into box $i - 1$. The solute accumulated in box i is thus proportional to $f_{i+1} - f_i$. The volume of box i is $b \theta r_i (r_{i+1} - r_i)$. The change of the solute concentration in box i during the time interval Δt_D is thus

$$\Delta \bar{C}_i = \frac{f_{i+1} - f_i}{\bar{r}_i (r_{i+1} - r_i)} \qquad (34)$$

The concentration in box i is now:

$$\bar{C}_i \text{ (new)} = \bar{C}_i \text{ (old)} + \Delta \bar{C}_i \qquad (35)$$

The new values are transferred to the $\bar{C}$ array, and the procedure is repeated.

Factors Bearing on Computer Time. The total duration of the experiment is divided by the simulation into a number of equal time intervals. In order to minimize the computer time required, it would seem desirable to use the smallest possible number of successive operations by selecting a relatively large value for Δt_D. However, there is an upper limit on acceptable values of Δt_D which is closely related to the interval between boundaries in the space grid and to the diffusion coefficient of the solute. [2,23,24] The problem is most simply understood considering the situation at the beginning of a simulation in which a sharp initial solute boundary is placed at r_i. The concentration is zero in box $i - 1$ and all boxes above it; the concentration is C_0 in box i and below. During the first operation of the diffusion routine, the calculated flow across boundary i from box i into box $i - 1$ will be, according to Eq. (33), proportional to $\bar{D}_i \Delta t_D g_i r_i$. Since there are no gradients anywhere else in the array, the flows at all other boundaries will be zero. In particular, the flow across boundary $i - 1$ will be zero, and so the change in concentration in box $i - 1$ will be [Eq. (34)]

$$\Delta \bar{C}_{i-1} = \frac{\bar{D}_i \Delta t_D g_i r_i}{r_{i-1}(r_{i+1} - r_i)} \tag{36}$$

$$= \frac{\bar{D}_i \Delta t_D r_i (\bar{C}_i - \bar{C}_{i-1})}{r_{i-1}(r_{i+1} - r_i)(\bar{r}_i - \bar{r}_{i-1})} \tag{37}$$

At the beginning of the simulation, the interval between the boundaries is uniform throughout the array, and so $(r_{i+1} - r_i) = (\bar{r}_i - \bar{r}_{i-1}) = \Delta r$. In addition, before the first transfer C_{i-1} is zero, and since $r_i/\bar{r}_{i-1}$ is not far from unity,

$$\Delta C_{i-1} \approx \frac{\bar{D}_i \Delta t_D}{(\Delta r)^2} C_i = \alpha C_i \tag{38}$$

Now suppose a value of Δt_D is chosen that makes α greater than unity. Then, after the first diffusion transfer, the calculated concentration in box $i - 1$ will be greater than the original concentration in box i. That would mean that more solute is transferred from box i to box $i - 1$ than was present in box i to begin with, which is obviously nonsense. If Δt_D is large enough to make α equal to 0.5, then half of the solute in box i will be transferred to box $i - 1$ in the first diffusion operation. The gradient at boundary i will drop immediately to zero, which is also unacceptable. Evidently, it is necessary to choose a time interval that will make α appreciably less than 0.5. It has been found in practice that the diffusion routine will behave properly if Δt_D is kept small enough to ensure that α will not exceed 0.2.

As will be seen below, the operation of the sedimentation routine may move adjacent boundaries progressively closer together as the simulation proceeds. The routine places a lower limit Δr_{min} on the distance between adjacent boundaries. In a self-associating solute the largest possible diffusion coefficient is that of the monomer D_1 which is equal to the average diffusion coefficient at infinite dilution. For a given choice of Δt_D, the highest value of the α that can occur during the simulation is $D_1 \Delta t_D / \Delta r_{min}^2$. Then, since Δr_{min} and D_1 are known at the beginning of the simulation, Δt_D can be selected so that α cannot possibly exceed 0.2 during the calculation:

$$\Delta t_D \leq (0.2 \Delta r_{min}^2)/D_1 \tag{39}$$

This time interval is generally quite short. For example, if D_1 is 8×10^{-7} cm^2/sec and a 1-cm solution column is divided into 100 boxes ($\Delta r = 0.01$ cm), Δt_D will be 25 sec. In order to simulate a centrifuge run lasting 50 min, at least 120 diffusion operations will be necessary.

The powerful effect on computing time of the original choice of the boundary interval Δr can now be understood. Compare, for example, the time required for a simulation in a 200-box array ($\Delta r = 0.005$ cm) with an otherwise identical computation in a 100-box array ($\Delta r = 0.01$ cm). The most obvious effect is that twice as many boundaries and boxes will need to be treated in the 200-box case. Still more important, if Δr is halved, Δt_D must be divided by four if α is to remain within the limit given by Eq. (39). In order to treat the same total experimental time in the 200-box model, four times as many diffusion operations must be carried out on twice the number of boxes. Halving the boundary interval will entail an 8-fold increase in the computer time required for the diffusion routine.

The diffusion routine is not the only computing operation involved in a simulation, and the time required for the other parts of the program is not so strongly dependent on Δr. Nevertheless, the simulation of diffusion does account for a large part of the computing time used by the simulation, and, in practice, halving the boundary interval increases the computing time by a factor of about 5.

Simulation of Sedimentation

Preliminary Considerations. The most obvious way to simulate sedimentation would be to compute the downward flow of mass across each boundary during a series of short time intervals. The solute moving across boundary i into box i during the time interval Δt_S would be:

$$f_i' = \bar{S}_i \omega^2 r_i \Delta t_S \bar{C}_{i-1} A_i \tag{40}$$

where ω is the angular velocity. One would then compute the mass accumulated in each box and then the change in concentration during the time interval in a way analogous to that used in the diffusion routine.

This simple procedure is, however, subject to several troublesome artifacts,[16] the most important of which can be understood by referring to Fig. 3. Consider, for simplicity, an infinitely steep gradient initially placed at the boundary at r_{i-1} (Fig. 3a). Flow for a time Δt_S sufficient to empty half the solute from box $i - 1$ will produce the situation shown in Fig. 3b. Sedimentation for a second time interval equal to the first should entirely empty box $i - 1$ (Fig. 3c).

The difficulty is that the simulation program deals with the distribution of solute in terms of the *mean* concentration in each box, which it obtains by dividing the total mass in each box by the volume of that box. The

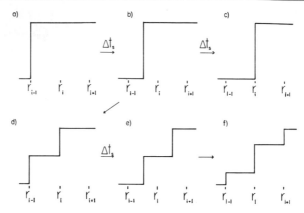

FIG. 3. Numerical dispersion produced by using Eq. (40) to simulate sedimentation of a nondiffusing boundary. See text for details.

distribution calculated by the computer will in fact resemble Fig. 3d, rather than 3b. Sedimentation for an additional time interval Δt_s will then give the result shown in Fig. 3e, which will be interpreted by the program as shown in Fig. 3f. The solute boundary, which should remain sharp in the absence of diffusion, spreads as a result of a computational artifact.

This problem, referred to as numerical dispersion, is a general one affecting finite-difference simulations of all kinds, and there are a number of ways of dealing with it. One particularly simple way involves moving the boundaries that set off the array of boxes instead of moving the solute across fixed boundaries.[11,12] Each boundary is moved at a rate equal to the local sedimentation coefficient, which depends on the local solute concentration. As a boundary moves, it does not pass and is not passed by any of the solute. The mass of solute in each box does not change. Usually, the boundaries move at different rates, since the solute concentration and the sedimentation coefficient vary from point to point in the cell. The boundaries that enclose a box will be moved closer together or farther apart during a short time interval. As a result, although the mass of solute in a box does not change, the concentration does change since the volume of the box changes. This scheme avoids the situation in which each box is partially emptied into the box below it and so eliminates the dispersion artifact.

The routine at this stage would result in moving the frame of observation downward relative to the real centrifuge cell. This is not necessarily a serious problem, but it can, in any case, be eliminated easily. Since the model is restricted to the infinite cell case, the concentration in the uppermost box must always be zero, and the sedimentation coefficient at r_1 is equal to the monomer sedimentation coefficient S_1. The time interval Δt_S can be selected at the beginning of the simulation so that the boundary at r_1 will move to the position originally occupied by boundary 2:

$$\Delta t_S = \frac{\ln(r_2/r_1)}{S_1\omega^2} \tag{41}$$

where ω is the rotor speed in radians per second. The boundaries and boxes are then reindexed: what was boundary 1 is now indexed as boundary 2, boundary 2 becomes boundary 3, and so on through the array. A new boundary 1 is inserted at the original position r_1, thus generating a new box 1 at the top of the array. A concentration of zero is assigned to the new box. The bottom box and boundary move out of the array and are discarded.

Simulation Procedure. The routine is provided with the boundary positions, r, and the mean solute concentrations in the boxes, $\overline{C}$, and begins by finding the concentration at each boundary [Eq. (4)]. The corresponding values of the sedimentation coefficient are extracted from the table. If a boundary initially at r_i moves with a local sedimentation coefficient $\overline{S}_i$ for a time interval Δt_S, it will move to a new position r_i^*:

$$r_i^* = r_i \exp(\overline{S}_i\omega^2\Delta t_S) \tag{42}$$

The old values of r_i are needed for the next step in the calculation and so must not be discarded at this point. The original volume of box i was:

$$V_i = b\theta\overline{r}_i(r_{i+1} - r_i) = [b\theta(r_{i+1}^2 - r_i^2)]/2 \tag{43}$$

The new volume of box i is $b\theta(r_{i+1}^{*2} - r_i^{*2})/2$, and, since the mass of solute in the box has not changed, the new concentration in the box is:

$$\overline{C}_i^* = \overline{C}_i[(r_{i+1}^2 - r_i^2)/(r_{i+1}^{*2} - r_i^{*2})] \tag{44}$$

The new boundary positions and concentrations are now reindexed and transferred to the r and $\overline{C}$ arrays:

$$\overline{C}_{i+1} = \overline{C}_i^* \tag{45}$$
$$r_{i+1} = r_i^* \tag{46}$$

The value of r_1 is left unchanged, and $\overline{C}_1$ is set equal to zero.

Control of Box Size. In all cases of interest, the sedimentation coefficient varies with concentration and thus differs from point to point in the centrifuge cell. During the operation of the sedimentation routine, adjacent boundaries move through different distances and thus the sizes of the various boxes change. In regions of the solute profile where the sedimentation coefficient rises with concentration, the boxes will be stretched; boxes will be compressed where the sedimentation coefficient decreases with concentration. Since a given box travels with the same general region of the solute profile throughout the simulation, particular boxes are likely to be repeatedly stretched or compressed.

The simulation scheme can tolerate considerable inhomogeneity in box size, but the compression and expansion of individual boxes cannot be allowed to continue without limit. Extreme compression of a few boxes is particularly troublesome, giving rise to serious artifacts in both the sedimentation and the diffusion routines. It is necessary, therefore, to set limits on box size and to eliminate from the array any boxes larger or smaller than the selected extremes. Noting the original box size established at the beginning of the simulation as Δr, it has been found sufficient to require that all boxes be kept larger than $0.8\Delta r$ and smaller than $2\Delta r$ throughout the calculation.

The array is inspected after each round of simulated sedimentation. When any box is found to be too large or too small, it is divided in half, and the solute is distributed between the two parts of the box (Fig. 4b). The solute is *not* equally divided; if it were, the concentration gradient would be reduced to zero at the boundary between the two halves of the box and a severe irregularity would be introduced into the simulated solute concentration profile. Instead, the solute is distributed so that the

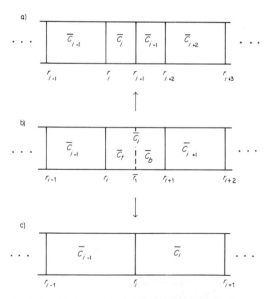

FIG. 4. Notation for box-splitting routine. (b) Solute in box i is distributed between upper and lower halves. (a) If box i is too narrow, upper and lower halves become new boxes. (c) If box i is too wide, upper and lower halves are merged with adjacent boxes.

concentration gradient between the two halves of the box is linearly interpolated between the gradients near the top and bottom boundaries of the original box. Suppose box i is to be divided by inserting a boundary at $\bar{r}_i$. The distance between the midpoints of the two halves of the box is $(r_{i+1} - r_i)/2$. If the concentrations in the top and bottom halves of the box are $\bar{C}_t$ and $\bar{C}_b$, the gradient at $\bar{r}_i$ is:

$$g_i^* = 2(\bar{C}_b - \bar{C}_t)/(r_{i+1} - r_i) \tag{47}$$

We require that g_i^* fall on a straight line between g_i', which lies at r_i' [Eqs. (5) and (6)] and g_{i+1}' at r_{i+1}':

$$g_i^* = g_i' + (g_{i+1}' - g_i')[(\bar{r}_i - r_i')(r_{i+1}' - r_i')] \tag{48}$$

Since the total mass of solute assigned to the two halves of the box must equal the mass originally present in the box,

$$\bar{C}_i(r_{i+1}^2 - r_i^2) = \bar{C}_t(r_i^2 - r_i^2) + \bar{C}_b(r_{i+1}^2 - \bar{r}_i^2) \tag{49}$$

Equations (47), (48), and (49) are solved for $\bar{C}_t$ and $\bar{C}_b$.

If the original box i was too large, the two halves of the box are now inserted separately into the array (Fig. 4a). The top half of the box becomes a new box i, the bottom half becomes box $i + 1$, and the former $\bar{r}_1$ becomes r_{i+1}. All subsequent boxes and boundaries in the array are reindexed. The former box $i + 1$ becomes the new box $i + 2$, box $i + 2$ becomes $i + 3$, and so on; the bottom box and boundary in the array are discarded. The reindexing must be done carefully; if, for example, one were to proceed as follows:

$$\begin{aligned}
\bar{C}_i &= \bar{C}_t \\
\bar{C}_{i+1} &= \bar{C}_b \\
\bar{C}_{i+2} &= \bar{C}_{i+1} \\
&\cdots\cdots \\
\bar{C}_n &= \bar{C}_{n-1}
\end{aligned} \tag{50}$$

then one would simply propagate $\bar{C}_b$ from box $i + 1$ to the bottom of the array. Either the values of $\bar{C}$ must be transferred into an intermediate array and then reindexed or the array must be reindexed from the bottom up:

$$\begin{aligned}
\bar{C}_n &= \bar{C}_{n-1} \\
&\cdots\cdots \\
\bar{C}_{i+2} &= \bar{C}_{i+1} \\
\bar{C}_{i+1} &= \bar{C}_b \\
\bar{C}_i &= \bar{C}_t
\end{aligned} \tag{51}$$

If the original box i was too small, the upper and lower halves are merged with the boxes above and below, and the contents of the new boxes are mixed (Fig. 4c). A new box $i - 1$ is formed, which includes the contents of the old box $i - 1$ and the top half of box i:

$$\bar{C}^*_{i-1} = \frac{\bar{C}_{i-1}(r_i^2 - r_{i-1}^2) + \bar{C}_t(r_i^2 - \bar{r}_i^2)}{(r_i^2 - r_{i-1}^2)} \tag{52}$$

Similarly, a new box i consists of the old box $i + 1$ and the bottom half of box i:

$$\bar{C}_i = \frac{\bar{C}_b(r_{i+1}^2 - \bar{r}_i^2) + \bar{C}_{i+1}(r_{i+2}^2 - r_{i+1}^2)}{(r_{i+2}^2 - \bar{r}_i^2)} \tag{53}$$

The old values of r_i and r_{i+1} are now eliminated. The former $\bar{r}_i$ becomes r_i and all subsequent boxes and boundaries are reindexed:

$$\begin{aligned}\bar{C}_j = \bar{C}_{j+1} \quad i + 1 < j < n - 1 \\ r_j = r_{j+1} \quad i + 1 < j < n \end{aligned} \tag{54}$$

Since the array now lacks one box, a new box is added to the bottom of the array. It is made identical in size and solute concentration with the box immediately above it.

Selection of Time Intervals

The complete simulation procedure consists of alternate applications to the concentration array of the routines that describe diffusion without sedimentation and sedimentation without diffusion. The time intervals used in each application of the sedimentation routine (Δt_S) and of the diffusion routine (Δt_D) are selected using criteria internal to the separate procedures: [Eqs. (39) and (41)]. It is evidently necessary to choose the *number* of applications of the sedimentation and diffusion routines so that the total times spent in sedimentation and in diffusion, first, are equal to each other and, second, are equal to the duration of the experiment to be simulated.

In virtually every case of interest, the sedimentation time interval defined by Eq. (41) is found to be considerably greater than the maximum acceptable diffusion time interval given by Eq. (39). That being so, it is necessary to alternate one sedimentation transfer with *several* successive applications of the diffusion routine.

In order to find n_S, the number of sedimentation transfers needed to simulate the entire experiment, one begins by using Eq. (41) to compute Δt_S. The ratio of the total duration of the experiment Δt_T and Δt_S is rounded off to the nearest integer; this integer n_S is the number of applications of the sedimentation routine. In general, $n_S \Delta t_S$ will not be precisely equal to Δt_T, but the error in the total simulated time may be acceptable. If it is not, Δt_S can be adjusted slightly. In this case, the new box generated at the top of the solution column by each round of simulated sedimentation will not be quite identical to the top box in the initial array, but this will cause no difficulty. In any case,

$$\Delta t_T = n_S \Delta t_S \tag{55}$$

where either Δt_T is a slightly erroneous value of the total experimental time or Δt_S differs slightly from the value given by Eq. (41).

Once the final value of Δt_S is in hand, Δt_D and n_D, the number of diffusion transfers per sedimentation transfer, can be calculated. The sedimentation time interval must be an integral multiple of Δt_D. One chooses the smallest integral value of n_D that will give an acceptably small value of Δt_D, according to Eq. (39). Since

$$\Delta t_D \leq (0.2 \Delta r_{min}^2)/D_1 \tag{56}$$

and

$$\Delta t_S = n_D \Delta t_D$$

the number of diffusion transfers per sedimentation transfer must be at least:

$$n_D \geq (D_1 \Delta t_S)/0.2 \Delta r_{min}^2 \tag{57}$$

The total number of diffusion transfers during the simulation is $n_D \cdot n_S$ and

$$n_S n_D \Delta t_D = n_S \Delta t_S = \Delta t_T \tag{58}$$

The complete simulation begins with n_D rounds of simulated diffusion followed by one round of simulated sedimentation. The combined calculation is repeated n_S times. At convenient intervals during the simulation, the program constructs the concentration gradient profile from arrays of g' and r' [Eqs. (5) and (6)] and prints or plots the result.

Summary Outline of the Procedure

It is worthwhile to recall at this point that the particular simulation procedure described here has certain definite limitations. First, it is applicable only to self-associating solutes that reequilibrate reasonably rapidly. Second, it cannot deal with cases in which the equilibrium constant varies from point to point in the centrifuge cell; pressure dependence is the most obvious case of this kind, but other instances are possible. Third, the solute boundary must be isolated from the ends of the solution column by regions of negligible concentration gradient at all times during the run. Keeping these limitations in mind, the essential steps of the computer simulation procedure are summarized in the following outline.

I. Preliminary procedures
   A. Read the parameters describing solute and experiment
      1. Aggregated species present, association constants (j, K_j)
      2. Sedimentation and diffusion coefficients of all species or equivalent information (Eqs. 19–22)
      3. Initial plateau concentration (C_o)
      4. Hydrodynamic concentration dependence of $\bar{S}$ $[k$, Eq. (10)]
      5. Rotor speed
   B. Define initial arrays of boundaries (r_i) and solute concentrations $(\bar{C}_i)$
      1. Select number of boxes and boundary interval (Δr)
      2. Specify meniscus position (r_1) and compute r_i array [Eq. (8)]
      3. Define initial concentration array [Eq. (8) or experimental profile]
   C. Compute time intervals and number of transfers for sedimentation and diffusion routines
      1. Read duration of experiment (Δt_T)
      2. Compute Δt_S [Eq. (41)]
      3. Compute n_S [Eq. (55)] and adjust either Δt_S or Δt_T so that $n_S \Delta t_S$-Δt_T
      4. Compute minimum acceptable value of n_D [Eq. (57)]
      5. Compute Δt_D [Eq. (58)]
   D. Table assembly
      1. Compute $\bar{S}$ and $\bar{D}$ at 100 values of C_T equally spaced between zero and C_o. For each value of C_T
         a. Find C_1 [Eq. (2), Fig. 2]
         b. Compute $\bar{S}$ [Eq. (9) or (10)] and $\bar{D}$ [Eq. (16)]
         c. Compute $\bar{S}_g$ [Eq. (31)] if no-diffusion result is needed
      2. Find maximum and minimum values of $\bar{S}$ in preliminary table
      3. Inspect preliminary table for gaps in $\bar{S}$ larger than $(\bar{S}_{max} - \bar{S}_{min})/100$
      4. Eliminate large gaps in table by inserting additional values of $\bar{S}$, $\bar{D}$, and $\bar{S}_g$ at intermediate values of C_T
II. Simulation
   A. Diffusion
      1. Find midpoint of each box $[\bar{r}_i$, Eq. (3)]
      2. Compute concentration at each *boundary* $[C_i$, Eq. (4)]
      3. Extract $\bar{D}_i$ for each boundary from table [Eq. (32)]
      4. Compute gradients midway between box centers $[g_i'$, Eq. (5)]
      5. Find points r_i' equidistant from box centers [Eq. (6)]
      6. Interpolate gradients at boundaries $[g_i$, Eq. (7)]
      7. Compute flow parameters $[f_i$, Eq. (33)]

      8. Compute $\Delta\overline{C}_i$ array [Eq. (34)]

      9. Compute adjusted concentration array [$\overline{C}_i$, Eq. (35)]

    10. Repeat steps 1–9 n_D times

  B. Sedimentation

      1. Find midpoint of each box [$\bar{r}_i$, Eq. (3)]

      2. Compute concentration at each boundary [C_i, Eq. (4)]

      3. Extract $\overline{S}_i$ for each boundary from tables [Eq. (32)]

      4. Compute new position of each boundary [r_i^*, Eq. (42)]

      5. Compute new concentration in each box [$\overline{C}_i^*$, Eq. (44)]

      6. Reindex boxes and boundaries and transfer r_i^* and $\overline{C}_i^*$ to r and $\overline{C}$ arrays [Eqs. (45) and (46)]

  C. Box splitting

      1. Inspect new r_i array for intervals larger than $2\Delta r$ or smaller than $0.8\Delta r$

      2. Split unacceptable boxes, distributing contents between top and bottom halves [Eqs. (47)–(49)]

      3. Merge and/or reindex boxes and boundaries as appropriate [Eq. (50) or Eqs. (52)–(54)]

  D. Repeat steps A, B, and C n_S times

III. Output

  A. Construct final gradient profile

      1. Compute $\bar{r}_i$ array [Eq. (3)]

      2. Compute g_i' array [Eq. (5)]

      3. Compute r_i' array [Eq. (6)]

      4. Compute $dC_T/d\overline{S}_g$ if no-diffusion solution is needed

  B. Report result

      1. Print input parameters: K_j; S_j, D_j for all species; C_0; k; r_1; rotor speed

      2. Print adjusted values of Δt_T, Δt_S, Δt_D, n_S, n_D

      3. Print tabulated values of C_T, $\overline{S}$, and $\overline{D}$

      4. Print or plot $d\overline{S}_g/dC_T$ and $\overline{S}_g$ if no-diffusion solution is needed

      5. Print or plot g_i' and r_i' arrays.

Applications and Extensions of the Simulation Procedure

The computational technique described here has been used to calculate representative sedimentation velocity profiles for a variety of classes of macromolecular solutes.[11,12,25–27] As a result, it is now possible to predict with some confidence the behavior of virtually any rapidly equilibrat-

[25] D. J. Cox, *Arch. Biochem. Biophys.* **142,** 514 (1971).

[26] D. J. Cox, *Arch. Biochem. Biophys.* **146,** 181 (1971).

[27] R. R. Holloway and D. J. Cox, *Arch. Biochem. Biophys.* **160,** 595 (1974).

ing, self-associating solute in the ultracentrifuge. The procedure has also been used successfully to mimic the behavior of at least two experimental systems, β-lactoglobulin[28] and tubulin.[29] The method has recently been extended to deal with mixtures of solutes that interact hydrodynamically but not chemically. The calculations are done by carrying two concentration arrays simultaneously and cross-referring between them as the simulation proceeds. Particular attention has been given to self-associating solutes contaminated with monomer that has been altered so as to be unable to participate in the association equilibrium.[30]

Zimmerman and Ackers have shown that the model developed for the ultracentrifuge can be adapted with little change to describe the gel permeation chromatography of self-associating solutes. The results of such simulations are particularly suitable for comparison with experimental profiles produced by the column scanning technique. The necessary modifications in the model have been described[31] and the predicted behavior of a number of rapidly equilibrating systems on various chromatographic media has been surveyed.[32,33] Zimmerman and Ackers have also used the simulation technique to consider the effect on chromatographic boundaries and zones of continuous variation in the packing of the gel matrix along the column. This complicating factor is a normal and unavoidable experimental feature of gel permeation chromatography, particularly with very porous gels.[34]

More recently, Zimmerman has very substantially enlarged the capabilities of the simulation technique as it is applied to gel permeation chromatography. He has described models that are capable of predicting the behavior of slowly equilibrating self-associating solutes by nesting a chemical relaxation routine within each iteration of the sedimentation–diffusion loop.[13,35] The procedure is applicable to the entire range of reaction rates from infinitely rapid to infinitely slow. In addition, Brown and Zimmerman have developed a technique for simulating the chromatography of active enzymes, monitoring the changing distribution in the column of the enzyme, substrate, and products as may be done experimentally with the column scanning system.[36] These recent models have

[28] L. M. Gilbert and G. A. Gilbert, this series Vol. 27, p. 273.
[29] S. N. Timasheff, R. P. Frigon, and J. C. Lee, *Fed. Proc., Fed. Am. Soc. Exp. Biol.* **35,** 1886 (1976).
[30] D. J. Cox, unpublished
[31] J. K. Zimmerman and G. K. Ackers, *J. Biol. Chem.* **246,** 1078 (1971).
[32] J. K. Zimmerman and G. K. Ackers, *J. Biol. Chem.* **246,** 7298 (1971).
[33] J. K. Zimmerman, D. J. Cox, and G. K. Ackers, *J. Biol. Chem.* **246,** 4242 (1971).
[34] J. K. Zimmerman and G. K. Ackers, *Anal. Biochem.* **57,** 578 (1974).
[35] J. K. Zimmerman, *Biophys. Chem.* **3,** 339 (1975).
[36] B. B. Brown and J. K. Zimmerman, *Biophys. Chem.* **5,** 351 (1976).

not yet been used to simulate the analogous experiments in the ultracentrifuge, but this extension of the work should be straightforward.

Alternative Models

It must be reemphasized that the model described here in detail is a representative example of a group of related simulation techniques that have appeared in the literature during the past 15 years. The model was designed to deal with the specific problem of rapidly equilibrating self-associating solutes, and, for these particular systems, it is probably the best compromise available between precision and computing efficiency. There are, however, a number of significant problems that lie beyond the scope of the technique outlined above. These cases include pressure-dependent equilibria in the centrifuge,[18-20] ligand-mediated association equilibria and other association reactions among unlike species $(nA + mB \rightleftharpoons A_nB_m)$, and approach to sedimentation equilibrium and other experiments in which solute concentration gradients at the ends of the solution column are not negligible. To simulate these cases, it is necessary to select among the several alternative simulation procedures available in the literature. These procedures fall into three general classes, and they differ in significant ways from each other and from the method considered above.

The method of Bethune and Kegeles was the first technique to incorporate the effect of diffusion into descriptions of the transport behavior of associating systems. The procedure is thoroughly described by Bethune and Kegeles[6] and certain important improvements are given by Bethune.[7] The computation is based on an ingenious analogy between countercurrent distribution and continuous transport processes, such as sedimentation and chromatography. In a countercurrent distribution apparatus, both the movement of a component through the array of tubes and the dispersion of the component within the array depend on a single variable, the partition coefficient. By appropriate scaling of time and position, the movement and spreading can be interpreted as sedimentation and diffusion coefficients, and so S and D are collapsed into a single parameter. The ratio of S and D, and so the solute molecular weight, can be given whatever value may be appropriate. For two solutes running simultaneously, the partition coefficients can be selected to give any desired ratio between the molecular weights of the two components. That is, not only are S_1/D_1 and S_2/D_2 correctly specified, but also S_2/S_1 and D_2/D_1 are correct; this last feature is the principal improvement described by Bethune.[7] For three components, all three sedimentation coefficients and the largest and smallest diffusion coefficients can be adjusted independently to give whatever ratios may be appropriate for the molecular weights of the solutes. The diffusion coefficient of the intermediate species is fixed by the values given to the other two components; however, the value specified in this way is usually close to the correct one.

The transport of an associating solute is described by alternate rounds of "sedimentation/diffusion" and chemical reaction—reequilibration for rapidly reacting solutes and incomplete relaxation for slow reactions. Pressure-dependent sedimentation can be dealt with by assigning different values of the association constant to the various "tubes" in the array which represent different depths in the solution column in an ultracentrifuge cell.

The countercurrent analog has been used to describe the transport of rapidly equilibrating self-associating solutes of various degrees of complexity[6,7,37,38] and for slowly relaxing monomer–dimer systems.[39] It has been particularly important in studies of associating solutes with pressure-dependent equilibrium constants, a number of which have been examined by Kegeles and his colleagues.[40–42]

The method of Cann and Goad is a powerful simulation technique that has been particularly useful in describing the transport behavior of interacting systems in which the association equilibria are influenced by low molecular weight ligands. The method was originally developed to deal with the electrophoresis of solutes that interact with a component of the electrophoresis buffer.[8] The model that has evolved from the procedure originally described is most completely described in the chapter by Goad in "Interacting Macromolecules."[2] In its general structure, the procedure of Cann and Goad resembles the technique outlined above. The solution column is divided into segments by a series of boundaries, and the concentration profile is carried as an array of mean concentrations in the segments. The simulation of diffusion is done by a finite difference model similar to the one described here. However, sedimentation is handled quite differently; solute is said to be driven across each boundary from each box into the one below using an expression similar to Eq. (40). The numerical dispersion problem (see Fig. 3) is handled by replacing $\bar{C}_{i-1}$ in Eq. (40) by an interpolated concentration at boundary i, C_i in the notation of Fig. 1. The interpolated concentration is obtained from the local concentration profile by fitting the mean concentrations in three boxes to a Taylor expansion truncated after the second derivative. The expression

[37] J. L. Bethune and P. J. Grillo, *Biochemistry* 6, 796 (1967).
[38] B. J. McNeil, L. W. Nichol, and J. L. Bethune, *J. Phys. Chem.* 74, 3846 (1970).
[39] D. F. Oberhauser, J. L. Bethune, and G. Kegeles, *Biochemistry* 4, 1878 (1965).
[40] G. Kegeles, L. Rhodes, and J. L. Bethune, *Proc. Natl. Acad. Sci. U.S.A.* 58, 45 (1967).
[41] G. Kegeles and M. L. Johnson, *Arch. Biochem. Biophys.* 141, 59 (1970).
[42] G. Kegeles and M. L. Johnson, *Arch. Biochem. Biophys.* 141, 63 (1970).

analogous to Eq. (40) uses the concentration at boundary i at the beginning of each sedimentation transfer, as though C_i remained constant throughout the finite time interval used for each step of the simulation. In fact, the concentration at each boundary changes during each time interval as the solute profile slides downward through the array of boundary positions. This problem is managed by two devices. First, the sedimentation transfer times are kept quite short. Instead of alternating several short diffusion steps with one long sedimentation transfer, the same short time interval is used for both sedimentation and diffusion. The combined flows due to the two processes can thus be computed simultaneously using a single expression. In addition, the entire boundary array is moved downward at a constant rate given by the velocity of the least rapidly sedimenting macromolecular species. Since the boundaries do not move with the local average sedimentation as they do in the method outlined above, the array of boxes is not distorted, and sedimentation flows between boxes are reduced but not eliminated. However, since the points in the boundary array move generally with the solute profile, the time variation of the solute concentration at each point is minimized.

For chemically interacting solutes, a new equilibrium distribution in every box is calculated after each round of simulated sedimentation and diffusion. Efficient procedures for this computation have been described by Goad.[2]

The Cann and Goad method can be used for self-associating solutes, but it is probably more demanding of computer time for these systems than the distorted grid model. However, the more general problem of ligand-mediated association equilibria is not easily handled by the latter procedure, and for these cases, the technique of Cann and Goad is the method of choice. The method has been applied to a wide variety of transport processes involving electrophoresis complicated by interaction with buffer components,[8,43] boundary[9] and zone[44] sedimentation of solutes undergoing rapid ligand-mediated association, and the behavior of ligand-mediated dimerizing solutes that reequilibrate slowly during boundary sedimentation.[14] The procedure has quite recently been extended to simulate Hummel–Dreyer chromatographic experiments[45] and the approach to isoelectric focusing equilibrium of solutes that interact with components of the ampholyte system.[46]

The method of Dishon, Weiss, and Yphantis is a sophisticated numerical solution of the flow equation for the ultracentrifuge. The most complete

[43] J. R. Cann and W. B. Goad, *J. Biol. Chem.* **240**, 1162 (1965).
[44] J. R. Cann, *Biophys. Chem.* **1**, 1 (1973).
[45] J. R. Cann and N. D. Hinman, *Biochemistry* **15**, 4614 (1976).
[46] D. I. Stimpson and J. R. Cann, *Biophys. Chem.* **7**, 115 (1977).

description of the procedure is given by Dishon et al.[10] In many respects, the method resembles the technique of Cann and Goad. The centrifuge cell is divided into segments and the initial solute distribution is defined by assigning values to the mean concentration in each segment. The calculation proceeds by computing the combined flow due to sedimentation and diffusion between adjacent segments during a succession of short time intervals. The principal difference between the two procedures is that the space grid remains stationary in the simulation of Dishon et al. and is not moved as it is by Cann and Goad. In consequence, the concentration at each grid point changes appreciably during the time interval used for the calculation. The resulting numerical dispersion problem is managed by an iteration procedure that has the effect of using the average over each time interval of the solute concentration at each grid point. This is the kind of approach described by Goad[2] in his discussion of "implicit" equations for the flow process. The advantages of the technique are, first, that the convergence of the calculated flows to stable values is rigorously tested at each stage of the calculation and, second, that the stationary space grid makes it particularly convenient to describe the course of events at the ends of the solution column. The most significant disadvantage of the procedure is that it requires substantial computer time. When the sedimenting boundary is sharp early in the experiment, the grid points must be closely spaced and short time intervals must be used. Several devices are used to reduce computer time as much as possible. First, the use of a stationary grid makes it possible to simplify the flow equations considerably by combining the boundary and box position for each segment into two constants that are carried without recalculation through the simulation. Second, since a sedimenting boundary normally becomes less steep during a centrifuge run, the time intervals, which must be quite short to begin with, are gradually lengthened as the computation proceeds. Finally, a variable grid spacing is used, closely spaced where the concentration rises steeply and more widely spaced where the concentration varies slowly. This last device is also suggested by Goad.[2]

The method of Dishon et al. is the only one thus far applied to problems involving the approach to sedimentation equilibrium. It has been used to consider the effect of concentration-dependent sedimentation on the time required to reach equilibrium[47] and the effect of continuous rotor slowing, in a magnetically suspended centrifuge rotor, on the steady state solute distribution.[48,49] The technique has also been used to describe the

[47] M. Dishon, G. H. Weiss, and D. A. Yphantis, *Biopolymers* **4**, 457 (1966).
[48] I. H. Billick, M. Dishon, M. Schulz, G. H. Weiss, and D. A. Yphantis, *Proc. Natl. Acad. Sci. U.S.A.* **56**, 399 (1966).
[49] I. H. Billick, M. Dishon, G. H. Weiss, and D. A. Yphantis, *Biopolymers* **5**, 1021 (1967).

approach to sedimentation equilibrium in a density gradient.[50] This last problem has also been considered by Sartory *et al.*[51] using a method that combines simulation of the development of the density gradient by the method of Dishon *et al.* with an analytical treatment of the migration of the macromolecular solute. The method has also been applied to several problems in sedimentation velocity: concentration-dependent sedimentation,[52] pressure-dependent sedimentation,[53] and sedimentation of concentration-dependent mixtures.[54]

All the simulation procedures described above have undergone continuous development, and most problems of interest can now be approached by more than one method. The choice among methods is often a matter of convenience and familiarity. There seem to be no instances in which the same problem has been dealt with by different models in the same computer, and so comparisons of computing efficiency and precision are no more than plausible estimates. However, some tentative recommendations can be made for broad classes of problems. For the concentration-dependent sedimentation of a single component or the sedimentation of a self-associating solute in rapid equilibrium, the distorted-grid model is accurate and appears to have the advantage in computing efficiency. For macromolecular solutes that interact with low-molecular weight ligands and especially for ligand-mediated association reactions, the procedure of Cann and Goad is the method of choice. The ligand-mediated association is a special case of interaction among unlike components ($nA + mB \leftrightarrows A_n B_m$) where B may be either a ligand or a second macromolecular component. For this reason, the method of Cann and Goad is probably the most suitable choice for mixed interactions of all kinds, although less has been done with association reactions between unlike subunits. The method of Dishon *et al.* is clearly preferable for problems involving the ends of the solution column: either early times in velocity sedimentation or approach to sedimentation equilibrium. The technique can also be used for pressure-dependent systems; indeed, this problem appears quite difficult in a moving coordinate system. The pressure-dependence problem can also be dealt with by the countercurrent analog, probably in less computer time. It is not clear whether the two procedures will give results of equal precision. Mixtures of solutes that interact hydrodynamically but not chemically have been described by

[50] M. Dishon, G. H. Weiss, and D. A. Yphantis, *Biopolymers* **10**, 2095 (1971).

[51] W. K. Sartory, H. B. Halsall, and J. P. Breillatt, *Biophys. Chem.* **5**, 107 (1976).

[52] M. Dishon, G. H. Weiss, and D. A. Yphantis, *Biopolymers* **5**, 697 (1967).

[53] M. L. Johnson, D. A. Yphantis, and G. H. Weiss, *Biopolymers* **12**, 2477 (1973).

[54] J. J. Correia, M. L. Johnson, G. H. Weiss, and D. A. Yphantis, *Biophys. Chem.* **5**, 255 (1976).

the distorted-grid model and by the numerical solution of Dishon *et al.* It is uncertain at the moment which of the two approaches is more advantageous in precision or efficiency. All four procedures can be modified to deal with slowly reacting solutes; here again, the choice among procedures is not clear-cut.

While the selection of the best method for a particular problem is not always obvious, it is clear that one method or another can be found to simulate nearly any transport experiment. It is to be expected that direct comparisons between simulation and experiment will be an increasingly useful procedure in defining interacting systems of every kind.

Acknowledgments

I wish to thank Dr. John R. Cann for informing me of new developments in his work before publication. This work was supported by the National Institutes of Health, project Grant No. GM-22243, by the Clayton Foundation Biochemical Institute, University of Texas, and by the Kansas Agricultural Experiment Station. This is contribution 192-B from the Department of Biochemistry, Kansas Agricultural Experiment Station.

[11] Measurements of Protein Interactions Mediated by Small Molecules Using Sedimentation Velocity[1]

By JOHN R. CANN

Previously (this series, Vol. 27 [12]) we described how ligand-mediated, macromolecular association–dissociation reactions can give rise to velocity sedimentation patterns (either moving-boundary or zonal) showing two well resolved peaks of large molecule despite instantaneous establishment of equilibrium. Of particular interest is the ligand-mediated dimerization reaction, $2M + nX \rightleftharpoons M_2X_n$, in which a macromolecule, M, associates into a dimer with the mediation of a small ligand molecule or ion, X, of which a fixed number, n, are bound into the complex, M_2X_n. Resolution of the reaction boundary or the reaction zone into two peaks is dependent upon generation of stable concentration gradients of unbound ligand across the sedimenting boundary or zone by reequilibration during

[1] Supported in part by Research Grant 5R01 HL13909-24 from the National Heart and Lung Institute, National Institutes of Health, United States Public Health Service. This publication is No. 652 from the Department of Biophysics and Genetics, University of Colorado Medical Center, Denver, Colorado 80262.

differential transport of M and M_2X_n. In general, the areas under the two peaks do not faithfully reflect the initial equilibrium composition. Nor do the peaks migrate with velocities per unit centrifugal field commensurate with the sedimentation coefficients of M and M_2X_n. Moreover, these quantities cannot be determined by extrapolation of the velocities of the peaks per unit field to infinite dilution of macromolecule at constant ligand concentration. The conditions for such sedimentation behavior require a sufficiently strong interaction of ligand with macromolecule to generate the gradients of ligand upon which resolution is contingent. These conditions are satisfied by two systems that provide experimental verification of theoretical predictions—namely, the reversible dimerization of New England lobster hemocyanin mediated by the binding of Ca^{2+} and H^+[2-4] and the dimerization of tubulin through the mediation of vinblastine.[5,6]

The foregoing theoretical predictions are for the limit in which chemical equilibration is very rapid, but the results of a recent investigation[7] admit the generalization that conclusions concerning the sedimentation behavior of ligand-mediated interactions in the limit of instantaneous establishment of equilibrium evidently are valid for kinetically controlled interactions characterized by half-times of reaction as long as 20–60 sec. Also of considerable practical importance is the finding that for half-time of dissociation of dimer less than about 200 sec, resolution of the reaction boundary into two peaks can occur only if dimerization is ligand-mediated (or presumably, ligand-facilitated; see below) as contrasted with the simple dimerization reaction, $2M \rightleftharpoons M_2$.

Although we have focused attention on macromolecular dimerization, higher-order ligand-mediated association (and dissociation[8]) reactions can also give well resolved bimodal reaction boundaries.[9,10] As to zone sedimentation it is now known[11] that, with the exception of association reactions highly cooperative in both macromolecule and ligand (e.g., $4M + 4X \rightleftharpoons$

[2] K. Morimoto and G. Kegeles, Arch. Biochem. Biophys. 142, 247 (1971).

[3] G. Kegeles and M. S. Tai, Biophys. Chem. 1, 46 (1973).

[4] M.-S. Tai and G. Kegeles, Biophys. Chem. 3, 307 (1975).

[5] R. C. Weisenberg and S. N. Timasheff, Biochemistry 9, 4110 (1970).

[6] J. C. Lee, D. Harrison, and S. N. Timasheff, J. Biol. Chem. 250, 9276 (1975).

[7] J. R. Cann and G. Kegeles, Biochemistry 13, 1868 (1974).

[8] As in the case of ligand-induced association, both ligand-mediated dissociation of a macromolecule into its hydrodynamically identical subunits ($M_4 + 4X \rightleftharpoons 4MX$) and ligand-facilitated dissociation (the set of sequential reactions $M_2 \rightleftharpoons 2M$, $M + X \rightleftharpoons MX$) can give sedimentation patterns showing well resolved, bimodal boundaries[9] or zones.[10,11]

[9] J. R. Cann, "Interacting Macromolecules," Figs. 88 and 89 on pp. 203–204 and text on p. 204. Academic Press, New York, 1970.

[10] J. R. Cann and W. B. Goad, Science 170, 441 (1970).

[11] J. R. Cann, Biophys. Chem. 1, 1 (1973).

M_4X_4 as contrasted to the sequential reaction $M + X \rightleftharpoons MX$, $4MX \rightleftharpoons M_4X_4$), rapidly equilibrating ligand-mediated interactions in general have the potentiality for showing bimodal zones irrespective of reaction mechanism. This generalization is subject to the provisos that overall ligand-binding is sufficiently strong to generate large gradients of unbound ligand along the sedimentation column, but that binding to the reactant itself is not so strong that in effect the system behaves like the analogous nonmediated interaction.

For weak interactions that require a large excess of ligand over macromolecule for the reaction, the concentration of unbound ligand along the centrifuge cell cannot be significantly perturbed by reequilibration during differential transport of the macromolecular species; and the system (e.g., $mM + nX \rightleftharpoons M_mX_n$) effectively collapses to the nonmediated, self-association reaction, $mM \rightleftharpoons M_m$. An example is the Mg^{2+}-mediated association of tubulin.[12,13] In this limit the moving-boundary sedimentation behavior is interpretable in terms of the Gilbert theory[14,15] (see also this series, Vol. 27 [11]) which, in contradistinction to the theory of relatively strong ligand-mediated interactions, predicts a unimodal reaction boundary for dimerization. In the case of higher-order polymerization ($m \geqslant 3$), the shape of the reaction boundary is dependent upon the concentration of macromolecule—unimodal at low concentrations and bimodal above a characteristic concentration determined by the value of m and the apparent equilibrium constant. An instructive elaboration of the classical Gilbert theory to include weak progressive self-association terminated by the highly favored formation of an m-mer has been applied by Frigon and Timasheff[12] to the forementioned Mg^{2+}-tubulin system.

It is anticipated that the new insights provided by these theoretical and experimental investigations will find application to a variety of biochemical reactions such as protein–drug interactions and the interaction of enzymes with cofactors and allosteric affectors. They also have important implications for the many analytical and preparative applications of velocity sedimentation in biochemistry and molecular biology. From this point of view the most provocative of the new insights is that an inherently homogeneous macromolecule can give sedimentation patterns showing two peaks due to ligand-mediated association-dissociation even when equilibration is rapid. Moreover, the peaks may be quite well resolved and could easily be misinterpreted as indicative of heterogeneity. Clearly, unambiguous interpretation of the sedimentation patterns depends upon

[12] R. P. Frigon and S. N. Timasheff, *Biochemistry* **14**, 4559 (1975).
[13] R. P. Frigon and S. N. Timasheff, *Biochemistry* **14**, 4567 (1975).
[14] G. A. Gilbert, *Discuss. Faraday Soc.* **20**, 68 (1955).
[15] G. A. Gilbert, *Proc. R. Soc. London, Ser. A* **250**, 377 (1959).

fractionation experiments. The following protocol incorporates the several precautions that must be taken (this series, Vol. 25 [11] and Vol. 27 [12]) to ensure the validity of the fractionation test for distinguishing between ligand-mediated interaction and inherent heterogeneity: Relatively concentrated solutions of protein are subject to ultracentrifugation several times. In the case of zone sedimentation in the preparative ultracentrifuge, several samples of protein from each peak are collected and pooled. For moving-boundary sedimentation in the analytical instrument a Yphantis–Waugh partition cell[16] is used to collect several samples of the slower moving peak, which are pooled. Each fraction is then reconcentrated by ultrafiltration; reequilibrated against buffer by rapid dialysis or by passage through a Sephadex G-25 column equilibrated with buffer; and analyzed in the ultracentrifuge. A comparison experiment is made with the unfractionated material at the same concentration. For interactions the fractions will behave like the unfractionated material and show two peaks, whereas for heterogeneity a single peak will be obtained. In practice this interpretation of the fractionation test is subject to the provisos that the protein preparation does not contain a significant amount of a noninteracting contaminant and that the reacting protein system itself is not microheterogeneous with respect to the conditions required for the interaction. If either of these latter circumstances should exist, even a rapidly interacting system may behave deceptively like a noninteracting, heterogeneous system in the sense that each fraction will show a single peak, as is the case for *Helix pomatia* α-hemocyanin at alkaline pH[17] or high salt concentration,[18] to be discussed in detail in this volume [12].

Returning to our discussion of ligand-mediated interactions, we note that when a ligand-mediated interaction is recognized as such, the fractionation test also serves to establish its reversibility. An independent method is then used to obtain information on the rates of reaction. For association–dissociation reactions, the rate of dissociation can be measured by means of simple dilution experiments monitored by light scattering as a function of time after dilution.[2] If the interaction happens to be pressure sensitive (this series Vol. 27 [13]), the rate of response of the shape of the sedimentation pattern to changes in rotor speed yields information on both the forward and reverse reaction rates.[2] For sufficiently slow equilibration, analysis of the sedimentation patterns is straightforward since the peaks correspond to separated monomer and polymer; and conventional analysis of the patterns may be used in conjunction with molecular weight determinations by the Archibald method or by light

[16] D. A. Yphantis and D. F. Waugh, *J. Phys. Chem.* **60**, 630 (1956).

[17] R. J. Siezen and R. van Driel, *Biochim. Biophys. Acta* **295**, 131 (1973).

[18] Y. Engelborghs and R. Lontie, *J. Mol. Biol.* **77**, 577 (1973).

scattering to determine thermodynamic parameters as function of ligand concentration. In the case of rapid equilibration, however, the patterns are not amenable to classical interpretation because the peaks cannot be placed into correspondence with individual reactant and product; and an entirely different method of analysis is required. Although there are several possible approaches depending upon the particular system and the individual style and tastes of the investigator, successful prosecution requires that cognizance be taken of the fundamental fact that generation of bimodal reaction boundaries and zones is not unique for any one reaction mechanism. Accordingly, it is imperative that appeal be made to the combined application of sedimentation with one or more other physical methods in order to elucidate the mechanism of reaction. The studies on the ligand-mediated association of hemocyanin and tubulin referred to above are exemplary.

One approach[5,6,12,13] combines comprehensive velocity sedimentation experiments with electron microscopy for visualization of the end product of self-association of the protein and with ligand-binding measurements. Where possible, a fitting prelude is the measurement of the gradients of ligand along the centrifuge column upon which resolution is dependent for sufficiently strong interactions. It is gratifying that, in the case of the vinblastine-tubulin system, these gradients have been measured[6] in the analytical ultracentrifuge using the photoelectric scanning absorption optical system (this series, Vol. 27 [1]). The fact that the change in concentration of vinblastine across the sedimenting boundary of tubulin compares favorably with theoretical prediction lends confidence in the interpretation of the bimodal patterns in terms of a rapidly equilibrating vinblastine-mediated dimerization of the protein. Ligand-binding measurements using two unrelated methods[6] indicate that two vinblastine molecules are involved in the reaction.

Quantitative interpretation of the patterns now hinges upon determination of the sedimentation coefficients of monomer and polymer. The sedimentation coefficient of the monomer is determined in a straightforward manner by sedimentation of the protein in the absence of ligand. Determination of the sedimentation coefficient of the dimer presents a more difficult problem. For strong interactions such as the Ca^{2+}- and H^+-mediated dimerization of hemocyanin, it can be determined by sedimentation in the presence of sufficiently high concentration of ligand to drive the reaction to completion, provided, of course, that further association into higher-order polymers does not ensue. For sufficiently weak interaction of ligand with protein resulting in collapse of the system to the Glibert limit, $m M \rightleftharpoons M_m$, even an indirect estimation of the sedimentation coefficient of M_m is problematical if m is small.[19] If, however, the stoichiometry of the reaction is high ($m > 10$) and the association constant large, the sedimentation of the fast peak in the reaction boundary may be regarded as approximating the sedimentation of M_m, at least at high protein concentration.[20] In that event, an estimate of the apparent sedimentation coefficient of M_m at infinite dilution and the hydrodynamic dependence of the apparent sedimentation coefficient on concentration can be obtained by conventional interpretation of a plot of the velocity of the fast peak per unit field against protein concentration sufficiently high that the velocity decreases linearly with increasing concentration. When applied to the Mg^{2+}-tubulin system this procedure indicated an end product of association hydrodynamically equivalent to a closed ring structure observed in the electron microscope at identical conditions.[12]

The sedimentation coefficients of monomer and polymer having been determined, there is in all likelihood already enough information at hand that qualitative reading of the sedimentation patterns will suggest a tentative model for association. Given such a model it is now possible to derive the stoichiometry and equilibrium constant(s) of reaction by analysis of the weight-average sedimentation coefficient as a function of protein concentration at constant ligand concentration. This is accomplished by nonlinear least-squares fitting of the experimental data to the theoretical equation which gives the weight average sedimentation coefficient as a function of protein concentration, sedimentation properties of the monomer and polymer, and the several parameters of the model[12] (this series, Vol. 27 [11]). The equilibrium constant(s) and other physical quantities are, in turn, used for computer simulation of the sedimentation patterns[12,21] (this series, Vol. 27 [11]). If the overall agreement between experimental and calculated patterns point to the validity of the general features of the model, the way is open for thermodynamic characterization of the interaction by analysis of weight-average sedimentation coefficients determined as a function of protein concentration, ligand concentration, and temperature. This is the procedure used to characterize the Mg^{2+}-tubulin interaction[12,13] and in principle is also applicable to strong interaction. While the approach is elegant, it is subject to a degree of uncertainty introduced by simplifying mechanistic assumptions in the models used to interpret sedimentation behavior and to potential complications inherent in the use

[19] J. R. Cann, "Interacting Macromolecules," Chapter 3. Academic Press, New York, 1970.
[20] Ibid, Chapter 4.
[21] J. R. Cann and W. B. Goad, Arch. Biochem. Biophys. 153, 603 (1972).

of sedimentation velocity for the study of self-association, e.g., pressure dependence of equilibrium constants and kinetics of reaction.

An alternative approach which has been applied to the hemocyanin system[2-4] circumvents these complications by focusing on direct molecular weight determinations and relaxation kinetics as detailed by Kegeles and Cann in this volume [12]. Briefly, the low-speed Archibald method is employed to obtain apparent molecular weights at the air–liquid meniscus as a function of protein and ligand concentrations. Analysis of these data yields the overall stoichiometry of reaction and the equilibrium constant at 1 atm pressure corrected for thermodynamic nonideality. This is valuable information, but it is a fundamental principle that thermodynamics per se is not revealing of mechanism. Thus, for example, Archibald molecular weight measurements are inherently incapable of distinguishing between ligand-mediated and ligand-facilitated dimerization of a protein. Whereas in the ligand-mediated mechanism obligatory binding of ligand molecules to macromonomer must precede dimerization, in the ligand-facilitated mechanism two macromonomer molecules associate to a dimer, which is then stabilized by the binding of ligand. In contrast to equilibrium measurements, relaxation kinetic studies are capable of making some distinctions between such mechanisms. Toward this end, Tai and Kegeles[4] have applied relaxation kinetics to the Ca^{2+}-hemocyanin system at pH 9.6. Concentration-jump dilution relaxation experiments in a stopped-flow apparatus were monitored by light scattering. The results rule out Ca^{2+}-facilitated dimerization and a "compromise" mechanism in which Ca^{2+} binds to both macromonomer and dimer in several stages of the overall process. On the other hand, they are in accord with the predictions of Ca^{2+}-mediated dimerization.

[12] Kinetically Controlled Mass Transport of Associating– Dissociating Macromolecules[1]

By GERSON KEGELES and JOHN R. CANN

The preceding chapter is concerned mainly with the sedimentation behavior of ligand-mediated, macromolecular associating–dissociating systems when equilibration is so rapid as to be considered instantaneous

[1] From the Section of Biochemistry and Biophysics U-125, The Biological Sciences Group, University of Connecticut, Storrs, Connecticut 06268 and the Department of Biophysics and Genetics, University of Colorado Medical Center, Denver, Colorado 80262. Supported in part by Research Grants NSF BMS75-07640 and NIH 5R01 HL13909-25. This publication is No. 679 from the Department of Biophysics and Genetics, University of Colorado Medical Center, Denver, Colorado 80262.

as far as mass transport is concerned. It is emphasized that accurate interpretation of the sedimentation patterns in terms of reaction mechanism requires the combined application of sedimentation with one or more other physicochemical methods. One such method is the relaxation technique for measuring the rates of rapid reactions, and mention is made of the use of relaxation kinetics to distinguish between possible mechanisms for the dimerization of New England lobster hemocyanin induced by the binding of Ca^{2+}. In the present chapter we develop the subject of kinetically controlled association–dissociation reactions in a format which first considers theoretical predictions as to the shape of the sedimentation pattern and then turns to the combined application of velocity sedimentation, thermodynamic measurements, and relaxation kinetics to the characterization of the association–dissociation of hemocyanins from different species.

Theory of Sedimentation for Kinetically Controlled Association–Dissociation

Irreversible Reactions. The simplest class of kinetically controlled macromolecular interactions includes irreversible dimerization

$$2 M \xrightarrow{k} M_2 \tag{1}$$

and irreversible dissociation of an m-mer into its hydrodynamically identical subunits

$$M_m \xrightarrow{k} mM \tag{2}$$

These reactions are taken to occur either in the absence of ligand-mediation or under conditions such that the system effectively behaves like a nonmediated interaction. Interest in the theory of mass transport for such irreversible interactions was aroused by discussions with colleagues concerning the electrophoresis, sedimentation, and gel-permeation chromatography of enzyme systems and other reacting macromolecules. In essence, the most frequently posed question was: "Will the mass-transport pattern of a macromolecule undergoing slow and irreversible reaction during the time course of the experiment show two peaks or a single, skewed peak with a long trailing or leading edge?" In order to answer this question, both moving-boundary and zone patterns were calculated for Reactions (1) and (2).[1a,2] The moving-boundary patterns apply

[1a] J. R. Cann and D. C. Oates, *Biochemistry* 12, 1112 (1973).
[2] J. R. Cann, *in* "Methods of Protein Separation" (N. Catsimpoolas, ed.), Vol. 1, Chapter 1, Fig. 8. Plenum, New York, 1975.

to analytical velocity sedimentation, while the results for the zonal mode of transport are valid for sedimentation through a preformed density gradient and apply qualitatively to gel-permeation chromatography.

As illustrated in Fig. 1 both moving-boundary and zone patterns typically show two well-resolved peaks for half-times of reaction ranging from about 0.3 to 2.5 times the duration of the transport experiment, depending upon the difference in velocity between product and reactant. The greater the difference in velocity, the better is the resolution for longer half-times. There is no indication of single skewed peaks with long trailing or leading edges as long as the reaction does not go to completion during the course of the experiment. It is characteristic of the bimodal patterns that the peak corresponding to a mixture of reactant and some product is sharp, whereas the one corresponding to product is broad and skewed in the direction from whence it was formed. The product peak grows with time of transport at the expense of the peak which contains the reactant, and the pattern never reaches the base line between the peaks. In other words, the two peaks constitute a single reaction boundary or zone. Nevertheless, rate constants for both dimerization and dissociation into subunits can be derived with reasonable accuracy (low by only 5–10%) from the rate of change of the area of the monomer peak in analytical sedimentation patterns recorded sequentially during the course of a single experiment. On the other hand, the zone patterns are not so readily interpreted. Thus, the rate constant for dimerization estimated from the amount of material in the monomer peak is subject to an unacceptable error; although in the case of dissociation into subunits, the rate constant estimated in this way is of sufficient accuracy (low by about 25%) to be useful in pilot experiments.

Inspection of the patterns indicates that, in practice, at least some of them could easily be misinterpreted in terms of heterogeneity. At the same time their general shape is virtually the same as shown by other classes of interaction; e.g., rapidly equilibrating ligand-mediated association–dissociation.[3-6] (see this series Vol. 27 [12]), kinetically controlled reversible dimerization[6] and the dissociation of certain macromolecular complexes.[1a,7] In general, fractionation experiments executed as described in this volume [11] provide an unambiguous method for distinguishing between interaction and heterogeneity (see [11] and the material in the present chapter on *Helix pomatia* hemocyanin, however; for peculiar circumstances under which the results of the fractionation

[3] J. R. Cann, "Interacting Macromolecules," p. 196. Academic Press, New York, 1970.
[4] J. R. Cann and W. B. Goad, *Arch. Biochem. Biophys.* **153,** 603 (1972).
[5] J. R. Cann, *Biophys. Chem.* **1,** 1 (1973).
[6] J. R. Cann and G. Kegeles, *Biochemistry* **13,** 1868 (1974).
[7] J. L. Bethune and G. Kegeles, *J. Phys. Chem.* **65,** 1755 (1961).

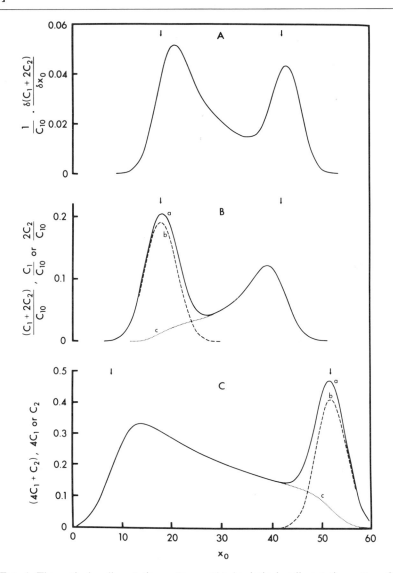

Fig. 1. Theoretical sedimentation patterns. (A) Analytical sedimentation pattern for irreversible dimerization [Reaction (1)]; sedimentation is from right to left. (B) Zone sedimentation pattern for irreversible dimerization [Reaction (1)]; sedimentation is from left to right: curve a, total concentration $(C_1 + 2C_2)/C_{10}$; curve b, concentration of monomer C_1/C_{10}; curve c, concentration of dimer $2C_2/C_{10}$. (C) Zone sedimentation pattern for irreversible dissociation of a tetramer into its identical subunits [Reaction (2) with $m = 4$]; sedimentation is from left to right: curve a, total concentration $4C_1 + C_2$; curve b, concentration of tetramer $4C_1$; curve c, concentration of subunit C_2. In each case the macromolecule exists entirely as reactant at the start of sedimentation. Reprinted with permission from J. R. Cann and D. C. Oates, *Biochemistry* **12**, 1112 (1973). Copyright by the American Chemical Society.

test may be deceptive), but not between different classes of interaction. Thus, in the case of heterogeneity each of the fractions will show a single sedimenting peak; for ligand-mediated association–dissociation, and kinetically controlled reversible dimerization, each fraction will give two peaks like the unfractionated material; while for irreversible dimerization or dissociation of an m-mer into subunits and for dissociation of certain complexes, the fraction comprised of product will sediment as a single peak whereas the one rich in reactant will exhibit two peaks. Analysis of both freshly prepared and aged fractions is suggested as a precautionary measure for distinguishing between kinetically controlled interactions and heterogeneity.

The foregoing is indicative of the interpretative difficulties that may be encountered in practice and underscores the need for the combined application of sedimentation with other physicochemical methods in order to elucidate the mechanism of an interaction.

Dissociation of a Complex. Calculation of theoretical zone sedimentation patterns for kinetically controlled dissociation of a complex[1a] were made in response to a colleague's question: "Will the slow (either irreversible or reversible) dissociation of an enzyme complex or a ribosome–enzyme complex during the course of zone sedimentation give two peaks of enzymatic activity or a single skewed peak which trails?" The kinetically controlled dissociation of either a large enzyme complex or ribosome–enzyme complex (C) to give a residual complex or ribosome (A) stripped of a particular enzyme (B), which now exists in free solution, is simulated by the reaction

$$C \underset{k_2}{\overset{k_1}{\rightleftharpoons}} A + B \tag{3}$$

for $k_2/k_1 \geq 0$ and $V_c = V_A > V_B$, where V_i is the sedimentation velocity of the ith species. As illustrated by the representative zone patterns presented in Fig. 2, both the profile of total material and of enzymic activity are typically bimodal. The faster migrating peak is a mixture of C, A, and B in the pattern of total material and a mixture of C and B in the pattern of enzymic activity, and in both cases the slower peak is comprised largely of B. Single skewed peaks with trailing edges are never predicted for patterns of total material, and only if the reaction goes to completion during the course of sedimentation will the profile of enzymic activity show a single peak that is skewed forward.

Zone sedimentation patterns similar to the theoretically predicted ones have been observed experimentally with enzymically active complexes between aminoacyltransferase I and proteinaceous cytoplasmic particles[8]

[8] E. Shelton, E. L. Kuff, E. S. Maxwell, and J. T. Harrington, *J. Cell Biol.* **45**, 1(1970).

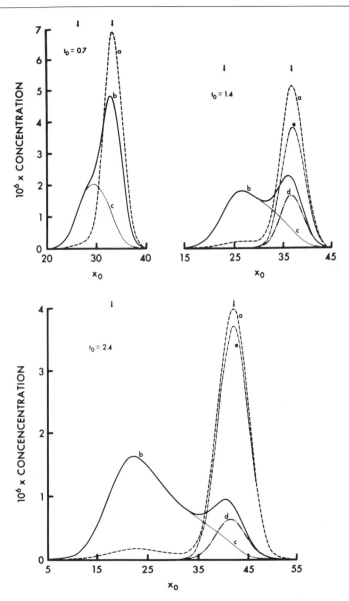

FIG. 2. Zone sedimentation patterns computed for the dissociation of a complex [Reaction (3) with $k_2/k_1 = 2 \times 10^5 \ M^{-1}$]; sedimentation is from left to right. Curves: a, total amount of material $C_4 + 0.9C_2 + 0.1C_3$; b, enzymic activity $C_1 + C_3$; curves c, molar concentration of B, C_3; d, molar concentration of C, C_1; and e, molar concentration of A, C_2. $t_0 = k_1 t$ is dimensionless time; pure complex at start of sedimentation. Reprinted with permission from J. R. Cann and D. C. Oates, *Biochemistry* **12**, 1112 (1973). Copyright by the American Chemical Society.

and between aminoacyl-tRNA synthetases and ribosomes.[9] In both cases, fractionation experiments have shown that the two peaks in the pattern of enzymic activity do not correspond to isozymes with different molecular weights, but instead constitute a bimodal reaction zone arising from dissociation of the complex during the course of sedimentation.

Fractionation per se does not, however, provide information as to the rates of reaction; and there are striking similarities between the bimodal patterns displayed in Fig. 2 for kinetically controlled (even irreversible) dissociation and those computed by Bethune and Kegeles[7] for very rapid equilibration. These investigators have computed zone patterns for several rapidly equilibrating systems, which are schematized by Reaction (3), but differ one from another in the relative mobilities of C, A, and B. A particularly timely case is for $V_C > V_A > V_B$, since it bears on the sedimentation behavior of ribosomes under conditions that cause their dissociation into subunits.

The dissociation according to the scheme of Eq. (3) of ribosomes themselves (in the case of ribosomes from *Escherichia coli* the ribosome, C, is referred to as the 70 S unit) into a larger, A (50 S), and a smaller, B (30 S), subunit is a complex process. This process is strongly controlled by the level of divalent cation (Mg^{2+}),[10,11] dependent on the level of uni-univalent electrolyte[11,12] and quite sensitive to hydrostatic pressure.[12,13] In addition, "vacant" ribosomes from *E. coli* (i.e., ribosomes devoid of peptidyl tRNA, messenger RNA, initiation factors, and peptidyl products) often show a microheterogeneity in which two distinct classes of ribosomes are possible.[14,15] These classes, designated as "tight couples" and "loose couples," are traditionally distinguishable by means of low speed zonal ultracentrifugal analysis on a stabilizing sucrose gradient;[14,15] the "tight couples" remain as 70 S units in the presence of 5 mM Mg^{2+}, while the "loose couples" dissociate into 30 S and 50 S subunits. "Tight" and "loose" couples are also distinguished by means of kinetic investigation of the subunit interaction, using such means as stopped-flow[16,17] or

[9] W. K. Roberts, personal communication, 1972.

[10] R. S. Zitomer and J. G. Flaks, *J. Mol. Biol.* **71**, 263 (1972).

[11] A. Spirin, B. Sabo, and V. A. Kovalenko, *FEBS Lett.* **15**, 197 (1971).

[12] A. A. Infante and R. Baierlein, *Proc. Natl. Acad. Sci. U.S.A.* **68**, 1780 (1971).

[13] J. G. Hauge, *FEBS Lett.* **17**, 168 (1971).

[14] H. Noll, M. Noll, B. Hapke, and G. van Dielien, *in* 24th Kolloquium der Gesellschaft für Biologische Chemie, p. 257, Springer-Verlag, Berlin and New York, 1973.

[15] O. P. van Diggelen and L. Bosch, *Eur. J. Biochem.* **39**, 511 (1973).

[16] A. D. Wolfe, P. Dessen, and D. Pantaloni, *FEBS Lett.* **37**, 112 (1973).

[17] A. Wishnia, A. Boussert, M. Graffe, P. Dessen, and M. Grunberg-Manago, *J. Mol. Biol.* **93**, 499 (1975).

pressure-jump[18-20,20a] relaxation, with a scattered-light detector. Although the kinetic processes are completed rapidly, i.e., in a few seconds or less when the Mg^{2+} ion level is high (perhaps, 8–10 mM),[16-20,20a] the half-time for reassociation at ribosome concentrations of 0.01% to several hundredths of 1% in the presence of 3 mM Mg^{2+} can also be as long as perhaps 10–20 sec; and at even lower Mg^{2+} ion levels (2 mM) the half-time for dissociation is as long as 75 sec.[17] These statements should be tempered by the knowledge now accumulating that both the kinetic and the earlier equilibrium results have frequently been obtained either on mixtures of "loose" and "tight" couples, or on different types of preparations from different laboratories. The major point being made here, however, is that the kinetic processes so far studied seem sufficiently rapid so that failure to reach local chemical equilibrium would not be expected[1a,6] (see also Figs. 2 and 4 and text discussion thereof) to contribute much to enhancing the resolution of ultracentrifuge patterns, except at Mg^{2+} levels as low as 2 mM. It should be noted that the total time of centrifugation is in the range from several hours to 10 hr.

According to the equilibrium light-scattering measurements of Zitomer and Flaks,[10] the overall formation constant of mixed 70 S ribosomes from their subunits approximates 10^9 liters/mole in buffers containing 10 mM Mg^{2+}. This estimate has been confirmed by kinetic studies on "tight couples" by Wishnia *et al.*[17] using stopped-flow scattered-light measurements and by Noll and Noll[21] on similar material using exchange measurements in subunits during zonal sedimentation. If one refers to the zonal calculations of Bethune and Kegeles[7] made on the basis of instantaneous reequilibration in the case where $V_C > V_A > V_B$, one would expect very little resolution for such a high formation constant. Moreover, if resolution should occur, this would develop because the region corresponding to the *slower* subunit, B (30 S), would become bimodal: there would be a 30 S zone moving essentially within a slowed down "70 S" zone, and a trailing 30 S zone of pure, resolved subunit. This situation would exist as long as the 30 S and 50 S subunits are present in equimolar amounts, as they would be if only 70 S ribosomes were originally present. Thus, one would never expect to see zonal patterns for rapidly reequilibrating systems that would contain a trailing 50 S zone, in addition to a 30 S zone and a 70 S zone.

[18] J. B. Chaires, C. Ke, M. S. Tai, G. Kegeles, A. A. Infante, B. Kosiba, and T. Coker, *Biophys. J.* **15**, 293a (1975).
[19] G. Kegeles and C. Ke, *Anal. Biochem.* **68**, 138 (1975).
[20] E. Schulz, R. Jaenicke, and W. Knoche, *Biophys. Chem.* **4**, 253 (1976).
[20a] J. B. Chaires, M. Tai, C. Huang, G. Kegeles, A. A. Infante, and A. J. Wahba, *Biophys. Chem.* **7**, 179 (1977).
[21] H. Noll and M. Noll, *J. Mol. Biol.* **105**, 111 (1976).

Predictions for zonal experiments become quite complicated,[22] however, showing such effects as time-dependent, and rotor-speed dependent sedimentation if the system being investigated has an appreciable volume of reaction and is subjected to large hydrostatic pressures, as in prolonged high speed ultracentrifugation. The direct application of these ideas to the sucrose gradient zonal ultracentrifugation of sea urchin ribosomes was made by Infante and Baierlein,[12] in their elegant experimental study (for a discussion, see also Harrington and Kegeles, this series Vol. 27 [13]). Infante and Baierlein found, in fact, in agreement with their predictions based on a volume of formation of 500 cm^3 per mole of ribosomes, that an original single zone of 75 S sea urchin ribosomes run at high speed would, in proceeding down the sucrose gradient into regions of increasingly higher hydrostatic pressure, gradually become dissociated into "35 S" and "56 S" subunits, the fast zone of complex, C, continually slowing down below 75 S and the slower zone speeding up in the process, because of coupling of the transport of species through the increasingly important reaction process. Infante and Baierlein did not find or predict three zones, however, and calculations for E. coli ribosomes confirm this conclusion over a wide range of parameters.[23] By use of low centrifugal fields and long times of centrifugation, the dissociating effects of pressure can be minimized. What has usually been found in E. coli ribosomes, on the other hand, under such circumstances, which correspond rather closely to the original infinitely rapid reequilibration conditions assumed by Bethune and Kegeles,[7] is that there are essentially three zones corresponding to 60–70 S ribosomes and both 30 S and 50 S subunits even at Mg^{2+} levels of 10 mM. Such a result can ensue only when there is microheterogeneity in the sample. This was the basis for the method developed by Noll et al.[14] and van Diggelen and Bosch[15] to distinguish between, and separate "tight" and "loose" couples. The procedure is to centrifuge a zone on a sucrose gradient at relatively low speed in a buffer containing 4–5 mM Mg^{2+}, a condition under which "loose" couples are practically completely dissociated, while "tight" couples, still having a very high overall formation constant,[17] are at most only slightly dissociated. The 70 S zone is then isolated, and is usually presumed to consist of whole ribosomes alone, of the "tight" couple type. If any dissociation of "tight" couples has occurred, however, predictions[7] show that this preparation would contain some molar excess of 50 S subunits, because of the deletion of some 30 S subunits in a trailing zone. Noll et al.[24] have, in fact, demon-

[22] G. Kegeles, L. Rhodes, and J. L. Bethune, Proc. Natl. Acad. Sci. U.S.A. 58, 45 (1967).
[23] J. B. Chaires and G. Kegeles, Biophys. Chem., 7, 173 (1977).
[24] M. Noll, B. Hapke, N. H. Schreier, and H. Noll, J. Mol. Biol. 75, 281 (1973).

strated that under some conditions this is what happens. If such a process is then repeated on such a preparation, it would be predicted[7] that the 50 S subunit might eventually be in sufficient excess so that the trailing zone, upon resolution, would become a true 50 S zone rather than a 30 S zone. This phenomenon has also been observed in direct experiments by Noll et al.[24]

In summary, the ultracentrifuge behavior of the fairly rapidly reequilibrating ribosome system is usually complicated by micro-heterogeneity and by sensitivity to hydrostatic pressure. However, low speed zonal experiments on purified "tight" couple systems sometimes conform rather closely[24] to expectations[7] developed on the basis of a model system that interacts extremely rapidly, and in which the controlling ligand (Mg^{2+}) is present at a sufficiently high level that no appreciable gradients of ligand are created by migration of the interacting system. Under these conditions the "tight couples," because of their high formation constant, sediment virtually as a single 70 S zone, leaving behind only a slight amount of material in a 30 S zone (see preceding paragraph).[23] Aside from its importance in molecular biology, elucidation of the ultracentrifuge behavior of the ribosome system illustrates the indispensable role of independent kinetic and equilibrium measurements for unambiguous interpretation of the sedimentation behavior of associating–dissociating systems in general.

Reversible Dimerization. As will be described shortly, the dimerization of New England lobster hemocyanin induced by Ca^{2+}-binding does not equilibrate instantaneously, the half-times of reaction being of the order of 40–100 sec, depending upon conditions. This finding prompted a comparative theoretical investigation[6] into the effect of chemical kinetics upon the shape of the reaction boundary for the nonmediated dimerization reaction

$$2\,M \underset{k_2'}{\overset{k_1'}{\rightleftharpoons}} M_2 \qquad (4)$$

and for the ligand-mediated reaction set

$$M + X \underset{k_2}{\overset{k_1}{\rightleftharpoons}} MX \qquad (5a)$$

$$2\,MX \underset{k_4}{\overset{k_3}{\rightleftharpoons}} M_2X_2 \qquad (5b)$$

where X is a small ligand molecule.

The family of curves displayed in Fig. 3 is for nonmediated dimerization [Reaction (4) with 52.7% dimerization] and traces the continuous transformation in the shape of the sedimentation pattern from the unimodal one in the limit of instantaneous reequilibration during differential

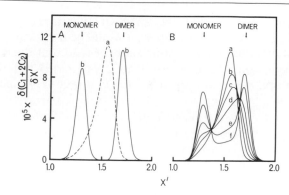

Fig. 3. Analytical sedimentation patterns computed for nonmediated dimerization [Reaction (4)]. (A) Calculated for the limiting case of instantaneous reequilibration during differential transport of monomer and dimer (curve a) and for the limit in which there is no interconversion during the course of the sedimentation experiment (curve b). (B) Calculated for kinetically controlled interaction: curve a, half-time of dissociation, 19 sec; b, 173 sec; c, 347 sec; d, 693 sec; e, 1733 sec; and f, 3465 sec. Time of sedimentation, 2000 sec. All patterns in this and the following figure are for 52.7% dimerization at an initial macromolecule concentration of 2.326×10^{-5} M. Reprinted with permission from J. R. Cann and G. Kegeles, *Biochemistry* **13**, 1868 (1974). Copyright by the American Chemical Society.

transport of monomer and dimer to the completely resolved one in the limit where a negligible amount of reaction occurs during the course of the sedimentation experiment. The patterns for half-times of dissociation as long as 60 sec are virtually the same as for instantaneous reequilibration. But, as the half-time is increased further, resolution ensues. Thus, for a half-time of 347 sec, the pattern shows a strong, centripetal shoulder; and for a half-time of 693 sec, the pattern is distinctly bimodal. Increasing the half-time still further enhances resolution of the two peaks with concomitant evidence of a third peak of intermediate velocity,[25,26] and it is apparent that the gradient will go to the base line between the two major peaks only when the half-time of reaction is manyfold greater than the time of sedimentation.

Sedimentation patterns for ligand-mediated dimerization [Reaction Set (5) with 52.7% dimerization] are displayed in Fig. 4. Those shown in Fig. 4A for instantaneous establishment of equilibrium are essentially the same as described previously[3-5] (see this series Vol. 27 [12] and this volume [11] preceding chapter) and require no further comment except that they serve as a point of reference for kinetically controlled interactions. The calculations for kinetically controlled interactions make the reasonable

[25] G. G. Belford and R. L. Belford, *J. Chem. Phys.* **37**, 1926 (1963).
[26] D. F. Oberhauser, J. L. Bethune, and G. Kegeles, *Biochemistry* **4**, 1878 (1965).

assumption that ligand-binding per se [Reaction (5a)] equilibrates rapidly (k_1 and k_2 in upper range observed for binding of small molecular to proteins) so that the monomer–dimer interconversion [Reaction (5b)] becomes rate controlling, the definitive kinetic parameter being the half-time of dissociation of the dimer. The family of patterns in Fig. 4B is for kinetic-control at a ligand concentration for which the pattern is unimodal when equilibration is instantaneous. As in the case of nonmediated dimerization, a half-time as long as 60 sec simply causes a slight distortion of the peak. Upon further increase in the half-time, resolution into two peaks ensues and becomes increasingly sharp as the limit is approached where negligible interconversion occurs during the course of sedimentation. This limit is approached asymptotically, however, and the gradient does not go to the base line between the well-separated peaks even for a half-time 5-fold greater than the time of sedimentation. As for the effect of

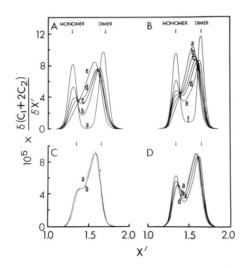

Fig. 4. Analytical sedimentation patterns computed for ligand-mediated dimerization [Reaction Set (5)]. (A) Instantaneous reequilibration during differential transport of the several species: curve a, initial equilibrium concentration of ligand, $C_3^0 = 10^{-7}\ M$; b, $5 \times 10^{-7}\ M$; c, $7.5 \times 10^{-7}\ M$; d, $10^{-6}\ M$; and e, $2 \times 10^{-6}\ M$. Time of sedimentation, 2000 sec. (B) Kinetically controlled interaction for $C_3^0 = 2 \times 10^{-6}\ M$: curve a, half-time of dissociation of dimer, 19 sec; b, 116 sec; c, 231 sec; d, 347 sec; e, 693 sec; and f, 6930 sec. Time of sedimentation, 1500 sec. (C) Comparison of the pattern for instantaneous reequilibration (curve a) with the pattern for kinetically, controlled interaction with half-time of 58 sec (curve b); $C_3^0 = 10^{-6}\ M$; time of sedimentation, 1500 sec. (D) Comparison of instantaneously reequilibrating and kinetically controlled interaction for $C_3^0 = 7.5 \times 10^{-7}\ M$: curve a, instantaneous reequilibration; b, half-time of 58 sec; c, 231 sec; and d, 462 sec. Time of sedimentation, 1500 sec. Reprinted with permission from J. R. Cann and G. Kegeles, *Biochemistry* **13**, 1868 (1974). Copyright by the American Chemical Society.

chemical kinetics at lower ligand concentration (percent dimerization held constant) where bimodal patterns obtain for instantaneous equilibration, decreasing the rate of the monomer–dimer interconversion merely enhances the resolution (Fig. 4D).

These results admit several generalizations. First, previous conclusions concerning the sedimentation behavior of ligand-mediated interactions in the limit of instantaneous equilibration[3-5] (see this series Vol. 27 [12] and this volume [11]) are valid for kinetically controlled interactions characterized by half-times as long as 60 sec and the same can be said for nonmediated dimerization. Also of considerable practical importance is the finding that for half-times less than about 200 sec, resolution of the reaction boundary into two peaks can occur only if dimerization is ligand-mediated. Finally, it bears repeating that cognizance must be taken of the fact that bimodal patterns for rapidly equilibrating ligand-mediated dimerization bear a strong resemblance to the patterns shown by sufficiently slow dimerization reactions whether ligand-mediated or not (cf Fig. 1A, 3 and 4).

The foregoing discussion has focused on a ligand-mediated mechanism of dimerization which assumes that obligatory binding of ligand to two monomer molecules must precede the formation of a molecule of dimer. Another possible mechanism termed ligand-facilitated assumes that two monomer molecules can react to form a dimer, which is then stabilized by binding of ligand

$$2 M \rightleftharpoons M_2$$
$$M_2 + X \rightleftharpoons M_2X \tag{6}$$
$$M_2X + X \rightleftharpoons M_2X_2$$

These two extreme mechanisms apparently cannot be distinguished by their sedimentation behavior, at least for instantaneous equilibration.[27] On the other hand, as we shall see below, relaxation kinetics is intrinsically capable of distinguishing between them.

Association–Dissociation of Hemocyanins

Already referred to above in the discussion of reversible dimerization is one example of considerable interest, because it has been studied in detail by a combination of several techniques. This is the association–dissociation reaction of the hemocyanin of *Homarus americanus,* the North American lobster. This protein, in contrast to the behavior of many other hemocyanins,[28] shows a completely reversible interaction at pH 9.6

[27] J. R. Cann, *Adv. Pathobiol.* J. R. Cann *in* "Differentiation and Carcinogenesis" (C. Borek, C. M. Fenoglis, and D. W. King, eds.) No. 6 of the series *Advances in Pathobiology,* 1977, p. 158.
[28] R. J. Siezen and R. van Driel, *Biophys. Biochim. Acta* **295,** 131 (1973).

in glycine–sodium hydroxide buffers, which can be explained by a single apparent overall equilibrium constant.[29] This apparent overall equilibrium constant is strongly dependent on the level of free calcium ion and on pH.

By means of Archibald ultracentrifuge measurements[30] at the liquid–air meniscus, these authors[29] were able to obtain data on the thermodynamic nonideality under these conditions, some 4 pH units above the isoelectric point, and to evaluate the thermodynamic formation constant of whole molecules from half molecules as a function of pH and free calcium ion concentration. By plotting the logarithm of this apparent overall equilibrium constant against the logarithm of the free calcium ion concentration, they were able to determine that the formation of each whole molecule of hemocyanin from two half molecules is accompanied by a net uptake of 5 ± 1 calcium ions. The sedimentation velocity patterns of this system under these conditions (pH 9.6, 0.1 ionic strength glycine–sodium hydroxide buffers containing from 0.0031 to $0.0053\,M$ free calcium ion) all show partial resolution, in disagreement with Gilbert theory[31] for a rapid simple protein dimerization process not controlled by ligand binding. The experimental patterns generally resemble a great deal many of those discussed above, in Figs. 3 and 4. Preliminary verification of the reversibility of the interaction was obtained by fractionation with a separation cell of the Yphantis–Waugh design.[32] The reconcentrated slower-moving material, recovered and pooled from several experiments and redialyzed, showed partial peak resolution when resubjected to sedimentation, with patterns nearly identical to those obtained by sedimenting the original hemocyanin sample diluted to the same concentration. Thus, the material appeared to have reequilibrated completely. Furthermore, rapid changes of rotor velocity from 42,000 rpm to 59,780 rpm and back resulted in reversible pattern shifts for a sample overlayed with mineral oil, which suggested a small but rapid, reversible dissociating effect on increase of hydrostatic pressure. Ultimately, this prediction of the effect of pressure was verified by means of pressure-jump kinetic observations.[33] Data were also obtained by temperature-jump light-scattering observation[34] of the apparent first collision step in a complex kinetic pathway. This pathway involves multiple ligand binding steps and the existence of a steady-state conformational intermediate of whole

[29] K. Morimoto and G. Kegeles, *Arch. Biochem. Biophys.* **142**, 247 (1971).
[30] W. J. Archibald, *J. Phys. Colloid Chem.* **51**, 1204 (1947).
[31] G. A. Gilbert, *Discuss. Faraday Soc.* **20**, 68 (1955).
[32] D. A. Yphantis and D. F. Waugh, *J. Phys. Chem.* **60**, 630 (1956).
[33] G. Kegeles, V. P. Saxena, and R. Kikas, *Fed. Proc., Fed. Am. Soc. Exp. Biol.* **33**, 1504 (1974).
[34] M. S. Tai and G. Kegeles, *Arch. Biochem. Biophys.* **142**, 258 (1971).

molecules[35,36] The most successful approach was to examine carefully the calcium ion concentration dependence of the relatively slow kinetic process,[35,36] which could be followed by "concentration-jump" dilution measurements[37] in a Durrum–Gibson stopped-flow apparatus[38] equipped for 90° light-scattering observation.[39] The advantage of examining the dependence of slower kinetic processes on prior ligand-binding equilibria was demonstrated in a study of magnesium ion-coupled conformational changes in tRNA.[40]

In the case of lobster hemocyanin at pH 9.6 in glycine–sodium hydroxide buffer, it was found that, at two closely spaced values of free calcium ion concentration, the overall recombination rate constant for half molecules, as determined[35] by concentration-jump measurements, appeared to carry most of the calcium ion dependence previously observed[29] in the equilibrium constant, while the overall dissociation rate constant for whole molecules appeared to be relatively independent of calcium ion concentration. This clue led to an extension of the kinetic measurements over the widest feasible range of calcium ion concentration under these conditions, and to a detailed mechanistic interpretation of the kinetics.[36] Two extreme mechanisms were postulated for the binding of four calcium ions and the associated dimerization of half molecules, similar to those discussed above in schemes (5a) and (5b) (ligand-mediated) and scheme (6) (ligand-facilitated). These are illustrated in reaction scheme (7), for ligand-mediated dimerization:

$$2M + 4X \underset{k_{21}}{\overset{k_{12}}{\rightleftharpoons}} 2MX + 2X \underset{k_{32}}{\overset{k_{23}}{\rightleftharpoons}} 2MX_2 \underset{k_{43}}{\overset{k_{34}}{\rightleftharpoons}} M_2X_4 \tag{7}$$

and in reaction scheme (VIII), for ligand-facilitated dimerization:

$$2M + 4X \underset{k_d}{\overset{k_a}{\rightleftharpoons}} M_2 + 4X \underset{k_{21}}{\overset{k_{12}}{\rightleftharpoons}} M_2X + 3X \underset{k_{32}}{\overset{k_{23}}{\rightleftharpoons}} M_2X_2 + 2X \underset{k_{43}}{\overset{k_{34}}{\rightleftharpoons}} M_2X_3 + X \underset{k_{54}}{\overset{k_{45}}{\rightleftharpoons}} M_2X_4 \tag{8}$$

The reciprocal of the slow relaxation time for these two reaction schemes is, under the assumption that ligand-binding steps are rapid compared to protein interaction,

$$1/\tau = 4k_{34}[\overline{M}]K_{12}^2 K_{23}^2 [\overline{X}]^4/(1 + K_{12}[\overline{X}] + K_{12}K_{23}[\overline{X}]^2) + k_{43} \tag{9}$$

[35] G. Kegeles and M. S. Tai, Biophys. Chem. 1, 46 (1973).
[36] M. S. Tai and G. Kegeles, Biophys. Chem. 3, 307 (1975).
[37] H. F. Fisher and J. R. Bard, Biochim. Biophys. Acta 188, 168 (1969).
[38] Q. H. Gibson and L. Milnes, Biochem. J. 91, 161 (1964).
[39] J. E. Stewart, Durrum Application Notes No. 7. Palo Alto, California, 1971.
[40] D. C. Lynch and P. R. Schimmel, Biochemistry 13, 1841 (1974).

for the ligand-mediated case [reaction scheme (7)], and

$$1/\tau = 4 k_a [\overline{M}] + k_d/(1 + K_{12} [\overline{X}] + K_{12} K_{23} [\overline{X}]^2 + K_{12} K_{23} K_{34} [\overline{X}]^3 \\ + K_{12} K_{23} K_{34} K_{45} [\overline{X}]^4) \quad (10)$$

for the ligand-facilitated case [reaction scheme (8)]. In (9) and (10), the symbols $[\overline{M}]$ and $[\overline{X}]$ denote equilibrium concentrations of the species, and capital K symbols denote equilibrium constants (i.e., $K_{12} = k_{12}/k_{21}$).

Additionally, it is necessary, in order to explain macromolecular interactions much slower than those governed by diffusion control, to assume a series of coupled macromolecular interaction steps, rather than a single-step reaction. The simplest scheme consistent with an overall bimolecular process appears to be a diffusion-controlled bimolecular collision process followed by an internal rearrangement.[41,42] If the primary product of the diffusion-controlled collision process is an intermediate conformer in a steady-state concentration, then the ensuing slow kinetics really appear pseudo-bimolecular.[36,43] These considerations were taken into account in developing the complete expressions for the relaxation time.[36] However, for the purpose at hand, of distinguishing between the ligand-mediated and the ligand-facilitated pathways, it is sufficient to examine the corresponding expressions (9) and (10) for the reciprocal relaxation times. It is noted that the experimentally accessible overall recombination rate constant is given by one-fourth of the coefficient of the sum of the concentrations of all liganded forms of monomer, $[\overline{M}_t]$. The overall dissociation rate constant is given by the term that does not contain the macromolecule concentration. Thus for scheme (9), the ligand-mediated mechanism, the recombination rate constant k_f and the dissociation rate constant k_r, since $[\overline{M}_t] = [\overline{M}] (1 + K_{12} [\overline{X}] + K_{12} K_{23} [\overline{X}]^2)$, are given by

$$k_f = k_{34} K_{12}^2 K_{23}^2 [\overline{X}]^4/(1 + K_{12} [\overline{X}] + K_{12} K_{23} [\overline{X}]^2)^2 \\ k_r = k_{43} \quad (11)$$

On the other hand, for scheme (10), the ligand-facilitated mechanism, k_f and k_r are, since $[\overline{M}_t] = [\overline{M}]$ in this scheme, given by

$$k_f = k_a \\ k_r = k_d/(1 + K_{12} [\overline{X}] + K_{12} K_{23} [\overline{X}]^2 + K_{12} K_{23} K_{34} [\overline{X}]^3 \\ + K_{12} K_{23} K_{34} K_{45} [\overline{X}]^4) \quad (12)$$

From (9), it is predicted that the reciprocal relaxation times for the ligand-mediated scheme, (7), will become larger as the equilibrium ligand concentration is raised, at least at small ligand concentrations. On the

[41] K. Šolc and W. H. Stockmayer, J. Chem. Phys. 54, 2981 (1971).
[42] K. Šolc and W. H. Stockmayer, Int. J. Chem. Kinetics 5, 733 (1973).
[43] K. Kirschner, private communication, 1974.

other hand, from (10) it is predicted that the reciprocal relaxation times for the ligand-facilitated reaction scheme, (8), will become smaller as the equilibrium ligand concentration is raised, which provides for one clear-cut distinction between these assumed mechanisms. Moreover, from (11), it is predicted that for the ligand-mediated scheme, (7), the recombination rate constant, k_f, obtained from the slope of a plot of $1/\tau$ versus concentration of monomeric forms, will increase as ligand concentration increases at least at small values of ligand concentration. Also, for the ligand-mediated scheme, (7), the dissociation rate constant, k_r, determined from the intercepts of these plots, should be constant, independent of the ligand concentration, according to relations (11).

However, if the ligand-facilitated reaction scheme, (8), is the mechanism of the reaction, then according to (12), the recombination rate constant, k_f, obtained from the slopes of plots of $1/\tau$ versus total concentration of monomeric forms, should be constant independent of ligand concentration, while the intercepts, giving the overall dissociation rate constants, should decrease progressively with increase in ligand concentration.

Figure 5 shows the results of the kinetics experiments[36] at four different molarities of calcium ion, 0.0031, 0.0036, 0.0047 and 0.0053. It is seen

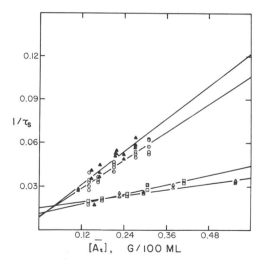

FIG. 5. Reciprocal relaxation times at 25° for concentration-jump light-scattering experiments at four different levels of free calcium. Five volumes of protein solution were mixed with two volumes of dialysis buffer, 0.1 ionic strength glycine–sodium hydroxide at pH 9.6, containing free calcium ion molarities equal to 0.0031 (△), 0.0036 (□), 0.0047 (○) and 0.0053 (▲). The horizontal axis represents total of all liganded forms of monomeric species (470,000 molecular weight). Reproduced from M. Tai and G. Kegeles, *Biophys. Chem.* **3**, 307 (1975), by permission of North-Holland Publishing Co.

that the intercepts are nearly independent of the level of free calcium ion, the reciprocal relaxation times increase at constant total monomer concentration as the calcium ion concentration increases, and the slopes increase progressively as the calcium ion concentration increases, leading to progressively higher recombination rate constants. All these findings are consistent with reaction scheme (7) and relations (9) and (11) for the ligand-mediated mechanism, and are inconsistent with reaction scheme (8) and relations (10) and (12) for the ligand-facilitated mechanism. Thus the data from relaxation kinetics provide a clear choice between these possible mechanisms in this case. It is to be noted that the apparent overall equilibrium constant derived from either mechanism can be reduced to the identical form, the result being wholly independent of the mechanism assumed. While this does not appear to be so when one refers to expressions (11) and (12), one sees immediately that this must be true by comparing the initial states, $2 M + 4 X$, and the final states, M_2X_4 in the reaction schemes (7) and (8). Since the initial states are identical in both schemes, and the final states are identical in both schemes, the total free energy change, and hence the overall equilibrium constant, is identical in both schemes. This is just a restatement of the truism that one cannot arrive at reaction mechanisms by equilibrium measurements. This is not to say, however, that one cannot arrive at an assessment of the population of the major species present at equilibrium, through a judicious choice of equilibrium and transport observations. The additional leverage afforded by resort to relaxation kinetics seems well illustrated here.

As mentioned above, the transport behavior of many hemocyanin systems cannot be analyzed as simply as is the case for lobster hemocyanin at pH 9.6. Even in the case of lobster hemocyanin, as the system is adjusted closer to the isoelectric point, and divalent metal ion is removed, the rate of interaction is tremendously slowed, and transport behavior becomes complicated.[44] In many other cases,[28] excellent resolution has been obtained in analytical ultracentrifugation, sometimes even complete resolution between peaks, which implies the extreme of a virtually noninteracting system.

One case which has been particularly puzzling, however, is that of the α-hemocyanin of *Helix pomatia,* the Roman orchard snail.[28,45-48] This fraction makes up approximately 75–80% of the total hemocyanin; the remaining 20–25%, termed β-hemocyanin, is characterized by inability to

[44] V. P. Saxena, G. Kegeles, and R. Kikas, *Biophys. Chem.* **5,** 161 (1976).
[45] S. Brohult, *J. Phys. Colloid Chem.* **51,** 206 (1947).
[46] K. Heirwegh, H. Borginon, and R. Lontie, *Biochim. Biophys. Acta* **48,** 517 (1961).
[47] W. N. Konings, R. J. Siezen, and M. Gruber, *Biochim. Biophys. Acta* **194,** 376 (1969).
[48] Y. Engelborghs and R. Lontie, *J. Mol. Biol.* **77,** 577 (1973).

dissociate into half molecules at pH 5.7 in the presence of 1 M NaCl.[45,46] From pH 7.6 to 7.85, and at somewhat more alkaline pH values in the presence of calcium ion, there appear to be a mixture of whole, half, and one-tenth size molecules for the isolated α-hemocyanin.[28,47] In analytical ultracentrifugation the relative concentrations remained constant over a very wide range of total protein concentration under a fixed set of buffer conditions.[28,47] Yet the base line between schlieren peaks representing whole and half molecules was always elevated, indicating some kind of reaction boundary. Pressure was shown to dissociate whole molecules,[49] and the elevation in the base line was ascribed[47] to progressive dissociation of whole molecules as a result of hydrostatic pressure; the area between peaks was therefore added to the area of the fast peak to estimate the fraction of material in the form of whole molecules.[47] Many virtually complete fractionations have been achieved[28] by differential sedimentation, or use of a moving-platform analytical ultracentrifuge separation cell,[32] or by gel filtration chromatography, in the alkaline region between pH 7.6 and pH 9.3. None of the fractions obtained by the different separation techniques showed any reequilibration if the conditions were kept constant, even after many days, as judged by resubjection to analytical ultracentrifugation, yet they responded immediately to changes in pH and ionic strength. This entire behavior was ascribed to microheterogeneity, in which it was postulated[28] that under any one set of buffer conditions only a tiny fraction of the whole sample was capable of undergoing reaction, the rest being either completely associated or completely dissociated. This picture was derived from analogy with the N-F transition of bovine plasma albumin at low pH studied by Foster and colleagues.[50] Moreover, both reversible boundary spreading electrophoresis experiments and fractionation at different pH values confirmed unequivocally the existence of microheterogeneity.

In a careful extension of equilibrium and transport studies, Engelborghs and Lontie[48] performed static light-scattering measurements on *H. pomatia* hemocyanin samples in 0.1 M acetate buffers at pH 5.7 containing various amounts of a series of univalent salts. In this buffer in the absence of added univalent salt, there is no dissociation of whole molecules. When the buffer is made 1 M in NaCl, the α-hemocyanin dissociates completely into half-molecules. In the presence of 0.4 M NaCl, velocity ultracentrifuge patterns show partial peak resolution, and analysis of the light-scattering data as a function of total protein concentration showed that the dissociation behavior could not be fit by a single equilibrium

[49] W. N. Konings, Doctoral Dissertation, University of Groningen, 1969, p. 31.
[50] J. F. Foster, M. Sogami, H. A. Petersen, and W. J. Leonard, *J. Biol. Chem.* **240**, 2495 (1965).

constant, confirming again the existence of microheterogeneity. However, changes in the salt concentration resulted in light-scattering changes that took place too rapidly to be followed in hand-mixing experiments.[48]

A reexamination of the light-scattering data of Engelborghs and Lontie at pH 5.7 in the presence of 0.4 M NaCl was undertaken,[51] to see whether a relatively simple model of microheterogeneity might fit. Two such models were tried, in which the system was postulated to be a mixture of nonreacting molecules, and reversibly interacting whole and half molecules with a single equilibrium constant. In one case, it was assumed that the inert or incompetent molecules were half molecules, and in the other case, it was assumed that the incompetent molecules were whole molecules. In either case, a closed solution of the problem exists, from which it is possible to extract the equilibrium constant for the reacting system and the percentage of material which is incompetent. The incompetent half molecule model resulted in a qualitative failure to represent the data. However, the incompetent whole molecule model fit the light-scattering data[48] very well, and resulted in an estimate of 35% of incompetent whole molecules and a formation constant of whole molecules from half molecules of 0.176 liters per gram, neglecting residual nonideality. The isoelectric point of the protein is at pH 5.3,[46] so that at such a high salt concentration the electrostatic contributions to nonideality at pH 5.7 should be negligible. The residual nonideality should derive from excluded-volume effects.

Samples of α-hemocyanin provided by Professor van Bruggen, Dr. van Driel, and Dr. Kuiper of the Biochemical Laboratory of the University of Groningen were then examined kinetically,[52] by stopped-flow dilution concentration-jump[34,37] experiments and by light-scattering pressure-jump experiments in a newly constructed apparatus.[19] While it was already known that the reaction would be driven by pressure, from previous studies,[47,49] it should have been predicted from the ultracentrifuge studies[28,47] that simple dilution should have no effect, since the relative peak areas in analytical ultracentrifugation were completely independent of total protein concentration, at least at alkaline pH values. On the contrary, reaction amplitudes for 1.4 times dilution in acetate buffer at pH 5.7 containing 0.4 M NaCl, and in alkaline buffers as well, were at least as large as those produced by shifting pressure by almost 100 atmospheres,[52] indicating that, indeed, a large fraction of the system was able to undergo a reversible whole molecule–half molecule interaction under constant buffer conditions. In both types of experiments at pH 5.7, relaxation times of the order of 1 sec were measured. Although the overall dis-

[51] G. Kegeles, *Arch. Biochem. Biophys*, **180**, 530 (1977).
[52] M. S. Tai, G. Kegeles, and C. Huang, *Arch. Biochem. Biophys*. **180**, 537 (1977).

sociation rate constant can be determined unequivocally in such relaxation experiments, it is impossible to evaluate the recombination rate constant without an independent estimate for the fraction by weight of material which is competent to react, the apparent rate constant being the product of the true rate constant multiplied by that fraction.[52] Thus, the estimate[51] that 65% of the α-hemocyanin is reactive at pH 5.7 was applied to the relaxation data to evaluate the overall recombination rate constant. The ratio of recombination rate constant to dissociation rate constant then provided values for the formation constant of whole molecules from half molecules, independently from concentration-jump and pressure-jump experiments. While the values for the individual rate constants agreed approximately from both sets of kinetics data, the equilibrium constants fortuitously agreed exactly, at 0.154 liters per gram, compared to 0.176 liters per gram derived[51] from the equilibrium light-scattering data.[48] Thus there seems to be a clear quantitative verification that a major part of the α-hemocyanin is reactive at pH 5.7 in the presence of 0.4 M NaCl.

It remains to explain why, at alkaline pH, relative concentrations estimated from corrected peak areas in sedimentation experiments are completely independent of total protein concentration, why fractionation is possible by use of ultracentrifuge partition cells, differential sedimentation, and gel filtration chromatography, and why such fractions, once obtained, do not appear to reequilibrate after long periods, as judged by resubjection to sedimentation. One explanation could be that behavior at alkaline pH is completely different from that at pH 5.7. However, the logical consequences of the model that successfully fits the light-scattering data at pH 5.7 show that it is unnecessary to postulate such differences in behavior.

Since the system is assumed to consist (at least at pH 5.7 in the presence of 0.4 M NaCl) of 35% of incompetent whole molecules, and 65% of a reversibly interacting half molecule–whole molecule system, how should it behave on ultracentrifugation? The incompetent whole molecules should form a leading schlieren peak, whose relative area, corrected for radial dilution and for the Johnston–Ogston effect,[53] should remain constant at 35% of the total area, independent of total protein concentration, since these faster-sedimenting molecules do not dissociate (except at high pressures). The remainder of the system should form a single trailing schlieren peak, since the whole molecule–half molecule reaction has a relaxation time (at pH 5.7) near 1 sec, and such a system should follow Gilbert theory[31] at such high buffer component concentrations, and should never resolve into two peaks. One set of exper-

[53] J. P. Johnston and A. G. Ogston, *Trans. Faraday Soc.* **42**, 789 (1946).

iments[48] in the acetate buffer system containing 0.4 M NaCl at pH 5.7 suggests that the fast peak is responsible for 41% of the total concentration, in good agreement with the estimate[51] of 35% of incompetent whole molecules. If the material is fractionated by separating between the two peaks, the slower moving material will be the purified interacting system. This fraction, when reexamined in analytical ultracentrifugation will always result in a single slower-moving peak, whose sedimentation velocity will be identical with that observed in the original mixture, if it is first reconcentrated to the same value it represented in the mixture. Thus, it might seem, according to this criterion, that this material "runs true" and consists of unreactive half molecules, which do not equilibrate at all, even after long standing. Only a very careful measurement of sedimentation velocity of this fraction as a function of concentration, or comparison with isolated half molecules whose reactivity has been destroyed, would show that the recovered fraction is indeed an interacting system. Also, any method such as gel filtration chromatography which might isolate the highest molecular weight material in pure form would produce a fraction which contains only incompetent whole molecules. If reconcentrated back to the original total protein concentration of the mixture, this fraction would sediment at about the same rate as the faster peak in the mixture. The material would, of course, not change its behavior after standing in a constant buffer system for long periods of time, because it has been isolated to be incompetent in that buffer. However, this does not preclude its dissociation if the salt concentration or pH are raised, since it can become partly competent to dissociate under such conditions.[28,48] Thus, most of the behavior which had already been described at alkaline pH[28,47] is consistent with what could be qualitatively predicted from the successful model at pH 5.7. This is not to say that every detail of the behavior of the α-hemocyanin system will be explained by this relatively simple model, but the major behavioral anomalies which have seemed so puzzling do find a reasonable explanation in this simple model.

One additional critique as to the use of fractionation criteria for interaction should now be stated as a result of these findings. A monomer–dimer system undergoing rapid reversible interaction not under the control of ligand-binding does not resolve in ultracentrifugation: except for concentration-dependence of transport rates, it can very easily be mistaken for a single inert component in a mixture. If such an interacting system is found in nature mixed with any resolvable inert species, the whole mixture will, on successful fractionation and resubjection to analysis by the original fractionation technique, pass all the qualitative tests for a nonreacting mixture. But only part of the system is nonreacting. The reacting portion may be reequilibrating in precisely that rapid interaction whose very existence the collected fractionation evidence might mistakenly deny. The easiest experiment which may resolve this possibility is stopped-flow dilution with a light-scattering detector. This can be performed either on the separated fractions or on the original sample. If no reaction amplitude is found, a bimolecular reversible interaction within the time range of the instrument may be ruled out. If there is some reaction amplitude, then a way should be found to obtain some quantitative information as to whether the interacting system is a major part, or possibly only a very minor part of the whole system. This could be conceivably achieved, if material is in plentiful supply, by using relaxation kinetic measurements of recombination rate constants to monitor the results of such fractionation experiments, however tedious and time-consuming this might be. Alternatively, detailed equilibrium measurements might be performed on fractions as they are purified, again a potentially exhausting project. The easiest experiments to perform are transport experiments such as sedimentation velocity or gel filtration chromatography. These, however, as seen above, are the experiments most difficult to interpret without chance of error.

[13] The Meaning of Scatchard and Hill Plots

By F. W. DAHLQUIST[1]

A number of interesting macromolecular questions are directly concerned with the binding of ligands (usually small molecules) to macromolecules. As a result, there has been a great increase in the number of publications primarily concerned with the measurement of such binding and its interpretation. A number of methods for graphical and computer-assisted analysis of the binding data have been employed. The purpose of this review is to consider two of the most commonly used graphical methods, the Scatchard plot[1a] and the Hill plot,[2] in some detail. The Hill plot, log [sites bound]/[sites free] versus log [free ligand], is also known as the Sips plot[3] in the immunological literature. When redox equilibria are considered as the binding of electrons, the Hill plot and Nernst plot are fully equivalent as well.[4] A second goal is to relate these common methods of data presentation to each other. It is clear that a given set of binding

[1] Alfred P. Sloan Foundation Fellow, 1975–1977.
[1a] G. Scatchard, *Ann. N. Y. Acad. Sci.* **51**, 660 (1949).
[2] A. V. Hill, *J. Physiol. (London)* **90**, iv–vii (1910).
[3] R. Sips, *J. Chem. Phys.* **16**, 490 (1949).
[4] See B. O. Malmström, *Quart. Rev. Biophys.* **6**, 389 (1974) for a discussion of the meaning of Nernst plots for cytochrome oxidase.

data contains the same information independent of the particular method used to interpret it. However, certain graphical methods show some aspects of this information more clearly than others. We shall demonstrate the quantitative relationships that allow one to directly convert from one method to the others. Finally, we will demonstrate how parameters for a number of models for apparent cooperative binding behavior may be extracted from the two plots without the need for extensive computer-dependent curve fitting. These models include those proposed by Koshland, Némethy, and Filmer[5] and by Monod, Wyman, and Changeux[6] to explain cooperative ligand binding to subunit containing proteins; a generalized model for ligand binding to a population of macromolecules with different affinities such as would apply to hapten–antibody interactions; and the binding of a ligand to a linear lattice of binding sites that would describe the nonspecific binding of a protein to a nucleic acid.

Most common methods for plotting experimental binding data are each derived from the various linearized forms of the binding equation derived for a single noninteracting binding site. This situation is described by an association constant, K, according to the scheme:

$$M + X \rightleftharpoons MX$$
$$K = [MX]/[M][X]$$

Here M is the macromolecule and X is the ligand. We shall now consider the Scatchard and Hill treatments of such a situation.

The Scatchard Plot.[1a] The fraction, $\bar{Y}$, of macromolecule sites occupied by the ligand is given by

$$\bar{Y} = MX/(M + MX)$$

Combination of this relationship with the equilibrium constant gives

$$\bar{Y}/X = K(1 - \bar{Y})$$

A plot of fraction of macromolecular sites bound divided by free ligand concentration versus fraction bound gives a straight line of slope equal to $-K$ and intercept equal to K. If the absolute concentration of binding sites is known, the fraction bound may be replaced by the concentration of bound sites, MX, to give the equation below:

$$\frac{MX}{X} = K(M_0 - MX)$$

where M_0 is the concentration of potential binding sites. In this variation, a plot of bound ligand concentration divided by free-ligand concentration versus the concentration of bound ligand gives a slope of $-K$ and the

---

[5] D. E. Koshland, Jr., G. Némethy, and D. L. Filmer, *Biochemistry* **5**, 365 (1966).
[6] J. Monod, J. Wyman, and J.-P. Changeux, *J. Mol. Biol.* **12**, 88 (1965).

abscissa intercept becomes M_0. This provides one of the most important aspects of the Scatchard plot. The total number of binding sites may be extrapolated from data in which complete saturation of the macromolecular binding sites is not observed. The intercept of the Scatchard plot in this form is KM_0, and, together with the value of M_0 obtained from the abscissa intercept, it provides another way of determining K. This will become important for the interpretation of nonlinear Scatchard plots.

Hill Plot.[2] Hill originally suggested that all forms of ligand binding can be represented by a single-step interaction between macromolecule and ligand molecules according to the scheme

$$M + nX \rightleftharpoons MX_n$$

This equilibrium can be defined by a phenomenological constant, K, such that

$$[MX_n]/[M][X]^n = K^n$$

The apparent value of K is just the inverse of the half-saturating ligand concentration.

The Hill binding equation can be linearized by taking logarithms:

$$\log [MX_n]/[M] = n \log K + n \log X$$

The binding data may then be plotted as log [sites bound]/[sites free] versus log [free ligand].

The slope of the plot gives n_H, the Hill coefficient. In the case of a single noninteracting site, the Hill coefficient is unity and the half-saturating ligand concentration is simply the apparent dissociation constant for the ligand–macromolecule complex.

It is important to recognize that the Hill procedure requires an independent knowledge of the concentration of free sites. This is usually obtained by difference, knowing the total number of potential sites and the number which are bound by ligand. In this sense it is less powerful than the Scatchard plot which does not require a knowledge of the total number of sites, but rather provides an extrapolated value for this parameter. However, the Hill coefficient provides a very useful quantitative measurement of the extent of cooperative interactions among the potential sites.

Figures 1A and 1B show Scatchard and Hill plots of the data of Dahlquist *et al.*[7] for the binding of chitotriose to egg white lysozyme. The binding is characterized by a single, noninteracting binding site with an association constant of 1×10^5 M^{-1}. The original data were obtained as fractional occupancy as a function of free ligand concentration. The

[7] F. W. Dahlquist, L. Jao, and M. A. Raftery, *Proc. Natl. Acad. Sci. U.S.A.* **56**, 26 (1966).

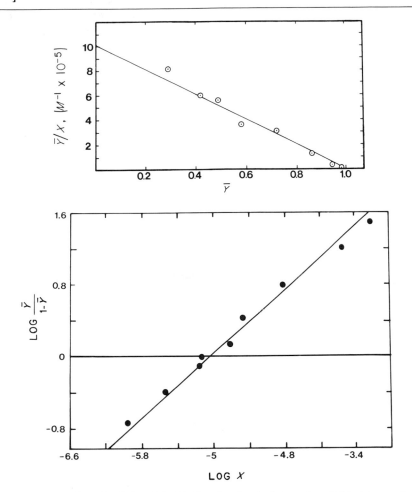

FIG. 1. (A) A Scatchard plot of the binding data for the interaction of chitotriose and egg white lysozyme at pH 5.5 and 25°. (B) A Hill plot of the same data as in Fig. 1A.

Scatchard plot gives a value of unity for the fractional occupancy at saturation. The Hill plot gives a value of 0.97 ± 0.05 for the Hill coefficient. Thus, all the potential binding sites are accounted for, and a single affinity constant describes the binding quite adequately. A more detailed account of the meaning of the Hill coefficient follows in later sections.

Qualitative Effects of Positive Cooperativity in Ligand Binding

The two methods of data treatment provide adequate procedures for the evaluation of the association constant for binding equilibria that are defined by a single, noninteracting binding site. For more complicated binding schemes, these simple arguments are no longer sufficient. The two procedures both give nonlinear plots.

Figures 2A and 2B show Scatchard and Hill plots for the binding of oxygen to hemoglobin.[8] This binding shows positive cooperativity. Phenomenologically, this implies that the first few ligands bind with a lower affinity than the subsequent ligands to the macromolecule. The most striking effect is seen in the Scatchard plot. This now shows a markedly nonlinear curve characterized by a well-pronounced maximum. A maximum in the Scatchard plot is characteristic of a system showing positive cooperativity. The position of the maximum shows the degree of the cooperativity to be evaluated. The cooperativity is more pronounced as the position of this maximum shifts along the abscissa to high degrees of saturation.[9]

The Hill plot is most nearly linear for the hemoglobin data from about 10% to 90% saturation. The slope of the Hill plot in this region has a value of 2.8. At very low and very high degrees of saturation, the slope of the Hill plot approaches unity. Under both these conditions the data is reflecting a single-site binding. At low degrees of saturation, this represents the binding of the first ligand to the macromolecule. At high degrees of saturation, the Hill plot represents the binding of the fourth ligand to hemoglobin molecules which have three ligands already bound. The first and fourth association constants can be estimated by extrapolation back to the abscissa intercept for each limiting slope of unity. These extrapolations are shown as dotted lines in Fig. 2B. Wyman[10] was the first to recognize this fact and has coined the term interaction energy, ΔG_{int}, to describe the free energy difference between the first and last ligand associations. Thus,

$$\Delta G_{int} = RT \ln (K_4/K_1)$$

For hemoglobin this corresponds to about 3.0 kcal/mol or 750 cal per site. While the interaction energy is a very useful quantity for defining the cooperativity of the binding, most binding data cannot be obtained with the accuracy required at very low or very high degrees of saturation. The most easily obtained data are in the region of half-saturation. In this region the Hill plot provides two parameters, the midpoint and the Hill coefficient.

The Hill coefficient is often used as a qualitative indicator of cooperative binding. Weber and Anderson[11] showed that the Hill coefficient can

[8] K. Imai, *J. Biol. Chem.* **249,** 7607 (1974).

[9] F. W. Dahlquist, *FEBS Lett.* **49,** 267 (1974).

[10] J. Wyman, *Adv. Protein Chem.* **19,** 223 (1964); J. Wyman, *Quart. Rev. Biophys.* **1,** 35 (1968).

[11] G. Weber and S. A. Anderson, *Biochemistry* **10,** 1942 (1965).

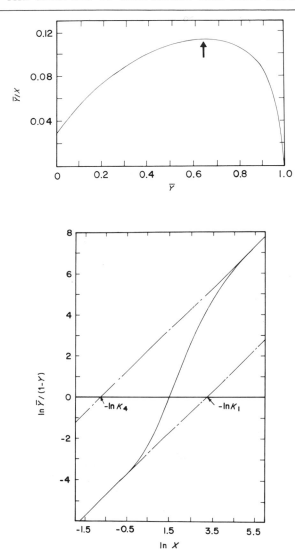

FIG. 2. (A) A Scatchard plot of the binding data for the interaction of oxygen with hemoglobin as determined by K. Imai [*J. Biol. Chem.* **249,** 7607 (1974)]. The solid line is deduced from the values of the four affinity constants deduced by Imai. The arrow shows the position of the maximum ($\bar{Y} = 0.64$). The concentration of free ligand is in partial pressure of oxygen, measured in Torr. (B) A Hill plot of the same data as in Fig. 2A. The solid line represents the binding data. The values of the affinity constants for the first and fourth ligand bound are shown as extrapolations from a very low and very high saturation.

never be greater than the number of binding sites per macromolecule, providing the macromolecule does not change its degree of polymerization due to ligand binding. It has been shown by Dahlquist[12] that if association–dissociation phenomena of the macromolecule are coupled to ligand binding, the Hill coefficient cannot exceed the largest number of binding sites present in the various macromolecular species. Thus, the Hill coefficient can vary between one and the largest number of binding sites when positive cooperativity is present.

Qualitative Effects of Site Heterogeneity or Negative Cooperativity

Many experimentally determined binding curves show a more gradual dependence of the number of ligands bound as a function of ligand concentration than would be predicted from a single binding affinity. There are many microscopic explanations for such behavior.

For example, the binding curves observed for hapten–antibody binding show this behavior. In such systems, populations of antibody molecules are present representing a range of different affinity constants. This binding-site heterogeneity requires more than one affinity to describe the binding isotherm. Similar behavior could be expected if a binding experiment were performed with a single macromolecular species that was partially denatured, with the partially denatured molecules binding the ligand with altered affinity.

The above situations are conceptually distinct from those systems that show negative cooperativity (or anticooperativity) in ligand binding. This latter class was first described by Conway and Koshland[13] for the binding of nicotine adenine dinucleotide to the enzyme glyceraldehyde-3-phosphate dehydrogenase from rabbit muscle. The enzyme contains four subunits, identical in amino acid sequence, each capable of binding one coenzyme. The binding data show that the first ligand bound has a higher apparent affinity than subsequent ligands. This could result from induced changes in the other binding sites resulting from the first ligand. An alternative explanation, originally proposed by Bernhard,[14] supposes that despite the fact that amino acid sequences are identical, the architecture of the tetrameric enzyme requires at least two distinct environments for the four binding sites. These two microscopic explanations cannot be distinguished by ligand binding isotherms since both predict the same mathematical expression for the binding equation. Site heterogeneity cannot be easily distinguished from negative cooperativity by binding measurements

[12] F. W. Dahlquist, *Eur. J. Biochem.* in press.
[13] A. Conway and D. E. Koshland, Jr., *Biochemistry* **7**, 4011 (1968).
[14] R. A. MacQuarrie and S. A. Bernhard, *J. Mol. Biol.* **55**, 181 (1971).

alone, although the algebraic form of the binding expression is different. However, a qualitative indicator of the phenomenon of negative cooperativity or site heterogeneity can be obtained from the Scatchard and Hill plots.

Figures 3A and 3B show the Scatchard and Hill plots for the binding of

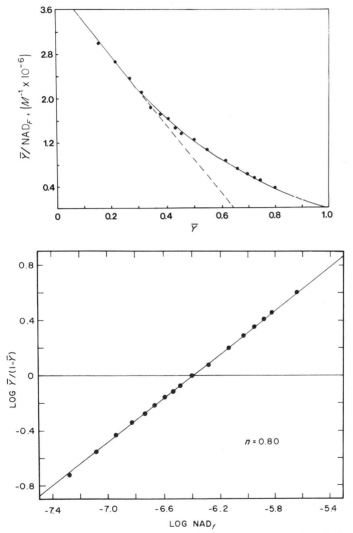

FIG. 3. (A) A Scatchard plot of the binding data for the interaction of nicotinamide adenine dinucleotide and sturgeon muscle glyceraldehyde-3-phosphate dehydrogenase at pH 7.0 and 25°. The dashed line represents the extrapolation of the initial part of the Scatchard plot to the $\bar{Y}$ axis. (B) A Hill plot of the same data as in Fig. 3A.

nicotinamide adenine dinucleotide to the glyceraldehyde-3-phosphate dehydrogenase isolated from sturgeon muscle. This binding is characterized by apparent negative cooperativity.

The Scatchard plot is concave and is decidedly nonlinear. The Hill plot of these data is quite linear, however. The slope of the Hill plot is less than unity. For the data shown in Fig. 3, the Hill coefficient is 0.80 ± 0.2. Thus negative cooperativity is characterized by a Hill coefficient less than unity.

Levitzki and Koshland[15] have pointed out that the qualitative shapes of the various plots provide an excellent indicator of positive and negative cooperativity in ligand binding. It is clear that either plot can be used to define the cooperativity in qualitative terms. The quantitative relationships are somewhat less clear and will be discussed in the following sections.

Quantitative Relationships between the Plots

It is clear that each of the two plotting methods contains the same intrinsic information concerning the binding equilibrium. This information is a function of the data itself, not the particular manner of plotting the data. What is less clear is the nature of the quantitative relationship between the curvature of the Scatchard plot and the slope of the Hill plot. This can be evaluated directly by comparison of the appropriate derivatives for each plotting procedure.

We shall consider the binding data to consist of a set of fractional occupancies of the binding sites, $\bar{Y}$, as a function of free ligand concentration, X.

The slope of the Scatchard plot can be evaluated as:

$$\frac{d(\bar{Y}/X)}{d\bar{Y}} = \frac{1}{X}\left[1 - \frac{\bar{Y}}{X}\frac{dX}{d\bar{Y}}\right] = \frac{1}{X}\left[1 - \frac{\bar{Y}\,d\ln X}{d\bar{Y}}\right] \tag{1}$$

The slope of the Hill plot:

$$\frac{d(\ln(\bar{Y}/1-\bar{Y}))}{d\ln X} = \frac{X}{Y}\left[\frac{1}{1-Y}\right]\frac{dY}{dX} \tag{2}$$

The Hill coefficient, n_H, is usually evaluated near half-saturation, and the Hill plot is usually linear in this region, so:

$$n_H = \frac{X}{4}\frac{d\bar{Y}}{dX} = \frac{1}{4}\left[\frac{d\bar{Y}}{d\ln X}\right] \tag{3}$$

[15] A. Levitzki and D. E. Koshland, Jr., *Proc. Natl. Acad. Sci. U.S.A.* **62**, 1121 (1969).

As expected, a simple relationship exists between the slopes of the two plots. Both are related to the derivative $d\bar{Y}/(d \ln X)$. Some rather important insights can be gained from these simple relationships.

A system showing positive cooperativity in ligand binding is characterized by a Scatchard plot showing a maximum. The position of the maximum can be related directly to the Hill coefficient.[9] For a maximum in the Scatchard plot:

$$0 = 1/X[1 - (\bar{Y}_m/X_m)(dX/d\bar{Y})]$$

or

$$[(d \ln \bar{Y})/(d \ln X)]_m = 1$$

where the subscript m refers to the quantity evaluated at the maximum.

Using the expression for the Hill slope at the fractional occupancy corresponding to the maximum [Eq. (2)]:

$$\text{Hill slope} = 1/(1 - \bar{Y}_m)$$

Since the Hill plot is usually quite linear in the region near half-saturation, we can say:

$$n_H = 1/(1 - \bar{Y}_m) \tag{4}$$

where $\bar{Y}_m$ is the position of the maximum in the Scatchard plot. Using the above relationships, a very good estimate of the Hill coefficient can be obtained from the position of the maximum in the Scatchard plot. This can be seen more clearly in Figs. 2A and 2B. The position of the maximum in Fig. 2A corresponds to a fractional saturation of about 0.65. From Eq. (4), this is equivalent to a Hill coefficient of about 2.8–2.9. This exactly corresponds to the value of the Hill coefficient obtained from the slope of the Hill plot in the region near the midpoint of Fig. 2B. Since the Hill plot is usually fairly linear in the region near the midpoint, the Hill coefficient can be somewhat more easily obtained than the position of the maximum in the Scatchard plot. For this reason, the Hill plot is the preferred method of plotting data showing positive cooperativity.

Quantitative Interpretations of the Scatchard and Hill Plots

Any quantitative interpretation of the binding data for systems not showing a single affinity requires more information than the binding curves themselves provide. As a result, quantitative conclusions require at least a minimal model for the molecular events controlling the binding. In many cases this can be quite general, or it may become rather specialized. The specific models for cooperative ligand binding to proteins proposed by Monod, Wyman, and Changeux[6] and Koshland, Némethy, and Filmer[5] are easily analyzed by the Hill and Scatchard plots.

As discussed above, the Hill plot is the preferred method for interpreting binding data showing positive cooperativity. For this reason we shall discuss the quantitative relationships of only the Hill coefficient to the various models for positive cooperativity. The quantitative interpretation of negative cooperativity or site heterogeneity will be explored for both the Scatchard and Hill plots. The Scatchard analysis is most useful in cases which show site heterogeneity or negative cooperativity.

Scatchard Plot

Binding Site Heterogeneity. This situation applies most directly to antigen–antibody interactions or other systems demonstrating a population of binding sites with a range of affinities distributed about some average value. If the fraction of sites with a particular association constant K_i is given by f_i, then $\bar{Y}$, the fractional saturation of all the sites, is the average fractional saturation of each kind of binding site[16]:

$$\bar{Y} = \sum_{\substack{\text{all} \\ \text{sites}}} f_i[K_i X/(1 + K_i X)] \tag{5}$$

If binding data can be obtained at low fractional saturation values, the intercept of the Scatchard plot can be obtained by extrapolation to $Y = 0$. Under these conditions we can evaluate the Scatchard plot intercept as:

$$\text{intercept} = \lim_{\bar{Y} \to 0} (\bar{Y}/X) = \lim_{X \to 0} \sum_{\substack{\text{all} \\ \text{sites}}} (f_i K_i)/(1 + K_i X)$$

$$= \sum_{\substack{\text{all} \\ \text{sites}}} f_i K_i = \bar{K}_{\text{form}} \tag{6}$$

Thus the Scatchard plot intercept gives the average association constant, $\bar{K}_{\text{form}}$.

We may evaluate the slope of the Scatchard plot by substitution of the expression for $\bar{Y}$ into the expression for the Scatchard plot slope at any position [Eq. (1)]:

$$\text{slope} = - \frac{\Sigma[(f_i K_i^2 X)/(1 + K_i X)^2]}{\Sigma[(f_i K_i X)/(1 + K_i X)^2]} \tag{7}$$

Figure 4 shows a Scatchard plot of the data of Nisonoff and Pressman[17] for the interaction of rabbit anti-p-azobenzoate with the

---

[16] R. N. Pinckard and D. M. Weir, *in* "Handbook of Experimental Immunology" (D. M. Weir, ed.), chap. 14, Blackwell, Oxford, Edinburgh 1967.

[17] A. Nisonoff and D. Pressman, *J. Immunol.* **80**, 417 (1958).

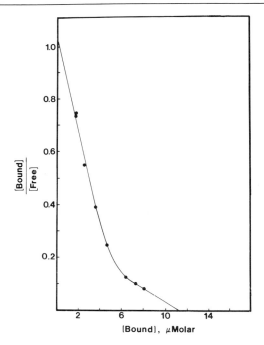

FIG. 4. A Scatchard plot of the binding data for the interaction of rabbit anti-*p*-azobenzoate with the hapten *p*-iodobenzoate [A. Nisonoff and D. Pressman, *J. Immunol.* **80**, 417 (1958).

hapten *p*-iodobenzoate. The Scatchard plot appears to be biphastic with different limiting slopes at low and high fractional saturation. Since the slope of the Scatchard plot for a system showing a single affinity is simply the negative of the association constant, it might seem reasonable that the limiting slopes represent the apparent affinity at low and high degrees of saturation. In fact we might define an apparent constant K_{app} as just the negative of Scatchard slope at any position. Thus:

$$K_{app} = \frac{\Sigma f_i [K_i^2 X/(1 + K_i X)^2}{\Sigma f_i [K_i X/(1 + K_i X)^2}$$ (8)

Providing the range in affinities is not too great, we may evaluate the limiting slopes at low and high fractional saturation. At low saturation $K_i X \ll 1$ for all sites and the apparent affinity from the Scatchard plot under these conditions becomes K_{app}^0:

$$K_{app}^0 = \frac{\Sigma f_i K_i^2}{\Sigma f_i K_i}$$ (9)

Thus the apparent limiting slope at low saturation gives the mean square affinity divided by the mean. This value will be weighted more strongly by the high affinity sites than by the low affinity ones.

At high saturation, the apparent affinity K_{app} can be evaluated since $K_i X \ll 1$ for all sites and:

$$K_{app} = \frac{1}{\Sigma f_i(1/K_i)}$$

For this situation it is more reasonable to consider an apparent dissociation constant, $1/K_{app}$:

$$1/K_{app} = \Sigma f_i(1/K_i) = \overline{K}_{diss} \tag{10}$$

or the apparent dissociation constant is the average dissociation constant, averaged over the total population of antibody molecules. It must be stressed that these limiting slopes should be evaluated only at very low ($\overline{Y} < 0.1$) and very high saturation (>0.9) for these relationships to apply properly.

We find a rather simple situation developing for the analysis of the binding data using the Scatchard plot. Extrapolation of the intercept to zero saturation gives the average formation constant for binding, while the limiting slope at high saturation gives the average dissociation constant for the interactions of the various binding sites with ligand. The initial slope provides a somewhat more complicated expression which relates to the second moment of the binding constant distribution.

Notice that if there is a single affinity for all the antibody molecules, K_{app}^0, $\overline{K}_{form}$, and $1/\overline{K}_{diss}$ are all equal and give a direct measure of the affinity and a straight line (constant slope) Scatchard plot will be observed. As the range of affinities becomes greater, the limiting slopes will become more different, and comparison of the limiting values will give some idea of the width and type of distribution of affinity constants.

In principle, the Scatchard plot can provide the average association constant, the mean square association constant and the mean dissociation constant. This can be seen more clearly in Fig. 4, where the binding of the hapten p-iodobenzoate and an antibody population isolated from a particular rabbit are presented. Extrapolation to zero binding gives a value for $\overline{K}_{assoc}$ of $9.3 \times 10^4 \, M^{-1}$. The limiting slope at high saturation gives a value for $(1/\overline{K}_{diss})$ of $2.5 \times 10^4 \, M^{-1}$. Thus there is about a 4-fold ratio in the two average parameters. Some feeling of the meaning of this factor of 4 can be gained if one considers only two equal populations of antibodies, each defined by a particular affinity. When $\overline{K}_{assoc} \cdot \overline{K}_{diss}$ equals four, the two populations have a 4-fold difference in intrinsic affinity as well. Thus $\overline{K}_{assoc} \cdot \overline{K}_{diss}$ is a measure of the range of affinities present.

To be more exact, one could assume a Gaussian distribution of binding energies[18] and use that relationship to calculate the relationships between the average dissociation constant, average association constant, and the mean-square association constant. Any two of these three parameters would determine the standard deviation of this Gaussian distribution of binding energies about the mean value. Thus the standard deviation can be determined twice from the three parameters. If these two determinations agree well, it would be reasonable to conclude that the binding energies are indeed Gaussian distributed.

Negative Cooperativity. Qualitatively, the Scatchard plot observed for a system showing negative cooperativity and that observed for site heterogeneity can be quite similar. The negative cooperativity of the binding shown in Fig. 3A gives a concave Scatchard plot and appears to consist of two parts. One might suppose that the limiting slope at low saturation represents the strongest binding while that observed at high saturation represents the weakest binding. This is not exactly correct, however. This can be seen more clearly if we examine the simplest case that can show negative cooperativity, a simple two-site binding. For simplicity we shall assume that the negative cooperativity is an induced phenomenon. Later we will give the general result that applies to either induced or preexisting asymmetry models.

Consider a macromolecule, M, with two identical sites for a ligand:

$$M + X \underset{}{\overset{K_1}{\rightleftharpoons}} MX_1 \underset{}{\overset{K_2}{\rightleftharpoons}} MX_2 \tag{11}$$

The binding of the first ligand is governed by an association constant K_1, while the binding of the second ligand is controlled by the association constant K_2. For such a scheme the fractional occupancy of the binding site is given by:

$$\bar{Y} = \frac{K_1 X + K_1 K_2 X^2}{1 + 2K_1 X + K_2 K_2 X^2} \tag{12}$$

for negative cooperativity to be seen in ligand binding $K_1 > K_2$.

The relationship of the slope of the Scatchard plot to the two constants K_1 and K_2 may be evaluated directly. If we define the apparent affinity as the slope of the Scatchard plot, then (using Eq. 1):

$$K_{app} = \frac{2K_1 - K_2 + 2K_1 K_2 X + K_1 K_2^2 X^2}{1 + 2K_2 X + K_1 K_2 X^2} \tag{13}$$

At low saturation, $1 \gg K_1 X > K_2 X$ and the limiting slope becomes:

$$K_{app}^0 = -(2K_1 - K_2) \tag{14}$$

[18] L. Pauling, D. Pressman, and A. L. Grossberg, *J. Am. Chem. Soc.* **66**, 784 (1944).

For negative cooperativity $K_1 > K_2$ and the Scatchard slope is negative. Important, however, is that the limiting slope does not give K_1 directly, but rather this more complicated form involving both K_1 and K_2. For highly negatively cooperative systems, K_2 would be small with respect to K_1 and the slope would differ from K_1 by a factor of 2.

For more than two binding sites, it is clear that a still more complicated expression for the initial slope of the Scatchard plot will result. It can be shown that the most negative value for the limiting slope at low degrees of saturation for a system with N identical binding sites is simply $-NK_1$. It is clear that if the identical binding sites are independent, a linear Scatchard plot of slope equal to $-K_1$ would result. From this we can set boundaries on the value of the initial slope of the Scatchard plot, provided no positive cooperativity is present:

$$-NK_1 < \text{limiting slope} < K_1 \tag{15}$$

where N is the number of identical binding sites. If the binding sites are nonidentical, then the most negative limit would become the first Adair[19] constant which is the sum of all the association constants for the first ligand being bound to the individual sites. The Adair treatment is discussed below.

The limiting slope of the Scatchard plot for the two identical binding site case at high fractional saturation is simply equal to $-K_2$ when the terms which are second order in ligand concentration dominate. For a system with N identical but interacting binding sites, the limiting slope at high saturation is simply $-K_N$, where K_N is the association constant of the ligand for the macromolecule when all but one of the binding sites are already occupied by ligand. If the sites are nonidentical and interact, the results are more difficult to interpret.

For a macromolecule with two binding sites with different affinities and no interaction between them, the limiting slopes at high and low saturation can be evaluated directly. For such a situation the fractional saturation of these nonequivalent binding sites is given by:

$$\bar{Y} = \frac{\frac{1}{2}(K_1 + K_1')X + K_1 K_1' X^2}{1 + (K_1 + K_1')X + K_1 K_1' X^2} \tag{16}$$

where K_1 and K_1' are the affinities of the nonequivalent, noninteracting binding sites.

The limiting slope at low fractional saturation becomes:

$$K_{\text{app}}^0 = (K_1^2 + K_1'^2)/(K_1 + K_1') \tag{17}$$

[19] G. S. Adair, *J. Biol. Chem.* **63**, 529 (1925).

while at high fractional saturation:

$$K^1_{app} = (2K_1K'_1)/(K_1 + K'_1) \tag{18}$$

or

$$\bar{K}_{diss} = 1/K^1_{app} = \frac{1}{2}(1/K_1 + 1/K'_1)$$

Once again, the Scatchard plot intercept at low saturation gives the average formation constant. The reciprocal limiting slope at high saturation can be interpreted as the mean dissociation constant. This is true, independent of the number of nonidentical noninteracting sites involved. This follows because this model of multiple binding sites without interaction between them is a special case of the general situation of binding site heterogeneity discussed above. The added restraint is that the fraction of sites of the various types are equal (or at least integer multiples of each other), since they correspond to the various types of subunits within the multimer.

Extrapolation of the Initial Scatchard Slope to the $\bar{Y}$ Axis. One of the great advantages of the Scatchard plot is that the number of total potential binding sites may be easily obtained by extrapolation. This extrapolation is valid only when a single affinity constant describes the binding. Many workers extend such extrapolations to binding data showing negative cooperativity. It is tempting to carry out these extrapolations because the Scatchard plot has a decidedly biphasic quality when negative cooperativity is present. Thus the extrapolation of the initial slope to the abscissa ($\bar{Y}$ axis) is thought to give the number of high-affinity sites present. Such an extrapolation is shown in Fig. 3A. As we shall now demonstrate, such interpretations are patently incorrect. In fact the qualitative interpretation of this extrapolation is quite similar to that of the Hill coefficient when positive cooperativity is present. It represents the *minimum* number of sites of that affinity.

Let us again consider the two identical site case with interaction between them. The intercept (when $\bar{Y} = 0$) of the Scatchard plot is given by the limit of $\bar{Y}/X$ as $\bar{Y}$ and therefore X approaches zero. Hence

$$\text{intercept} = \lim_{x \to 0} \frac{K_1 + K_1K_2X}{1 + 2K_1X + K_1K_2X^2} = K_1 \tag{19}$$

The initial slope is given by $-(2K_1 - K_2)$ so extrapolation to the abscissa of the initial slope gives:

$$\bar{Y}_{extrap} = K_1/(2K_1 - K_2) \tag{20}$$

For a situation showing negative cooperativity $K_1 > K_2$ and $0.5 < \bar{Y}_{extrap} < 1.0$. This clearly shows that the position of the extrapolated value of $\bar{Y}$ depends on the cooperativity, not directly on the number of sites.

More insight can be gained if we consider the general case of N binding sites that are not necessarily identical. The binding expression for this general situation is given by the general Adair expression:

$$\bar{Y} = (1/N) \sum_{i=1}^{N} i\psi_i X^i/(1 + \Sigma\psi_i X^i) \tag{21}$$

The parameters ψ_i are phenomenological constants describing the binding of i ligands to the macromolecule.

The intercept of the Scatchard plot is given by ψ_1/N. Evaluation of the initial slope of the Scatchard plot using this general expression gives a limiting value of the initial slope as $-\psi_1$. As a result, the extrapolated value of $\bar{Y}$ obtained from the initial slope must be $1/N$ or larger. This provides very important information about the degree of anticooperative binding expressed. The smallest the extrapolated value of $\bar{Y}$ can be is $1/N$. This would correspond to the maximum anticooperative binding possible when all the macromolecular species present have only one ligand bound. As the degree of anticooperative interaction weakens and individual species exist with more than one ligand bound when an average of one ligand is bound per mole of macromolecule, the extrapolated value of $\bar{Y}$ will increase from its limiting value of $1/N$.

These results suggest that the value of $\bar{Y}$ extrapolated from the initial slope of the Scatchard plot when anticooperativity is present provides the same kind of information as the Hill coefficient when positive cooperativity is present. The value of the extrapolated value of $\bar{Y}$ can vary between $1/N$ and unity when negative cooperativity is present. The reciprocal of this extrapolated value therefore represents the minimum number of binding sites present. Comparison of this value with the actual number of binding sites present is a good qualitative measure of the degree of anticooperativity present.

Hill Plot

The Hill (Sips or Nernst) plot of log [sites bound]/[sites free] versus log [free ligand] is really a plot of the difference in chemical potential between the bound and free macromolecular sites as a function of the chemical potential of the binding ligand. In principle, these chemical potentials contain the important thermodynamic information. In fact the statistical mechanical meaning of the Hill plot has been investigated by Wyman[10] and Heck.[20] Rather sophisticated, model independent, conclusions about

[20] H. deA. Heck, J. Am. Chem. Soc. **93**, 23 (1971).

the information content of the Hill plot have been drawn by these workers. The interested reader is referred to their elegant work for such discussions.

We shall be concerned with the two parameters that may be obtained from Hill plots of the majority of experimental data, the slope of the region of half-saturation or Hill coefficient n_H, and the half-saturating ligand concentration X_M. With extremely good data, the Hill plot may be extended to very low and very high saturation, and other parameters such as the interaction energy may be obtained. Since one can obtain such precise binding data in only a limited number of systems, we shall restrict our attention to the Hill coefficient and half-saturating ligand concentration. The important question becomes the nature of the relationship between these experimental parameters and the thermodynamics of the binding itself.

Binding-Site Heterogeneity. As discussed above, the fractional saturation for this case is given by:

$$\bar{Y} = \sum_{\substack{all \\ sites}} (f_i K_i X)/(1 + K_i X)$$

Using Eq. (3), the Hill coefficient may be evaluated as:

$$n_H = \frac{1}{4}[d\bar{Y}/(d \ln X)] \qquad \text{evaluated at half saturation}$$

Taking the derivative gives:

$$n_H = \frac{1}{4} \sum_{\substack{all \\ sites}} (f_i K_i X_m)/(1 + K_i X_m)^2$$

and the condition for half saturation gives:

$$\frac{1}{2} = \sum_{\substack{all \\ sites}} (f_i K_i X_m)/(1 + K_i X_m)$$

Now these two expressions are not particularly useful unless one makes some assumption concerning the fraction of molecules, f_i, with affinity, K_i, and the affinity itself. Again this may involve the supposition of a Gaussian distribution in binding energy, and then the Hill coefficient can be related to the spread in binding energy.

It seems clear the Scatchard plot is a better way to analyze such a situation.

Cooperativity. The simplest case which can show cooperativity is the two site situation. It is often used as a simple model for negative cooperative effects using a "functional dimer" model when there is an even number of binding sites.

For a simple two-site case (or functional dimer model), the most general expression for the fractional saturation of those binding constants is given by an Adair equation:

$$\bar{Y} = (\tfrac{1}{2}\psi_1 X + \psi_2 X^2)/(1 + \psi_1 X + \psi_2 X^2) \qquad (22)$$

where ψ_1 and ψ_2 are phenomenological constants representing the formation of mono- and di-bound macromolecules from the unbound state. This is quite generally correct provided the macromolecule does undergo dissociation–association phenomena and equilibrium is reached. For this situation use of Eq. (1) and evaluation of Eq. (22) at-half saturation gives

$$n_H = 4/(\psi_1/\sqrt{\psi_2} + 2) \qquad (23)$$

and

$$X_m = 1/\sqrt{\psi_2} \qquad (24)$$

where x_m is the concentration of ligand at half-saturation.

Note that both phenomenological constants ψ_1 and ψ_2 can be determined from the Hill slope and midpoint.

If negative cooperativity results from identical sites with induced interactions ($\psi_1 = 2K_i$ and $\psi_2 = K_1 K_2$), then:

$$n_H = 2/(\sqrt{K_1/K_2} + 1) \qquad (25)$$
$$X_m = 1/\sqrt{K_1 K_2} \qquad (26)$$

Both microscopic constants K_1, the association constant for the first ligand, and K_2, the association constant for the second ligand, can now be obtained from the Hill plot. Note that the Hill coefficient is less than unity when $K_1/K_2 < 1$ and is equal to unity when $K_1 = K_2$. When $K_2 > K_1$, the Hill coefficient is greater than 1 but less than 2, and positive cooperativity is observed. The half-saturating ligand concentration is a geometric mean of the affinity constants K_1 and K_2. The reciprocal of the half-saturating ligand concentration has the interpretation of an affinity representing the average binding energy for the formation of the mono- and di-bound species.

One could also interpret the two site binding case in terms of two different and noninteracting sites described by affinity constants K_1 and K_1'. This would lead to the expressions:

$$n_H = 4/(\sqrt{K_1/K_1'} + \sqrt{K_1'/K_1} + 2) \qquad (27)$$

and

$$X_m = 1/\sqrt{K_1' K_1} \qquad (28)$$

For such a model, the Hill coefficient varies from unity to values less than unity when K_1/K_1' is much different from 1.

These models for two-site binding predict *identical* binding isotherms and hence cannot be distinguished by binding measurements alone. However, the parameters of these models can be evaluated directly from the Hill plot without the necessity of computer-aided curve-fitting procedures.

These arguments may be employed to evaluate the Hill plot for an N site case. For a general situation of N sites on a macromolecule in which association-dissociation phenomena are not present:

$$\bar{Y} = 1/N \left[\left(\sum_{i=1}^{N} i\psi_i X^i \right) \Big/ \left(1 + \sum_{i=1}^{N} \psi_i X^i \right) \right]$$

and the Hill coefficient becomes:

$$n_{\mathrm{H}} = 1/4N \left[\left(\sum_{i=1}^{N} i^2 \psi_i X^i \right) \Big/ \left(1 + \sum \psi_i X^i \right) - (N/2)^2 \right] \qquad (29)$$

Wyman[10] has pointed out that this general expression has an important meaning. The fraction $\psi_i X^i/(1 + \Sigma\psi_i X^i)$ is simply the fraction of molecules with i ligands bound. Then the term $\Sigma i^2 \psi_i X^i/(1 + \Sigma\psi_i X^i)$ is the average value of the square of the number of ligands bound. The term $(N/2)^2$ is the square of the average number of ligands bound at half saturation. So the Hill coefficient is a measure of the mean square deviation of the number of ligands bound at half-saturation. Consider a situation with four equivalent and equal binding sites. At half-saturation the fraction of molecules with 0, 1, 2, 3, and 4 ligands bound is given by a normal distribution with values $1/16$, $1/4$, $1/2$, $1/4$, and $1/16$, respectively. For positive cooperativity, the Hill coefficient is larger than unity and the observed distribution is broader than this. Thus there are proportionally more zero- and mono-bound and tri- and tetra-bound than di-bound species at half-saturation. Similarly, negative cooperativity has a Hill coefficient less than 1, and there is a correspondingly larger fraction of di-bound species present at half-saturation. Thus the Hill coefficient really is a measure of the second moment of the bound-state distribution.

Weber and Anderson[11] were able to show that Eq. (29) can never be larger than N for any values of the Adair constants. So the Hill coefficient can be used as a measure of the minimum number of binding sites.

In order to gain more specific information about the macroscopic constants, it is necessary to relate the phenomenological Adair constants to each other with the aid of some model. For systems showing cooperativity, two such models are often considered: those of Monod, Wyman, and Changeux[6] (MWC) and of Koshland, Neméthy, and Filmer[5] (KNF).

In its simplest form the MWC model supposes that a protein of N sites exists in a form with identical, but very poor, affinity. This form is in equilibrium with a minority species again with N identical binding sites, but now of much greater affinity. In order to bind the first ligand, the energetically unfavorable step of conversion of the protein from the low-affinity to the high-affinity form is required. Subsequent ligands bind without the necessity of performing this conversion and therefore have higher affinity and positive cooperativity results. The model is specified by the allosteric constant, L, the equilibrium constant between the low and high affinity forms of the enzyme; and K_r, the affinity of the high affinity sites. In the model's simplest form, the low-affinity sites are never occupied. The expression for $\bar{Y}$ becomes:

$$\bar{Y} = [LK_rX(1 + K_rX)^{N-1}]/[1 + L(1 + K_rX)^N] \tag{30}$$

and substitution into Eq. (3) and appropriate rearrangement gives:

$$L = (\alpha + 1)^{N-1} (\alpha - 1) \tag{31}$$

where

$$\alpha = (N + n_H - 2)/(N - n_H)$$

and

$$X_m = \alpha/K_r \tag{32}$$

This shows very clearly that the allosteric constant L determines the value of the Hill coefficient and K_r and L together establish the half-saturating ligand concentration. Thus the Hill plot may be used to evaluate K_r and L without the need for computer fitting.

The relationship between L and n_H is shown in Fig. 5 for the four-site case. Rather small values of L are sufficient to cause n_H to deviate substantially from unity. However to generate high cooperativity, a large value of L is necessary. Thus to achieve a Hill coefficient of 3.0 for this tetrameric case, L must have a value of 864.

It is clear than the Hill plot has two easily measured parameters, the slope and midpoint, and hence models for cooperativity with two *independent* parameters may be evaluated directly using the Hill plot. The simplest KNF model for cooperativity supposes that ligand binding causes conformational changes in subunits and that the energetics of the cooperativity are related to the relative energies of interaction of the subunits depending on whether ligands are bound to those subunits. Cooperativity is determined by the relative stability of a liganded subunit–unbound subunit interaction as compared to the interactions of two bound subunits or two free ones. Clearly the number of subunit contacts depends on the architecture of the subunits. For a tetrameric protein in a tetrahedral geometry, each subunit contacts three others and the expression for $\bar{Y}$ is:

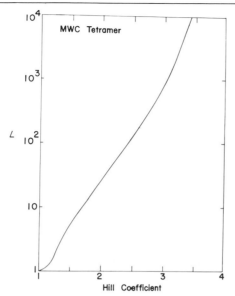

FIG. 5. A plot of the relationship between L, the allosteric constant, and the Hill coefficient for a tetramer according to the MWC model [J. Monod, J. Wyman, and J.-P. Changeux, *J. Mol. Biol.* **12**, 88 (1965)] for cooperative binding behavior.

$$\bar{Y} = \frac{1}{4} \left[\frac{\begin{array}{c} 4K_{AB}^{3}[K_{s}K_{t}(X)] + K_{AB}^{4}K_{BB}[K_{s}K_{t}(X)]^{2} \\ + 12K_{AB}^{3}K_{BB}^{3}[K_{t}K_{s}(X)]^{3} + 4K_{BB}^{6}[K_{t}K_{s}(X)]^{4} \end{array}}{\begin{array}{c} 1 + 4K_{AB}^{3}[K_{s}K_{t}(X)] + 6K_{AB}^{4}[K_{s}K_{t}(X)]^{2} \\ + 4K_{AB}^{3}[K_{s}K_{t}(X)]^{3} + K_{BB}^{6}[K_{s}K_{t}(X)]^{4} \end{array}} \right] \tag{33}$$

A square planar geometry is also possible such that each subunit is in contact with two others, then $\bar{Y}$ becomes:

$$\bar{Y} = \frac{1}{4} \left[\frac{\begin{array}{c} 4K_{AB}^{2}[K_{s}K_{t}(X)] + 4(K_{AB}^{4} + 2K_{AB}^{2}K_{BB})[K_{s}K_{t}(X)]^{2} \\ + 12K_{AB}^{2}K_{BB}^{2}[K_{s}K_{t}(X)]^{3} + 4K_{BB}^{4}[K_{s}K_{t}(X)]^{4} \end{array}}{\begin{array}{c} 1 + 4K_{AB}^{2}[K_{s}K_{t}(X)] + (2K_{AB}^{4} + 4K_{AB}^{2}K_{BB})[K_{s}K_{t}(X)]^{2} \\ + 4K_{AB}^{2}K_{BB}^{2}[K_{s}K_{t}(X)]^{3} + K_{BB}^{4}[K_{s}K_{t}(X)]^{4} \end{array}} \right] \tag{34}$$

Substitution of these expressions into Eq. (3) and evaluation at half-saturation gives:

Tetrahedral

$$n_{H} = \frac{4[(K_{AB}^{2}/K_{BB})^{3/2} + 1]}{3(K_{AB}^{2}/K_{BB})^{2} + 4(K_{AB}^{2}/K_{BB})^{3/2} + 1} \tag{35}$$

$$X_{m} = 1/(K_{BB}^{3/2}K_{s}K_{T}) \tag{36}$$

Square

$$n_H + \frac{4[(K_{BB}^2/K_{BB}) + 1]}{(K_{AB}^2/K_{BB})^2 + 6(K_{AB}^2/K_{BB}) + 1} \tag{37}$$

$$X_m = 1/(K_{BB}K_S K_T) \tag{38}$$

Again the striking result is that the Hill coefficient is simply related to only *one* independent parameter, the quantity K_{AB}^2/K_{BB}. This controls the cooperativity. When K_{AB}^2/K_{BB} is equal to unity, the Hill coefficient is unity. When K_{AB}^2/K_{BB} is greater than 1, both geometries predict negative cooperativity, while values of K_{AB}^2/K_{BB} less than 1 generate positive cooperativity. These quantitative relationships between the Hill coefficients and K_{AB}^2/K_{BB} are shown in Figs. 6 and 7.

It should be stressed that one usually cannot distinguish between the various models using this simple analysis of the Hill plot. More extensive and careful binding measurements are required, particularly at very high and very low saturation, to distinguish mechanisms. The models are essentially indistinguishable for binding data taken near half-saturation with the usual precision.

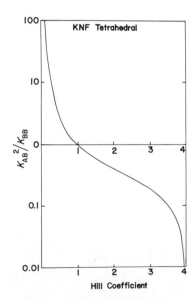

FIG. 6. A plot of the relationship between K_{AB}^2/K_{BB} and the Hill coefficient for a tetrameric protein with tetrahedral geometry in the KNF model [D. E. Koshland, Jr., G. Némethy, and D. L. Filmer, Biochemistry 5, 365 (1966)] for cooperative binding behavior.

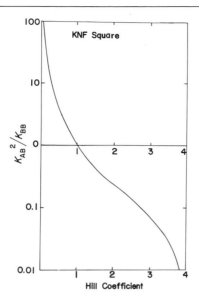

FIG. 7. A plot of the relationship between K_{AB}^2/K_{BB} and the Hill coefficient for a tetrameric protein with square geometry in the KNF model [D. E. Koshland, Jr., G. Némethy, and D. L. Filmer, *Biochemistry* **5**, 365 (1966)] for cooperative binding behavior.

These analyses depend on the fact that the majority of experimental binding data give approximately linear Hill plots near half-saturation. Occasionally this is not observed, and the Hill plot has "bumps" and the slope is difficult to determine. Cornish-Bowden and Koshland[21] have investigated this problem and concluded that it is usually the result of a system showing a mixture of positive and negative cooperativity. When such curves are observed, one must resort to curve-fitting procedures to establish the functional form of the binding isotherm.

Recently, Henis and Levitzki[22] have considered the results of these models for positive cooperativity on the Scatchard plot. They show that useful information may be obtained at very low and very high degrees of saturation. This information is related to the values of the Adair constants derived for the various models, and may be used to evaluate the parameters of the various models. Their results are summarized in the table.

As can be seen, there are rather complicated relationships between the Scatchard slopes at low and high saturation and the parameters of the models. Again these parameters also control the Hill plot and are perhaps more easily related to the Hill coefficient when positive cooperativity is present.

[21] A. Cornish-Bowden and D. E. Koshland, Jr., *J. Mol. Biol.* **95**, 201 (1975).
[22] Y. I. Henis and A. Levitzki, *Eur. J. Biochem.* **71**, 529 (1976).

THE EXPRESSION FOR SCATCHARD SLOPES ACCORDING TO DIFFERENT MODELS[a]

Model	ψ_1	ψ_2	Slope when $[X] \to 0$	Slope when $[X] \to \infty$
Koshland et al.[c]				
(a) Dimer[b]	$2K_{AB}K_SK_t$	$K_{BB}(K_SK_t)^2$	$\left(\frac{K_{BB}}{K_{AB}} - 2K_{AB}\right)K_SK_t$	$-\frac{K_{BB}}{K_{AB}}K_SK_t$
(b) Tetramer–tetrahedral case	$4K_{AB}^3K_SK_t$	$6K_{AB}^4K_{BB}(K_SK_t)^2$	$(3K_{BB} - 4K_{AB}^2)K_{AB}K_SK_t$	$-\left(\frac{K_{BB}}{K_{AB}}\right)^3K_SK_t$
(c) Tetramer–square case (no diagonal interactions)	$4K_{AB}^2K_SK_t$	$2(2K_{BB} + K_{AB}^2)K_{AB}^2(K_SK_t)^2$	$(2K_{BB} - 3K_{AB}^2)K_SK_t$	$-\left(\frac{K_{BB}}{K_{AB}}\right)^2K_SK_t$
Monod et al.[d]				
(a) Dimer-exclusive[b] binding ($K_T = 0$)	$\frac{2}{(1 + L)K_R}$	$\frac{1}{(1 + L)K_R^2}$	$\frac{L - 1}{(1 + L)K_R}$	$-\frac{1}{K_R}$
(b) Dimer-nonexclusive[b] binding	$\frac{2(K_T + LK_R)}{(1 + L)K_R}$	$\frac{K_T^2 + LK_R^2}{(1 + L)K_R^2K_T^2}$	$\frac{(L - 1)K_T^2 + (1 - L)LK_R^2 - 4LK_RK_T}{(1 + L)(K_T + LK_R)K_RK_T}$	$-\frac{K_T^2 + LK_R^2}{(K_T + LK_R)K_RK_T}$
(c) Tetramer-exclusive binding	$\frac{4}{(1 + L)K_R}$	$\frac{6}{(1 + L)K_R^2}$	$\frac{3L - 1}{(1 + L)K_R}$	$-\frac{1}{K_R}$
(d) Tetramer-nonexclusive binding	$\frac{4(K_T + LK_R)}{(1 + L)K_RK_T}$	$\frac{6(K_T^2 + LK_R^2)}{(1 + L)K_R^2K_T^2}$	$\frac{(3L - 1)K_T^2 + (3 - L)LK_R^2 - 8LK_RK_T}{(1 + L)(K_T + LK_R)K_RK_T}$	$-\frac{K_T^4 + LK_R^4}{K_RK_T(K_T^3 + LK_R^3)}$
Preexistent asymmetry				
(a) Dimer[b]	$K_1 + K_1'$	K_1K_1'	$-\frac{K_1^2 + K_1'^2}{K_1 + K_1'}$	$-\frac{2K_1K_1'}{K_1 + K_1'}$
(b) Tetramer (dimer of dimers)	$2(K_1 + K_1')$	$(K_1 + K_1')^2 + 2K_1K_1'$	$-\frac{K_1^2 + K_1'^2}{K_1 + K_1'}$	$-\frac{2K_1K_1'}{K_1 + K_1'}$

[a] The parameters used for the Koshland model are as defined originally by Koshland *et al.* L is the allosteric equilibrium constant and K_T and K_R are the intrinsic dissociation constants of the ligand to the R and T states, respectively, as defined originally by Monod *et al.*

[b] It can be shown that a multisubunit assembly composed of isologous dimers, and in which the subunit interactions occur only within the dimers, behaves as a dimer, and thus the equations applicable to the dimer case can be used.

[c] D. E. Koshland, Jr., G. Némethy, and D. L. Filmer, *Biochemistry* **5,** 365 (1966).

[d] J. Monod, J. Wyman, and J.-P. Changeux, *J. Mol. Biol.* **12,** 88 (1965).

Ligand Binding to a Linear Lattice of Binding Sites: Protein– Nucleic Acid
   Interactions

The interaction of ligands with polymers without discrete binding sites
may also be analyzed using the Scatchard or Hill approach. The clearest
example of such interactions involve the binding of proteins to a DNA (or
RNA) molecule. If there is no specificity of base sequence involved in the
binding, the DNA may be considered to represent a linear array of a large
number of binding sites. The total number of binding sites would simply
be the ratio of the number of bases in the DNA molecule divided by the
number of bases occupied by the protein. If there are no cooperative
interactions which either attract or repel protein molecules on the DNA,
one might expect to see a linear Scatchard plot or Hill plot of unit slope.
This is *not* the case, however. At low degrees of saturation, each protein
will be presented with a large number of possible binding sites, each of
which is independent and equivalent. However, as the degree of saturation increases, it is necessary that there be a sufficient number of free
bases arranged sequentially along the DNA to form a binding site. Thus
the number of binding sites available becomes *less* than the total number
of free bases divided by the number of bases occupied by the protein,
since some of the free bases are in linear arrays not large enough to form a
binding site. As a result, an apparent negative cooperativity in protein
binding occurs at high degrees of saturation. The Scatchard plot would be
concave, and the Hill coefficient would be less than unity.

McGhee and von Hippel[23] analyzed this situation in detail for both
noninteracting and interacting binding of a protein along a linear DNA
chain. Their analysis is quite complete, and we shall only quote the important relationships derived in that work. If we let K represent the association constant of the protein for a free binding site in the absence of
cooperative ligands, and define a number, m, as the number of bases
covered by the binding of the protein, and the number of protein
molecules bound per base as v, then the Scatchard binding equation
becomes:

$$v/X = K(1 - mv) \left[\frac{1 - mv}{1 - (m - 1)v} \right]^{m-1} \tag{39}$$

where X is the free concentration of protein.

When the protein covers only one site, the equation reduces to the
usual Scatchard equation when no cooperativity is present. However, as
m becomes large, greater and greater deviations from a linear Scatchard
plot are observed. This can be seen in Fig. 8, where various theoretical

[23] J. D. McGhee and P. H. von Hippel, *J. Mol. Biol.* **86**, 469 (1974).

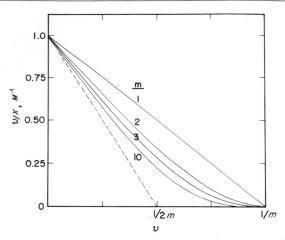

Fig. 8. Theoretical Scatchard plots of the binding of protein to a nucleic acid for various values of m, the number of bases required for a single protein binding site. These theoretical curves do not include any interactions between adjacent protein molecules. The value of K is fixed at $1\ M^{-1}$

curves have been constructed with m varying from 1 to 10 and have been plotted using an arbitrary value of $K = 1\ M^{-1}$.

McGhee and von Hippel[23] demonstrated that the limiting slope of the Scatchard plot at low degrees of saturation is:

$$\lim_{\nu \to 0} \text{limiting slope} = -K(2m - 1) \tag{40}$$

So at low saturation, a value of K can be estimated if m has been determined. The value of m may be estimated by the curvature of the Scatchard plot. For example, the limiting slope may be evaluated at saturation ($\nu = 1/m$):

$$\lim_{\nu \to 1/m} \text{limiting slope} = 0$$

This implies that the overlapping binding essentially inhibits fully the last ligand binding if the DNA chain is infinitely long. As a result, it is difficult to estimate the total number of binding sites present. If one only considers data taken at low degrees of saturation, use of Eq. (40) gives an extrapolated value of ν equal to $1/(2m - 1)$, or, if m is large, the actual number of binding sites will be underestimated by a factor of 2. This extrapolation is shown as the dashed line in Fig. 8.

The Hill approach is also a valuable one for these binding situations. Of course this requires a knowledge of the total number of bases present so that the number occupied and the number free may be estimated.

Taking logarithms of both sides of Eq. (39), and then derivatizing both sides with respect to v, one obtains:

$$\frac{1}{v} - \frac{d \ln L}{dv} = \frac{-m}{1 - mv} + (m - 1) \left[\frac{-m}{1 - mv} + \frac{m - 1}{1 - (m - 1)v} \right] \quad (41)$$

This derivative may be evaluated at half-saturation ($v = \frac{1}{2}m$ to give at half saturation:

$$(d \ln L)/dv = 4m[2m/(m + 1)]$$

This may be directly related to the Hill coefficient since at half-saturation

$$(d \ln L)/dv = (m \ d \ln L)/d\overline{Y} = 4m/n_H$$

Thus a very simple relationship is obtained:

$$n_H = (m + 1)/2m \quad (42)$$

This may be used to estimate the number of lattice positions occupied by the ligand. In the limit of high m, the Hill coefficient approaches a limiting value of 0.5. Of course, this approach assumes that there is no cooperativity in ligand binding. If negative cooperativity is present, the value of m will be overestimated, while positive cooperativity will lead to an underestimate of m.

McGhee and von Hippel[23] considered the effects of cooperativity in their model by including a parameter, ω, such that the intrinsic affinity of protein for a site adjacent to an already bound protein molecule would be $K\omega$. If a binding site were bounded by two sites with bound protein, the affinity would become $K\omega^2$. The model considers only interactions between adjacent proteins in a pairwise fashion. For $\omega > 1$, positive cooperativity results, and for $\omega < 1$, negative cooperativity results. The Scatchard equation becomes:

$$\frac{v}{X} = K(1 - m\overline{Y}) \left[\frac{(2\omega - 1)(1 - mv) + v - R}{2(\omega - 1)(1 - mv)} \right]^{m-1} \left[\frac{1 - (m + 1)v + R}{2(1 - mv)} \right]^2 \quad (43)$$

where R is given by:

$$R = \{[1 - (m + 1)v]^2 + 4\omega v(1 - mv)\}^{1/2}$$

In Figure 9, a series of Scatchard plots are shown for various values of ω in which $m = 1$ and $K = 1 \ M^{-1}$. As expected, one obtains maxima in the Scatchard plots, indicating positive cooperativity.

Using our rule of thumb that the Hill approach is more appropriate for systems showing positive cooperativity, we may evaluate the Hill coefficient for this more complete model by a procedure similar to that employed for the simpler model involving only binding site interactions. A

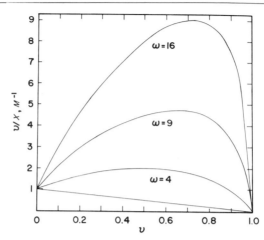

FIG. 9. Theoretical curves for the binding of a protein to a nucleic acid when cooperative interactions between sites are present. The parameter ω measures the relative affinity for a protein to a site adjacent to a bound protein as compared to a site with no adjacent protein. For these curves $K = 1\ M^{-1}$ and $m = 1$.

particularly useful result is obtained if we consider the case of $m = 1$. Then, by the sequence of taking logarithms, derivitization, evaluation at the midpoint, and relating $(d \ln L)/d\nu$ to the Hill coefficient, one obtains the remarkably simple result:

$$n_H = \sqrt{\omega} \qquad (44)$$

Thus, if negative cooperativity exists ($\omega < 1$), the Hill coefficient can vary between zero and one. Note that if ω is very small, the lattice will only hold one-half the number of protein molecules as compared to binding sites.

If positive cooperativity exists, the Hill coefficient increases; for large values of ω, the Hill coefficient can be very large indeed. The model used by McGhee and von Hippel[23] essentially assumes that the DNA chains are of infinite length. Thus n_H may increase without bound. Of course, for real molecules the upper limit of n_H is the number of binding sites available. This can be quite large and the transition from the free to the bound state can become essentially all or none in its character. For values of m greater than 1, the actual cooperativity will be less than that observed at the same value of ω when m equals unity. Thus the value of ω obtained as $(n_H)^2$ is the minimum interaction between sites required to obtain a particular cooperativity.

The Hill coefficient can be evaluated for any value of m and ω if we take the appropriate derivatives in Eq. (43) and evaluate those derivatives at the midpoint. The analytical forms of these solutions are not particularly illuminating and are not presented here.

Conclusions

The results presented here show that the curvature of Scatchard plots or the slope of the Hill plot contains considerable information concerning the cooperativity of the interactions, and this information may be directly evaluated if a model that controls the degree of saturation as a function of ligand concentration is applied to the problem. These techniques are not particularly well suited to distinguishing models for cooperative binding, but do allow a direct evaluation of the important parameters of the various models from a knowledge of the detailed shape of the Scatchard plot or the slope of the Hill plot.

[14] Ligand-Binding by Associating Systems[1]

By John R. Cann

Recently, two groups of investigators[2-4] have addressed themselves to the theory of ligand-binding by systems in which macromolecular association is induced by the small ligand molecule in question. While their particular emphases differ somewhat, they complement each other: On the one hand, there is concern regarding the hazard of ascribing mechanistic significance to the shape of the ligand-binding curve merely on the grounds that it exhibits apparent positive or negative cooperativity.[2] On the other hand, emphasis is placed on determination of the intrinsic binding constant and the stoichiometry of the binding reaction per se[3] and the possibility of distinguishing between the extreme mechanisms of ligand-mediated and ligand-facilitated association.[4] The thrust of this communication is to present a unified view maximally effective for the analysis of biological systems as diverse as the interaction of enzymes with cofactors and allosteric affectors, the interaction of vinca alkaloids with tubulin and the binding of lectins to cells. The following format will be used in the

[1] Supported in part by Research Grant 5 RO1 HL 13909-25 from the National Heart and Lung Institute, National Institutes of Health, United States Public Health Service. This is publication No. 657 from the Department of Biophysics and Genetics, University of Colorado Medical Center, Denver, Colorado 80262.
[2] L. W. Nichol and D. J. Winzor, *Biochemistry* **15**, 3015 (1976).
[3] J. R. Cann and N. D. Hinman, *Biochemistry* **15**, 4614 (1976).
[4] J. R. Cann in "Differentiation and Carcinogenesis" (C. Borek, C. M. Fenoglis, and D. W. King, eds.) No. 6 of the series *Advances in Pathobiology*, p 158, 1977.

presentation: First, the binding characteristics of two model systems illustrating the ligand-mediated mechanism of dimerization will be discussed and contrasted with a model for the ligand-facilitated mechanism of dimerization. The complications introduced by possible extraneous binding of ligand to sites not specifically involved in the dimerization process per se will then be examined. Finally, guidelines will be drawn for unambiguous interpretation of binding curves when it is known or suspected that the ligand induces association of the protein.

In the ligand-mediated mechanism of dimerization, obligatory binding of ligand to macromonomer must precede formation of the dimer. Ligand-binding data have been generated by computer simulation of two such reaction schemes:

$$M + X \rightleftharpoons MX, \ k_0 \tag{1a}$$
$$M + MX \rightleftharpoons M_2X, \ K \tag{1b}$$

in which M is the monomeric form of the macromolecule and X, the small ligand molecule or ion; and

$$M + X \rightleftharpoons MX, \ k_0 \tag{2a}$$
$$2MX \rightleftharpoons M_2X_2, \ K \tag{2b}$$

The results are expressed as the mean number of moles of ligand bound per mole of constituent macromolecule, ν, at the equilibrium concentration of unbound ligand, $[X]$. The constituent concentration, $\overline{C}$, is the total concentration of macromolecule in all its forms; e.g., for Reaction Scheme (1), $\overline{C} = [M] + [MX] + 2[M_2X]$ and $\nu = ([MX] + [M_2X])/\overline{C}$. Representative results are displayed in Figs. 1A, C, and D as Scatchard plots[5] of $\nu/[X]$ vs ν. It can be shown analytically that the intercept of the Scatchard plot with the ordinate for the two model systems is: (a) Reaction Scheme (1), $K_{app} = k_0 + k_0 K\overline{C}$, where the apparent binding constant, K_{app}, of ligand to macromolecule extrapolates to the intrinsic binding constant, k_0, at infinite dilution of macromolecule; (b) Reaction Scheme (2), k_0. It can also be shown analytically that for Reaction Scheme (1) the binding curves are stationary at their midpoint ($\nu = 0.5$) where $1/[X] = k_0$ independent of K and $\overline{C}$.

The simulated data are analyzed by means of Scatchard plots rather than double-reciprocal plots[6] of $1/\nu$ vs $1/[X]$, because the latter are very difficult to interpret for associating systems.[3] For example, the double-reciprocal plots for Reaction Scheme (2) are concave toward the abscissa and extrapolate to the ordinate with a very flat slope. Whereas the intercept with the ordinate and, thus through reciprocity, the number of binding sites on the monomer are readily determined, the limiting slope is

[5] G. Scatchard, *Ann. N. Y. Acad. Sci.* **51**, 660 (1949).
[6] I. M. Klotz and D. L. Hunston, *Biochemistry* **10**, 3065 (1971).

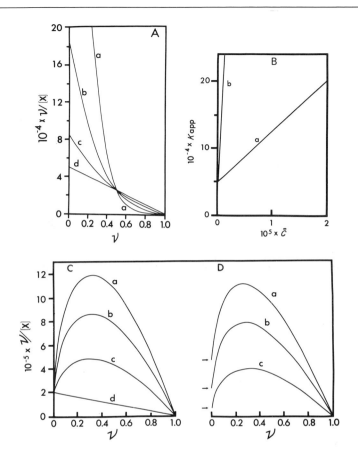

FIG. 1. Theoretical binding curves for the ligand-mediated and ligand-facilitated mechanisms of macromolecular dimerization; their shape and dependence upon macromolecule concentration. (A) Scatchard plots for ligand-mediation via reaction scheme (1), $k_0 = 5 \times 10^4$ M^{-1} and $K = 1.5 \times 10^5 M^{-1}$: curve a, $\bar{C} = 7.27 \times 10^{-5} M$; b, $1.80 \times 10^{-5} M$; c, $4.55 \times 10^{-6} M$; and d, limit of infinite dilution of macromolecule. (B) Extrapolation of apparent binding constant given by the intercept of the Scatchard plot with the ordinate to infinite dilution of macromolecule, ligand-mediation via reaction scheme (1): curve a, $k_0 = 5 \times 10^4 M^{-1}$ and $K = 1.5 \times 10^5 M^{-1}$ corresponding to Fig. 1A; b, $k_0 = 5 \times 10^4 M^{-1}$ and $K = 4.5 \times 10^6 M^{-1}$. (C) Scatchard plots for ligand-mediation via reaction scheme (2), $k_0 = 2 \times 10^5 M^{-1}$ and $K = 7.5 \times 10^5 M^{-1}$: curve a, $\bar{C} = 1.4 \times 10^{-4} M$; b, $7 \times 10^{-5} M$; c, $1.8 \times 10^{-5} M$; and d, limit of infinite dilution of macromolecule. A similar set of curves was obtained when K was varied from $7.5 \times 10^5 M^{-1}$ to $10^2 M^{-1}$ at fixed values of $k_0 = 2 \times 10^5 M^{-1}$ and $\bar{C} = 1.4 \times 10^{-4} M$. (D) Scatchard plots for ligand-facilitation via reaction scheme (3), $K = 10^2 M^{-1}$ and $k_0 = 1.732 \times 10^7 M^{-1}$ (similar picture for $k_0 = 2 \times 10^5 M^{-1}$): curve a, $\bar{C} = 1.4 \times 10^{-4} M$; b, $7 \times 10^{-5} M$; and c, $1.8 \times 10^{-5} M$. Arrows indicate intercepts with the ordinate.

not the intrinsic binding constant (k_0) but rather some weighted average of k_0 and the dimerization constant, K. In principle, extrapolation to infinite dilution of the apparent binding constants determined at progressively lower macromolecule concentration should eliminate the role of dimerization to give k_0, but the extrapolation would be uncertain in practice because of the sigmoid shape of the plot of apparent constant vs concentration. In contrast, interpretation of the Scatchard plots is straightforward. Thus, the first guideline to emerge from these calculations is to avoid double-reciprocal plots of binding data when it is known or suspected that the ligand induces association of the macromolecule.

The Scatchard plots for Reaction Scheme (1) (Fig. 1A) are concave toward the abscissa. This shape is reminiscent of nonassociating systems characterized either by inherent heterogeneity of binding sites with respect to their intrinsic affinity for ligand or by binding to multiple sites with negative cooperativity.[7] In contrast to heterogeneity and negative cooperativity, however, the extent of binding is dependent upon macromolecule concentration, the plots for different $\bar{C}$ showing the analytically predicted stationary point. These features can be understood as follows: At ligand concentrations less than the midpoint concentration, dimerization [Reaction (1b)] enhances binding so that for given $\nu < 0.5$, [X] decreases (i.e., $\nu/[X]$ increases) with increasing $\bar{C}$. On the other hand, at higher ligand concentrations dimerization inhibits binding by removing unliganded monomer from the reaction arena, and very high concentrations of ligand are required to reverse the dimerization reaction via binding to the released unliganded monomer thereby driving the binding reaction (1a) to completion. Accordingly, for given $\nu > 0.5$, $\nu/[X]$ decreases with increasing $\bar{C}$. As predicted analytically, the apparent binding constant as given by the intercept with the ordinate increases with increasing $\bar{C}$, but extrapolation to infinite dilution of macromolecule (Fig. 1B) erases the role of dimerization in determining the extent of binding at finite concentrations and yields k_0.

In contrast to Reaction Scheme (1), Reaction Scheme (2) gives Scatchard plots (Fig. 1C) that are open downward in shape because dimerization [Reaction (2b)] enhances binding at all ligand concentrations. Here the shape of the plot is reminiscent of nonassociation systems characterized by binding to multiple sites with positive cooperativity, but in the case of ligand-mediated dimerization the plots are dependent upon macromolecule concentration. The higher $\bar{C}$, the greater is the deviation from that expected for simple binding in the absence of dimerization. But,

---

[7] Although our model system assumes a single binding site on the monomer, calculations for a system with two binding sites having the same intrinsic affinity for ligand substantiate this observation.

in agreement with the analytical prediction for this reaction scheme, the intercept with the ordinate gives k_0 irrespective of $\bar{C}$ and $\bar{K}$.

Let us now consider the ligand-facilitated mechanism of dimerization in which two macromonomers associate to a dimer, which is then stabilized by the binding of ligand as schematized by the set of reactions

$$2M \rightleftharpoons M_2, \; K \tag{3a}$$
$$M_2 + X \rightleftharpoons M_2X, \; 2k_0 \tag{3b}$$
$$M_2X + X \rightleftharpoons M_2X_2, \; (\tfrac{1}{2})k_0 \tag{3c}$$

It can be shown analytically that, for this reaction scheme, the intercept of the Scatchard plot with the ordinate is

$$K_{app} = k_0[(1 + 8\bar{C}K)^{1/2} - 1]/\{2 + [(1 + 8\bar{C}K)^{1/2} - 1]\}$$

The open-downward shape of the Scatchard plots (Fig. 1D) reflects the cooperativity conferred by stabilization of the dimer via the binding reaction, 3b. The increase in cooperativity with increasing $\bar{C}$ derives from a mass action effect on Reaction (3a); the higher the concentration of unliganded dimer, the greater is the extent of ligand-binding by the system. In agreement with analytical prediction, K_{app} decreases with decreasing $\bar{C}$ and extrapolates to zero at infinite dilution. This is so because the extrapolation erases the dimerization upon which ligand-binding is dependent at finite $\bar{C}$, which raises the question of how k_0 can be determined. In principle, it can be obtained by extrapolation of K_{app} to infinite $\bar{C}$, but in practice this may be difficult. The procedure of choice would appear to be nonlinear least-squares fitting of the experimental data to the appropriate theoretical equation.

The foregoing results indicate that analysis of binding data by means of Scatchard plots can distinguish between the extreme mechanisms by which ligand-binding may induce dimerization of a macromolecule. Thus, the open-downward shape of the plots for the ligand-facilitated mechanism schematized by reaction set (3) serves to distinguish it from the ligand-mediated Reaction Scheme (1); while the dependence of K_{app} upon $\bar{C}$ distinguishes it from the ligand-mediated Reaction Scheme (2). Likewise, the two ligand-mediated schemes are readily distinguishable one from another. This presupposes, however, that there is no extraneous binding of ligand to sites not specifically involved in the dimerization process per se. Obviously, a heavy overlay of nonspecific binding will mask the underlying mechanism of dimerization; but a priori it is not clear how the binding of only a few extraneous ligand molecules will affect the binding characteristics of the system. Thus, our calculations were extended to include two different models that postulate the binding of a single extraneous ligand molecule per monomeric unit. These calculations[8] proved to be quite revealing and are summarized below.

In the first model the underlying mechanism of dimerization is the ligand-mediated one given by Reaction Scheme (2). The sequential set of reactions is

$$
\begin{array}{lll}
M + X \rightleftharpoons XM & , & k & \text{extraneous binding to monomer} \\
M + X \rightleftharpoons MX & , & k_0 \\
XM + X \rightleftharpoons XMX & , & k_0 & \text{specific binding to monomer} \\
2MX \rightleftharpoons M_2X_2 & , & K \\
MX + XMX \rightleftharpoons XM_2X_2 & , & 2K & \text{dimerization} \\
2XMX \rightleftharpoons X_2M_2X_2 & , & K \\
M_2X_2 + X \rightleftharpoons XM_2X_2 & , & 2k \\
XM_2X_2 + X \rightleftharpoons X_2M_2X_2 & , & (1/2)k & \text{extraneous binding to dimer}
\end{array}
\qquad (4)
$$

in which $k < k_0$; the apparent dimerization constant, $Kk_0^2[X]^2/(1 + k_0[X])^2$, is independent of k and the same as for reaction scheme (2); the total number of binding sites per monomeric unit is $n = 2$; and the intercept of the Scatchard plot with the ordinate is $k + k_0$.

In the second model the underlying mechanism of dimerization is the ligand-facilitated one given by Reaction Scheme (3). In this case the set of reactions is

$$
\begin{array}{llll}
M + X \rightleftharpoons XM & , & k & & \text{extraneous binding to monomer} \\
2M \rightleftharpoons M_2 & , & K \\
M + XM \rightleftharpoons XM_2 & , & 2K & & \text{dimerization} \\
2XM \rightleftharpoons X_2M_2 & , & K \\
M_2 + X \rightleftharpoons XM_2 & , & 2k \\
XM_2 + X \rightleftharpoons X_2M_2 & , & (1/2)k & & \text{extraneous binding to dimer} \\
X_iM_2 + X \rightleftharpoons X_iM_2X & , & 2k_0, & i = 0,1,2 \\
X_iM_2X + X \rightleftharpoons X_iM_2X_2 & , & (1/2)k_0, & i = 0,1,2 & \text{specific binding to dimer}
\end{array}
\qquad (5)
$$

in which $k < k_0$; the apparent dimerization constant, $K(1 + k_0[X])^2$, is independent of k and the same as for Reaction Scheme (3); $n = 2$; and the intercept of the Scatchard plot with the ordinate is

$$
K_{\text{app}} = k + k_0[(1 + 8\bar{C}K)^{1/2} - 1]/\{2 + [(1 + 8\bar{C}K)^{1/2} - 1]\}
$$

which extrapolates to k at infinite dilution of macromolecule.

Binding data were generated systematically for each of these models using a range of values for the several parameters. The Scatchard plots displayed in Fig. 2 are quite representative of Reaction Scheme (4). Clearly, the binding of even 2 or 3 extraneous ligand molecules would not mask the underlying mechanism of dimerization. Likewise, as evidenced by Fig. 3A and curve a in Fig. 3B, the underlying ligand-facilitated mechanism of Reaction Scheme (5) would not be masked in situations where

---

[8] J. R. Cann, previously unpublished.

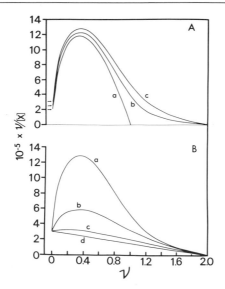

FIG. 2. Theoretical Scatchard plots for ligand-mediated dimerization with extraneous binding to a site not specifically involved in the dimerization per se, reaction scheme (4). (A) Dependence of shape upon the value of k at fixed values of $k_0 = 2 \times 10^5 M^{-1}$, $K = 7.5 \times 10^5 M^{-1}$ and $\bar{C} = 1.4 \times 10^{-4} M$; curve a, $k = 0$; b, $5 \times 10^4 M^{-1}$; and c, $1 \times 10^5 M^{-1}$. Arrows indicate intercepts with the ordinate. (B) Dependence upon the value of K at fixed values of $k = 1 \times 10^5 M^{-1}$, $k_0 = 2 \times 10^5 M^{-1}$, and $\bar{C} = 1.4 \times 10^{-4} M$: curve a, $K = 7.5 \times 10^5 M^{-1}$; b, $1 \times 10^5 M^{-1}$; c, $1 \times 10^4 M^{-1}$; and d, $1 \times 10^2 M^{-1}$. Similar pictures were obtained for $\bar{C} = 1.8 \times 10^{-5} M$.

$k_0 \gg k$. But, when the value of k_0 is progressively decreased, other things held constant, the shape of the calculated Scatchard plot changes profoundly and progressively from an open-downward shape for $k_0 \gg k$ (curve a of Fig. 3B) to a concave-upward shape for k_0 of the same order of magnitude as k (curve c). Thus, the underlying mechanism of dimerization is obscured when $k_0 \approx k$. However, it can still be ascertained that the ligand induces dimerization, since the Scatchard plot remains sensitive to macromolecule concentration (compare curves c and d).

In view of the several theoretical results described above, it is of foremost importance in practice to distinguish between cooperative binding to a nonassociating macromolecule and ligand-induced association. This can be achieved either by making binding measurements over a wide range of macromolecule concentration or preferably by direct observation in the ultracentrifuge. The studies of Gerhart and Schachman[9] and Changeux and Rubin[10] on the allosteric interactions in aspartate

[9] J. C. Gerhart and H. K. Schachman, *Biochemistry* **7**, 538 (1968).
[10] J.-P. Changeux and M. M. Rubin, *Biochemistry* **7**, 553 (1968).

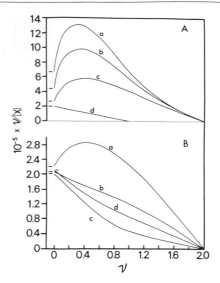

FIG. 3. Theoretical Scatchard plots for ligand-facilitated dimerization with extraneous binding to a site not specifically involved in the dimerization per se [Reaction Scheme (5)]. (A) Dependence of shape upon macromolecule concentration at fixed values of $k = 2 \times 10^5$ M^{-1}, $K = 1 \times 10^2 M^{-1}$ and $k_0 = 2 \times 10^5 M^{-1}$: curve a, $\bar{C} = 1.4 \times 10^{-4} M$; b, $7 \times 10^{-5} M$; c, 1.8 $\times 10^{-5} M$; and d, limit of infinite dilution of macromolecule. (B) Dependence of shape upon the value of k_0 at fixed values of $k = 2 \times 10^5 M^{-1}$ and $K = 1 \times 10^2 M^{-1}$ with $\bar{C} = 1.8 \times 10^{-5} M$ for curves a, b, and c and $1.4 \times 10^{-4} M$ for curve d: curve a, $k_0 = 6 \times 10^6 M^{-1}$; b, $2 \times 10^6 M^{-1}$; c and d, $4 \times 10^5 M^{-1}$. Arrows indicate intercepts with the ordinate.

transcarbamylase and of Levitzke and Koshland[11] on the ATP- or UTP-induced dimer to tetramer transformation in cytosine triphosphate synthetase are exemplary of the application of ultracentrifugation to this end.

The following guidelines can now be delineated for interpretation of experimental binding data once it is known or suspected that the ligand induces association of the protein or other macromolecule:

1. Double-reciprocal plots of the data must be avoided in favor of Scatchard plots.

2. Binding measurements, which define as precisely as possible the shape of the Scatchard plots, should be made at different protein concentrations.

3. The results should be scrutinized so as to avoid fitting to a straight-line data points, which may actually describe a curve concave toward the abscissa as in Fig. 1A and in Fig. 3B, curves c and d. Not only will linearization underestimate the intercepts with the abscissa *(n)* and the

[11] A. Levitzki and D. E. Koshland, Jr., *Biochemistry* **11**, 247 (1972).

ordinate (apparent average binding constant to the first site), but more important, it can be conceptually misleading.

4. The average intrinsic binding constant to the first site on the monomer (or the intrinsic binding constant in the less complicated cases) is determined by extrapolation of the apparent constant to infinite dilution of protein. The extrapolation need not necessarily be linear; indeed it will, in general, be nonlinear if for no other reason than the usual thermodynamic nonideality. Although this guideline was arrived at by computer simulation of model systems, a model is not required for determination of the intrinsic binding constant. This is so because extrapolation to infinite dilution erases the role of association in determining the extent of ligand binding at finite protein concentration.

5. It is expected that the extrapolation to infinite dilution together with the shape of the Scatchard plots will provide some mechanistic insight into the interaction. Scatchard plots at different protein concentrations can then be computer simulated using the mathematical relationships admitted by plausible mechanisms. When done in conjunction with independent ultracentrifugal,[11-14] gel-permeation chromatographic[15] or rapid kinetic[16] measurements, such simulations should permit quantitative distinctions between these mechanisms.

Finally, a few words concerning the methods for measurement of ligand-binding by associating systems are in order. Recently it has been shown[3] that the gel-chromatographic procedure of Hummel and Dreyer[17] has distinct advantages over static equilibrium methods, such as equilibrium dialysis,[18] for quantification of systems of this kind. These advantages are conferred by the dilution that results from axial dispersion as the zone of protein moves down the chromatographic column. As a consequence, association is reduced by mass action so that its role in determining the extent of ligand-binding is decreased. Thus, the behavior of the system tends to approach the limit in which association no longer enhances ligand-binding, and the apparent binding constant extrapolates to the intrinsic constant at infinite dilution with a flatter slope. The limit will be approached still closer if binding can be measured by fluorometric titration of the protein with the ligand,[12] because the high sensitivity of fluorometry permits measurements at much lower protein concentrations.

[12] J. C. Lee, D. Harrison, and S. N. Timasheff, *J. Biol. Chem.* **250**, 9276 (1975).

[13] R. P. Frigon and S. N. Timasheff, *Biochemistry* **14**, 4559, 4567 (1975).

[14] K. Morimoto and G. Kegeles, *Arch. Biochem. Biophys.* **142**, 247 (1971).

[15] D. J. Winzor and H. A. Scheraga, *Biochemistry* **2**, 1263 (1963).

[16] M. S. Tai and G. Kegeles, *Biophys. Chem.* **3**, 307 (1975).

[17] J. P. Hummel and W. J. Dreyer, *Biochim. Biophys. Acta* **63**, 530 (1962).

[18] I. M. Klotz, F. M. Walker, and R. B. Pivan, *J. Am. Chem. Soc.* **68**, 1486 (1946).

[15] Pressure-Jump Light-Scattering Observation of Macromolecular Interaction Kinetics

By GERSON KEGELES

Historical Survey

The specific use of light-scattering measurements to follow protein interaction kinetics was detailed in an elegant, classical paper[1] on the dimerization of bovine plasma mercaptalbumin with mercurial coupling reagents. With more rapid kinetic processes, too fast to follow by ordinary hand-mixing experiments, continuous flow,[2] stopped flow,[3] and perturbation methods[4-7] have been employed.

Bimolecular processes involving macromolecular reactants are followed in a particularly straightforward way by observation of light scattered at 90°,[1] and the present author may have been the first to suggest such an application[8] to the temperature-jump method of Czerlinski and Eigen.[4] Development of an apparatus designed specifically for that purpose was begun in 1963, using the microcell design described by Czerlinski[9] for fluorescence recording. The principles and use of this light-scattering apparatus for following a macromolecular dimerization were presented in 1970 and described in detail in 1971.[10] Prior to this, turbidity measurements in the transmission mode were used to study the formation of micelles by the temperature-jump method,[11] and 90° light scattering was used to monitor stopped-flow changes in red blood cells.[12] Another 90° light-scattering temperature-jump apparatus was developed during this

[1] H. Edelhoch, E. Katchalski, R. H. Maybury, W. L. Hughes, and J. T. Edsall, *J. Am. Chem. Soc.* **75**, 5058 (1953).

[2] H. Hartridge and F. J. W. Roughton, *Proc. R. Soc. London, Ser. B* **94**, 336 (1923).

[3] Q. H. Gibson and L. Milnes, *Biochem. J.* **91**, 161 (1964).

[4] G. Czerlinski and M. Eigen, *Z. Elektrochem.* **63**, 652 (1959).

[5] M. Eigen and L. de Maeyer, *in* "Technique of Organic Chemistry" (A. Weissberger, ed.), Vol. 8, Part 2, Chap. 18. Wiley (Interscience), New York, 1963.

[6] S. Ljunggren and O. Lamm, *Acta Chem. Scand.* **12**, 1834 (1958).

[7] H. Strehlow and M. Becker, *Z. Elektrochem.* **63**, 457 (1959).

[8] L. W. Nichol, J. L. Bethune, G. Kegeles, and E. L. Hess, *in* "The Proteins" (H. Neurath, ed.), Vol. 2, Chap. 9, p. 329. Academic Press, New York, 1964.

[9] G. Czerlinski, *Rev. Sci. Instrum.* **33**, 1184 (1962).

[10] M. S. Tai and G. Kegeles, *Arch. Biochem. Biophys.* **142**, 258 (1971).

[11] G. C. Krescheck, E. Hamori, G. Davenport, and H. A. Scheraga, *J. Am. Chem. Soc.* **88**, 246 (1966).

[12] R. I. Sha' Afi, G. T. Rich, V. W. Sidel, W. Bossert, and A. K. Solomon, *J. Gen. Physiol.* **50**, 1377 (1967).

period and used to study human red blood cells[13] and lauryl sulfate micelle formation.[14] The assembly of tobacco mosaic virus was studied by temperature-jump turbidity measurements in the transmission mode,[15] and the sickling of red blood cells was followed by turbidity changes induced by pressure.[16] The stopped-flow apparatus was used to observe kinetics of macromolecular interaction by means of scattered light as well[17,18] and the "concentration-jump" relaxation technique was developed.[18] Commercial stopped-flow equipment was modified[19] so as to become an efficient detection and measuring apparatus for fluorescence and for 90° light-scattering. The "concentration-jump" procedure was employed with 90° light-scattering in one example, in this apparatus, to obtain records with high signal-to-noise ratio for kinetics in the range of 20 sec to several minutes, which were used to examine detailed mechanisms of ligand-controlled association.[20–22] The same basic apparatus has been used in kinetic studies of ribosomal subunit interaction.[23,24] Recently another apparatus has been described for measurements of temperature-jump by means of fluorescence,[25] and a detailed kinetics study has been published on the self-assembly of glutamate dehydrogenase, which was based on 90° light-scattering measurements in temperature-jump and stopped-flow experiments.[26] A laser has been introduced into the stopped-flow apparatus for the purpose of making 90° light-scattering measurements,[27] which appears to be effective at considerably lower macromolecule concentrations than are convenient to study with the commercial design.[19]

The pressure-jump technique has a particularly high potential for the

[13] J. D. Owen, B. C. Bennion, L. P. Holmes, E. M. Eyring, M. W. Berg, and J. L. Lords, Biochim. Biophys. Acta 203, 77 (1970).
[14] B. C. Bennion, L. K. Tong, L. P. Holmes, and E. M. Eyring, J. Phys. Chem. 73, 3288 (1969).
[15] R. B. Scheele and T. M. Schuster, Biophys. Soc. Abstr. No. 236a (1970).
[16] M. Murayama and F. Hasegawa, Fed. Proc., Fed. Am. Soc. Exp. Biol. 28, 536 (1969).
[17] C. Huang and C. Frieden, Fed. Proc., Fed. Am. Soc. Exp. Biol. 28, 536 (1969).
[18] H. F. Fisher and J. R. Bard, Biochim. Biophys. Acta 188, 168 (1969).
[19] J. E. Stewart, Durrum Application Notes No. 7. Durrum Instruments Corp., Palo Alto, California, 1971.
[20] G. Kegeles and M. Tai, Biophys. Chem. 1, 46 (1973).
[21] M. S. Tai and G. Kegeles, Biophys. Chem. 3, 307 (1975).
[22] G. Kegeles and J. R. Cann, this volume [12].
[23] A. D. Wolfe, P. Dessen, and D. Pantaloni, FEBS Lett. 37, 112 (1973).
[24] A. Wishnia, A. Boussert, M. Graffe, P. Dessen, and M. Grunberg-Manago, J. Mol. Biol. 93, 499 (1975).
[25] R. Rigler, C.-R. Rable, and T. M. Jovin, Rev. Sci. Instrum. 45, 580 (1974).
[26] D. Thusius, P. Dessen, and J. M. Jallon, J. Mol. Biol. 92, 413 (1975).
[27] D. Riesner and H. Buenemann, Proc. Natl. Acad. Sci. U.S.A. 70, 890 (1973).

study of macromolecular interactions, because it is the *molar* volume of reaction that determines how well pressure drives a reaction.[28-31] Thus, contrary to earlier expectations, it now appears that this technique will be applicable to a very wide range of protein–protein interactions. The original pressure-jump techniques[6,7] were applied to aqueous electrolyte solutions, using electrical conductivity changes to follow kinetic processes. One light absorption apparatus had been employed[32] for studies in the 2 μsec to 2 msec time range, using shock waves to initiate the pressure jump, up to 500 and 1000 atmospheres, previous to the time of the last review in this series[33] of the pressure-jump method. Another light-absorption pressure-jump apparatus was published[34] while publication of that review was pending, which studied the ferric thiocyanate complex at pressures of several hundred atmospheres up to almost 1400 atmospheres. Still another light-absorption pressure-jump apparatus was reported[35] at about the same time as the previous review. This device was used to study the interaction of a dye with bovine plasma albumin, and still another, similar description of a pressure-jump light-absorption apparatus was published in 1976.[36] One closely related technique has been described very recently, in which repeated pressure pulses of small amplitude are used to force minor changes in the composition of the sample, being monitored optically. In this apparatus, signal averaging over a very large number of pulses has been accomplished,[37] thereby greatly increasing the signal-to-noise ratio. A brief description of two separate types of equipment designed specifically for pressure-jump studies using light scattered at 90° as monitor was published in 1974.[38] The second design was redescribed in detail recently,[39] and this has been used to observe the interaction of a variety of proteins and to measure rate constants for the interaction of ribosomal subunits[22,40,52] and the interaction of subunits of *Helix*

[28] G. Kegeles, L. Rhodes, and J. L. Bethune, *Proc. Natl. Acad. Sci. U.S.A.* **58**, 45 (1967).
[29] L. F. TenEyck and W. Kauzmann, *Proc. Natl. Acad. Sci. U.S.A.* **58**, 888 (1967).
[30] R. Josephs and W. F. Harrington, *Proc. Natl. Acad. Sci. U.S.A.* **58**, 1587 (1967).
[31] W. F. Harrington and G. Kegeles, this series Vol. 27 [13].
[32] A. Jost, *Ber. Bunsenges. Phys. Chem.* **70**, 1057 (1966).
[33] M. T. Takahashi and R. A. Alberty, this series Vol. 16, [2].
[34] K. R. Brower, *J. Am. Chem. Soc.* **90**, 5401 (1968).
[35] D. E. Goldsack, R. E. Hurst, and J. Love, *Anal. Biochem.* **28**, 273 (1969).
[36] W. Knoche and G. Wiese, *Rev. Sci. Instrum.* **47**, 220 (1976).
[37] R. M. Clegg, E. L. Elson, and B. W. Maxfield, *Biopolymers* **14**, 883 (1975).
[38] G. Kegeles, V. P. Saxena, and R. Kikas, *Fed. Proc., Fed. Am. Soc. Exp. Biol.* **33**, 1504 (1974).
[39] G. Kegeles and C. Ke, *Anal. Biochem.* **68**, 138 (1975).
[40] J. B. Chaires, C. Ke, M. S. Tai, G. Kegeles, A. A. Infante, B. Cosiba, and T. Coker, *Biophys. J.* **15**, 293a (1975).

pomatia α-hemocyanin.[22,41] This 90° light-scattering pressure-jump apparatus is capable of observing changes of only a few percent in the molecular composition of reacting systems having molecular weights as small as 48,000.

Pressure-Jump Equipment Useful for Transmitted Light

The basic principles of the use of scattered-light observation for following macromolecular interaction kinetics have been described[10,26] and are presented in more detail in a current review.[42] The general practical application of relaxation kinetics, in particular the temperature-jump method, is described in a very detailed recent review.[43] This covers practical applications of theory as well as problems involved in methodology, equipment, data acquisition, and data processing that are common to all relaxation methods. Two reviews[33,44] have treated the subject of nonoptical pressure-jump apparatus. Therefore, the remainder of the present review will be devoted particularly to conveying information on the optical applications of the pressure-jump technique itself, some recent modifications of the basic technique, a comparison where possible of the capabilities and limitations of existing apparatus, and a brief outline of the equipment and procedure for light-scattering pressure-jump experiments, with some assessment of the probable capabilities of this method for following interactions of macromolecular systems.

The two earliest reports[6,7] of the pressure-jump method outlined two basically different techniques for obtaining rapid shifts of pressure, both of which were applied to electrolyte solutions using electrical conductivity as the monitor. In one procedure,[6] commercial gas cylinders were used to increase quickly the pressure on the system under observation, from an original pressure of 1 atm. In the other procedure,[7,45] arrangements were made to build up very slowly the pressure on the system under observation, and then a plummet with a sharp tip was used to initiate the bursting of a thin membrane separating that system from atmospheric pressure, producing a sudden release of pressure. In the procedure[6] employing rapid pressurization, the gas supply cylinder was simply opened to the pressure autoclave containing the sample, the pressure rising to its final

[41] M. S. Tai, G. Kegeles, and C. Huang, *Arch. Biochem. Biophys.* **180,** 537 (1977).
[42] D. Thusius, *in* "Applications of Chemical Relaxation to Molecular Biology" p. 339, (I. Pecht and R. Rigler, eds.), Springer-Verlag, Berlin and New York, 1977.
[43] A. F. Yapel, Jr., and R. Lumry, *in* "Methods of Biochemical Analysis" (David Glick, ed.), Vol. 20, p. 169. Wiley (Interscience), New York, 1973.
[44] W. Knoche *in* "Techniques of Chemistry" (G. G. Hammes, ed.), Vol. 6, Part 2, p. 187. Wiley (Interscience), New York, 1973.
[45] H. Strehlow and H. Wendt, *Inorg. Chem.* **2,** 6 (1963).

value in about 50 msec. In the apparatus using the bursting membrane,[7] the system could be depressurized as rapidly as 64 μsec. Oil was used to surround the sample and transmit pressure to it. A thin plastic membrane separated the sample from direct contact with the surroundings, while allowing transmission of pressure. There was an appreciable adiabatic temperature drop in the oil at depressurization, which could affect the temperature of the sample. For this reason, a more recent version of the bursting-membrane technique[46] uses water as the pressurization fluid, the adiabatic temperature drop being about one-fifth to one-tenth as large.[33,46] The bursting-membrane technique tends to generate sonic disturbances above the sample and in the room, which have been reduced with a muffler.[33] In the recent version,[46] obviation of the sonic disturbances has been accomplished by evacuation of the cell above the bursting membrane. Since considerable time was required to replace circular membranes, this more recent modification also employs a strip of metal film, which can be advanced rapidly to an unbroken area, after each experiment. While the original publication[7] indicated stability after about 64 μsec, the more recent reports in which the conductivity cell of the bursting membrane apparatus is replaced by transmitted-light optical detection indicate overall vibration-limited instrument deadtime from about 100 μsec[35] to about 0.5 msec.[36] The hydraulically pressurized high-pressure, light-absorption, pressure-jump apparatus referred to earlier[34] employs direct valving of pressurization oil and reports a minimum instrument dead-time of 6 msec. This arrangement made use of a deformable sample cell with plastic windows, inserted inside oil in a commercial high pressure transmitted-light cell, which might turn out to be disadvantageous when turbidity is to be measured. This apparatus is the only one which has been arranged to easily shift the pressure by an arbitrary amount in either direction, between arbitrary initial and final pressures. The shock-wave procedure referred to earlier[32] is a pressurization technique which employs transmitted light detection. It is the only application described that is capable of measuring relaxation times down to 2 μsec.

Pulsed Pressure-Perturbation Apparatus

In addition to the single-step relaxation methods,[4-7] perturbation methods have included interactions of chemically interacting systems with periodic forcing functions, such as sound waves in the case of pressure-volume energy. Dispersion and absorption effects are associated with passage of sound through such systems. In their extensive review, Eigen and de Maeyer[5] have provided theoretical and experimental back-

[46] W. Knoche and G. Wiese, *Chem. Instrum.* 5 (2), 91 (1973–1974).

ground for this type of observation. An oscillating sinusoidal pressurization method was described by Wendt,[47] which is closely related to the acoustical techniques. The recently reported signal-averaging method of Clegg et al.[37] for small pressure-induced perturbations is a little more akin to such procedures than it is to the conventional pressure-jump method itself, but it shares features of both methods. Pressure pulses are produced periodically by piezoelectric transducers, and observation may be optical. Excellent signal-to-noise ratio output has been achieved by the averaging of over 3600 signals. The method has been reported to go down to the 100 μsec time range.[37] Considerable care will be required if the method is applied to systems with multiple relaxation times, since only a portion of the total relaxation curve is observed with a pulse method if the forcing pressure pulse is short.[48] This method may turn out to be more useful than the conventional pressure-jump technique, and it will be interesting to follow the course of its further development and application.

Pressure-Jump Apparatus for 90° Light-Scattering

The first device of this kind described[38] was based on a modification of the analytical ultracentrifuge cell. A 90° viewing hole was cut in the cell housing, and the centerpiece was replaced by a transparent one made of polished Lucite. The sample was introduced through a small circular cylindrical filling hole into a chamber which also had the shape of a circular cylinder and which was polished inside. No air space was allowed to remain in contact with the sample; instead the filling hole was filled nearly full with Dow-Corning light silicone oil, DC 200. Schumaker and colleagues[49] have described a nitrogen-pressurized loading chamber for use with analytical ultracentrifuge cells. A modification of this device[50] was used to pressurize this cell and then seal it under pressure. The cell was then removed, inserted in a photomultiplier optical system, and manually depressurized in a darkened room by loosening the filling-hole screw, while a recording oscilloscope was triggered manually, slightly in advance. Although it took two people to set off the experiment, we found it possible with practice to coordinate well enough to measure at oscilloscope scanning speeds of 20 msec per division. The depressurization times varied, but could be reduced to about 20 msec, as judged by calibration with light transmitted through a filter and a cell filled with phosphate

[47] H. Wendt, Ber. Bunsenges. Phys. Chem. 70, 556 (1966).
[48] Eigen and de Maeyer,[5] p. 922.
[49] V. N. Schumaker, A. Wlodawer, J. T. Courtney, and K. M. Decker, Anal. Biochem. 34, 359 (1970).
[50] V. P. Saxena, G. Kegeles, and R. Kikas, Biophys. Chem. 5, 161 (1976).

buffer and a pH indicator. The major difficulty with this apparatus in addition to its awkwardness was the introduction of occasional optical artifacts caused by reflections of light from the oil-aqueous solution meniscus, even when masks were used to try to hide this from view. With this crude equipment, it was possible to make pressure-jump relaxation observations on interactions in lobster hemocyanin and even in chymotrypsin solutions under conditions where the weight-average molecular weight is below 48,000.

As a result of the encouragement provided by this limited success with relatively simple equipment, we developed a stainless steel cell[38,39] fitted with three standard quartz analytical ultracentrifuge cell windows in standard window holders. A pressure sensor is also embedded in the cell, with its stainless steel pressure transmitting cap in direct contact with the liquid sample. The sample is introduced through the hole which is also used for pressurization by gas from a commercial cylinder. A protective layer of Dow-Corning light silicone oil DC200 is again placed over the sample, to prevent dissolving of the pressurization gas in the sample.

Figure 1 shows a photograph of the cell with its windows, window holders, and gaskets. A delrin cap is inserted into the bottom of the cell before it is fitted onto a mounting rod, in order to reduce heat transfer. The cell is also bored through with a water conduit, through which is

FIG. 1. Photograph of the pressure-jump 90° light-scattering cell and its window assemblies. The pressure sensor (not seen) is mounted on the opposite side of the cell from the 90° window. Photograph by courtesy of Michael C. Daily.

FIG. 2. Photograph of the solenoid-value and pressure-jump cell assembly, with cell windows removed.

passed a constant flow of thermostatted water, for temperature control. The opening at 90° is elongated, so as to enlarge the optical aperture at which scattering can be observed. The cell holds a sample of just under 2 ml. An adjustable high-pressure reducing valve is fitted to the gas cylinder, and the valve is fitted in turn with ¼-inch stainless steel seamless high-pressure tubing, to communicate through a manual shutoff valve to a high-speed high-pressure solenoid valve, which is normally closed. This valve is connected above to a manually operated pressure release valve, and below through standard fittings and ¼-inch stainless steel tubing to the cell.

Figure 2 shows a photograph of the solenoid valve and cell assembly. The published detailed description lists specifications for valves, pressure sensor, amplifiers, and commercial sources of supply.[39]

Figure 3 shows a schematic diagram of the entire apparatus, and Fig. 4 is a photograph of the apparatus, including the optical components, but not showing the commercial gas cylinder and high-pressure reduction valve or the recording electronic equipment. A PEK 110-W mercury-arc lamp housing and projection system has been modified to hold a 12-V, 50-W tungsten iodine lamp, which is much more stable in the time range from several milliseconds to several minutes. The projected light passes through an iris diaphragm, which is closed down to about 1 mm diameter, and a multielement short focal length lens then serves to focus an image of the iris diaphragm in the center of the cell, care being taken to assure that no light is reflected from the inner cell walls. To prevent unwanted light reflections, the pressure sensor is also mounted away from the cell center, and out of sight of the 90° photomultiplier, as shown in the machine drawing of the cell (Fig. 5). Two matched end-window photomultipliers are rigidly fastened to the steel supporting track, one directly in line to receive transmitted light, and one arranged for observation at 90°. The transmitted light is used as a reference beam in scattered-light experiments, and for this purpose is passed through a two-stage optical at-

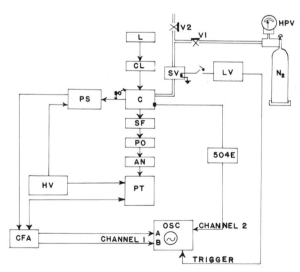

FIG. 3. Schematic diagram of pressure-jump assembly. N_2, high-pressure N_2 cylinder; HPV, high-pressure reduction valve; V_1, "valving-off" valve 1; V2, pressure release valve 2; SV, high-speed solenoid valve; L, tungsten-iodine lamp; CL, condenser lens; C, reaction cell; SF, step filter; PO, polarizer; AN, adjustable analyzer; PT, transmitted-light photomultiplier; PS, scattered-light photomultiplier; LV, low-voltage power supply; HV, high-voltage power supply; CFA, dual cathode-follower amplifier; OSC, oscilloscope; 504E, Kistler pressure transducer amplifier. From G. Kegeles and C. Ke, *Anal. Biochem.* **68**, 138 (1975) by permission of Academic Press, Inc.

FIG. 4. Photograph of pressure-jump assembly, by courtesy of Michael C. Daily. From G. Kegeles and C. Ke, *Anal. Biochem.* **68,** 138 (1975) by permission of Academic Press, Inc.

tenuator, described below, before being received by the in-line photomultiplier.

To operate this apparatus, a volume of gas at a preset high pressure is valved off from the commercial cylinder, and suddenly introduced into the

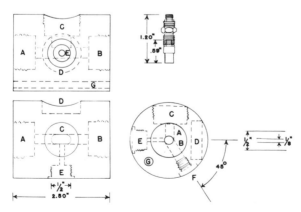

FIG. 5. Scale mechanical drawing of the cell. A, B, C, window seats for three quartz ultracentrifuge windows in standard window holders; D, mounting hole; E, filling hole; F, hole to contain Kistler 601B pressure sensor insert, details shown on the top right; G, thermostat water conduit. From G. Kegeles and C. Ke, *Anal. Biochem.* **68,** 138 (1975) by permission of Academic Press, Inc.

cell by switching open the high-speed solenoid valve. The same switching voltage triggers the sweep circuits of the oscilloscope, which records photomultiplier outputs of scattered and transmitted light. In using the apparatus for successive experiments on the same solution, the entire operation consists of opening and closing two valves and resetting the data-collection system prior to triggering the next experiment. To date we have used photographic recording of an oscilloscope screen with open camera shutter. The output of the pressure sensor can be recorded, if desired, together with that of the photomultipliers by use of a multichannel oscilloscope.

With the aid of this pressure sensor, which is designed for microsecond recording in shock-wave experiments, we have determined that failure to close off the high-pressure gas supply allows pressurization to continue building up through the high-pressure reduction valve for about 40 msec, using nitrogen gas. If the manual shutoff valve is closed between the solenoid valve and the gas supply, the dead-time of the instrument is immediately reduced to 8 msec, using nitrogen. If helium is substituted for nitrogen, the time for complete pressurization is further reduced to 2 msec, and the slight tendency of the cell to ring after pressurization with nitrogen disappears if helium is used.

Optically, this equipment is a dual-channel differential-double beam instrument,[10,39] which can also be used in the electronically offset single-beam mode for light-absorption measurements. In the double-beam scattering mode, we have used the transmitted light photomultiplier output to balance the 90° scattered-light photomultiplier output. It is generally desirable to use white light[10,13,14] rather than monochromatic light for such work, to avoid confusion from possible shifts of optical density in narrow spectral regions. In the pressure-jump apparatus, a two-stage optical attenuator is placed in the transmitted beam exiting from the cell, before it reaches the in-line photomultiplier. This consists of a four-unit neutral density-filter system, followed by a fixed polarizer and an adjustable analyzer. Electronic balance is achived by connection of the same power supply in parallel to both photomultipliers, which are of identical type, and by the use of balanced electronic amplifiers. Final balance is then achieved by optical attenuation. This technique appreciably improves the overall signal-to-noise ratio. Ordinarily, a reaction amplitude corresponding to a shift of several oscilloscope divisions in the photomultiplier difference signal is not even detectable as a reaction amplitude if turbidity alone is measured by transmitted light at the same instrument settings.

Features of the present arrangement that differ slightly from those published[39] include the following: a glass lens has been permanently fastened in the magnetic lens nosepiece of the end-window photomultiplier at 90° to focus the image of the light-scattering cell volume approximately on the photomultiplier cathode. A new oscilloscope and amplifier system have been installed, to permit both slow and fast recording of balanced optical output and of pressure-sensor output within a single experiment. The configuration consists of a Tektronix 5444 double-beam oscilloscope equipped with two 5A38 dual-trace amplifiers and a 5B44 dual-time base. The two optical signals are first subtracted, and the difference is processed through an AM 502 differential amplifier, which allows for adjustable RC filtering of the output signal. Photography is done with a simple C-5A camera with open shutter. At current prices, automatic digital data collection and processing should be considered as a viable alternative to an elaborate oscilloscope system.

A few words might be said here about the virtues of pressurization versus depressurization. Although the instrument dead-times in the literature are somewhat shorter for depressurization if one excludes the shock wave pressurization technique, the danger of incorrectly recording scattered light for slow processes, i.e., from several seconds to several minutes, is appreciably greater in the case of depressurization. This is because of the possibility of slow evolution of small gas bubbles that will scatter light, after the system returns to 1 atm, a possibility that does not exist after pressurization. The derived kinetic data for pressurization experiments represent values at the elevated pressure,[6] which are nearly the same as those at 1 atm unless one is concerned with volumes of reaction of several hundred milliliters per mole and pressure-jumps of 100 atm or more. However, in some macromolecular systems, some corrections to 1 atm may have to be considered. In all pressure-jump work, degassing of solutions is important. For sudden depressurization, pressure oscillations or even cavitation could be caused by sizable gas bubbles still present under pressure. For sudden pressurization, the existence of sizable gas bubbles in optical view at 1 atm will cause optical artifacts, especially during the initial pressurization period, as they shrink to tiny bubbles. It is desirable to have a cell-filling arrangement which makes the trapping of bubbles impossible, a feature not present in our current design:[39] considerable care is required to fill the cell with protein solutions free of all air bubbles. It is cautioned that in any design, all filling holes must be absolutely leak-proof at high pressures.

This review might now be concluded with some guesses as to the usefulness and applicability to the study of macromolecular interactions of the light-scattering pressure-jump method. It has been found feasible to calibrate optical pressure-jump equipment by light absorption using an indicator in Tris buffer at pH 8.2,[35] or with an indicator in phosphate buffer[38] having a volume of ionization of −28 ml per mole.[33] It would be expected that most macromolecular interactions would show molar volume changes on reaction far in excess of this.[31] For a monomer–dimer reaction at $\frac{2}{3}$ dissociation by weight,[10] there would be a 5% change in weight-average molecular weight for a volume of reaction of 100 ml per mole and a pressure-jump of 100 atm. If 90° light-scattering changes are monitored in a well balanced double-beam instrument, it should be straightforward to follow a 5–10% change in weight-average molecular weight if the weight-average molecular weight is 50,000 or higher and the reacting macromolecule system is present at a concentration of several tenths of 1%. If the conditions for appreciable interaction require dilution to 0.1% by weight or lower, however, then one should no longer expect to be able to measure the relaxation time in any but a very high molecular weight system as a function of concentration, a procedure necessary for assessing reaction mechanism. With this type of restriction in mind, the technique is potentially widely applicable under current design, with a lower time limit slightly longer than 2 msec and an upper time limit governed only by light source and electronics stability. Application of the shock-wave technique[32] might possibly effectively extend the time range of the light-scattering pressure-jump method down to several microseconds, it being noted that transmission of pressure at sound velocity requires 7 μsec per centimeter in water.

While this review was in press, a publication by Davis and Gutfreund[51] appeared, which described another optical cell designed for 90° observation of pressure-jump by light-scattering or fluorescence. Their depressurization times are conservatively estimated by the present author to be at least ten times smaller than the corresponding pressurization times with our own equipment, owing in part to their small cell volume. Their report confirms our assessment above, of the potential usefulness of the light-scattering pressure-jump technique for the detailed study of macromolecular reactions, which, in fact, we have already reported in two separate kinetic studies.[41,52]

Acknowledgments

The development of the pressure-jump light-scattering method described by Kegeles *et al.*[38,39] was made possible in large part by support from the National Science Foundation. I wish to thank Mr. Stanley Manter and Mr. Theodore Swol of the Instrument Shop of the Institute of Materials Science for construction of the stainless steel cell shown in Figs. 1 and 2. I am indebted to Mr. Michael Daily for the photographs in Figs. 1 and 4. Support by the University of Connecticut Research Foundation under Grant 35-787 for the oscilloscope assembly described is gratefully acknowledged.

[51] J. S. Davis and H. Gutfreund, *FEBS Lett.* **72,** 199 (1976).
[52] J. B. Chaires, M. S. Tai, C. Huang, G. Kegeles, A. A. Infante, and A. J. Wahba, *Biophys. Chem.,* **7,** 179 (1977).

[16] Measurement of Protein Dissociation Constants by Tritium Exchange

By ALFRED D. BARKSDALE and ANDREAS ROSENBERG

The chemical reactivity of a group in a macromolecular subunit can be strongly influenced by the extent of the subunit's incorporation into the polymer. The difference in reactivity when the subunit resides in the polymer or in the dissociated state can, therefore, afford a measure of the degree of dissociation of the subunit–subunit complex if the system's kinetics are studied as a function of macromolecular concentration. The method is not new, having been used sporadically over the last 10 years. For example, Klapper and Klotz[1] have studied the dissociation of hemerythrin by ligand-affinity measurements.

An overall reaction scheme may be represented

$$
\begin{array}{ccc}
2L + P_2 & \overset{K_d}{\leftrightarrow} & 2P + 2L \\
k_1 \uparrow\downarrow k_{-1} & K_d^1 & k_2 \uparrow\downarrow k_{-2} \\
PL_2 & \overset{}{\leftrightarrow} & 2PL
\end{array}
$$

The relationships between equilibrium constants in the above scheme can be extended to the rate constants. We can consider the reaction between P_2 and P molecules and ligand L. Over the initial course of the reaction the contribution to the kinetics of the ligand loss reaction (k_{-1}, k_{-2}) may be neglected, as may the dissociation of the liganded polymer. If the rates for association and dissociation are very much larger than the rates of ligand binding to either form, P and P_2 are always in equilibrium proportion and the observed kinetics are second order with the observed rate constant

$$k_{obs} = (1 - \alpha)k_1 + \alpha k_2$$

with α = extent dissociation as it appears in the expression for the dissociation constant

$$K_d = 2\alpha^2 P^0/(1 - \alpha)$$

where P^0 = total concentration of subunits = (P)+2(P_2). Thus by changing P^0 so as to change the extent of dissociation, and measuring k_{obs} as a function of P^0, K_d can be estimated. Other reaction schemes and their kinetics have been developed.[2]

The difference reactivity method has not received wide attention in

[1] M. H. Klapper and I. M. Klotz, *Biochemistry* **7**, 223 (1968).
[2] R. A. Alberty and W. G. Miller, *J. Chem. Phys.* **26**, 1231 (1957).

protein studies because of experimental difficulties. First, the reaction may modify the protein so as to expose further groups or to alter the avidity to reagent, thus complicating the kinetics. Second, it may be impossible to measure (e.g., by spectroscopy) the kinetics over a sufficient concentration range (four or more orders of magnitude typically) to precisely determine K_d. Finally, the rates of ligand binding may be faster or comparable to the rates of form interconversion.

Independently, these authors[3] and Schreier and Baldwin[4] have developed a radioisotope method for determining protein dissociation constants, a method that circumvents the difficulties encountered in conventional kinetic experiments or interposed by protein alteration. That is, trace amounts of tritium and carbon-14 are incorporated into the protein. Because the amounts of isotope are trace, the protein retains its natural properties. The protein tritium is exchanged with solvent hydrogen at rates that are sensitive to the state of the protein (folded, unfolded, associated or dissociated), which range over several orders of magnitude so as to be very much less than form interconversion rates, and which are readily measurable over a wide range of concentrations, including very dilute solutions. The carbon-14 remains affixed to the protein, thereby providing "by its radioactivity" a measure of the protein concentration, even at less than nanomolar concentrations.

Before discussing the double-label technique, we should review some of the history of hydrogen-exchange kinetics and methods in protein systems.

A Brief Survey of Pertinent Hydrogen Exchange Theory and Methods

Hydrogen exchange kinetics were introduced into protein chemistry by Linderstrøm-Lang.[5,6] One can measure either the uptake (inexchange) or loss (outexchange) of a hydrogen isotope from a site (group) on the protein or a polypeptide. The process may be most simply illustrated as

$$H^*OH + GH \underset{k_{out}}{\overset{k_{in}}{\rightleftharpoons}} GH^* + HOH \tag{1}$$

where H* indicates a hydrogen isotope (deuterium or tritium) and G represents a group having an exchangeable hydrogen. Under conditions of moderate temperature and pH, G = —SH, —OH, —NH$_2$, or —CONH—.

[3] G. J. Ide, A. D. Barksdale, and A. Rosenberg, *J. Am. Chem. Soc.* **98**, 1595 (1976).

[4] A. A. Schreier and R. L. Baldwin, *J. Mol. Biol.* **105**, 409 (1976).

[5] K. U. Linderstrøm-Lang, *Chem. Soc. Spec. Publ.* **2**, 1 (1955).

[6] K. U. Linderstrøm-Lang and J. A. Schellman, *in* "The Enzymes," 2nd ed. (P. D. Boyer, H. Lardy, and K. Myrbäck, eds.), Vol. 1, p. 443 Academic Press, New York, 1959.

The kinetics of hydrogen exchange by sulfhydryl, hydroxyl, or amino groups are generally too fast for measurement by the conventional techniques described below. The hydrogen exchange rates for the peptide group in proteins, however, are characterized by times of seconds to days, for which reason the hydrogen exchange properties of peptide amino moieties have given the most information about "structure" in polypeptides and proteins.[7]

Let us concern ourselves with the outexchange step, since it is the process that will provide the means for determining protein dissociation constants. The outexchange occurs not only by direct exchange [Eq. (1)] but also by acid and base catalysis via the intermediates[8,9]

$$
\begin{array}{cc}
\text{O} - & \text{O} + \\
\| & \| \\
-\text{C}-\text{N}- & -\text{C}-\text{N}-
\end{array}
\tag{2}
$$

Therefore, the apparent rate constant for isotope loss is comprised of three terms, one each for the three possible pathways,

$$
k_{obs} = k_{out} + k_H(H^+) + k_{OH}(OH^-)
\tag{3}
$$

Present evidence indicates that direct exchange makes very little contribution to k_{obs} so that the last two terms predominate and, moreover, interact to produce a minimum outexchange rate at "pH min" at about pH 3. The k_{obs} in all cases increases monotonically on either side of the minimum.

The rate constant for outexchange is pseudo first order, since the concentration of labeled sites (mM) is vastly exceeded by the concentration of water sites (111 M). Furthermore, the backexchange process can be neglected so that first-order kinetics are observed over the entire time course of reaction.

The outexchange kinetics of simple polypeptides behave quite regularly as a function of pH and temperature, since all peptide sites are identical. For example, k_{obs} for poly(DL-alanine) is given by[10,11]

$$
k_{obs} = 50(10^{-pH} - 10^{pH-6})10^{0.05(T-20)}min^{-1}
\tag{4}
$$

The kinetics of random-coil heteropolypeptides cannot as a rule be described by a single rate constant, indicating that outexchange from a peptide group is modified by nearest-neighbor interactions.[12] Equations

[7] S. W. Englander, N. W. Downer, and H. Teitelbaum, *Annu. Rev. Biochem.* **41**, 903 (1972).
[8] A. Berger, A. Loewenstein, and S. Meiboom, *J. Am. Chem. Soc.* **81**, 62 (1959).
[9] B. H. Leichtling and I. M. Klotz, *Biochemistry* **5**, 4026 (1966).
[10] A. Hvidt, *C. R. Trav. Lab. Carlsberg* **34**, 299 (1964).
[11] S. W. Englander and A. Poulsen, *Biopolymers* **7**, 379 (1969).
[12] C. K. Woodward and A. Rosenberg, *Proc. Natl. Acad. Sci. U.S.A.* **66**, 1067 (1970).

such as (4) and the rates measured in small peptides have been used to predict with reasonable accuracy the outexchange characteristics of the random-coil protein oxidized ribonuclease A on the basis of amino acid content.[13]

The peptide groups in a polypeptide or protein having structure of the secondary (α−helix, β−sheet), tertiary (chain folding) or quaternary (subunit interactions) types are often greatly attenuated relative to random coil rates. For example, native ribonuclease A loses tritium at observed rates which are spread out over seven or so orders of magnitude.[14] Whether the rate attentuation observed in protein hydrogen exchange studies results from hydrogen bonding or from restricted solvent accessibility,[14,15] it is nevertheless clear that each peptide group is affected to a greater or lesser degree by the overall solution structure of the protein. If the structure changes, then rates change.

We define H_{rem} as the number of hydrogen atoms remaining per molecule at time t. Since the outexchange rate of each peptide site in the molecule is in principle different, H_{rem} is customarily written as a sum of parallel first-order reactions. That is,

$$H_{rem} = \sum_{sites} e^{-k_i t} \tag{5}$$

Equation (5) represents a general statement of protein hydrogen isotope loss kinetics. Experimentally, however, one finds that only a portion of total isotope exchange can be measured under a particular set of conditions. That is, the time window for observation (e.g., minutes to days) permits kinetic examination during perhaps three of the seven or more decades of time over which outexchange occurs. Consequently, some hydrogens are lost before measurement begins; others can exchange so slowly as to appear kinetically inactive.[16] Thus, phenomenologically, Eq. (5) can take a form

$$H_{rem} = \Sigma A_j e^{-kt} + B \tag{6}$$

where each A_j, k_j pair characterizes the apparent size and reactivity of an apparent "class" of sites, and B = the "background" contribution from very slowly exchanging hydrogens.

We will conclude our survey of hydrogen exchange history with these comments:

1. Protein hydrogen isotope exchange kinetics can be measured precisely over large time periods and under a variety of conditions (pH,

[13] R. S. Molday, S. W. Englander, and R. G. Kallen, *Biochemistry* **11**, 150 (1972).
[14] C. K. Woodward and A. Rosenberg, *J. Biol. Chem.* **245**, 4105 (1971).
[15] S. W. Englander and R. S. Staley, *J. Mol. Biol.* **45**, 277 (1969).
[16] C. K. Woodward, L. M. Ellis, and A. Rosenberg, *J. Biol. Chem.* **250**, 432 (1975).

temperature, protein concentration, solvent composition). The resultant data can be treated quantitatively.

2. The rates of isotope loss from a protein are often exquisitely sensitive to the conformational state or other structural features of the protein.

3. That hydrogen exchange properties of a protein do disclose quantitative structural information about the protein permits us to use the results to monitor protein structure as a function of variables, much as one might use ultraviolet (UV) or optical rotatory dispersion (ORD) spectra to evaluate the state of a protein. In this light, knowledge of the microscopic details of the exchange process become of lesser consequence.

Some discussion of hydrogen exchange methods is now in order. The first technique to appear was the density gradient method of Linderstrøm-Lang.[17] A fully deuterated protein would be dissolved in water, at which point outexchange begins. Some time later an aliquot of solution is removed, lyophilized to remove solvent, which now contains deuterium, and H_{rem} is determined from the solvent density. Over the last 15 years, the density gradient method, which is time consuming, as well as fraught with potential sources of error in the cryosublimation steps[18,19] (i.e., isotope loss during freeze-drying and subsequent storage of cryosublimate) has generally fallen into disfavor (although the cryosublimation step has been revived elegantly and recently by Schreier and Baldwin[4] as discussed in the Results section below).

The newer infrared technique championed by Hvidt and Willumsen involves measuring the decrease in the amide II infrared absorption band as a function of time after an undeuterated protein is exposed to D_2O (the absorption decrease resulting from replacement of peptide hydrogen by deuterium).[20,21] The absorption bands are weak and consequently the infrared technique is limited to high concentrations. Moreover, both the density gradient and the infrared methods suffer from the criticism that a partially deuterated protein molecule may differ conformationally from either the undeuterated or fully deuterated species.

The tritium loss technique of Englander[22] has, in the present context, decided advantages that make it the method of choice for determining protein dissociation parameters:

1. Only trace amounts of label enter the protein.

2. The experiments are relatively simple.

[17] I. M. Krause and K. U. Linderstrøm-Lang, C. R. Trav. Lab. Carlsberg Ser. Chim. **29,** 367 (1955).
[18] B. E. Hallaway and E. S. Benson, Biochim. Biophys. Acta **107,** 154 (1965).
[19] R. H. Byrne and W. P. Bryan, Anal. Biochem. **33,** 414 (1970).
[20] E. R. Blout, C. deLoze, and A. Asadourian, J. Am. Chem. Soc. **83,** 1895 (1961).
[21] L. Willumsen, C. R. Trav. Lab. Carlsberg **38,** 223 (1971).
[22] S. W. Englander, Biochemistry **2,** 798 (1963).

3. The protein concentration can be very low (e.g., nanomolar) yet H_{rem} can be measured precisely.

An experiment consists of five steps: inexchange, first separation, outexchange, second separation, determination of H_{rem}. To achieve inexchange the protein is incubated in tritiated water for a sufficient length of time under appropriate conditions of pH and temperature (usually chosen to partially unfold the protein without denaturing it and to accelerate chemical steps to achieve a sufficient level of statistical tritium incorporation). By statistical we mean that a given site will be labeled, but not in all molecules. Totally trace amounts of tritium are inexchanged such that if only certain molecules were to receive all the tritium, then these fictitious fully labeled chains would amount to no more than about 1% of total molecules present.

The first separation, removal of excess solvent tritium so that outexchange can begin, is accomplished by filtration of the inexchange solution through Sephadex G-25. The protein band emerges well ahead of the solvent band, such that the zero time (beginning of outexchange) can be precisely defined. Since the gel column has been pre-equilibrated with the desired solvent at the desired temperature, the first separation brings the protein to the desired set of conditions. Outexchange continues for a predetermined length of time, at which point an aliquot is removed and subjected to a second gel filtration (second separation), which separates from the protein the solvent isotope that had accumulated during outexchange. The eluate is partitioned, part for scintillation counting, part for determination of protein concentration, customarily by spectrophotometry. The desired quantity, H_{rem}, is determined by

$$H_{rem} = 111 \times CPM/CPM^0 \times P^0 \qquad (7)$$

where cpm^0 = counts per minute per unit volume in the inexchange sample, and P^0 = protein concentration.

An alternative method, due to Englander and Crowe, replaces the gel filtration in the first and second separation with rapid dialysis.[23] The tritiated protein solution is placed in a dialysis bag with thin walls. The bag is suspended in a large volume of buffer (buffer volume approximately 100-fold greater than bag volume) and spun rapidly long enough to achieve equilibrium distribution across the membrane. As a method of first separation the dialysis technique may prove cumbersome in that several changes of dialyzate may be needed to rid the system of the initial solvent tritium, resulting in an ill-defined zero time. As a method of second separation, the dialysis technique has a decided advantage over filtration.

[23] S. W. Englander and D. Crowe, *Anal. Biochem.* **12**, 579 (1965).

The protein does not undergo the relatively indeterminate dilution which occurs during filtration. Rather, the tritium activity and the protein concentration in the outexchange sample are determined directly from bag contents or with precisely known dilutions involved. Such considerations become quite important if H_{rem} is a sensitive function of protein concentration, as is the case when a protein dissociation constant is being determined by hydrogen exchange kinetics.

Recently Schreier[24] has introduced phosphocellulose filtration as a third method of second separation. Briefly, the aliquot of outexchange sample is vacuum-filtered through a disk of phosphocellulose paper. At low pH and ionic strength the filter paper can quantitatively bind the protein while the solvent and its radioisotopes pass through. The pH must be adjusted to about 3 (pH min) and the temperature kept low to minimize protein tritium loss during the filtration. The bound protein is subsequently separated from the filter paper by washing in a concentrated salt solution prior to scintillation counting. Provided the initial protein–paper interaction and the later protein–paper separation can be made quantitative, the phosphocellulose method is superior to dialysis for two reasons: more points can be taken since a bag need not be opened and closed to remove sample, and very dilute protein concentrations, with radioactivity levels near solvent background, can be used, since the filtration also acts as a concentration step.

The weak link in any variant of the tritium loss technique is the determination of P^0. P^0 can be measured, however, by labeling a small proportion of molecules with ^{14}C, whose radioactivity can be determined in very dilute solutions. Independently the authors[3,25] and Schreier and Baldwin[4,24,26] have shown that carbon-14 activity can be nearly correlated with protein or peptide concentration in solutions that have optical densities too large for measurement down to optical densities too low for spectral determination. By labeling dissociation reaction partner or partners with both tritium and carbon-14 (double label), both tritium activity and protein concentration, hence H_{rem} itself, can be determined in one step. Details of the double-label method will be given in the Results section.

The Mathematics of Protein Dissociation Measured by Tritium Exchange

Consider a site on a subunit. We label that site with tritium. If the protein itself undergoes dissociation, the site can have two environments (polymer or free), in each of which it can have a unique rate constant for

[24] A. A. Schreier, *Anal. Biochem.*, In press.
[25] A. D. Barksdale and A. Rosenberg, submitted.
[26] A. A. Schreier and R. L. Baldwin, *Biochemistry* **16**, 4203 (1977).

tritium loss. The system can be represented as

$$(A^*)n \underset{k_b}{\overset{k_f}{\rightleftharpoons}} nA^*$$
$$\downarrow k_0 \qquad \downarrow k_1$$

where k_f, k_b are the rate constants for interconversion of polymer to subunits and back, and k_0, k_1 are the rate constants for tritium loss from the site when it is located in the polymer and subunits, respectively. We assume that k_f, $k_b \gg k_0$, k_1 such that in any adjustment of concentration, equilibrium is established prior to tritium loss.[2] We define the extent of dissociation

$$\alpha = \text{sites in form } A^*/\text{total sites labeled} = [A^*]/C^* \qquad (8)$$

with $C^* = n[(A^*)_n] + [A^*]$. The dissociation constant is given by

$$K_d = [A^*]^n/[A^*)_n] = (\alpha C^*)^n / \left[\frac{C^*(1 - \alpha)}{n} \right] = n(\alpha C^*)^n/C^*(1 - \alpha) \qquad (9)$$

The observable in this simple system is the amount of label remaining as a function of time (analogous to H_{rem}, as defined in the preceding section), whose rate law is

$$\frac{dC^*}{dt} = \frac{nd[(A^*)_n]}{dt} + \frac{d[A^*]}{dt} \qquad (10)$$

The rate laws for the polymer and subunit can be written

$$\frac{d[(A^*)_n]}{dt} = -k_f[(A^*)_n] + k_b[A^*]^n - nk_0[(A^*)_n] \qquad (11a)$$

$$\frac{d[A^*]}{dt} = k_f[(A^*)_n] - k_b[A^*]^n - k_1[A^*] \qquad (11b)$$

Combining Eqs. (11a) and (11b) and the definitions of concentrations employed in Eq. (10), there results

$$dC^*/dt = -k_0(1 - \alpha)C^* - k_1\alpha C^* = -k_{obs}C^* \qquad (12)$$

That is, although apportioned between the two states of the system, the decrease in C^* with time is a first-order process with integrated rate law,

$$\ln(C^*/C^0) = -k_0(1 - \alpha)t - k_1\alpha t = -k_{obs}t \qquad (13)$$

A schematic representation of the behavior of $\ln(C^*/C^0)$ with extent of dissociation is given in Fig. 1. We note that if three samples of initial concentrations C_0^0 ($\alpha = 0$, no dissociation), C_α^0 ($\alpha = \alpha$, dissociation), and C_1^0 ($\alpha = 1$, complete dissociation) are allowed to simultaneously undergo tritium loss and are then sampled simultaneously at time t_s

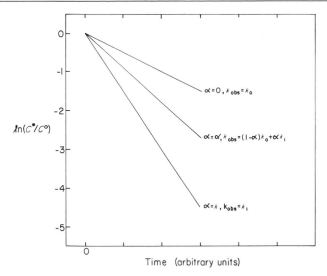

FIG. 1. First-order outexchange from a singly labeled subunit as a function of the degree of dissociation.

after exchange has begun, then the extent dissociation in the sample of intermediate concentration is given by

$$\alpha = \frac{\ln (C_0^*/C_0^0) - \ln (C_\alpha^*/C_\alpha^0)}{\ln (C_0^*/C_0^0) - \ln (C_1^*/C_1^0)} = (k_0 - k_{obs})/(k_0 - k_1) \qquad (14)$$

In practice, one allows a number of samples to simultaneously outexchange, thereby yielding a number of k_{obs}, C_α^0 pairs from which K_d is statistically evaluated by Eq. (9). Two forms of statistical evaluation are given in the Results section below.

Let us next consider the subunit with N sites labeled with tritium and proceed to examine the relationships between the time dependence of tritium loss (H_{rem}) and the extent of dissociation. We should, at the outset, state that multisite kinetics as they occur in protein hydrogen exchange are quite complex and that a full discussion of them is beyond the intent of this chapter.

Rather, for purposes of these discussions, simple and straightforward mathematics will be set forth. That these simple forms accurately describe the data comes from a blessing of nature: over limited to fairly wide time ranges, a portion of the tritium loss from a protein or a polypeptide can often be represented by a single expression[27]

$$H_{rem} = A_{obs}\exp(-k_{obs}t) \qquad (15)$$

[27] S. W. Englander, Ann. N.Y. Acad. Sci. 244, 10 (1975).

where, as discussed in the preceding section, the preexponential A represents the size of a "class" of hydrogens of rate constant k. Generally, one observes behavior describable by Eq. (15) toward the end of the outexchange process, such that the sites undergoing tritium loss are somewhat removed from solvent and thus presumably occupy more nearly mutually similar environments than may sites with greater solvent accessibility or other factors that promote encounter with solvent.

Given that the subunit tritium loss can be described by Eq. (15) (which must be proved experimentally, as shown in the Results section IV), then a relationship analogous to Eq. (13) exists:

$$\ln H_{rem} = k_{obs}t + \ln A_{obs} \qquad (16)$$

with the slope and intercept ($-k_{obs}$, $\ln A_{obs}$, respectively) related to the fraction dissociation α, as

$$k_{obs} = k_0 (1 - \alpha) + \alpha k_1 \qquad (17a)$$
$$\ln A_{obs} = (1 - \alpha)\ln A_0 + \alpha \ln A_1 \qquad (17b)$$

The subscripts 0 and 1 indicate, as before, zero and complete dissociation.

Consequently, the hydrogen exchange in a protein sample of partial dissociation is given by

$$\ln H_{rem,\alpha} = -(1 - \alpha)k_0 t - \alpha k_1 t + (1 - \alpha)\ln A_0 + \alpha \ln A_1 \qquad (18)$$

For purposes of later reference we set down the expressions equivalent to Eq. (18) for the free and associated subunits,

$$\ln H_{rem,0} = -k_0 t_0 + \ln A_0 \qquad (19a)$$
$$\ln H_{rem,1} = -k_1 t_1 + \ln A_1 \qquad (19b)$$

Following the procedure outlined for the single-site case, we allow protein solutions to simultaneously undergo outexchange, having prepared the solutions such that their concentration range is wide enough so that there are examples of $\alpha = 0$, $\alpha = \alpha$, $\alpha = 1$.

There are two ways to use the data obtained (H_{rem}, t pairs) to calculate the extent of dissociation. The first is to examine all the samples at a particular point in the outexchange (constant H_{rem}) and to relate the extent of dissociation to the time required to reach the chosen H_{rem}. Then, combining Eqs. (18), (19a), (19b) with $\ln H_{rem,\alpha} = \ln H_{rem,0} = \ln H_{rem,1}$)

$$\alpha = (k_1 t_1 - k_0 t_0 + \ln A_0 - \ln A_1)/(k_1 t_\alpha + \ln A_0 - \ln A_1) \qquad (20)$$

Use of Eq. (20) requires very precise data over long periods of time, since it is virtually impossible to sample each solution at the exact moment it has reached the chosen point in outexchange. That is, there must be sufficient and precise data points to permit least-square or regression analysis of $\ln H_{rem}$ vs time so that (1) the $\ln A$ term (intercept) can be determined with accuracy and (2) the time to reach the desired $\ln H_{rem}$ can be accurately interpolated or extrapolated.

It is much simpler to examine the outexchange in the array of solutions at a fixed time, relating $\ln H_{rem}$ at that time to the extent of dissociation. Combining Eqs. (18), (19a), and (19b) with $t_0 = t_1 = t_\alpha$, we have

$$\alpha = (\ln H_{rem,0} - \ln H_{rem,\alpha})/(\ln H_{rem,0} - \ln H_{rem,1}) \qquad (21)$$

The advantage accruing from constant time, rather than constant H_{rem}, determination of α are these:

1. Fewer time points need be taken, so that more samples can be run in parallel.

2. Changes in total protein concentration between time points (an occurance when dialysis is used) will affect the α in that sample but not the dissociation constant calculated from statistical analysis of the $\ln H_{rem}$, concentration pairs.

3. The time points can be fairly close together, since the extrapolation of $\ln H_{rem}$ vs t to $t = 0$ to determine the $\ln A$ terms is unnecessary. To reiterate, it is necessary that the length of time to the first time point be long enough that chemical equilibrium be established between the free and associated subunits and that outexchange in all solutions obey Eq. (15), which must be carefully verified.

Results

To date the dissociation constants of two proteins, carboxyhemoglobin A_0[3] and ribonuclease S[4] have been measured by tritium exchange. Both investigations have also employed trace amounts of carbon-14 affixed to the polypeptide chains as a very sensitive monitor of concentration.

The tetramer–dimer Dissociation of Human Carboxyhemoglobin A_0

The authors have investigated the tetramer–dimer dissociation of carboxyhemoglobin A_0 (HbCO) as a function of pH, temperature, and buffer composition.[4,25] The reaction scheme may be represented as $(\alpha\beta)_2 = 2(\alpha\beta)$, where $\alpha\beta$ indicates the hemoglobin dimer whose two heme groups are ligated with CO. The dissociation constant is expressed by

$$K_{4,2} = [\alpha\beta]^2/[(\alpha\beta)_2] = 4\alpha^2[\text{HbCO}]/(1 - \alpha) \qquad (22)$$

where $[\text{HbCO}]$ = total concentration of protein in moles of tetramer per liter.

It was desired at the outset of our investigation to have an accurate and precise method of determining the hemoglobin concentration at levels where standard spectrophotometric methods become difficult (e.g., $[HbCO] < 10^{-7}, > 10^{-3}$). Accordingly, we allowed concentrated samples of HbCO to react with one of two ^{14}C-labeled compounds to place a $^{14}CH_3$ group at lysine and N-terminus groups. The radioisotopes remain affixed under all conditions. A general discussion of alkylation of proteins is given by Means and Feeney.[28]

The reagent we employed initially was $^{14}CH_3I$, which reacts as

$$-NH_2 + {}^{14}CH_3I \rightarrow -NH^{14}CH_3 + HI$$

The method of Link and Stark[29] was followed. Briefly, a sample of methyl iodide, a gas, at an activity of 50 μC_i (New England Nuclear), was introduced by vacuum transfer to a cell containing the protein solution (HbCO, 300 mg/ml, pH 9.8, 0.1 M borate buffer). The cell was sealed off, wrapped with foil, and placed in the dark at 2° for 3 days. Approximately 20% of the $-NH_2$ groups were labeled, amounting to 1.7% of chains, there being 12 $-NH_2$ groups per chain.

A far less cumbersome technique for alkylating $-NH_2$ groups in proteins is the formaldehyde method of Rice and Means.[30] Schreier and Baldwin,[4] whose work will be discussed in the next section, have used $^{14}CH_2O$ successfully in their study of the dissociation of ribonuclease S. We have turned to this method in our more recent hemoglobin dissociation work.

Based on the procedure of Rice and Means,[30] we have followed this scheme to label HbCO with $^{14}CH_2O$: 0.06 ml of 300 mg/ml HbCO, prepared as described earlier,[31] is mixed with 0.04 ml of 0.5 M, pH 9.5 borate buffer in a 1-ml disposable syringe. The ampoule of $^{14}CH_2O$ (50 μCi, New England Nuclear) is frozen in Dry Ice to crystallize the reagent, after which the neck of the ampoule is removed. The HbCO solution is injected into the ampoule, which is then swirled in an ice bath for 30 sec. At that point, 10 μl of freshly prepared $NaBH_4$ (Fisher, 1 mg/ml) are added to quench the reaction. A second borohydride aliquot is added after another 30 sec. The reaction mixture is sucked into a 1-ml disposable syringe and applied to a Sephadex G-25 column (0.5 × 6 cm), preequilibrated with pH 9.2, 0.2 M borate buffer at 0°. The eluted protein, now free of small

[28] G. E. Means and R. E. Feeney, "Chemical Modification of Proteins," Chap. 6. Holden-Day, San Francisco, 1971.

[29] T. P. Link and G. R. Stark, *J. Biol. Chem.* **243**, 1082 (1968).

[30] R. H. Rice and G. E. Means, *J. Biol. Chem.* **246**, 831 (1971).

[31] A. D. Barksdale, B. E. Hedlund, B. E. Hallaway, E. S. Benson, and A. Rosenberg, *Biochemistry* **14**, 2695 (1975).

molecular products, has about 40% of $-NH_2$ groups labeled, amounting to 3.3% of HbCO chains. The ^{14}C activity at this stage works out to be 1 mCi/mmole of chains.

To correlate ^{14}C activity with hemoglobin concentration, a fixed volume of ^{14}C HbCO was added to a number of volumes of pH 7.0 buffer (0.1 M $NaPO_4$) such that the final calculated hemoglobin concentrations ranged from 8×10^{-4} to 2×10^{-8} M in heme groups (i.e., chains). Aliquots were removed for scintillation counting. Several samples were converted to the cyanmet[32] derivate for spectral determination of heme concentration. Knowing the [Fe] in these samples and their ^{14}C activities, we then calculated what the ^{14}C activity should be in the remainder of the solutions, based on the dilutions used. We found a 1:1 correlation between [Fe] and ^{14}C activity over the entire concentration range.

The level of ^{14}C activity obtained by the alkylation procedure can be converted to the observable counts per minute as (CPM)

$$(1 \text{ mCi/mmol chain}) \times (2.2 \times 10^9 \text{ CPM/mCi})$$
$$= 2.2 \times 10^9 \text{ CPM/mmol chain} \quad (23)$$

We want as little modified protein in the system as possible, because the [^{14}C]HbCO might be different structurally from unlabeled HbCO (although Means and Feaney[28] argue otherwise). On the other hand, there must be sufficient ^{14}C present for accurate and precise scintillation counting. In our experience, about 300–500 CPM are the lower limit for accuracy. Therefore, before introducing tritium, we calculate how much [^{14}C]HbCO must be mixed with cold hemoglobin to have the desired CPM in the most dilute samples.

A typical inexchange mixture might contain 0.7 ml of unlabeled HbCO (5×10^{-3} M in tetramer), 0.1 ml [^{14}C]HbCO, 0.1 ml ^3H_2O (1 Ci/ml, New England Nuclear), and 0.1 ml pH 9.5, 0.5 M borate buffer. The two HbCO solutions and the buffer were flushed with CO prior to mixing. The inexchange mixture, now having less than 1% modified chains, would be incubated at 32° for 16 hr. We arrived at this set of conditions by trial and error to balance the extent of ^3H incorporation against ferric hemoglobin formation. These conditions yield about 70% of maximum statistical ^3H incorporation and no more than 7% ferric hemoglobin. Complete statistical inexchange (discussed above in the section giving a survey of theory and methods) can be accomplished by longer incubation at higher pH and/or temperature, but at a price of 20% or more ferric forms.

Measurement of the dissociation constant, $K_{4,2}$, of HbCO follows this scheme (Fig. 2): first separation, dilution, outexchange, sampling, scintilla-

---

[32] E. Antonini and M. Brunori, "Hemoglobin and Myoglobin in Their Reactions with Ligands," p. 45. North-Holland Publ., 1971.

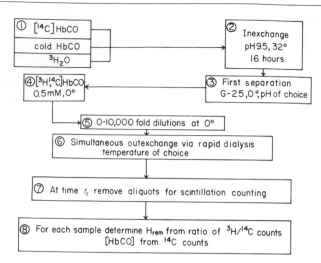

FIG. 2. Schematic representation of the double label radioisotope method for determining the dissociation constant of carboxyhemoglobin.

tion counting, calculation of H_{rem} and [HbCO], and calculation of $K_{4,2}$. Each step will be discussed in some detail.

First separation—in which the vast excess solvent tritium is removed from the system so that outexchange can begin—starts with the application of 0.2 ml of inexchange mixture (about $4 \times 10^{-3}\ M$ in tetramer) to a 1×10 cm Sephadex G-25 column, preequilibriated at $0°$ with CO saturated buffer (pH 6–8.6 in our work). The protein band, about 3 ml in volume, is collected on ice, and a few crystals of sodium dithionite (Fisher, final concentration <1 mg/ml) are added to reduce the 4–7% ferric hemoglobin in the inexchange mixture to about 1%. The dithionite is diluted substantially very quickly in the subsequent steps; therefore, no measurable sulfhemoglobin is formed. The HbCO concentration is now about $3 \times 10^{-4}\ M$ in tetramer.

The HbCO eluate is next apportioned into 17 samples, which are diluted 1- to 10^4-fold with CO-saturated buffer at $0°$. The exact dilution factors are designed to give a smooth logarithmic span in [HbCO]. The final volumes are large enough to fill the dialysis bag, (1–10 ml). The precision of these dilutions need not be great, since the protein concentration will change during dialysis. Moreover, the [HbCO] is determined exactly by ^{14}C activity at the sampling step. The total time for making these dilutions is about 3 min.

Except for the short time for dilutions, all HbCO molecules have had a common history up to the beginning of the dialysis step. Tritium loss began when the protein descended the column, but because the [HbCO] remains sufficiently high to be entirely tetrameric, all HbCO molecules will have lost the same amount of tritium.

The beginning of the dialysis step (which lasts 2–24 hr) initiates the unique outexchange behavior of each sample. Dialysis bags (⅜-inch width for concentrated samples, ¾-inch width for the dilute samples), boiled in frequently distilled water for several hours and cut to a length of about 12 inches, are mounted on Englander–Crowe racks[21] (rods of plastic with a magnet at the top, and with clamps below the magnet and at the bottom of the rod to hold the dialysis bag). For the concentrated samples, about 1–2 ml in volume, 3 bags are mounted per rack; for the dilute samples (5–10 ml), 2 bags occupy a rack. Each rack is inserted into a glass cylinder, filled with CO-saturated buffer, and placed in a waterbath at the desired temperature. The cylinders, except for bag filling and sampling, are kept tightly sealed.

Each dialysis bag, mounted before the first separation step, is filled with the appropriate HbCO solution in the order most concentrated to least concentrated. This order is strictly adhered to so that each sample will have been in its bag, at the desired temperature, for the same length of time when sampled. Filling the 17 bags (at $t = 0$) requires 7 min. Sampling ($t = t_s$) also requires 7 min. After filling, the racks are spun magnetically to establish diffusion equilibrium across the membrane.

We should add that the buffer in the cylinder containing the most concentrated samples ($\sim 3 \times 10^{-4}$ M in tetramer) sometimes receives 20 mg/ml of Dextran T-40 (Pharmacia or Sigma) to counteract the protein osmotic gradient, which would otherwise swell the bags and dilute their contents.

Sampling occurs at a predetermined time after bag filling. The length of time outexchange is allowed to proceed is based on two criteria: the desire to sample when the $\ln H_{rem}$, time curve obeys Eq. (15), and the need to have completed sampling before ferric hemoglobin formation and/or protein denaturation, become appreciable. These events occur more readily in dilute solutions. If either has occured appreciably, the values of $\ln H_{rem}$ at low concentration are inconsistent with the rest of the $\ln H_{rem}$, log [HbCO] curve (see Fig. 4). On the basis of our experience, sampling at 25° should start after 90 min and be finished before 240 min have passed. On the other hand, sampling at 5° may begin as much as 12 hr after filling and be concluded even 24 hr later.

Some comments about the sampling procedure are in order. For statistical purposes we customarily take three sets of data (i.e., at three times) for each outexchange. Before extracting aliquots from each bag, we also sample the dialyzate to determine the solvent ^3H and ^{14}C activities, which must be subtracted from those of the protein solution in the dialysis bags. Finally, color quenching must be accounted for.[33] That is, the reddish color of hemoglobin can shift the ^{3}H and ^{14}C energy spectra, thus altering the number of counts observed. We have overcome the problem by making all samples for scintillation counting have the same low hemoglobin concentration. Thus, an aliquot at $t = t_s$ is removed in the proper order from each bag and diluted with buffer. The volume of the solution extracted (5 μl to 0.5 ml) and the volume of buffer for the dilution are chosen so as to give a final [HbCO] $\sim 10^{-7}$ M in tetramer, a concentration high enough for accurate and precise determination of counts per minute but low enough to eliminate color quenching concerns.

Scintillation counting is performed on a sample containing 10.0 ml of scintillation cocktail and 0.5 ml of solution (protein or dialyzate). Choice of the proper scintillation cocktail again involved some trial and error. A relatively inexpensive Triton X 100–xylene solution (see below) was found wanting, in that the amount of quenching was quite sensitive to the pH, salt, and protein composition of the buffer. The following cocktail gives pH- and salt-independent results: 108 g (100 ml) of BBS-3 Biosolvent (Beckman), 900 ml of toluene (spectral grade), 0.2 g of POPOP (Packard), and 4 g of PPO (Packard) to make 1 liter of cocktail, which is mixed overnight and stored in a brown glass bottle.

For reasons discussed below, each sample is counted twice for 20 min.

Calculation of $H_{rem}/[Fe]$ requires four standard scintillation samples. The first two establish the spillover factors. That is, the liquid scintillation counter (Beckman LS-233) has two channels, A and B. The A channel is adjusted to monitor 99% of the ^{3}H activity, 1% appearing in (i.e., spilling over into) the B channel. About 30% of the ^{14}C CPM are recorded in the A channel with the balance appearing in the B channel. The spillover factors are the fractions of an isotope's total activity appearing in a particular channel. To determine the hemoglobin concentration it is necessary to know the total ^{14}C activity; hence, the total counts per minute in each channel must be broken into their ^{3}H and ^{14}C components. Mathematically

$$CPM_A = aCPM_H + (1 - b)CPM_C \qquad (23a)$$
$$CPM_B = (1 - a)CPM_H + bCPM_C \qquad (23b)$$

where the subscripts A, B, H, and C indicate A and B channels, ^{3}H, and ^{14}C, respectively. The symbol a is the fraction of total ^{3}H CPM appearing in the A channel, and b represents the fraction of total ^{14}C activity appear-

[33] D. L. Horrocks, "Applications of Liquid Scintillation Counting," Chap. 11. Academic Press, New York, 1974.

ing in the B channel. Simultaneous solution of Eqs. (23a) and (23b) yields

$$CPM_H = [bCPM_A - (1 - b)CPM_B]/c \qquad (24a)$$
$$CPM_C = [aCPM_B - (1 - a)CPM_A]/c \qquad (24b)$$

where $c = a + b - 1$.

The factor a, for tritium spillover, is determined by counting a sample containing ^3H_2O in buffer plus about $10^{-7} M$ in tetramer cold HbCO, the hemoglobin being added to provide the constant amount of color quenching, as discussed above. Therefore, from the activity of the standard tritium sample,

$$a = CPM_A/(CPM_A + CPM_B) \qquad (25)$$

The ^{14}C spillover factor b is measured by counting a sample containing only buffered $10^{-7} M$ tetramer $[^{14}C]HbCO$ from whose activity

$$b = CPM_B/(CPM_A + CPM_B) \qquad (26)$$

A third sample is needed to determine the extent of ^3H labeling in the inexchange. A small aliquot (say 10.0 μl) is extracted from the inexchange mixture and diluted precisely by about 10^5. A drop of cold hemoglobin is added to provide the proper color quenching. The extent of labeling [see Eq. (7)] is calculated, after spillover correction, as

$$E = CPM_H \times D/111 \qquad (27)$$

where D = dilution factor from inexchange to counting sample.

The fourth standard sample establishes the ratio of heme concentration to ^{14}C activity. After completion of the sampling step (so that tritium loss will be maximal), an aliquot of one of the more concentrated dialysis solutions is removed and precisely diluted. The [Fe] in this solution is determined by the cyanmet method.[32] A further dilution is then prepared for scintillation counting. After spillover corrections, the concentration factor F, is calculated as

$$F = [Fe] \times D_1/CPM_C \times D_2 \qquad (28)$$

where [Fe] = moles of heme per liter in the spectral sample, D_1 = dilution factor from bag to spectral sample, and D_2 = dilution factor from bag to scintillation sample.

The four standard vials are counted along with the vials from an experiment so as to minimize day-to-day variations in the scintillation counter's behavior.

Calculation of [Fe] and H_{rem} in the solutions from the dialysis experiments follows this protocol (spillover corrections being applied in all cases):

$$CPM_H^B = CPM_H \times D - B_H \qquad (29a)$$
$$CPM_C^B = CPM_C \times D - B_C \qquad (29b)$$

where the CPM_H^B, CPM_C^B are respectively the net ^3H and ^{14}C activities in the dialysis bags, CPM_H and CPM_C are the total activities in the counting sample, D = dilution factor from bag to counting sample, and B_H, B_C are the background (solvent) tritium and carbon 14 activities. Continuing,

$$H = \text{extent of outexchange} = CPM_H^B/E \qquad (30)$$
$$[Fe]^B = [Fe] \text{ in dialysis bag} = CPM_C^B/F \qquad (31)$$
$$H_{rem} = H/[Fe]^B \qquad (32)$$
$$[HbCO] = [Fe]^B/4 \qquad (33)$$

We now have 17 pairs of H_{rem}, $[HbCO]$ from which to statistically estimate the dissociation constant $K_{4,2}$. Rearrangement of Eqs. (10) and (22) yields

$$\alpha_i = \frac{-K_{4,2} + \sqrt{(K_{4,2})^2 + 16[HbCO]_i K_{4,2}}}{8[HbCO]_i} \qquad (34)$$
$$\ln H_{rem,i} = (1 - \alpha_i) \ln H_{rem,0} + \alpha_i \ln H_{rem,1} \qquad (35)$$

where the subscript i indicates the ith of the 17 data pairs.

In practice there are three unknowns: $K_{4,2}$, $\ln H_{rem,0}$, and $\ln H_{rem,1}$. The latter two behave effectively as unknowns because the range of $[HbCO]$ covers four orders of magnitude, rather than six, which if symmetrically distributed about $K_{4,2}$, would yield the outexchange behavior of tetramer and dimer independently.

To solve for the three unknowns we turn to established statistical methods. The chi-square[34b]

---

* Errors can be propagated stepwise by successive use of these equations.[34a] Let y be a function of variables u, v, x, . . .

$$y = f(u, v, z, \ldots)$$

The variance of y is given, to a good approximation in our experience, as

$$\sigma_y^2 = (df/du)^2 \sigma_u^2 + (df/dv)^2 \sigma_r^2 + (df/dz)^2 \sigma_z^2 + \ldots$$

where df/du, df/dv, . . . are partial derivatives of y with respect to u, v, ... For example, let us consider Eq. (27)

$$E = cpm_H \times D/111$$

The variance

$$\sigma_E^2 = (cpm_H/111)^2 \sigma_D^2 + (D/111)^2 \sigma_{CPM_H}^2$$

where σ_D^2 can be estimated by calibration of glassware and $\sigma_{cpm_H}^2$ estimated from counting errors. Progressing to Eq. (30), the next step in which E appears

$$H = CPM_H^B/E$$

we find $\sigma_H^2 = (1/E)^2 \sigma_{CPM_H^B}^2 + (CPM_H^B/E^2)^2 \sigma_E^2$

$$\chi^2 = \Sigma[(\ln H_{rem,i})_{calc} - (\ln H_{rem,i})_{obs}]/\sigma_i^2 \qquad (36)$$
$$= \Sigma[(1 - \alpha_i) \ln H_{rem,0} + \alpha_i \ln H_{rem,1} - (\ln H_{rem,i})_{obs}]/\sigma_i^2$$

is differentiated with respect to $\ln H_{rem,0}$ and $\ln H_{rem,1}$, and the derivatives are set equal to zero (χ^2 minimization). There results

$$\ln H_{rem,0}\Sigma(1 - \alpha_i)^2/\sigma_i^2 + \ln H_{rem,1}\Sigma\alpha_i(1 - \alpha_i)/\sigma_i^2$$
$$= \Sigma(1 - \alpha_i)(\ln H_{rem,i})_{obs}/\sigma_i^2 \quad (37)$$
$$\ln (H_{rem,0})\Sigma\alpha_i(1 - \alpha_i)/\sigma_i^2 + \ln (H_{rem,1})\Sigma\alpha_i^2/\sigma_i^2 = \Sigma\alpha_i(\ln H_{rem,i})_{obs}/\sigma_i^2 \quad (38)$$

which can be solved simultaneously for $\ln H_{rem,0}$ $\ln H_{rem,1}$ in terms of the α_i's. To know the α_i's, one should know $K_{4,2}$ or be able to solve for $K_{4,2}$. Unfortunately, Eq. (36) cannot be differentiated with respect to $K_{4,2}$ in a straightforward manner.

Therefore, a regression algorithm, discussed in detail by Bevington[4a,4c] to statistically evaluate $K_{4,2}$ and its standard deviation, was adopted.

1. Guess at $K_{4,2}$.
2. Calculate α_i by Eq. (34).
3. Calculate $\ln(H_{rem})_0$, $\ln(H_{rem})_1$ from the solutions to Eqs. 37 and 38.
4. Calculate χ^2.
5. Refine the guess at $K_{4,2}$ until a minimum in χ^2 (χ^2_{min}) is achieved. The value of $K_{4,2}$ at $\chi^2_{min} = \hat{K}_{4,2}$, the most probable value.
6. Adjust $K_{4,2}$ to lesser or greater values until $\chi^2 = \chi^2_{min} + 1$.
7. The standard deviation of the dissociation constant is defined as

$$\sigma_{K_{4,2}} = |\hat{K}_{4,2} - K_{4,2} \text{ at } \chi^2_{min} + 1| \qquad (39)$$

It makes no difference if $K_{4,2}$ is adjusted to values higher or lower than $\hat{K}_{4,2}$ to determine $\sigma_{K_{4,2}}$. That is, the well in χ^2 space is symmetrical about χ^2_{min}.

To this point we have made no allusion to the weighting factors, σ_i^2 = variance of $\ln (H_{rem})_{i,obs}$. These can be arrived at by propagating errors in scintillation counting and dilution [Eqs. (23) through (32)]. The absolute magnitude of σ_i^2 has no effect on the χ^2 minimization. That is, the same $\hat{K}_{4,2}$ is obtained if $\ln H_{rem}$ is in error by 1% or by 10%. The absolute magnitude of σ_i^2 becomes critical, however, in determining $\sigma_{K_{4,2}}$. For example, if $\sigma_i^2 \cong 0.001$ (corresponding to 1% error in H_{rem}) then $\chi^2_{min} \cong 10$. If $\sigma_i^2 \cong 0.11$, then $\chi^2_{min} \cong 0.1$. In the latter case, $K_{4,2}$ must be adjusted by about a 10-fold greater amount than in the former case to determine the standard deviation of $K_{4,2}$.

[4a] P. R. Bevington, "Data Reduction and Error Analysis for the Physical Sciences." McGraw-Hill, New York, 1969, p 60.
[4b] P. R. Bevington, Op. Cit., p 84.
[4c] P. R. Bevington, Op. Cit., Ch. 11.

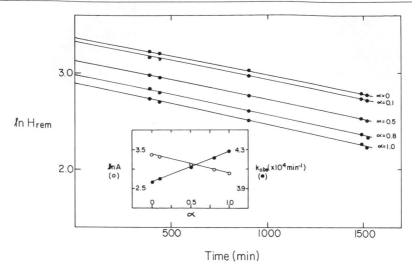

Time (min)

FIG. 3. Outexchange of HbCO as a function of dissociation at 20°, pH 7.0, sodium phosphate buffer. Solutions were prepared to have proper concentrations to give the desired extents of dissociation from the data of Crepeau *et al.* (see text footnote 35) for HbCO under identical conditions. ●, experimental points; ——, linear regression of ln H_{rem} upon t. Inset: variation of k_{obs} (negative of slope in ln H_{rem}, t plot) and ln A (intercept of ln H_{rem}, t plot) with extent of dissociation.

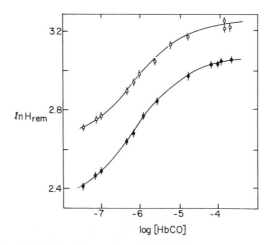

log [HbCO]

FIG. 4. ln H_{rem} at constant time vs log [HbCO]. Conditions: pH 7.4, 0.1 M sodium phosphate, 25°. ——, Weighted least-squares fit to determine $K_{4,2}$. ○, points taken 110 min after beginning the dialysis step, $K_{4,2} = (2.1 \pm 0.2) \times 10^{-6}$; ●, points taken 195 min after beginning dialysis, $K_{4,2} = (1.8 \pm 0.2) \times 10^{-6}$. Error bars: ±1 SD of ln H_{rem}.

TABLE I

COMPARISON OF $K_{4,2}$ FOR CARBOXYHEMOGLOBIN AS DETERMINED BY
TRITIUM EXCHANGE AND BY OTHER METHODS

References[a]	Conditions	Method[b]	$K_{4,2}$	$K_{4,2}$ by tritium exchange[c]
1	pH 7.0, 20°	UC	$(1.1 \pm 0.20) \times 10^{-6}$	$(1.6 \pm 0.2) \times 10^{-6}$ (20°)
2	pH 7.0, 22°	UC	1.5×10^{-6}	—
3	pH 7.0, 12°	LS	1×10^{-6}	$(1.4 \pm 0.2) \times 10^{-6}$ (10°)

[a] Key to references:

1. R. H. Crepeau, C. P. Hensely, Jr., and S. J. Edelstein, *Biochemistry* **13**, 4860 (1974).
2. J. O. Thomas and S. J. Edelstein, *J. Biol. Chem.* **247**, 7870 (1972).
3. I. Noren, D. A. Bertoli, C. Ho, and E. F. Casassa, *Biochemistry* **13**, 1683 (1974).

[b] UC, ultracentrifugation; LS, light scattering.

[c] Barksdale and Rosenberg, present study.

It is our practice to propagate all errors to arrive at the standard deviation of ln (H_{rem}). In the initial phases of the work, these included 0.5 to 1.0% in dilution factors (based upon manufacturers glassware specifications) and 1–3% error in scintillation counting (Beckman LS-233 specifications). These gave a 2–4% error in H_{rem}, which translates to a constant error of 0.02 to 0.04 in ln (H_{rem}). Subsequently, $\hat{K}_{4,2}$ would have a calculated error of about 25–35%. We later refined our errors by calibrating the volumetric pipettes and other glassware, which reduced the dilution errors to 0.2–0.5%. The counting errors were also reduced by double countings of each sample, the results of which were averaged. The average counts had a standard deviation well below the error predicted by the Beckman literature. The outcome of our error refinement was to reduce the standard deviation of H_{rem} to 0.5–1.0% (an error in ln H_{rem}) of 0.005–0.01). This, in turn, reduced $\sigma_{K,2}$ to 10–20% of $\hat{K}_{4,2}$.

We noted in the Mathematics section that if Eq. (35) is to be used to evaluate $K_{4,2}$ [Eq. (22)] then at all α (i.e., all concentrations) Eq. (15) as well as Eqs. (17a) and (17b) must be obeyed. To test the hypothesis we prepared solutions which at pH 7.0, 0.1 M sodium phosphate, 20° would have proper concentrations of HbCO to give $\alpha = 0, 0.1, 0.5, 0.8$, and 1.0. We could calculate the concentrations needed from the $K_{4,2}$ evaluated under the above conditions by Crepeau *et al.*[35] The data fitted to Eqs. (17a) and (17b) are given in Fig. 3.

The results of a typical experiment are shown in Fig. 4. Comparisons between our work and that of others is given in Table I. The full scope of our findings is the subject of another publication.[25] Our purpose here is to

[35] R. H. Crepeau, C. P. Hensely, Jr., and S. J. Edelstein, *Biochemistry* **13**, 4860 (1974).

illustrate how the double-label technique can be used to determine protein dissociation constants precisely and with relative simplicity.

The Dissociation of Ribonuclease S

Schreier and Baldwin have elegantly investigated the dissociation of ribonuclease S[4,24,26] into the S-peptide and S-protein as a function of concentration, pH, and temperature by $^3H/^{14}C$ double labeling. Although the principles are the same as those in the determination of the tetramer–dimer association of hemoglobin, the experimental methods and data analysis are quite different. Therefore, their work will be discussed in detail.

The ribonuclease S (RNase S) dissociation scheme may be represented in the following manner:

$$\text{RNase S*} \underset{\downarrow k_0}{\overset{K_d}{\rightleftharpoons}} \text{S-protein} + \text{S*peptide} \downarrow k_1$$

where the dissociation constant K_d is given by

$$K_d = [\text{S-protein}][\text{S-peptide}]/[\text{RNase S}] \tag{40}$$

Since the S-protein and S-peptide can be separated readily, it is possible in this system to introduce tritium into only one reaction partner—the S-peptide—after which the two species can be rejoined to form the intact protein, which will dissociate to whatever degree conditions dictate. Thus the kinetics of tritium loss from the peptide in its two environments provides the means of measuring K_d. Moreover, only the S-peptide bears the ^{14}C concentration marker.

Let us consider the experimental methods of Schreier and Baldwin. First, $^{14}CH_3$ groups are affixed to the S-peptide chain by the method of Rice and Means[30] discussed above. Inexchange is carried out by incubating the S-peptide (with about 1% ^{14}C-labeled chains) in 20 mCi/ml ^3H_2O at 50° for 30 min. The first separation by cryosublimation follows immediately (0° under a CO_2 atmosphere). The [^3H,^{14}C]S-peptide powder is mixed at 0° with a precisely determined 30% excess of S-protein at pH 3.3 (pH min for RNase), and outexchange is allowed to occur for 1–24 hr. The purpose of this step is to allow the rapidly exchanging hydrogens of the S-peptide to be lost from the complex, but at a relatively slow and controllable rate, so as to leave only the 10 or so more slowly exchanging protons. The efficacy of allowing rapid exchange to be completed before final adjustment of pH and/or concentration will become clear later. At the end of the waiting period the system is diluted to the desired concentration with buffer of the desired pH. Schreier and Baldwin, as did Barksdale and Rosenberg, prepared a number of samples of varying concentrations, which simultaneously undergo outexchange and are sampled simultaneously.

At time t_s following initiation of outexchange, an aliquot is removed from each of the samples for evaluation of H_{rem} and concentration. The second separation is accomplished by cryosublimation or by phosphocellulose filtration. In both cases, the aliquot is precisely diluted with pH 3 buffer (to return the system to pH min). If cryosublimation is used, the sample is frozen at $-78°$, a temperature low enough to terminate outexchange, and at some later, more convenient time, the solvent is removed by freeze-drying at $-35°$ under CO_2, conditions that eliminate ^3H loss from the peptide during the second separation. The powder is then dissolved in a standard pH 8 buffer (regardless of the pH at which outexchange was carried out), and H_{rem} and total concentration of peptide are determined by scintillation counting, with spillover and quench corrections, as described above.

If phosphocellulose filtration is used, the aliquot is added to the well of a stainless steel vacuum filter. The Whatman P-81 filter paper has been pretreated with $0.1 M$ NaOH, deionized H_2O, $0.1 M$ HCl, deionized H_2O, and 50 mM sodium formate buffer at pH 3. The well already contains 5 ml of the formate buffer. The funnel is encased in a collar that is filled with ice so that the entire filter and its contents are quite close to $0°$.

The protein solution is filtered slowly (20–30 sec). There follow rapid rinses with $0°$ pH 3 buffer. The moist paper (2.4 cm in diameter) is placed in a glass scintillation vial containing about 0.75 ml of 2.5 M NaCl in 50 mM pH 3 formate buffer. The salt solution elutes the protein and peptide from the paper (although Schreier[24] hastens to add that different salt concentrations and other conditions may be necessary to separate other proteins from the paper). After rinsing in the salt solution for about 1 hr, 10.0 ml of a 75% xylene, 25% Triton X-100, PPO, POPOP scintillation fluid is added, and scintillation counting begins. Standard samples, prepared with the same high salt concentrations to minimize quench corrections, are counted along with the experimental samples. The binding of the protein and peptide to the paper is not always complete, but by virtue of the quantitation afforded by the ^{14}C internal concentration standard, complete binding is unnecessary, provided sufficient ^3H and ^{14}C activity are present in the scintillation vial.

To demonstrate the efficacy of the phosphocellulose filtration, Schreier[24] ran parallel experiments on the S-peptide in which second separations by both cryosublimation and phosphocellulose filtration were compared to the standard gel filtration technique of Englander.[22] The results (Fig. 5) show the three methods to give identical results.

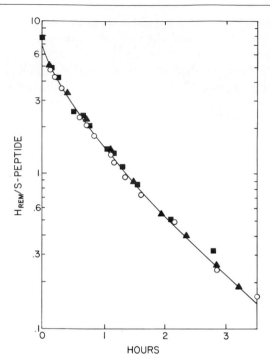

FIG. 5. Comparison of three assay methods for the extent of tritium exchange of the S-peptide in 50 mM sodium acetate, pH 3.0, 0°. Symbols: second separation by gel chromatography (▲), freeze drying (■), and phosphocellulose filtration (○). Data of A. A. Schreier *Anal. Biochem.* In press (1977).

The data analysis in the RNase S system is made relatively simple by virtue of these facts:

1. The slowly exchanging hydrogens of the S-peptide behave as though they were identical with a single rate constant [Eqs. (13) and (15)] at long times for the free and complexed S-peptide (Fig. 6), the linearity in lnH_{rem} time having been enhanced by letting the rapidly exchanging hydrogens "get out of the picture." Moreover, propagation of errors in lnH_{rem} is less critical because of the very large differences between the outexchange characteristics of the bound and free S-peptide ($k_1 \sim 1.5$ hr^{-1}, $k_0 \sim 0.005$ hr^{-1} at pH 3.3, 0°).

2. Since only the S-peptide bears tritium (cf. all subunits in hemoglobin) and since the difference in rates as a function of S-peptide incorporation into the complex is so large,[4] one need not work at very low dilution (i.e., $\alpha \to 0$) to obtain an accurate and precise estimate of K_d.

That is, following Schreier and Baldwin[4] but with the symbols introduced in Section III, we relate K_d to k_{obs} in the following manner.

If $\alpha \ll 1$, then,

$$\alpha = [\text{S-peptide}]/\{[\text{RNase S}] + [\text{S-peptide}]\}$$
$$\cong [\text{S-peptide}]/[\text{RNase S}] \quad (41)$$

whence, recalling Eq. (40),

$$\alpha = K_d/[\text{S-protein}] \quad (42)$$

Since $\alpha \ll 1$, then, by Eq. (14) rearranged:

$$k_{obs} = k_0 + K_d k_1/[\text{S-protein}] \quad (43)$$

where $[\text{S-protein}] = [\text{S-protein}]^0 + [\text{S-peptide}]^0 = 1.3[\text{S-peptide}]^0$ by experimental design, and $[\text{S-peptide}]^0$ is determined by ^{14}C activity, standardized against amino acid analysis. Thus, at low degrees of dissociation, there exists a linear relationship between the k_{obs} for tritium exchange in the S-peptide and $1/[\text{S-protein}]$, the slope being $K_d k_1$. The exactness of the linearity is demonstrated in Fig. 5. At $pH \leq 4.3$, Schreier and Baldwin were able to measure k_1 precisely and thus to calculate K_d. At $pH > 4.3$,

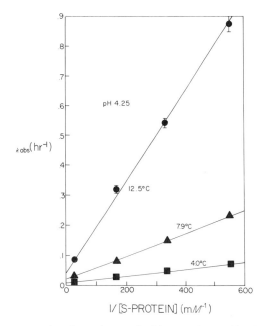

$1/$ [S-PROTEIN] (mM^{-1})

FIG. 6. The concentration dependence of tritium exchange kinetics of the S-peptide bound in RNase S at three temperatures. Conditions: 50 mM sodium acetate, pH 4.25. The samples had been preoutexchanged at 0°, pH 4.25, before being diluted and brought to the temperatures given (see text for explanation). Curves: weighted least squares to Eq. (43) in text. From A. A. Schreier and R. L. Baldwin, *Biochemistry* **16**, 4203 (1977).

TABLE II

COMPARISON OF K_d FOR RNase S AS DETERMINED BY TRITIUM
EXCHANGE AND OTHER METHODS

References[a]	Conditions	Method[b]	K_d	K_d by tritium exchange[c]
1	pH 2.7, 0°	UV	1×10^{-6}	1.1×10^{-6}
2	pH 7.0, 0°	Cal	1.5×10^{-10}	2×10^{-10}

[a] Key to references:
   1. Schreier and Baldwin, unpublished observations.
   2. R. P. Hearns, F. M. Richards, J. M. Sturtevant, and G. D. Watt, *Biochemistry* **10**, 806 (1971).
[b] UV, ultraviolet absorption; Cal, calorimetry.
[c] A. A. Schreier and R. L. Baldwin, *J. Mol. Biol.* **105**, 409 (1976).

the kinetics of the free S-peptide hydrogens are too rapid for the techniques described; consequently, k_1 was estimated from a relationship similar to Eq. (4). In Table II are listed Schreier and Baldwin's values for K_d of RNase S, together with other literature values. The agreement is seen to be excellent.

Summary

The tritium exchange method for measuring protein dissociation constants offers several advantages over other methods, especially when coupled with the ^{14}C concentration tag. The most obvious advantage is that of determining a protein property in very dilute, or very concentrated solutions. A second is the relative simplicity of the equipment needed. Indeed, the only expensive item is the liquid scintillation counter, which has become a stock feature in many laboratories.

Acknowledgments

The authors wish to thank Drs. R. L. Baldwin and A. A. Schreier for sending unpublished data and allowing its appearance in these pages. Thanks are also due Dr. A. A. Schreier, Professor Ben E. Hallaway, and Mr. G. J. Ide for helpful discussions. Finally, the authors acknowledge support for these studies from the National Science Foundation (Grant BMS 74-22686), the National Institutes of Health (Grant HL 16833) and the Minnesota Medical Foundation.

[17] The Use of Singlet– Singlet Energy Transfer to Study Macromolecular Assemblies

By Robert H. Fairclough and Charles R. Cantor

Fluorescence spectroscopy has assumed an increasingly active role in the study of the dynamics and structure of biological macromolecular assemblies. Intrinsic or extrinsic fluorescent emission, being extremely sensitive to local environment, can be used to monitor the kinetics and thermodynamics of the incorporation of a particular subunit or substrate into an assembly. Nevertheless, the most definitive information available from fluorescence measurements is the distance between pairs of loci in the assembled structure. The distance is determined from the efficiency of nonradiative singlet–singlet energy transfer between pairs of fluorescent dyes. The rate of this transfer is proportional to the inverse sixth power of the distance between a given pair of chromophores. One may determine, by appropriate choice and placement of fluorescent dyes, a distance that corresponds either to the specific distance between two chemically unique points in the assembled structure or to the center-to-center distance between two globular components in the structure.

With this information potentially available from fluorescence measurements, it is encouraging to observe the recent improvements in commercial instruments available for measuring fluorescence spectra and excited singlet-state lifetimes. During the past 5 years, an arsenal of covalent fluorescent labels and noncovalent probes has also become commercially available. In addition, refinements in the analysis of singlet–singlet energy transfer data have been presented,[1,2] and advances in the analysis of excited singlet lifetime data have been made.[3] In comparison to other physical techniques, fluorescence spectroscopy is a relatively simple and inexpensive method for studying self-assembling structures. The following discussion concentrates on singlet–singlet energy transfer. Note, however, that many of the samples that must be prepared as controls for an energy transfer experiment can provide useful information themselves apart from the transfer experiment.

[1] R. E. Dale and J. Eisinger, *in* "Biochemical Fluorescence: Concepts" (R. F. Chen and H. Edelhoch, eds.), Vol. I, p. 115. Dekker, New York, 1975.
[2] Z. Hillel and C.-W. Wu, *Biochemistry* **15,** 2105 (1976).
[3] I. Isenberg, *in* "Biochemical Fluorescence: Concepts" (R. F. Chen and H. Edelhoch, eds.), Vol. I, p. 43. Dekker, New York, 1975.

Introduction to Fluorescence Spectroscopy

The Fluorescence Time Scale

Weber has elegantly described fluorescence spectroscopy as the observation of a relaxation process that begins with the absorption of light and terminates with its emission.[4] Only those processes that occur with a frequency equal to or greater than the inverse of the time interval between absorption and emission have a chance of altering the relaxation process under observation, namely: fluorescence.

For many common fluorescent, substituted aromatic systems, the excited singlet-state lifetime is of the order of 10 nsec. Thus, for a process to alter fluorescent emission from this system, the process must occur with a rate of the order of 10^8 sec^{-1} or faster. Three processes that occur with this rate are particularly interesting. Elementary considerations show that translational diffusion of the fastest moving species in solution allows only particles within 10 Å of a fluorophore any chance to interact with the fluorophore in 10 nsec.[4] Hence, one of the most important factors governing the fluorescent emission of an excited system is the *immediate* environment of the fluorophore. A second process that can occur on the time scale of the first excited singlet state lifetime is rotational diffusion. Both absorption and emission of light by a fluorophore are highly orientation dependent. Observation of the polarization of the fluorophore's emission as a function of time or viscosity allows analysis of the rotational freedom enjoyed by the fluorophore. A third process known to occur on the time scale of the first excited singlet lifetime is long-range nonradiative singlet–singlet energy transfer. The rate of transfer was predicted theoretically by Förster[5] and verified experimentally by Stryer and his collegues.[6,7] The rate of transfer for many pairs of fluorophores is on the order of 10^7 to 10^9 sec^{-1}. In the next three sections each of these three processes is considered in more detail to see how it influences fluorescence spectroscopic measurements.

Influence of Local Environment on Fluorescence

Fluorescence is light emitted as a chromophore radiatively relaxes from the ground vibrational level of the first excited electronic singlet

[4] G. Weber, *in* "Spectroscopic Approaches to Biomolecular Conformation" (D. W. Urry, ed.), p. 23. American Medical Association, Chicago, Illinois, 1970.

[5] T. Förster, *in* "Istanbul Lectures" (O. Sinanouglu, ed.), Part III, p. 93. Academic Press, New York, 1965.

[6] L. Stryer and R. P. Haugland, *Proc. Natl. Acad. Sci. U.S.A.* **58**, 719 (1967).

[7] J. Yguerabide and L. Stryer, *Biophysical Journal—Proceedings of Los Angeles National Meeting*, 1969.

state to the ground electronic state. The first excited singlet state, S_1, is characterized experimentally by its lifetime, τ. This is just the reciprocal of the sum of the rates of all modes of deexcitation:

$$\tau = (k_f + \Sigma k_i)^{-1} \qquad (1)$$

where k_f is the rate of fluorescence and Σk_i is the sum of the rates of all

other competing processes that deactivate the state, such as intersystem crossing, internal conversion, solvent quenching. The lifetime is evaluated from the exponential decay of emission from a population of pulse-excited fluorophores.

The quantum efficiency, ϕ, of the emission provides additional characterization of the first excited singlet state. It is the ratio of the rate of emission of fluorescence to the sum of the rates of all modes of deexcitation:

$$\phi = k_f/(k_f + \Sigma k_i) \qquad (2)$$

The spectral distribution $f(\nu)$ or $f(\lambda)$ of the emission provides characterization of the energy separation between the ground vibronic level of the first excited electronic state and the entire vibronic envelope of the ground electronic state.

The immediate environment of the fluorophore can affect all the parameters used to characterize the first excited singlet state. Changes in polarity or rigidity in the immediate vicinity of the fluorophore can alter the relative spacing between the S_1 and the S_0 states, giving rise to "red shifts" or "blue shifts" in the fluorescence emission spectrum. Changes in the viscosity of the immediate environment or changes in the rotational freedom about covalent bonds in the fluorophore, or changes in rates of protonation or deprotonation with solvent can affect the relative rates of the various processes competing with fluorescence to deexcite the first excited singlet state. These changes can alter the singlet lifetime, τ, as well as the fluorescence quantum efficiency, ϕ.

No matter what the origin of these spectral changes, the important point to bear in mind is that the changes themselves allow one to monitor the kinetics and thermodynamics of processes giving rise to the changes in the microenvironment of the fluorophore so long as the rates of these processes are slower than 10^8 sec^{-1}. For more detailed description of environmental effects, the reader is referred to articles by Stryer,[8] Weber,[4] and Brand.[9,10]

[8] L. Stryer, *Science* **162**, 526, (1968).
[9] L. Brand and B. Witholt, see this series Vol. 11, p. 776.
[10] L. Brand and J. R. Gohlke, *Annu. Rev. Biochem.* **41**, 843 (1972).

Influence of Rotational Diffusion on Polarization of Emission

To see how rotational diffusion influences the polarization of emission, consider a solution of fluorophores at the origin of a Cartesian coordinate system (Fig. 1). The dyes and the transition dipoles that determine preferred absorption directions are at random orientations throughout the solution. Suppose one excites the sample with an instantaneous pulse of light propagating along the X axis and plane polarized in the XZ plane. Those fluorophores with absorption transition dipoles oriented in the Z direction are preferentially excited since the probability of excitation is proportional to $\langle \mathbf{u}_{oa} \cdot \mathbf{E} \rangle^2$ where $\mathbf{u}_{oa}$ is the transition dipole moment and $\mathbf{E}$ is the electric field vector of the light. This preferential excitation is called photoselection. For many substituted fluorescent aromatics, the longest wavelength absorption transition dipole and the fluorescence emission transition dipole are nearly parallel. Suppose absolutely no rotation of the excited fluorophores occurs. The fluorescence intensity observed on the nanosecond time scale along the Y axis through a polarizer

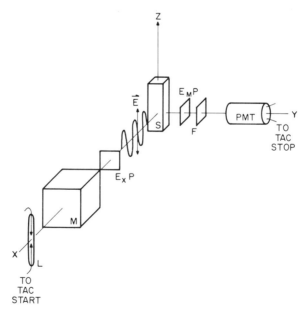

Fig. 1. Schematic diagram of the optical arrangement for studying the nanosecond depolarization of fluorescence. L is a pulsed light source; M is a monochromator; E_x P is the excitation polarizer; E is the direction of the electric field vector of the exciting light; S is the sample solution; F is a filter excluding exciting light, but passing fluorescent emission; E_m P is the emission polarizer; PMT is a photomultiplier; and TAC is a time to amplitude converter.

oriented in the Z direction, f_z, should decay exponentially with the lifetime, τ, of the excited singlet state:

$$f_z(t) = f_z(0)e^{-t/\tau} \tag{3}$$

If the fluorophores are at liberty to rotate isotropically during the excited state lifetime, f_z has an additional time dependence, $r_z(t)$, contributed by the rotation of the fluorophore emission transition dipoles:

$$f_z(t) = f_z(0)r_z(t)e^{-t/\tau} \tag{4}$$

To separate the time dependence of the rotational motion, $r_z(t)$, from the excited-state lifetime decay, one also measures the time dependence of fluorescence along the Y axis with the emission polarizer oriented parallel to the X axis, $f_x(t)$. Then two functions can be computed:

$$S(t) = f_z(t) + 2f_x(t) \tag{5}$$
$$r(t) = [f_z(t) - f_x(t)]/S(t) \tag{6}$$

Both $f_z(t)$ and $f_x(t)$ contain time dependence from rotational motion and from excited singlet decay. $S(t)$ is the time dependence of the total emission and is generally observed as a single exponential decay with a lifetime, τ, of the first excited singlet state of the fluorophore[11]:

$$S(t) = S_0 e^{-t/\tau} \tag{7}$$

$r(t)$, the emission anisotropy, is defined such that the excited state lifetime decay factor, $e^{-t/\tau}$, in each term of the numerator, $f_z(t)$ and $f_x(t)$, is removed by the $e^{-t/\tau}$ from $S(t)$ in the denominator.[11] Thus the behavior of $r(t)$ reflects only rotational motion.[11] At very short times after excitation, the emission dipoles largely retain their excitation distribution about the Z axis. One thus expects f_z to be greater than f_x and hence, $f_z(t) - f_x(t)$ to be large for small t. As time progresses, the excited dipoles redistribute randomly so that more lie along the X and Y axes: $f_x(t)$ increases and $f_z(t)$ decreases. At very long times these two functions should become equal and $f_z(t) - f_x(t) = 0$ for $t = \infty$. $r(t)$ is thus a decreasing function of time, and for rigid spheres, $r(t)$ decays exponentially[11]:

$$r(t) = r(0)e^{-t/\theta} \tag{8}$$

where θ is the rotational relaxation time. It is related to the solvent viscosity, η, the hydrated volume, V, and the absolute temperature, T, by $\theta = V\eta/kT$, where k is Boltzmann's constant. For a more complete discussion of and interpretation of polarized nanosecond experiments, the

[11] J. Yguerabide, see this series Vol. 26, p. 498.

reader is referred to articles by Yguerabide,[11] Wahl,[12] and Dale and Eisinger.[1]

Frequently the steady-state emission anisotropy, r, is more accessible than the time-resolved anisotropy discussed previously:

$$r = (I_z - I_x)/(I_z + 2I_x)$$

In measuring I_z and I_x the excitation is continuous rather than pulsed, but the polarizer orientations are the same as described for nanosecond studies. r is related to the rotational motion of the excited fluorophores through the equation:

$$1/r = 1/r_0 + 1/r_0(\tau/\theta) \qquad (9)$$

where τ and θ are the same as in Eqs. (7) and (8), and r_0 is the limiting anisotropy at high viscosity and low temperature. To evaluate both r_0 and θ, one can measure $1/r$ as a function of viscosity, η; prepare a Perrin plot of $1/r$ vs T/η (see Fig. 6); observe its linearity (to show τ is independent of viscosity); then determine r_0 and θ from the intercept and slope, knowing τ from an independent measurement. In principle, r_0 should be the same as $r(0)$ obtained from time-dependent measurements.

Two parameters characterize the effect of rotational diffusion on polarized emission. The limiting anisotropy r_0, or $r(0)$, is associated with the viscosity-independent rapid reorientation of dipoles; while θ is associated with the viscosity-dependent, slow reorientation of the dipoles. Here "rapid" and "slow" are relative to the nanosecond time scale.

Nonradiative Energy Transfer

Theoretical Background

Light energy absorbed by one fluorophore, the donor, can be transferred nonradiatively over substantial molecular distances to a second chromophore, the acceptor. Generally the wavelengths of light used to monitor this effect range from 3000 Å to 6000 Å, and the distances over which the energy is nonradiatively transferred range from 15 Å to 70 Å. Although the distances are substantial compared to the dimensions of many biological macromolecules, they are small relative to the wavelength of light. Thus, direct coupling of the emission transition dipole of the donor to the absorption transition dipole of the acceptor is a reasonable mechanism for the transfer process. Using this mechanism, Förster

---

[12] P. Wahl, in "Biochemical Fluorescence: Concepts" (R. F. Chen and H. Edelhoch, eds.), Vol. I, p. 1. Dekker, New York, 1975.

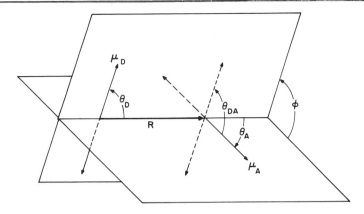

FIG. 2. Diagram of the angles used in the definition of κ^2. μ_D is the emission transition dipole vector of the donor; μ_A is the absorption transition dipole vector of the acceptor. R is the vector joining the centers of the two chromophores.

has calculated the rate of nonradiative transfer[5]:

$$k_t = 1/\tau(8.79 \times 10^{-5}) \cdot \kappa^2 \cdot n^{-4} \cdot \phi_D \cdot J_{DA} \cdot R^{-6} \qquad (10)$$

where τ and ϕ_D are the excited state lifetime and quantum efficiency of the donor in the absence of the acceptor; n is the refractive index of visible light in the intervening medium; and R is the distance in Å between the donor and the acceptor. κ^2, the orientation factor is defined as:

$$\kappa^2 = (\cos \theta_{DA} - 3 \cos \theta_D \cos \theta_A)^2$$
$$= (\sin \theta_D \sin \theta_A \cos \phi - 2 \cos \theta_D \cos \theta_A)^2 \quad (11)$$

where the angles involved are shown in Fig. 2. J_{DA}, the spectral overlap integral is defined as[13]:

$$J_{DA} = [\int_0^\infty f_D(\lambda)\epsilon_A(\lambda)\lambda^4 d\lambda]/[\int_0^\infty f_D(\lambda)d\lambda] \qquad (12)$$

where $f_D(\lambda)$ is the corrected emission spectrum of the donor in quanta per unit wavelength interval; $\epsilon_A(\lambda)$ is the molar extinction coefficient of the acceptor in $M^{-1}cm^{-1}$; λ is the wavelength in nanometers. It is convenient

[13] The overlap integral has been alternatively expressed as:

$$J_{DA} = [\int_0^\infty f_D(\nu)\epsilon^{-4}d\nu]/[\int_0^\infty f_D(\nu)d\nu] \qquad (a)[14]$$
$$J_{DA} = [\int_0^\infty f_D(\bar\nu)\epsilon_A(\bar\nu)\bar\nu^{-4}d\bar\nu]/[\int_0^\infty f_D(\bar\nu)d\bar\nu] \qquad (b)[1]$$
$$J_{DA} = [\int_0^\infty f_D(\lambda)\epsilon_A(\lambda)\lambda^2 d\lambda]/[\int_0^\infty f_D(\lambda)\lambda^{-2}d\lambda] \qquad (c)[15]$$

(a) and (b) are both used by Förster in his derivation of the rate of transfer.[5] Form (c) is incorrect (R. E. Dale, personal communication); the correct form for wavelength-dependent spectra is Eq. (12) in the text.

[14] R. B. Gennis and C. R. Cantor, *Biochemistry* **11**, 2509 (1972).

[15] K. H. Huang, R. H. Fairclough, and C. R. Cantor, *J. Mol. Biol.* **97**, 443 (1975).

to abbreviate Eq. (10) as:

$$k_t = 1/\tau(R_0/R)^6 \qquad (13)$$

where

$$R_0^6 = (8.79 \times 10^{-5}) \cdot \kappa^2 \cdot n^{-4} \cdot \phi_D \cdot J_{DA} \qquad (\text{in } \text{Å}^6) \qquad (14)$$

R_0 is called the "Förster critical distance." It is the distance at which the rate of nonradiative energy transfer equals the sum of the rates of all the other modes of depopulating the first excited singlet state of the donor:

$$k_t = k_f + \Sigma k_i \qquad \text{at } R = R_0 \qquad (15)$$

R_0^6 is related to the spectral properties of the donor and acceptor through the factors $n^{-4} \cdot \phi_D \cdot J_{DA}$, and to the relative orientation of the donor's emission transition dipole and the acceptor's absorption transition dipole through κ^2. κ^2, which may vary between 0 and 4, is not easily determined experimentally. At the current state of the art, the uncertainty in κ^2 usually limits the precision of any distances determined through detection of nonradiative energy transfer. Methods to cope with this difficulty use emission anisotropy to determine the rotational freedom of the donor and acceptor, and then use this knowledge to set limits upon R_0 as a function of various possible extreme models of donor and acceptor transition dipole geometries.[1,2] For purposes of comparison, we shall use $R_0(\frac{2}{3})$ to denote the R_0 calculated for a particular donor-acceptor pair by setting κ^2 equal to $\frac{2}{3}$.

Measurement of Energy Transfer

1. Donor Quenching. In practice one does not experimentally measure k_t directly, but rather an efficiency of transfer, E, which is defined analogously to the quantum efficiency in terms of the rates of elementary processes:

$$E = k_t/(k_t + k_f + \Sigma k_i) \qquad (16)$$

Since the presence of an energy acceptor in the vicinity of an excited energy donor provides an additional mode of deexcitation of the first excited singlet state of the donor, the efficiency of transfer can be effectively monitored by observing the effect of the acceptor on either the singlet lifetime or the quantum efficiency of the donor. To evaluate E one must compare the donor lifetime, τ_D, or the quantum efficiency, ϕ_D, in the absence of the acceptor to the lifetime, τ_{DA}, or the quantum efficiency, ϕ_{DA}, in the presence of the acceptor. Combining Eqs. (1) and (16) or (2) and (16), one can express E as a function of the lifetimes or the quantum efficiencies:

$$E = 1 - \tau_{DA}/\tau_D \tag{17}$$
$$E = 1 - \phi_{DA}/\phi_D \tag{18}$$

2. *Sensitized Emission.* The presence of an energy donor in the vicinity of a fluorescent energy acceptor provides additional excitation to the acceptor in the region of donor absorption. This sensitized emission of the acceptor can be used to measure the efficiency of the transfer process.

One monitors the emission at λ_A and scans the corrected excitation spectrum in the region of donor excitation, λ, for samples with and without the donor. The observed intensities,[16] $F_{DA}(\lambda,\lambda_A)$ and $F_A(\lambda,\lambda_A)$* respectively, along with the corresponding absorbances,[16] $A_{DA}{}^A(\lambda), A_{DA}{}^D(\lambda)$, and $A_A{}^A(\lambda)$ can be related to the efficiency of transfer, E.

In absence of donor: $F_A(\lambda,\lambda_A) \propto A_A{}^A(\lambda)$

In presence of donor: $F_{DA}(\lambda,\lambda_A) \propto A_{DA}{}^A(\lambda) + A_{DA}{}^D(\lambda) \cdot E$

$$E = \left[\frac{F_{DA}(\lambda,\lambda_A)}{F_A(\lambda,\lambda_A)} - \frac{A_{DA}{}^A(\lambda)}{A_A{}^A(\lambda)} \right] \cdot \frac{A_A{}^A(\lambda)}{A_{DA}{}^D(\lambda)} \tag{19}$$

Another way to monitor sensitized emission is to set the excitation monochromator at the λ_{max} of the donor, λ_D, and scan the emission spectrum of samples containing (1) donor and acceptor, DA; (2) acceptor alone, A; and (3) donor alone, D. The observed intensities at acceptor emission wavelengths, λ, are $f_{DA}(\lambda_D,\lambda)$, $f_A(\lambda_D,\lambda)$, and $f_D(\lambda_D,\lambda)$, respectively.[16] If the relative fluorescence yield at λ for species i, $\phi_i(\lambda)$, is defined as $\phi_i(\lambda) = [f_i(\lambda_D,\lambda)]/[A_i{}^i(\lambda_D)]$, one can write:

1. $f_{DA}(\lambda_D,\lambda) = kA_{DA}{}^D(\lambda_D)E\phi_A(\lambda) + kA_{DA}{}^D(\lambda_D)(1 - E)\phi_D(\lambda) + kA_{DA}{}^A(\lambda_D)\phi_A(\lambda)$
2. $f_A(\lambda_D,\lambda) = kA_A{}^A(\lambda_D)\phi_A(\lambda)$
3. $f_D(\lambda_D,\lambda) = kA_D{}^D(\lambda_D)\phi_D(\lambda)$

where k is an instrumental constant.

If care is taken to ensure that $A_A{}^A(\lambda_D) = A_{DA}{}^A(\lambda_D)$ and $A_D{}^D(\lambda_D) = A_{DA}{}^D(\lambda_D)$, one can show that:

$$\frac{f_{DA}(\lambda_D,\lambda) - f_A(\lambda_D,\lambda)}{f_D(\lambda_D,\lambda)} = E \frac{\phi_A(\lambda)}{\phi_D(\lambda)} + 1 - E \tag{20}$$

which, solved for E, gives:

$$E = \frac{[f_{DA}(\lambda_D,\lambda) - f_A(\lambda_D,\lambda) - f_D(\lambda_D,\lambda)]}{[A_D{}^D(\lambda_D)]/[A_A{}^A(\lambda_D)] \cdot f_A(\lambda_D,\lambda) - f_D(\lambda_D,\lambda)} \tag{21}$$

---

[16] $F(\lambda,\lambda_2)$ represents an excitation spectrum varying λ and monitoring emission at λ_2. $f(\lambda_1,\lambda)$ represents an emission spectrum exciting at λ_1 and monitoring emission as a function of λ. $A(\lambda)$ is the absorbance at λ. Subscripts denote the sample. Superscripts identify the fluorescent or absorbing species as either the donor, D, or the acceptor, A.

Relation of Transfer Efficiency to Distance

To establish the relationship between E and the distance, R, between the donor and acceptor, one combines Eqs. (1), (13), and (16) to give:

$$E = R_o^6/(R_0^6 + R^6) \tag{22}$$

and solving this for R one finds:

$$R = (1/E - 1)^{1/6}R_0 \tag{23}$$

It is Eq. (23) that enables one to interpret transfer efficiencies in terms of distances between fluorescent dyes on macromolecules or on macromolecular assemblies. Equation (22) calls forth a new interpretation of R_0, namely if $R = R_0$, one sees that $E = \frac{1}{2}$. Hence R_0 is the distance at which the efficiency of transfer between a donor and acceptor is 0.5.

Application of Energy Transfer to Biological Assemblies

The problem of determining the chemical structure of a biological assembly with a particle weight of up to several million is staggering. To do so one would have to determine $3N - 6$ atomic coordinates, where $N \sim 10^5$ or larger. Consequently structural studies have concentrated upon the organization of components in the assembly with the hope that eventually the components will be studied individually, their shapes determined, and the structure of the assembly deduced in the fashion of a three-dimensional jigsaw puzzle. The basic questions asked in these studies are: Is a given component near to or far from another? How near? How far? Where in the overall structure is the given component? On the surface? Buried?

This section outlines the use of nonradiative energy transfer to study the organization of assemblies of biological macromolecules. The basic approach is to label one region of the assembly with a fluorescent energy donor, label a second region with an energy acceptor, measure the efficiency of nonradiative energy transfer between the donor and acceptor, and then convert this efficiency to a distance between the two labeled regions.

Approaches to Fluorescent Labeling of Assembled Structures

Two approaches to structural information about assemblies from nonradiative energy transfer measurements are discussed. The first approach is designated as the specific-site labeling scheme. With this scheme one maps the distances between functionally important loci on the assembly. A fluorescent energy donor is placed at the first site, and an energy acceptor at the second site. The efficiency of transfer is measured, enabling the distance between the two sites to be calculated.

The second approach is designated as the random labeling scheme. In this scheme, one covalently labels as randomly and extensively as feasible the surface of a single purified component with a fluorescent energy donor. A second purified component is likewise labeled with an energy acceptor. The two labeled components are reincorporated into the assembly either through partial or total reconstitution. The average efficiency, $\bar{\bar{E}}$, of nonradiative energy transfer is measured, and this efficiency is converted to a center-to-center distance between the two labeled components. Center-to-center distances are similarly determined for a sufficient number of component pairs so that the organization of the whole assembly can be deduced. For N (≥ 4) components, a minimum of $4N - 10$ distance measurements is sufficient to determine the relative three-dimensional arrangement of the components. Obviously one may combine the two approaches to get distances from loci to components.

Many factors influence the choice of approach. Consider the specific site scheme. One may choose to label the sites covalently. The best situation ensues if the labeling reagents used are so specific that the intact assembly can be exposed to the reagent, and only the site of interest is labeled. This occurrence can arise since "active" sites of macromolecular assemblies frequently have unusually reactive side chains at or near them for their biological activity. If substantial nonspecific reaction does occur, it may be possible to partially or totally disassemble the structure, specifically label a purified component known to be at or near the site of interest and then perform a partial or total reconstitution with the labeled species.

In assemblies with labeled components, one must ask whether the geometrical arrangement of components is comparable to the arrangement in the unlabeled assembly. A comparison of the labeled and unlabeled assemblies on a sucrose gradient or in sedimentation velocity experiments is useful in detecting gross structural abnormalities. A more stringent comparison can be made if the assembly has an assayable activity. Generally if the labeled assembly behaves as well as the unlabeled assembly in these comparative studies, one argues that the functional similarity is consistent with structural similarity of the two assemblies.

Another option is open for the specific labeling of functionally important regions. These sites may strongly bind antibiotics, hormones, cofactors, or other biological ligands that exhibit intrinsic fluorescence or can be fluorescently modified without altering their biological activity. These fluorescent "ligands" specifically bound to the assembly may be used to label the particular site. This approach has the advantage of not altering the macromolecular assembly in any foreign fashion and can yield kinetic and thermodynamic data on ligand binding as well as energy transfer-distance measurements.

The chemical considerations to be made in applying the site-specific labeling approach are dictated by the desire for very specific reaction. In all cases it is necessary to verify the specificity of the labeling reaction. Reactive thiol groups are a good first target to aim for, since they are few in number and a variety of relatively —SH specific reagents exist. Carboxamide groups of protein-bound glutaminyl residues can be fluorescently labeled enzymatically using transglutaminase and fluorescent amino compounds.[17] The 3′ end of ribonucleic acids is subject to relatively specific modification via periodate oxidation and reduced Schiff base formation.[18] Also proteins with N terminal serine or threonine can be chemically oxidized and modified via reduced Schiff base formation at the N-terminus.[19] Specific modification of antibiotics, hormones, and biological ligands depend upon their unique chemical structure and the experimenter's resourcefulness in specific nonlethal modification of these ligands.

Now consider the random labeling scheme. Three prerequisites exist for the use of the random labeling approach. First, the assembly must consist of components that are approximately spherical, since the method for converting $\bar{\bar{E}}$ to center-to-center distances is strictly valid only for spheres. Second, the components must be fractionated in order to randomly label the surface of a selected protein. And finally the assembly must reconstitute from its components.

If these prerequisites are met, the next consideration is how to nonspecifically label the surface of particular protein. An excellent discussion of fluorescent labeling of proteins is presented by Dandliker and Portmann.[20] Many protein components of macromolecular assemblies have a substantial number of lysine residues on their surfaces. The ε-amino groups of these lysines can serve as points for "nonspecific" covalent attachment of fluorescent dyes. Some chemical functionalities that can be used to label ε-amino groups are isothiocyanates, (—N=C=S); sulfonyl chlorides, —(O=)S(=O)—Cl; activated car-

[17] A. Dutton and S. J. Singer, *Proc. Nat. Acad. Sci. U.S.A.* **72**, 2568 (1975).
[18] S. A. Reines and C. R. Cantor, *Nucleic Acids Res.* **1**, 767 (1974).
[19] H. B. F. Dixon and R. Fields, see this series Vol. 25, p. 409.
[20] W. B. Dandliker and A. J. Portmann, *in* "Excited States of Proteins and Nucleic Acids" (R. F. Steiner and I. Weinryb, eds.), p. 199. Plenum, New York, 1971.

boxylic acids; mixed anhydrides

$$
\underset{\text{O}}{\overset{\text{O}}{\underset{\|}{\text{C}}}}-\text{O}-\underset{\text{O}}{\overset{\text{O}}{\underset{\|}{\text{S}}}}-\text{O}^{-}, \quad \underset{}{\overset{\text{O}}{\underset{\|}{\text{C}}}}-\text{O}-\underset{}{\overset{\text{O}}{\underset{\|}{\text{C}}}}-\text{OEt}
$$

hydroxysuccinimide esters

$$
\underset{}{\overset{\text{O}}{\underset{\|}{\text{C}}}}-\text{O}-\text{N}
$$

paranitrophenolic esters, $-\text{C}(=\text{O})-\text{O}-\text{ØNO}_2$; α-halocarbonyl compounds, $\text{XCH}_2\text{C}(=\text{O})$; and aldehydes followed by borohydride reduction
of resultant Schiff bases.

Once the components are labeled, one must verify the randomness of
the labeling by peptide mapping of the labeled proteins. Next one must
ascertain the ability of the modified species to reincorporate into the
assembly. One usually finds that a lightly labeled component (≤ 1 dye per
protein) reconstitutes adequately, but a more heavily labeled one (≥ 4
dyes per protein) reconstitutes less efficiently. And, finally, one must
compare the hydrodynamic and functional behavior of modified to un-
modified assemblies.

The Choice of Dyes for Fluorescent Labeling

In addition to the above practical and chemical considerations, one
must choose a donor and acceptor on the basis of the known spectral
properties of the dyes and any estimates of the distance to be measured.
The first consideration derives from the functional dependence of the
efficiency, E, on the distance, R, between the donor and acceptor:

$$
E = R_0^6/(R_0^6 + R^6) \tag{22}
$$

A plot of Eq. (22) is shown in Fig. 3. This shows that for a given donor–
acceptor pair, E is a sensitive function of R only over the range 0.5
$R_0 < R < 1.9 R_0$. Hence, E can be used only to measure distances within
this range. The most sensitive distance measurements are made when R_0
for the chosen donor–acceptor equals the distance to be measured.

In many assemblies one wants to measure distances greater than 30 Å,
which is an approximate center-to-center distance between two small

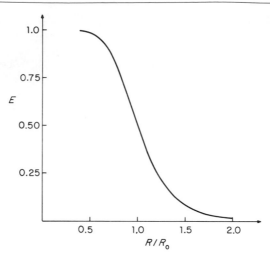

Fig. 3. Graph of the transfer efficiency, E, vs the distance, R, between the donor and acceptor in units of R_0.

globular proteins of molecular weights 12,000–15,000. Thus, one needs donor–acceptor pairs with R_0's of 30 Å or longer for measuring distances on assemblies of macromolecules. The factors that determine R_0 are given in Eq. (14). By inspection of this equation, one can see that the largest R_0's are obtained when ϕ_D and J_{DA} are as large as possible. ϕ_D is made large (>0.2 or 0.3) by appropriate choice of the donor. J_{DA} is maximized for donor–acceptor pairs with (1) good spectral overlap of donor emission and acceptor absorption, (2) large acceptor extinction ($>20,000 M^{-1}cm^{-1}$), and (3) the region of overlap as far to the red as spectrally possible (large λ).

The next considerations derive from experimental optimization of conditions for monitoring nonradiative energy transfer. One should choose donor–acceptor pairs for which the fluorescence emission spectra of the two dyes are well resolved. In using sensitized emission to follow transfer, the acceptor's emission is frequently contaminated by the long-wavelength tail of the donor fluorescence. By choosing a donor with a narrow emission envelope, one can minimize this contamination. Likewise, in using donor quenching to follow transfer, the blue edge region of the donor's emission can be freed of acceptor contamination by appropriate choice of donor and acceptor.

In addition to good spectral overlap of donor emission and acceptor absorption, and good resolution of the emission bands, one finds that having a large wavelength interval (70–100 nm) between ϵ_{max} and f_{max}

(Stoke's shift) for each of the dyes minimizes the spectral contamination of fluorescence emission by stray light scattered from the very large assemblies of macromolecules. Furthermore if the spectral sensitivity, $\epsilon_{max} \cdot \phi$, for the donor and for the acceptor are nearly equal, the emission of the one will not "swamp" the emission of the other. Finally, in using sensitized emission, donor–acceptor pairs should be chosen so that ϵ_{max} of the donor occurs in the region of a relative minimum of $\epsilon(\lambda)$ for the acceptor. At λ_{ex}, the exciting wavelength, $\epsilon_A(\lambda_{ex})/\epsilon_D(\lambda_{ex})$, should be as small as possible for maximum sensitivity of the technique.

In addition to the general spectral properties mentioned above, the use of donor singlet lifetime to monitor the transfer process is facilitated by a donor with a reasonably long lifetime (>10 nsec) and an acceptor with a very short lifetime (<2 nsec). A nonfluorescent acceptor may be used, but then information about the rotational mobility from polarization of fluorescence emission is lost.

The final considerations derive from the critical dependence of the measured transfer efficiency upon knowledge of the amounts of donor and acceptor present in different samples. If absorption measurements are used to quantitate dye stoichiometries, donor–acceptor pairs with good spectral resolution of their absorption bands must be chosen.

The quantitation of donor stoichiometry is generally more difficult than that for the acceptor, since the donor absorption is frequently heavily contaminated by acceptor absorption and the donor frequently has a low ϵ_{max}. The acceptor, on the other hand, generally has a large ϵ_{max} and an absorption free of donor contamination. Radioactive dyes permit quantitation of dye stoichiometry without the concomitant difficulties and artifacts that enter into absorption measurements, namely: (1) the difficulty of measuring low absorbances (routinely <0.01 and frequently <0.001) in the presence of large light-scattering particles, and (2) the difficulty of sorting out simultaneous absorption of both donor and acceptor with uncertain extinction coefficients for the bound dyes. If only one dye can be made radioactive, it is clear that one should try to have it be the donor, since the donor frequently has the lower ϵ_{max} and the problem of simultaneous acceptor absorption. We have found tritiated dyes with a specific activity of 30 mCi/mmol to be adequate for most detection purposes.

While all of these desirable characteristics of the donor–acceptor pair should be kept in mind, one must realize that the deficiencies of a given donor–acceptor pair rarely prevent performance of good experiments within the limitations imposed by the particular choice. Listed in Table I are some donor–acceptor pairs that have been, or could be, used in energy-transfer experiments on macromolecular assemblies. In Fig. 4 the spectral properties and structures of the dyes are summarized.

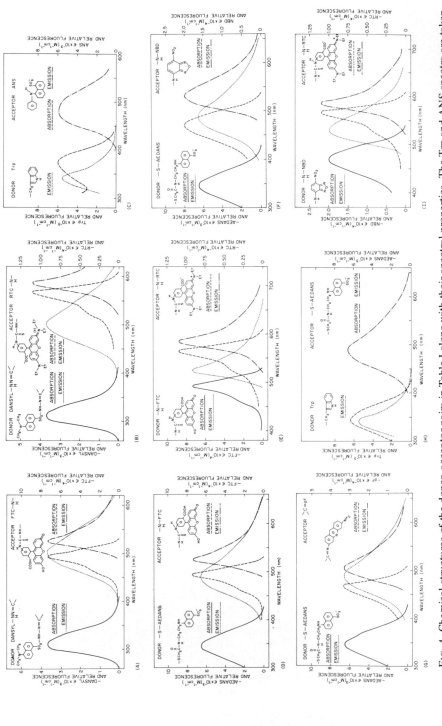

FIG. 4. Chemical structures of the donor-acceptor pairs in Table I along with their spectral properties. The Trp and ANS spectra are taken from S. Matsumoto and G. G. Hammes [Biochemistry **14**, 214(1975)] and G. G. Hammes and G. B. Shepherd and G. G. Hammes [Biochemistry **15**, 311 (1976)], and the RTC spectra are taken from R. F. Chen [Arch. Biochem. Biophys. **133**, 263 (1969)]. The rest of the spectra are from our laboratory and are either corrected excitation or corrected emission spectra.

TABLE I

DONOR–ACCEPTOR PAIRS USEFUL FOR ENERGY TRANSFER MEASUREMENTS

	Donor[a]	ϕ_D	τ (nsec)	Acceptor	$\epsilon_{max} \times 10^{-4}$ ($M^{-1}cm^{-1}$)	$J_{DA} \times 10^{-15}$ ($M^{-1}cm^{-1}nm^4$)	$R_o(2/3)$ (Å)
A.	Dansyl	0.1–0.2	15–20	FTC	4.2–8.5	0.79–1.59	33–41
B.	Dansyl	0.1–0.2	15–20	RTC	1.2	0.31	28–31
C.	Trp	0.10	2	ANS	0.6	—	22
D.	AEDANS	0.1–0.5	13–17	FTC	4.2–8.5	0.79–1.60	33–48
E.	FTC	0.5	4	RTC	1.2	0.50	40
F.	AEDANS	0.1–0.5	13–17	NBD	2.0	0.46	30–39
G.	AEDANS	0.1–0.5	13–17	pf	3.3	0.39	29–38
H.	Trp	0.10	2	AEDANS	0.65	0.07	22
I.	NBD	0.1–0.5	—	RTC	1.2	0.48	30–39

[a] Abbreviations as in Fig. 4.

Preparation of Samples for Energy Transfer Measurements

After selecting a labeling approach and a pair of dyes, one must construct an assembly containing the donor and the acceptor. The specifics of this task necessarily depend on the idiosyncrasies of the particular assembly. Once able to produce donor–acceptor labeled assemblies, one is ready to construct samples for energy-transfer experiments.

Consider assemblies with two distinguishable sites, schematically represented as 1 2. Energy-transfer experiments between dyes placed in sites 1 and 2 require preparation of at least four different samples:

    I. D Y
    II. D A
    III. X A
    IV. X Y

where D is the donor, A is the acceptor, X is a nonabsorbing, nonfluorescing donor analog, and Y is an equivalent acceptor analog. If the presence of X and Y have no direct effect on the spectra of A and D, samples I, III, and IV can be replaced by D __, __ A, and __ __, respectively. Each labeled sample requires 2 ml of solution with dye concentrations of 10^{-8} to 10^{-7} M to conveniently monitor the fluorescence. If assemblies are labeled nearly stoichiometrically, one needs ~200 pmol of assembly for each sample. If UV-visible absorption measurements are also required, one should prepare at least 1 nmol of assemblies for each sample. This guarantees an optical density at the dye λ_{max} of 0.01 using 1-cm pathlength cells requiring 1 ml of sample. However, it is important to keep the sum of the optical densities of D, A, and assemblies in any sample used in fluorescence measurements less than 0.03 to avoid the complications of inner filter effects and emission reabsorption. The easiest procedure is to dilute concentrated samples after absorption measurements and to use these for the fluorescence measurements.

In constructing the $\underline{D}$ $\underline{A}$ sample, one should minimize the number of assemblies having donors but lacking acceptors. These assemblies have donors unable to participate in the transfer process. If the two sites incorporate label with different efficiencies, one places the acceptor in the site with the highest incorporation efficiency and the donor in the other site. This minimizes the number of acceptor-free donors, and thus maximizes the sensitivity of the experiment.

Fluorescence Spectroscopy of the Labeled Samples

After preparing the four samples and measuring the concentration of donor, acceptor, and assemblies in each sample, one is ready to study the samples fluorometrically via one or more of the three modes outlined previously: (1) donor's quenched emission; (2) acceptor's sensitized emission; (3) donor's shortened lifetime.

Donor Quenching

The excitation monochromator of the spectrofluorometer is set at λ_D near the absorption λ_{max} of the donor, and the emission monochromator is scanned for each of the four samples. The spectra from samples I, II, and III are corrected for the light scattered from the large assemblies by subtracting the spectrum of sample IV from each to give $f_I(\lambda_D,\lambda)$,[21] $f_{II}(\lambda_D,\lambda)$, and $f_{III}(\lambda_D,\lambda)$, respectively. $f_I(\lambda_D,\lambda)$ and $f_{II}(\lambda_D,\lambda)$ are compared to ensure that the presence of the acceptor does not alter the shape or the position of the emission maximum of the donor spectrum. Such alterations are indicative of processes other than Förster transfer, and interpretation of spectral changes observed in terms of the Förster mechanism is not valid. Barring any such spectral alterations, one compares $f_I(\lambda_D,\lambda)$ and $f_{III}(\lambda_D,\lambda)$, the spectra from the donor-labeled and the acceptor-labeled assemblies alone, and selects a wavelength interval, $\Delta\lambda$, containing donor emission but no acceptor emission. Selecting wavelengths, λ_i, within $\Delta\lambda$, one evaluates the ratio $f_{II}(\lambda_D,\lambda_i)/f_I(\lambda_D,\lambda_i)$ for a number of λ_i's and averages the results: $[f_{II}(\lambda_D,\lambda_i)/f_I(\lambda_D,\lambda_i)]_{av}$. Knowing the relative concentrations of donor in samples I and II, $[D]_I/[D]_{II}$, one evaluates the ratio of quantum efficiencies, ϕ_{DA}/ϕ_D, using the equation:

$$\frac{\phi_{DA}}{\phi_D} = \left[\frac{f_{II}(\lambda_D,\lambda_i)}{f_I(\lambda_D,\lambda_i)}\right]_{av} \frac{[D]_I}{[D]_{II}} \tag{24}$$

[21] $f_I(\lambda_D,\lambda)$ is the scatter corrected emission spectrum of sample I excited at λ_D, etc.

This ratio is used to calculate E from Eq. (18). Equation (18) applies to the case of two samples of assemblies each labeled one-to-one with the acceptor. If the acceptor stoichiometry in less than one-to-one with assemblies, one must correct E in Eq. 18 for the fraction of donors not quenched by an acceptor (see page 368).

Donor quenching is perhaps the most straightforward assay for energy transfer. But one must ensure that the decrease in the donor quantum efficiency is in fact the result of energy transfer, not the result of a conformational change induced by the acceptor, nor the result of direct radiative absorption of donor emission by the acceptor. The use of nonabsorbing, nonfluorescing analogs of the acceptor protects against the first possibility, and the use of low optical densities of acceptor (≤ 0.03) protects against the second.

Sensitized Emission

Two different experimental arrangements that give equivalent information can be used to measure sensitized emission. With the first and most widely described configuration, one monitors emission at a wavelength, λ_A, rich in signal from the acceptor and devoid of signal from the donor. The corrected excitation spectrum through the absorption bands of the donor and the acceptor is recorded for all four samples. If λ_A is devoid of emission from the donor, the spectra from samples I and IV should be identical. If they are not, one can correct for the donor emission by using the second experimental arrangement described later. The spectrum from sample IV is subtracted from the spectra of samples II and III to give the scatter corrected $F_{II}(\lambda, \lambda_A)$ and $F_{III}(\lambda, \lambda_A)$. From these spectra one evaluates $F_{II}(\lambda_D, \lambda_A)/F_{III}(\lambda_D, \lambda_A)$ for λ_D in the region of donor excitation maximum. Next one must evaluate $A_{II}^D(\lambda_D)$, $A_{II}^A(\lambda_D)$, and $A_{III}^A(\lambda_D)$, the absorbances of the donor in sample II, the acceptor in II, and the acceptor in III at λ_D, respectively. This is performed by dual-beam absorption spectroscopy with sample IV as a blank, or alternatively by liquid scintillation counting of radioactive-dye labeled assemblies and known extinction coefficients of bound dyes. One calculates the transfer efficiency from Eq. (19):

$$E = \left[\frac{F_{II}(\lambda_D, \lambda_A)}{F_{III}(\lambda_D, \lambda_A)} - \frac{A_{II}^A(\lambda_D)}{A_{III}^A(\lambda_D)} \right] \cdot \frac{A_{III}^A(\lambda_D)}{A_{II}^D(\lambda_D)} \tag{19'}$$

In the second experimental configuration, useful primarily when the donor contributes signal at λ_A, one excites at λ_D near the λ_{max} of the donor. The emission spectra of all four samples are recorded well through the acceptor emission. The spectrum of sample IV is subtracted from those of samples I, II, and III to give the scatter-corrected spectra $f_{\mathrm{I}}(\lambda_{\mathrm{D}},\lambda)$, $f_{\mathrm{II}}(\lambda_{\mathrm{D}},\lambda)$, and $f_{\mathrm{III}}(\lambda_{\mathrm{D}},\lambda)$, respectively. $f_{\mathrm{I}}(\lambda_{\mathrm{D}},\lambda)$ and $f_{\mathrm{III}}(\lambda_{\mathrm{D}},\lambda)$ are next corrected for any differences in concentration of donor or acceptor, respectively, compared to the corresponding concentrations in sample II to give $f_{\mathrm{I}}'(\lambda_{\mathrm{D}},\lambda)$ and $f_{\mathrm{III}}'(\lambda_{\mathrm{D}},\lambda)$. Selecting wavelengths in the region of acceptor emission, λ_{A}, one calculates E from Eq. (21):

$$E = \frac{[f_{\mathrm{II}}(\lambda_{\mathrm{D}},\lambda_{\mathrm{A}}) - f_{\mathrm{III}}'(\lambda_{\mathrm{D}},\lambda_{\mathrm{A}}) - f_{\mathrm{I}}'(\lambda_{\mathrm{D}},\lambda_{\mathrm{A}})]}{\left[\dfrac{A_{\mathrm{II}}{}^{\mathrm{D}}(\lambda_{\mathrm{D}})}{A_{\mathrm{II}}{}^{\mathrm{A}}(\lambda_{\mathrm{D}})}\right] \cdot f_{\mathrm{III}}'(\lambda_{\mathrm{D}},\lambda_{\mathrm{A}}) - f_{\mathrm{I}}'(\lambda_{\mathrm{D}},\lambda_{\mathrm{A}})} \tag{21'}$$

One can see directly from $[f_{\mathrm{II}}(\lambda_{\mathrm{D}},\lambda) - f_{\mathrm{III}}'(\lambda_{\mathrm{D}},\lambda)]$ compared to $f_{\mathrm{I}}'(\lambda_{\mathrm{D}},\lambda)$ whether any sensitized emission is occurring by a "bump" on the $[f_{\mathrm{II}}(\lambda_{\mathrm{D}},\lambda) - f_{\mathrm{III}}'(\lambda_{\mathrm{D}},\lambda)]$ spectrum in the region of acceptor emission (see Fig. 5).

Sensitized emission is generally harder to monitor than donor quenching. This is largely the result of contaminating donor fluorescence in the acceptor's emission envelope. The sensitivity of the fluorescence mea-

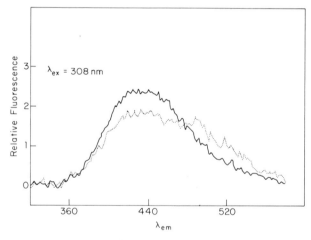

FIG. 5. Raw data of an energy transfer experiment. The donor in this case is the Y base of *yeast tRNA*$_{\mathrm{f}}^{\mathrm{Phe}}$. The acceptor is proflavine (pf) of the modified yeast tRNA$_{\mathrm{pf}}^{\mathrm{Phe}}$ [W. Wintermeyer and H. G. Zachau, *FEBS Lett.* **18**, 214 (1971)]. The solid line is the emission from a sample of poly(U)-programmed 70 S ribosomes to which an equivalent of tRNA$_{\mathrm{Y}}^{\mathrm{Phe}}$ has been added and then an additional equivalent of the nonfluorescent, nonabsorbing *Escherichia coli* tRNA^{Phe}. A background of 70 S ribosomes is subtracted to arrive at the solid line. The dotted spectrum arises from poly(U)-programmed 70 S prepared as for the first sample, only now tRNA$_{\mathrm{pf}}^{\mathrm{Phe}}$ is substituted for *E. coli* tRNA^{Phe}, and the background subtracted is 70 S · poly(U) · *E. coli* tRNA^{Phe} · tRNA$_{\mathrm{pf}}^{\mathrm{Phe}}$. Comparing the dotted spectrum to the solid one, one notes both quenching of the donor in the region 390–460 nm and sensitized emission of the acceptor in the region of 480–580 nm.

surements $[F_{II}(\lambda_D,\lambda_A)]/[F_{III}(\lambda_D,\lambda_A)]$ or $[f_{II}(\lambda_D,\lambda_A)]/[f'_{III}(\lambda_D,\lambda_A)]$ is maximized by minimizing the ratio $\epsilon_{II}^A(\lambda_D)/\epsilon_{II}^D(\lambda_D)$.

Donor Lifetime Quenching

Nonradiative energy transfer can also be observed by the decrease in the excited singlet lifetime of the donor in the presence of the acceptor. Two schemes are used for measuring excited singlet lifetimes: (1) the phase shift technique described by Weber,[22] and (2) the single photon counting technique described by Tao,[23] Yguerabide,[11] Ware[24] and Isenberg.[3] The single-photon counting technique offers the advantage of multiple lifetime decay characterization.

Using a single-photon counting apparatus as diagrammed in Fig. 1 without the excitation polarizer, one examines the time course of emission of the donor in samples I and II above and uses sample IV, if nonfluorescent, as a "scatter solution" to calibrate the instrumental response function. In these measurements the monochromator is set at the λ_{max} of the donor, and the emission is first passed through a polarizer oriented 54.7° to the vertical to separate lifetime decay from anisotropy decay. The emission is then filtered with a narrow band pass or interference filter passing the donor's emission but excluding exciting light and emission from the acceptor.

The data are routinely collected to a preset value in the peak channel of a pulse-height analyzer, and the data are deconvoluted using the method of moments,[3] Laplace transforms,[25] or nonlinear least squares fitting.[26] An excellent recent review of the analysis of single photon counting data has been prepared by Isenberg.[3] Using the lifetimes τ_I and τ_{II} obtained by the deconvolution of data from samples I and II, respectively, E is calculated from Eq. (17):

$$E = 1 - \tau_{II}/\tau_I \tag{17'}$$

This equation is impervious to small variation in labeling stoichiometry, but low acceptor incorporation makes lifetime analysis difficult. Donor lifetime analysis is best reserved for the specific site labeling approach, since the random labeling approach generates a population of assemblies with different donor–acceptor distances resulting in a spectrum of lifetimes making analysis for all but the longest lifetime impossible.

[22] G. Weber, see this series Vol. 16, p. 380.

[23] T. Tao, *Biopolymers* **8**, 609 (1969).

[24] W. R. Ware, *in* "Creation and Detection of the Excited State" (A. A. Lamola, ed.), Vol. I, Part A, p. 213. Dekker, New York, 1971.

[25] A. Gafni, R. L. Modlin, and L. Brand, *Biophys. J.* **15**, 263 (1975).

[26] A. Grinvald and I. Z. Steinberg, *Anal. Biochem.* **59**, 583 (1974).

Lifetime measurements used in conjunction with steady-state measurements frequently aid in the interpretation of the measured E in terms of a distance. Consider the case in which the steady state E is 0.10. This steady-state measurement cannot distinguish whether the efficiency is the result of a single donor-acceptor separation for which $E = 0.10$ or whether the efficiency is the result of the total quenching of donor fluorescence by 10% of the donor–acceptor pairs and the other 90% are so far apart there is no transfer at all. Lifetime measurements on such a system allow these two extremes to be distinguished. In the first situation donor lifetime measurements in the absence and in the presence of the acceptor will give rise to $E = 0.10$ also. In the second situation, the donor lifetime measured in the presence of the acceptor will be the same as the lifetime in the absence of the acceptor and $E = 0$. Thus, from the lifetime measurements, one forms a better idea of the distance distribution between donor and acceptor.

Interpretation of E as a Distance in the Specific-Site Labeling Case

Suppose the labeled assemblies have the donor at one unique site and the acceptor at a second unique site. After measuring the concentration of donors, acceptors, and assemblies in each sample and the efficiency of nonradiative transfer, one converts these measurements to a distance between the donor and the acceptor in the following manner. First one corrects the measured efficiency for acceptor stoichiometry if necessary. Next one determines the range of R_0's for the labeled system under study. And finally one uses the corrected efficiency and the range of R_0's to calculate a range of distances.

Correction of E for Acceptor Stoichiometry

One must correct the observed transfer efficiency, E_{obs}, for the acceptor stoichiometry if the transfer efficiency is measured by steady-state techniques and the acceptor labeling stoichiometry is less than one-to-one. If the fraction of assemblies with acceptor is f_a, the corrected efficiency, E_c, which would be observed if all assemblies had an acceptor, is given by:

$$E_c = E_{obs}/f_a \tag{25}$$

This correction must be employed when E is calculated from raw spectral data using Eqs. (18), (19), or (21), since these formulas are derived for the case of each donor transferring energy to an acceptor with efficiency E. Lifetime data need not be corrected for acceptor stoichiometry. Knowing the stoichiometry, though, may help one to deconvolute the lifetime data. No correction is necessary at this stage for donor stoichiometry.

The Range of R_0's

The next step is to determine the R_0 for the donor–acceptor pair. The method for doing this is described at length by Dale and Eisinger[1] in "Biochemical Fluorescence" and is briefly discussed here. Recall Eq. (14) which relates R_0 to the spectral properties and relative orientation of the donor and acceptor:

$$R_0^6 = (8.79 \times 10^{-5}) \; \kappa^2 n^{-4} \phi_D J_{DA} \tag{14}$$

where $J_{DA} = [\int_0^\infty f_D(\lambda)\epsilon_A(\lambda)\lambda^4 d\lambda]/[\int_0^\infty f_D(\lambda)d\lambda]$. ϕ_D, the quantum efficiency of the donor, is determined by comparison of the integrated corrected emission spectrum of the donor-labeled assembly to that of a known standard, such as quinine bisulfate (0.70 in 0.1 N H_2SO_4)[27] or fluorescein (0.85 in 0.1 N NaOH).[28] The refractive index for visible light, n, ranging from 1.33 for water to 1.49 for an 80% sucrose solution, is usually taken as a rough average of these extreme values, 1.4. $f_D(\lambda)$ is taken from a corrected emission spectrum of the donor-labeled assembly, and $\epsilon_A(\lambda)$ is evaluated from an absorption spectrum (or a corrected excitation spectrum) of the acceptor-labeled assembly and the known extinction coefficient of the bound dye. By assigning κ^2 a value of ⅔ in Eq. (14) one can calculate an estimate of R_0, called $R_0(⅔)$.

A κ^2 of ⅔ occurs only if both the donor and acceptor transition dipoles are free to assume all orientations during the excited-state lifetime of the donor. This situation rarely arises in actual practice. In theory κ^2 may be any value between 0 and 4 depending on the mutual orientation of the donor and acceptor transition dipoles. Generally, these dipoles are free to sample all orientations in a restricted region of space following excitation. This restricted region can be modeled as the surface of a cone if the dye is allowed to reorient about a single covalent bond, or it can be modeled as the volume of a cone if the dye is free to reorient about several bonds. Dale and Eisinger have dealt quantitatively with the constraint on κ^2 of rapid restricted reorientation of donor and acceptor transition dipoles relative to the macromolecule following excitation.

In the Dale–Eisinger procedure, one measures separately the decay of the emission anisotropy, $r(t)$, for the donor-labeled species and for the acceptor-labeled species, each excited in the longest wavelength absorp-

[27] T. G. Scott, R. D. Spencer, N. G. Leonard, and G. Weber, *J. Am. Chem. Soc.* **92**, 687 (1970).
[28] C. A. Parker and W. T. Rees, *Analyst* **85**, 587 (1960).

tion band. A plot of $\ln[r(t)]$ vs t is extrapolated from the linear decay region back to $t = 0$ to get the limiting emission anisotropy $r_m(0)$. Instead one may measure the steady-state emission anisotropy as a function of viscosity. From these data one prepares a Perrin plot and extrapolates from the linear high temperature, low viscosity region back to infinite viscosity to get a corresponding limiting anisotropy, r_{om}, at infinite viscosity (see, for example, Fig. 6). These limiting anisotropies are independent of the rotational motion of the macromolecular matrix and can be considered to reflect rapid reorientation of the dye transition dipoles relative to the matrix following excitation. If after excitation no reorientation of the dipoles occurs, one obtains the fundamental emission anistropy, $r_f = 0.4$.[1]

The ratio of the limiting anisotropy, r_{om} or $r_m(0)$, to the fundamental anisotropy, r_f, is defined as the average depolarization factor, $\langle d' \rangle_d$[1]:

$$r_m(0)/r_f = r_{om}/r_f = \langle d' \rangle_d \qquad (26)$$

If no reorientation occurs following excitation, $\langle d' \rangle_d = 1$. If complete randomization of orientation occurs, $\langle d' \rangle_d = 0$. It is evident that $\langle d' \rangle_d$

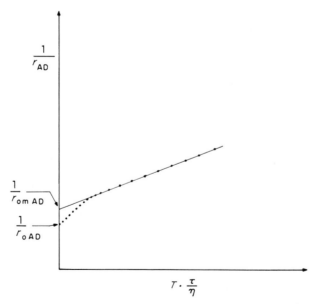

FIG. 6. Schematic diagram of a Perrin plot indicating the two limiting reciprocal anisotropies $1/r_{omAD}$ and $1/r_{oAD}$. The diagram is specifically for the case of the contribution of transferred energy to the emission anisotropy of the acceptor, r_{AD}, but in exactly analogous fashion one can evaluate the corresponding reciprocal limiting emission anisotropies, $1/r_{om}$ and $1/r_o$, for any fluorescently labeled macromolecule.

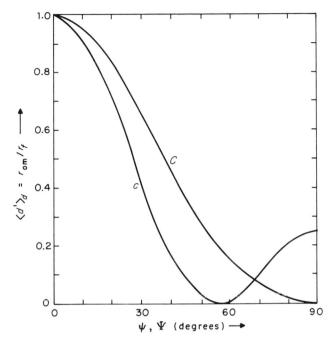

FIG. 7. Dynamic depolarization factors, $\langle d' \rangle_d$, calculated by Dale and Eisinger (see text footnote 1) for rapid reorientation of dipoles: (1) over the surface c of a cone with half-angle, ψ; or (2) through the volume C of a cone with half-angle, Ψ. Adapted from p. 231, Dale and Eisinger, by courtesy of Dekker, New York.

between these extreme values corresponds to reorientation over some restricted geometry characterized by Dale and Eisinger[1] as either surfaces or volumes of cones. The average depolarization factor, $\langle d' \rangle_d$, is functionally related to the half-angle ψ, of the cone surface over which the dipole may reorient; or it is functionally related to the half-angle, Ψ, of the cone volume through which the dipole may reorient (see Fig. 7).

Dale and Eisinger[1] have calculated $\langle \kappa^2 \rangle$ for the rapid reorientation of the donor and acceptor transition dipoles over the combinations of (1) cone surfaces, ψ_D and ψ_A, (2) cone surface $\psi_{D(A)}$ and cone volume $\Psi_{A(D)}$, and (3) cone volumes Ψ_D and Ψ_A for various models of relative cone axes alignment. Knowing the cone half-angles for the donor and acceptor, one examines the graphical results of the Dale–Eisinger calculations for each relative orientation of cone axes and makes note of the possible values of $\langle \kappa^2 \rangle$. Several orientations including those which generate maximum and minimum values of $\langle \kappa^2 \rangle$ are depicted in Fig. 8. From the largest and

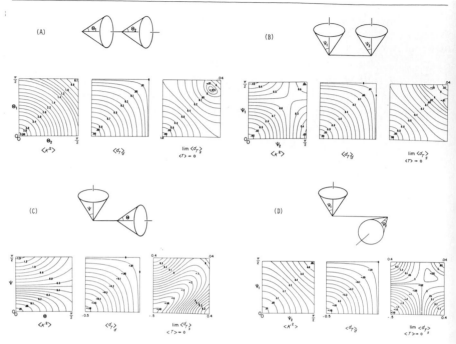

FIG. 8. Several relative orientations of cone axes and the values of $\langle \kappa^2 \rangle$, $\langle d_T \rangle_d$, and $\langle d_T \rangle_s$ as functions of the cone half-angles from Dale and Eisinger. Included here are the orientations (A) and (D) giving rise to the largest value and smallest value, respectively, of $\langle \kappa^2 \rangle$ for a given set of cone half-angles. (Θ and Ψ are used interchangeably here). Adapted from Dale and Eisinger (see text footnote 1), pp. 158–159, 170–171, 186–187, 200–201, by courtesy of Dekker, New York.

smallest values of $\langle \kappa^2 \rangle$ one calculates a range of R_0's:

$$R_0(\text{min}) = \left[\frac{\langle \kappa^2 \rangle_{\text{min}}}{\frac{2}{3}} \right]^{1/6} R_0(\frac{2}{3})$$

$$R_0(\text{max}) = \left[\frac{\langle \kappa^2 \rangle_{\text{max}}}{\frac{2}{3}} \right]^{1/6} R_0(\frac{2}{3})$$

where $R_0(\frac{2}{3})$ is the estimated R_0 calculated using $\kappa^2 = \frac{2}{3}$ in Eq. (14).

These represent the extreme range of R_0 values for a system. In some cases it may be possible to narrow this range. Suppose one can measure the contribution to the acceptor's emission anisotropy resulting from energy transferred from the donor, r_{AD}. In this case, one can study r_{AD} as a function of viscosity. Preparation of a Perrin plot of $1/r_{\text{AD}}$ vs $T\tau/\eta$ allows one to extract two limiting anisotropies: r_{omAD} and r_{oAD}. The first is obtained by extrapolation of the linear high-temperature, low-viscosity region of the Perrin plot back to infinite viscosity; and the second is obtained by extrapolation of the low-temperature, high-viscosity data to infinite viscosity (see Fig. 6). These two limiting anisotropies are used to define two new depolarization factors:

 1. The dynamic transfer depolarization factor, $\langle d_T \rangle_d = r_{\text{om AD}}/r_f$.
 2. The static transfer depolarization factor, $\langle d_T \rangle_s = r_{\text{oAD}}/r_f$.

These experimentally determined depolarization factors are compared to those calculated by Dale and Eisinger for the various relative orientations of cone axes. Certain orientations may give rise to $\langle d_T \rangle_d$ or $\langle d_T \rangle_s$ inconsistent with the experimentally determined factors. One may then rule out these orientations and thereby further restrict $\langle \kappa^2 \rangle$ and hence R_0.

The Range of Distances, R

The corrected transfer efficiency, E_c, and the range of R_0's from the Dale–Eisinger analysis are used in Eq. (23) to calculate a range of distances between the two sites:

$$R \text{ (min)} = [(1/E_c) - 1]^{1/6} R_0 \text{ (min)}$$
$$R \text{ (max)} = [(1/E_c) - 1]^{1/6} R_0 \text{ (max)}$$

The range of distances generated by the Dale–Eisinger analysis is a conservative estimate. Generally it is unlikely that the actual distance lies at the extremes of this range. A more liberal interpretation of energy-transfer measurements based upon statistical weighing of relative donor–acceptor geometries has been described by Hillel and Wu.[2] Their procedure has the general effect of significantly lowering the upper limit, $R(\text{max})$, of the Dale–Eisinger analysis because of its low statistical probability.

Interpretation of $\bar{\bar{E}}$ as a Center-to-Center Distance in Random Labeling Case

For samples prepared via the random labeling scheme, the measured transfer efficiency, $\bar{\bar{E}}$, is converted to a center-to-center distance using the method described by Gennis and Cantor.[14] Recall that the random labeling scheme generates one globular component of radius R_D, with its surface nonspecifically labeled with μ_D donors per protein. The scheme also generates a second globular component of radius R_A, with its surface nonspecifically labeled with μ_A acceptors per protein. These labeled components are then used to prepare, by reconstitution, the various labeled and unlabeled assemblies needed for energy-transfer measurements discussed on pages 363–64. Gennis and Cantor have shown if donor–donor transfer is unlikely, the observed transfer efficiency, $\bar{\bar{E}}$, is a function of: (1) the donor–acceptor average $R_0(\frac{2}{3})$; (2) the radii of the globular components, R_D and R_A; (3) the acceptor labeling ratio, μ_A; (4) the distance between the centers of the two components, R_{c-c}.

What is done in the Gennis–Cantor procedure is to calculate $R_0(\frac{2}{3})$ from the average spectral properties of the donor-labeled and of the acceptor-labeled assemblies. Then R_D and R_A are calculated from the known molecular weights of the labeled proteins and an average partial specific volume for proteins. Allowing for a Poisson distribution of labeling stoichiometries, one calculates $\bar{\bar{E}}$ as a function of μ_A for a reasonable range of R_{c-c}'s.[14] The calculated $\bar{\bar{E}}$'s are plotted vs μ_A for the range of R_{c-c}'s, yielding a grid of values of $\bar{\bar{E}}$ as a function of μ_A and R_{c-c} (see Fig. 9). The experimentally measured $\bar{\bar{E}}$ is placed on the grid at the exper-

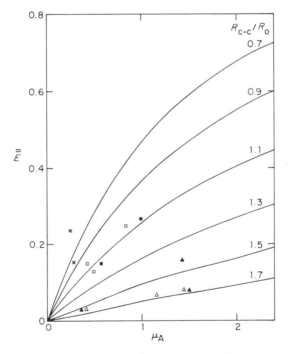

FIG. 9. Analysis of typical random-labeling energy-transfer results to yield distance estimates. The data represent measurements performed between labeled proteins of the 30 S subunit of the *Escherichia coli* ribosome. The experimentally determined $\bar{\bar{E}}$ is plotted against μ_A, the mean number of acceptors in the assembly. The grid is calculated as described in the text for globular components 1 and 2 with $R_1 = 17$ Å and $R_2 = 13$ Å and $R_0 = 43$ Å. Each symbol on the grid represents two different fluorescent labeled components. The ■ data, for example, represent two measurements with two values of μ_A and indicate an average center-to-center distance of 1.1 R_0, or 47 Å, between the two components. Adapted from K. H. Huang, R. H. Fairclough, and C. R. Cantor, *J. Mol. Biol.* **97**, 464 (1975), by courtesy of Academic Press, New York.

imentally determined μ_A, and the value of R_{c-c} is determined by interpolation from the grid. It is useful to have determinations of $\bar{\bar{E}}$ at several μ_A's for maximum reliability of the procedure.

The Reconstruction of Structure from Distance Measurements

After measuring the distances between the centers of N spherical subunits or N loci on a macromolecular aggregate, one can construct a three-dimensional model incorporating the information gained from the distance measurements. If the model is to consist of N points (sphere centers or chemical loci), $3N$-6 coordinates must be specified in order to build the three-dimensional model. Six distance determinations are required to specify the relative configuration of the first four noncoplanar points. There remains an unresolvable ambiguity in the handedness of this structure. For each additional point, four distance measurements enable unique specification of the three Cartesian coordinates of the point in the model. Thus for $N \geq 4$ points, the total number of distances sufficient to specify the $3N$-6 coordinates is just $4N$-10. More distance determinations further increase one's confidence in the model.

Frequently one can only measure $4N$-11 or $4N$-12 distances instead of the $4N$-10. This may be the result of protein scarcity, an inability to fluorescently modify a component, or an inability to reconstitute a labeled component. Nevertheless, with a fortuitous set of $4N$-11 distances, one can limit the arrangement of the points to two possible models in which the unknown coordinates are specified in each of the models by solution of a quadratic equation derived from the known distance measurements and the algebraic distance formula. Similarly, $4N$-12 distances result in constraining the structure to one of four possible models (see Fig. 10).

A Possible Future Use of Energy Transfer

In addition to pure structural studies, energy transfer offers a new approach to the study of assemblies with multiple binding sites for a particular subunit or ligand. Most standard techniques give the average incorporation of subunit or ligand, but no information about the distribution of subunit or ligand in a population of assemblies under nonsaturating conditions. Consider, for example, assemblies that can bind four copies of a particular subunit. Suppose one asks: At one-quarter saturation (in which) an average of one copy of subunit is incorporated per assembly), what is the nature of the population of assemblies? Does one have four copies of subunit on one quarter of the assemblies and none on the other three quarters? Or does one have a distribution of assemblies with no subunits,

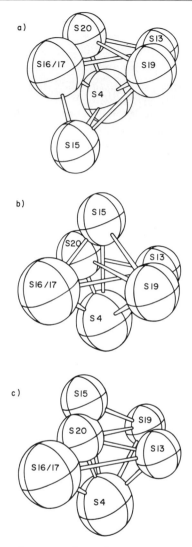

Fig. 10. Reconstruction of structure from distance measurements. In the course of the random labeling and reconstitution of the 30 S subunit of *Escherichia coli* ribosomes, distance estimates accumulated between six of the twenty-one proteins, which enabled some model building to be undertaken as described in the text. In this case twelve distances were in hand but fourteen were required for unique specification of a single model for the arrangement of the six proteins. Nevertheless, it was possible to limit the possible models to one of four, which when the equations were solved algebraically gave rise to two of the four models being equivalent. The three remaining models are illustrated here. They were constructed with an ORTEP thermal ellipsoid program kindly provided by Dr. Stephen J. Lippard. We thank him, John Gill, and C. T. Lamn for their advice and assistance in its use.

one subunit, two, three, and four subunits? If all the subunits bind near one another on the assembly, fluorescence energy transfer can allow determination of the distribution as well as the average incorporation of subunit.

To see how this might work, suppose one labels one portion of subunits with a donor and a second portion with an acceptor, and keeps a third portion unlabeled. Mixing the donor-labeled subunits with an excess (10:1) of (1) unlabeled and (2) acceptor-labeled subunits, one incubates to form assemblies under saturating conditions so that four subunits are present on all assemblies. One measures nonradiative energy transfer via donor quenching and calls it E_{sat}. Ideally E_{sat} is large ≥ 0.5). One next prepares a population of assemblies with one subunit per assembly, i.e., $\mu_{sub} = 1$. Some possible results of an energy-transfer experiment performed on this population of assemblies are listed:

1. If each assembly has one subunit, $E_{\mu=1} = 0.$
2. If one quarter has four subunits, and
   three quarters have none, $E_{\mu=1} = E_{sat}$
3. If one half have two subunits, and one half
   have none, $E_{\mu=1} = \frac{1}{3}E_{sat}$
4. If population has a binomial distribution
   of subunits, $E_{\mu=1} = 0.155E_{sat}$

Thus from the results of the experiment one can learn about the distribution of subunits in the population of assemblies.

Examples and Conclusion

Nonradiative energy transfer has been used to study a variety of macromolecular assemblies. Table II lists some of the systems and the results of energy transfer experiments using different donor-acceptor pairs. The reader's attention is drawn to the energy transfer study of Langlois *et al.* on 50 S subunits of ribosomes[29] for a particularly detailed conversion of energy transfer data to a range of distances using the Dale–Eisinger procedure.

From the results of the studies shown in Table II, it is clear that singlet-singlet energy transfer has the ability to generate structural details of complex macromolecules in solution. While this technique has its limitations and is often tedious because distances must be measured one at a time, singlet–singlet energy transfer offers one major advantage over other techniques that have been applied to the study of macromolecular as-

[29] R. Langlois, C. C. Lee, C. R. Cantor, R. Vince, and S. Pestka, *J. Mol. Biol.* **106**, 297 (1976).

TABLE II
SURVEY OF ENERGY TRANSFER MEASUREMENTS IN ASSEMBLED SYSTEMS[a]

System	Labeling scheme	Donor–acceptor	$R_o(2/3)$ (Å)	Distance measured (Å)	References
Chymotrypsin(C)–trypsin(T) in complex with black-eyed pea trypsin inhibitor (BEPTI)	R	DANSYL FTC	33	64	1
(C)—(BEPTI)—(T)	S to R	DANSYL FTC	33	33–43	1
Trypsin–trypsin inhibitor	R	DANSYL FTC	33	30–40	2
	R	DANSYL RTC	24	30–40	2
Aspartate transcarbamylase	S	Trp Pp	20	27	3
	S	Trp ANS	22	27	3
	S	Pp Mnp	33	26	3
	S	Pp ANS	25	26	3
	S	Pp Cp_4F	26	42	3
Microtubules to membranes	R	FTC RTC	—	40	4
30 S Subunits of ribosomes	R	AEDANS FTC	43	30–80	5
RNA Polymerase	R to S	DANSYL rif	38	27–38	6
	S	AEDANS rif	31	42–85	6
Pyruvate dehydrogenase	S	ANS FAD	40	58	7
	S	ϵCoA FAD	36	50	8

TABLE II (*Continued*)

System	Labeling scheme	Donor–acceptor	$R_0(2/3)$ (Å)	Distance measured (Å)	References
50 S Subunit of ribosomes	S	AEDANS FTC	46	58–84	9
Cell lectin receptors	R	DANSYL RTC	—	—	10

[a] Abbreviations used: R represents random and S represents specific labeling scheme. DANSYL, FTC, Trp, AEDANS, RTC, ANS, are as in Fig. 4. Pp is pyridoxamine phosphate; Mnp is mercurinitrophenol; Cp_4F is

rif is rifampicin; FAD is flavin adenine dinucleotide; ϵCoA is $1,N^6$-etheno coenzyme A. R_0's listed are as used by the authors.

[b] Key to references:

1. L. S. Gennis and C. R. Cantor, *J. Biol. Chem.* **251**, 769 (1976).
2. L. S. Gennis, R. B. Gennis, and C. R. Cantor, *Biochemistry* **11**, 2517 (1972).
3. S. Matsumoto and G. G. Hammes, *Biochemistry* **14**, 214 (1975).
4. J. S. Becker, J. M. Oliver, and R. D. Berlin, *Nature (London)* **254**, 152 (1975).
5. K. H. Huang, R. H. Fairclough, and C. R. Cantor, *J. Mol. Biol.* **97**, 443 (1975).
6. C.-W. Wu, L. R. Yarbrough, F. Y.-H. Wu, and Z. Hillel, *Biochemistry* **15**, 2097 (1976).
7. G. B. Shepherd and G. G. Hammes, *Biochemistry* **15**, 311 (1976).
8. G. B. Shepherd, N. Papadakis, and G. G. Hammes, *Biochemistry* **15**, 288 (1976).
9. R. Langlois, C. C. Lee, C. R. Cantor, R. Vince, and S. Pestka, *J. Mol. Biol.* **106**, 297 (1976).
10. S. M. Fernandez and R. D. Berlin, *Nature (London)* **264**, 411 (1976).

semblies. Unlike X-ray diffraction, neutron scattering, electron microscopy, and cross-linking, fluorescent techniques can be carried out on dilute solutions of assemblies that maintain full biological activity.

Acknowledgment

This work was supported by grants from the National Institutes of Health (GM 14825, GM 19843) and the National Science Foundation (BMS 14329).

[18] Fluorescence Methods for Measuring Reaction Equilibria and Kinetics[1]

By W. B. DANDLIKER, J. DANDLIKER, S. A. LEVISON, R. J. KELLY, A. N. HICKS, and JOHN U. WHITE

The present basic goal of biological science is to reduce the description of biological systems and events to the terms of organic and physical chemistry. In this way it should be possible to utilize what exists in the way of underlying theory for these latter sciences, to predict features of structure and behavior for biological systems. A major stride in this direction has been in the development of molecular biology, where the transfer of heritable characteristics has been reduced to the existence of perferential reactions in polynucleotides between adenine and thymine and between guanine and cytosine rather than other combinations. While the biochemistry of small molecules is well known at the organic chemical level, changes involving macromolecules usually are understood in much less detail and still have to be dealt with at a more descriptive level, e.g., in terms of the concentrations of "active sites" of unknown or partially known structure or of the macromolecules themselves. In these areas simple thermodynamic and kinetic analyses often afford the best descriptions that are accessible.

Once the stoichiometry of a reaction has been tentatively established, thermodynamic and kinetic descriptions often can encapsulate large amounts of unwieldy data into a few constants and, at the same time, test the form of mass or rate laws over a wide range of reactant and product concentrations. The importance of this test can hardly be overemphasized both with respect to obtaining meaningful constants and with respect to inferences to be drawn concerning reaction mechanisms. The behavior of the reaction at different temperatures and in the presence or the absence of various inhibitors or catalysts gives additional information, often of crucial value in characterizing a reaction. The rates and equilibria in different solvents and in the presence or the absence of various ions may give valuable clues about reactions, e.g., as to whether or not they are diffusion controlled or perhaps involve gain or loss of solvent molecules or ions. Wherever possible, the reaction should be studied utilizing different means for measuring the extent of reaction as, for example, by chemical

[1] Supported by Research Contract No. N01-CB-43905 from the National Cancer Institute, Research Grant No. GB-31611 from the National Science Foundation, Research Grant No. R-803885 from the Environmental Protection Agency and by Cordis Corporation, Miami, Florida.

analysis with or without separation of reactants and products, titration, pH shifts, radiolabeled techniques, light absorption, and fluorescence intensity or polarization. In assessing the adequacy of equilibrium and kinetic data obtained on a reaction, one of the most elementary internal checks that can be applied consists in the comparison of the equilibrium constant with the ratio of rate constants for the forward and backward reactions. Agreement between these quantities together with constancy of the form of the mass and rate laws is reasonable assurance that equilibrium and kinetic approaches have been utilized to their full advantage.

In order to obtain data of the range and quality needed to satisfy the above type of criteria, it is usually necessary to utilize in conjunction a number of experimental techniques and instruments. These include simple equilibrium measurements, kinetics of the forward reaction at the level of manual mixing and of stopped flow, relaxation measurements by temperature or pressure jumps and measurements on the back reaction, e.g., by dilution jump. In this chapter we have described some instruments and techniques pertinent to these approaches especially where changes in fluorescence intensity or polarization are utilized as indices of the extent of reaction.

Since reactions of interest rarely show a usuable, innate fluorescence change, we have made extensive use of fluorescence labeling of one component to make it possible to follow reactions generally. The advantages of utilizing fluorescence intensity or polarization changes to monitor reactions include high sensitivity, high speed, and the capability of measurement without physically separating reactants and products (or bound and free). The disadvantage of fluorescence labeling lies in the possible alteration of the properties of the molecule being labeled by the mere presence of the fluoresence label (or more trivially, by the effects of too drastic conditions chosen for labeling). However, experience has shown that many large and small molecules may be labeled with fluorescent dyes and still retain much of the native reactivity characteristic of the unlabeled molecule. At any rate, objective measurements can be made to assess the effects of labeling and to compensate for them if they are significant.

The physical basis underlying the appearance of fluorescence changes during a reaction are for the most part quite different for intensity as compared to polarization changes. Alteration of observed fluorescence intensities (at constant incident intensity and at constant wavelengths of excitation and emission) may arise either from spectral shifts or changes in quantum yield and usually imply some rather direct involvement, in the reaction being monitored. Changes of this type are to be expected if the natural fluorescence of a reactant or product is being monitored but is unlikely and probably undesirable if the fluorescence of an added label is being observed. In this latter case the influence of the label is hopefully outside the immediate sphere of reaction enabling the label to maintain a passive role. In this situation only polarization changes due to changes in rotational brownian motion consequent to the reaction are to be expected. If both intensity and polarization changes are simultaneously present in a reaction, its course can be followed by either effect. The use of fluorescent molecules as environmental probes is obviously only a slightly different way of viewing changes in intensity and polarization as a result of or during a reaction.

Theory

Stoichiometric Expressions

The physically simplest reaction to be considered is that in which a univalent fluorescent molecule (or fluorescent labeled molecule) thought of as the ligand, for convenience, in a one-step reaction binds to a single type of receptor molecule, which need not be fluorescent. The case in which there is also present a nonfluorescent ligand that competes with the fluorescent one for the same receptor sites is only slightly more complicated. In both of the above cases the pertinent optical properties of the fluorescent molecule can be completely specified in terms of the fluorescence intensity and polarizations of the "free" and "bound" states. In the equations developed below it will be assumed that this is the case and that the parameters Q_f, Q_b, P_f and P_b furnish a proper description.

If however, there is more than one type of receptor site present, or if the reaction proceeds in distinct steps, additional attention may have to be given to characterizing the absorption and emission properties of the fluorescent molecule in its different possible "bound" states. In this context there is a definite advantage in utilizing changes in polarization together with intensity changes, if present, since in this way different types of binding may possibly be distinguished. An additional route to similar information is to discriminate between different types of sites on a kinetic basis, i.e., to possibly observe the binding to one type of site before the subsequent reaction has proceeded to an appreciable extent. In the first and simplest case mentioned above the reaction can be represented by:

$$\mathfrak{F} + \mathfrak{R} \rightleftarrows \mathfrak{F}\mathfrak{R} \tag{1}$$

in which $\mathfrak{F}$ stands for the fluorescent ligand, $\mathfrak{R}$ for the nonfluorescent receptor, and $\mathfrak{F}\mathfrak{R}$ for the complex. If a nonfluorescent inhibitor $\mathfrak{N}$ is present, an additional reaction can be represented by

$$\mathfrak{N} + \mathfrak{R} \rightleftarrows \mathfrak{N}\mathfrak{R} \tag{2}$$

If two types of receptor sites are present the following reactions are possible

$$\mathfrak{F} + \mathfrak{R}_1 \rightleftarrows \mathfrak{F}\mathfrak{R}_1 \tag{3}$$
$$\mathfrak{F} + \mathfrak{R}_2 \rightleftarrows \mathfrak{F}\mathfrak{R}_2 \tag{4}$$
$$\mathfrak{N} + \mathfrak{R}_1 \rightleftarrows \mathfrak{N}\mathfrak{R}_1 \tag{5}$$
$$\mathfrak{N} + \mathfrak{R}_2 \rightleftarrows \mathfrak{N}\mathfrak{R}_2 \tag{6}$$

The notation in Eqs. (1) through (6) lends itself to the expression of concentration of univalent ligand as molarity and of receptor as molarity of binding sites. The magnitudes of equilibrium and kinetic constants derived from expressions below reflect this convention.

Symbols

The nomenclature used here assumes that any particular molecule of univalent fluorescent ligand $\mathfrak{F}$, is in one of two distinct chemical states, viz., either "free" or "bound," and that these states can be fully characterized optically by four constants, Q_f, Q_b, p_f, and p_b.

a_F,	heterogeneity index, Eq. (23)
a_N,	heterogeneity index, Eq. (24)
b,	subscript indicating "bound"
e,	subscript indicating value at equilibrium
F,	molar concentration of $\mathfrak{F}$
$\mathfrak{F}$,	chemical symbol for the fluorescent ligand
f,	subscript indicating "free"
$F_{b,max}$,	the maximum value of F_b taken to be equal to the molar concentration of receptor sites
$\left.\begin{array}{l} F_{b,max.1} \\ F_{b,max.2} \end{array}\right\}$	similar to $F_{b,max}$ except for two discrete classes (1 and 2) of binding sites
$\mathfrak{F}\mathfrak{R}$,	chemical symbol for the fluorescent ligand–receptor complex
k,	empirical rate constant, Eq. (25)
k',	empirical rate constant, Eq. (25)
k_1,	second-order rate constant for the forward reaction
k_{-1},	first-order rate constant for the backward reaction
K_F,	association constant for the reaction between $\mathfrak{F}$ and $\mathfrak{R}$
K_N,	association constant for the reaction between $\mathfrak{N}$ and $\mathfrak{R}$
$\left.\begin{array}{l} K_{F.1} \\ K_{F.2} \\ K_{N.1} \\ K_{N.2} \\ K_{F.S} \\ K_{N.S} \end{array}\right\}$	association constants similar to K_F and K_N except for nonuniform receptor sites: 1 and 2 for two discrete classes of receptor sites, S for Sips distribution
M,	the total molar concentration of $\mathfrak{F}$ in both free and bound forms
N,	molar concentration of $\mathfrak{N}$
$\mathfrak{N}$,	chemical symbol for unlabeled ligand
N_1,	order of reaction with respect to receptor
N_2,	order of reaction with respect to ligand
N_3,	order of reaction with respect to complex

$\mathfrak{RR}$, chemical symbol for the nonfluorescent ligand–receptor complex
$_0$, subscript indicating value at zero time
p, the polarization of the excess fluorescence, i.e., $p = (\Delta v - \Delta h)/(\Delta v + \Delta h)$, where Δv and Δh are the intensities in arbitrary units of the components in the excess fluorescence (above that of the blank) polarized in the vertical and horizontal directions, respectively.
Q, molar fluorescence of a mixture of free and bound forms of $\mathfrak{F}$ as they exist in a solution under observation, i.e., $Q = (\Delta v + \Delta h)/M$.
$\mathfrak{R}$, chemical symbol for the receptor
t, time
W, the total molar concentration of $\mathfrak{R}$ in both free and bound forms
x, the ratio F_b/F_f

Equations Relating Fluorescence Parameters to Species Concentrations[1a–3]

The measured polarization, p, furnishes a direct measure of the ratio of "bound/free"

$$\frac{F_b}{F_f} = \frac{Q_f}{Q_b}\left(\frac{p - p_f}{p_b - p}\right) \tag{7}$$

Since $M = F_b + F_f$,

$$F_b = \frac{MQ_f(p - p_f)}{Q_b(p_b - p) + Q_f(p - p_f)} \tag{8}$$

and

$$F_f = \frac{MQ_b(p_b - p)}{Q_b(p_b - p) + Q_f(p - p_f)} \tag{9}$$

Differentiation of the expression for F_f yields:

$$\frac{dF_f}{dt} = \frac{-MQ_bQ_f(p_b - p_f)}{[Q_b(p_b - p) + Q_f(p - p_f)]^2}\left(\frac{dp}{dt}\right) \tag{10}$$

For initial rates if $p \approx p_f$ then

$$\left(\frac{dF_f}{dt}\right)_0 = \frac{-MQ_f}{Q_b(p_b - p_f)}\left(\frac{dp}{dt}\right)_0 \tag{11}$$

If the reaction produces quenching or enhancement of fluorescence then the following relationships are useful:

$$\frac{F_b}{F_f} = \frac{Q_f - Q}{Q - Q_b} \tag{12}$$

[1a] W. B. Dandliker, H. C. Schapiro, J. W. Meduski, R. Alonso, G. A. Feigen, and J. R. Hamrick, Jr., *Immunochemistry* **1,** 165 (1964).
[2] W. B. Dandliker and S. A. Levison, *Immunochemistry* **5,** 171 (1968).
[3] S. A. Levison and W. B. Dandliker, *Immunochemistry* **6,** 253 (1969).

and

$$\frac{dF_f}{dt} = \frac{F_f^2}{M} \left(\frac{Q_f - Q_b}{[Q - Q_b]^2} \right) \frac{dQ}{dt} \tag{13}$$

As $t \to 0$, if $Q \approx Q_f$,

$$\left(\frac{dF_f}{dt} \right)_0 = \frac{M}{Q_f - Q_b} \left(\frac{dQ}{dt} \right)_0 \tag{14}$$

Evaluation of the Optical Constants Q_f, Q_b, p_f and p_b[1a]

The molar fluorescence, Q_f and the polarization, p_f for the "free" fluorescent ligand are determined from measurements on the fluorescent ligand without added receptor, other conditions being the same. The units in which fluorescence intensity are expressed are immaterial in applying them to the equations. For purposes of comparison and standardization it is desirable to employ a secondary standard, e.g., a fluorescent glass block, and to standardize the secondary standard in terms of known concentrations of a pure fluorescent dye, e.g., fluorescein, in a standard solvent.[4]

The determination of Q_b and p_b necessitates an extrapolation to the hypothetical completely bound state. A means for obtaining Q_b and p_b involves first extrapolating several p vs M curves (run at different receptor concentrations) to $M = 0$. These intercepts, p', are then plotted vs $(p' - p_f)$ divided by the receptor concentration in any convenient units. The intercept as (1/receptor concentration) $\to 0$ is equal to p_b. Q_b is found in an analogous way by plotting Q' vs $(Q_f - Q')/($receptor concentration).[1a]

Equilibrium Equations

Classical Mass Law, Uniformly Binding Sites, No Site–Site Interactions

Here the equations applicable to reactions (1) and (2) are:

$$F_b/F_f = K_F(F_{b,max} - F_b - N_b) \tag{15}$$
$$N_b/N_f = K_N(F_{b,max} - F_b - N_b) \tag{16}$$

These equations assume that $\mathfrak{F}$ and $\mathfrak{N}$ compete for the same sites but bind with different association constants. The constants K_F and $F_{b,max}$ are evaluated from Scatchard plots of data obtained in the absence of inhibitor, $\mathfrak{N}$.

[4] R. J. Kelly, W. B. Dandliker, and D. E. Williamson, *Anal. Chem.* **48**, 846 (1976).

The constant, K_N is then obtained from the equation.[5]

$$K_N = \frac{K_F[F_{b,max} - [Mx/(x + 1)] - x/K_F]}{x[W - (F_{b,max} - [Mx/(x + 1)] - x/K_F)]} \tag{17}$$

Two Distinct Types of Uniformly Binding, Noninteracting Sites

This case is represented by reactions (3) to (6). If no inhibitor, $\mathfrak{N}$, is present, the separate binding equations can be written as:

$$F_{b,1}/F_f = K_{F,1}(F_{b,max,1} - F_{b,1}) \tag{18}$$

and

$$F_{b,2}/F_f = K_{F,2}(F_{b,max,2} - F_{b,2}) \tag{19}$$

The experimental data consist of pairs of values of $(F_{b,1} + F_{b,2}) \equiv X$ and $(F_{b,1} + F_{b,2})/F_f \equiv Y$. Adding Eqs. (18) and (19), and substituting X and Y gives:

$$K_{F,1}K_{F,2}X^2 + (K_{F,1} + K_{F,2})XY + Y^2 - K_{F,1}K_{F,2}(F_{b,max,1} + F_{b,max,2})X$$
$$- (K_{F,1}F_{b,max,1} + K_{F,2}F_{b,max,2})Y = 0 \tag{20}$$

This is the equation of a hyperbola with asymptotes

$$Y = K_{F,1}F_{b,max,1} - K_{F,1}X \tag{21}$$

and

$$Y = K_{F,2}F_{b,max,2} - K_{F,2}X \tag{22}$$

Hence evaluation of the slopes and intercepts of the two asymptotes gives K_{F1}, K_{F2}, $F_{b,max,1}$ and $F_{b,max,2}$ separately.

A Large Number of Distinct Types of Uniformly Binding, Noninteracting Sites

This case has been discussed analytically by Hart[6] and graphically by Rosenthal[7]; see also Thieulant et al.[8]

Binding Sites Characterized by a Sips Distribution of Binding Affinities

A modified form of mass law in which the distribution of binding affinities follows Sips equation (see Nisonoff and Pressman[9]) has been

[5] W. B. Dandliker, R. J. Kelly, J. Dandliker, J. Farquhar, and J. Levin, Immunochemistry 10, 219 (1973).
[6] H. E. Hart, Bull. Math. Biophys. 27, 87 (1965).
[7] H. E. Rosenthal, Anal. Biochem. 20, 525 (1967).
[8] M. L. Thieulant, L. Mercier, S. Samperez, and P. Jouan, J. Steroid Biochem. 6, 1257 (1975).
[9] A. Nisonoff and D. Pressman, J. Immunol. 80, 417 (1958).

widely used in immunochemistry. In linearized form, the equations for the binding of $\mathfrak{F}$ and $\mathfrak{R}$ can be expressed as:

$$\log F_f = \frac{1}{a_F} \log \left(\frac{F_b}{F_{b,max} - F_b - N_b} \right) - \log K_{F,S} \tag{23}$$

$$\log N_f = \frac{1}{a_N} \log \left(\frac{N_b}{F_{b,max} - F_b - N_b} \right) - \log K_{N,S} \tag{24}$$

Methods for the numerical computation of results in terms of these equations are available.[1a,10-12]

Kinetic Equations[2,3]

General Formulation

The analysis given here applies to the simple reaction shown in Eq. (1). A general form of the rate equation for that reaction may be written as:

$$-dF_f/dt = dF_b/dt = k(F_{b,max} - F_b)^{N_1}(F_f)^{N_2} - k'(F_b)^{N_3} \tag{25}$$

where k and k' are empirical rate constants.

Concentrations have been expressed in all cases either as molarity of fluorescent ligand, $\mathfrak{F}$, or as molarity of receptor sites, regardless of the number of binding sites present on the kinetic unit of the receptor. In the special case where $N_1 = N_2 = N_3 = 1$, Eq. (25) can be expressed in terms of the second order forward rate constant k_1 and the first-order backward rate constant k_{-1}:

$$-dF_f/dt = dF_b/dt = k_1(F_{b,max} - F_b)(F_f) - k_{-1}(F_b) \tag{26}$$

Determination of Reaction Order

In Eq. (25) the constants N_1 and N_2, being the order of reaction with respect to receptor and ligand, respectively, can be evaluated by studying initial rates over a range of concentrations of ligand and receptor. For initial rates Eq. (25) becomes

$$(dp/dt)_0 = Q_b/Q_f(p_b - p_f)k(F_{b,max})^{N_1}(F_{f,0})^{N_2-1} \tag{27}$$

If the receptor is present in such large excess that its concentration may be regarded as constant, then Eq. (27) can be written as:

$$\log (dp/dt)_0 = (N_2 - 1) \log (F_{f,0}) + a \text{ constant} \tag{28}$$

[10] F. Kierszenbaum, J. Dandliker, and W. B. Dandliker, *Immunochemistry* **6**, 125 (1969).
[11] W. B. Dandliker, *in* "Methods in Immunology and Immunochemistry" (C. A. Williams and M. W. Chase, eds.), Vol. III, p. 435. Academic Press, New York, 1971.
[12] D. Mavis, H. C. Schapiro, and W. B. Dandliker, *Anal. Biochem.* **61**, 528 (1974).

which allows easy evaluation of N_2 from a linear plot of data at varying $F_{f,0}$. Conversely, if the ligand is present in very large excess, then Eq. (27) becomes

$$\log (dp/dt)_0 = N_1 \log (F_{b,max}) + \text{a constant} \qquad (29)$$

which allows the determination of N_1.

If the reaction between ligand and receptor produces quenching or enhancement of fluorescence, then these changes can also be used to follow the reaction. In this case Eqs. (13) and (25) would be combined to give the analog of Eq. (27). An important feature of the determination of reaction order from Eqs. (28) and (29) is that only relative, not absolute, concentrations need be known.

Determination of Rate Constants from Initial Rate and Integrated Rate Measurements

N_1 and N_2 having been determined as above, the empirical rate constant, k, can be determined with the aid of Eq. (27).

If $N_1 = N_2 = 1$ then the second-order forward rate constant, k_1, can be found from

$$(dp/dt)_0 = Q_b/Q_f(p_b - p_f)k_1(F_{b,max}) \qquad (30)$$

If $N_2 = N_3 = 1$ and if the receptor concentration is sufficiently high that it may be regarded as constant throughout the reaction, then both k_1 and k_{-1} can be found from

$$\log \left(\frac{F_{b,e}}{F_{b,e} - F_b} \right) = \left(\frac{k_1[F_{b,max}]^{N_1} + k_{-1}}{2.3} \right) t \qquad (31)$$

If there is no significant quenching or enhancement so that $Q_f \approx Q_b$, then the following relationship from Eqs. (7) and (31) holds:

$$\log (p_e - p) = \log (p_e - p_f) - \left(\frac{k_1[F_{b,max}]^{N_1} + k_{-1}}{2.3} \right) t \qquad (32)$$

Hence, a plot of $\log (p_e - p)$ vs, time should be linear with a slope equal to $(-1/2.3) (k_1[F_{b,max}]^{N_1} + k_{-1})$. Furthermore, a plot of this latter quantity vs $F_{b,max}$ will be linear if the order with respect to receptor concentration also is 1. The rate constants k_1 and k_{-1} are obtained as the slope and intercept, respectively, of this latter plot.

Rate constants can also be determined from the half-time for reaction. If in Eq. (32), $[F_{b,max}]^{N_1} \gg k_{-1}/k_1$ and if the half-time is defined as the time when $[(p_e - p)/(p_e - p_f)] = \frac{1}{2}$, then Eq. (32) becomes:

$$0.692 = k_1[F_{b,max}]^{N_1} t_{1/2} \qquad (33)$$

and

$$\log t_{1/2} = -N_1 \log (F_{b,max}) + \log [0.692/k_1] \tag{34}$$

Equation (25) is a general form of rate equation applicable to Eq. (1). As shown above, N_1 and N_2 are readily determined from initial rate measurements. Under conditions where $N_1 = N_2 = N_3 = 1$ [Eq. (25)] and where the reaction starts with $F_{b,0} = 0$, the rate equation can be integrated analytically as follows. The expression for equilibrium can be written

$$k_{-1} = \frac{k_1(F_{b,max} - F_{b,e})(F_{f,0} - F_{b,e})}{F_{b,e}} \tag{35}$$

Substitution of Eq. (35) into Eq. (26) gives the rate equation as

$$\frac{-dF_f}{dt} = k_1[F_f - F_{f,e}] \left[(F_f - F_{f,e}) + \frac{F_{f,0}F_{b,max} - (F_{f,0} - F_{f,e})^2}{F_{f,0} - F_{f,e}} \right] \tag{36}$$

or as the equivalent expression

$$\frac{-dF_f}{dt} = k_1[F_{b,e} - F_b] \left[(F_{b,e} - F_b) + \frac{F_{b,max}F_{f,0} - (F_{b,e})^2}{F_{b,e}} \right] \tag{37}$$

After forming partial fractions, integration gives

$$\ln \left\{ \frac{F_{f,0}F_{b,max}[F_f - F_{f,e}]}{[F_{f,0} - F_{f,e}][(F_f - F_{f,e})(F_{f,0} - F_{f,e}) + F_{f,0}F_{b,max} - (F_{f,0} - F_{f,e})^2]} \right\}$$
$$= -k_1 t \left[\frac{F_{f,0}F_{b,max} - (F_{f,0} - F_{f,e})^2}{F_{f,0} - F_{f,e}} \right] \tag{38}$$

or the equivalent expression

$$\ln \left\{ \frac{F_{b,max}F_{f,0}[F_{b,e} - F_b]}{F_{b,e}[(F_{b,e} - F_b)(F_{b,e}) + F_{b,max}F_{f,0} - F_{b,e}^2]} \right\}$$
$$= -k_1 t \left[\frac{F_{b,max}F_{f,0} - F_{b,e}^2}{F_{b,e}} \right] \tag{39}$$

In the limit when the receptor is present in very large excess, Eqs. (38) and (39) reduce to

$$\ln \left[\frac{F_f - F_{f,e}}{F_{f,0} - F_{f,e}} \right] = \ln \left[\frac{F_{b,e} - F_b}{F_{b,e}} \right] = -k_1 t \, F_{b,max} \tag{40}$$

In the limit of very high ligand concentrations, Eqs. (38) and (39) become

$$\ln \left[\frac{F_f - F_{f,e}}{F_{f,0} - F_{f,e}} \right] = \ln \left[\frac{F_{b,e} - F_b}{F_{b,e}} \right] = -k_1 t \, F_{f,0} \tag{41}$$

It may be noted that a set of similar equations containing k_{-1} could be obtained by solving for k_1 instead of k_{-1} in Eq. (35).

Determination of Dissociation Rate Constants by Direct Observation of the Back Reaction (Dilution Jump)

Information on the backward rate constant, k_{-1}, can be obtained by monitoring the rates of dissociation of preformed complexes following rapid dilution (dilution jump). Depending upon the rate constants involved, the jump may have to be very fast (a few milliseconds) or for slow reactions a slower jump made by manual mixing may be adequate. Attention must also be given to the dilution factor, which is limited to perhaps 1000 by practical considerations. The greater the dilution factor, the less important the recombination reaction will be, enabling the dissociation reaction to go nearer to completion before equilibrium effects become significant. Investigation of a large fraction of the dissociation reaction is important in revealing any heterogeneity in binding affinities. In the dilution-jump technique the dissociation of complexes differing in stability are partially resolved in time. If the dissociation is first order, as would be expected if it is unimolecular, then half of the complex present at time zero will have disappeared by the half-time regardless of its initial concentration. Thus, as time proceeds the reaction mixture becomes relatively enriched in the more stable complexes, which then can be observed after the weaker complexes have largely disappeared. An experimental way to simulate the effect of very large dilution factors, approaching infinite dilution, is to determine the behavior at several different dilutions and then to make an empirical extrapolation to infinite dilution. If in Eq. (1) the reaction starts with $\mathfrak{FR}$ present only, and if $N_3 = 1$ then the integrated rate equation in terms of k_{-1} the rate constant for dissociation, and in terms of $F_{b,e}$ the equilibrium concentration of complex (see Frost and Pearson,[13] p. 187) is

$$\ln \left\{ \frac{F_{b,max}^2 - F_{b,e}F_b}{(F_b - F_{b,e})(F_{b,max})} \right\} = k_{-1} \left\{ \frac{F_{b,max} + F_{b,e}}{F_{b,max} - F_{b,e}} \right\} t \qquad (42)$$

An alternative form is:

$$\ln \left\{ \frac{F_{b,max}^2 - (F_{b,max} - F_{f,e})(F_{b,max} - F_f)}{[(F_{b,max} - F_f) - (F_{b,max} - F_{f,e})]F_{b,max}} \right\}$$
$$= k_{-1} \left\{ \frac{2F_{b,max} - F_{f,e}}{F_{b,max} - (F_{b,max} - F_{f,e})} \right\} t \qquad (43)$$

Equation (43) may also be expressed as

$$\ln \left\{ \frac{F_{b,max}(F_{f,e} + F_f) - F_{f,e}F_f}{F_{b,max}(F_{f,e} - F_f)} \right\} = -k_1 t \left(\frac{2F_{b,max} - F_{f,e}}{F_{f,e}} \right) \qquad (44)$$

[13] A. A. Frost and R. G. Pearson, "Kinetics and Mechanism." Wiley, New York, 1961.

If the dilution is so large that the dissociation goes practically to completion then Eqs. (42) to (44) take on their limiting first-order forms:

$$\ln \left(\frac{F_{b,max}}{F_b}\right) = \ln \left(\frac{F_{b,o}}{F_b}\right) = \ln \left(\frac{M}{F_b}\right) = k_{-1}t \tag{45}$$

In terms of the polarization, p, Eq. (45) becomes:

$$\ln \left\{\frac{Q_b(p_b - p) + Q_f(p - p_f)}{Q_f(p - p_f)}\right\} = k_{-1}t \tag{46}$$

If $Q_f = Q_b$ then

$$\ln \left(\frac{p_b - p_f}{p - p_f}\right) = k_{-1}t \tag{47}$$

which permits the evaluation of k_{-1} from a linear plot.

Determination of Rate Constants from Chemical Relaxation

Chemical systems at equilibrium (or in which only slow changes are occurring) can be perturbed in a number of ways so that they relax to a new state adjacent to the initial one; the rate of relaxation serves as a means to determine the reaction rate constants k_1 and k_{-1} [Eq. (26)]. Relaxation methods lend themselves to the study of fast reactions, but suffer from the disadvantage that only a small concentration region is usually accessible to experimentation. Hence, results from relaxation methods should be supplemented by data from other types of kinetic experiments whenever possible. The most usual means of perturbing the equilibrium is by a temperature jump, produced by a capacitor discharge. In systems where the ΔH of reaction is small, it is sometimes possible to couple the reaction to another one with a large ΔH, e.g., by a pH change. Other means of perturbation include rapid changes in pressure or electric field.

The general concept of relaxation time arose originally in considering the rate of response of a physical system after it has been strained or deformed by some external agency. The relaxation time is the time required for the system to traverse all but $1/e$ of its path to the new equilibrium state. To apply these ideas to the reaction represented by Eq. (1), let the concentrations of $\mathfrak{F}, \mathfrak{R}$ and $\mathfrak{F}\mathfrak{R}$ be denoted by α, β, and γ, respectively. Then, the rate equation can be written as

$$-d\alpha/dt = -d\beta/dt = d\gamma/dt = k_1\alpha\beta - k_{-1}\gamma \tag{48}$$

After the perturbation, the conditions for equilibrium have changed so that the existing concentrations differ from the equilibrium values by the amounts $\Delta\alpha$, $\Delta\beta$, and $\Delta\gamma$. Hence, the rate of approach to the new equilibrium state is

$$-d\alpha/dt = k_1(\alpha_e + \Delta\alpha)(\beta_e + \Delta\beta) - k_{-1}(\gamma_e + \Delta\gamma) = k_1(\alpha_e\beta_e + \alpha_e\Delta\beta + \beta_e\Delta\alpha + \Delta\alpha\Delta\beta) - k_{-1}(\gamma_e + \Delta\gamma) \quad (49)$$

Neglecting the higher-order term ($\Delta\alpha\,\Delta\beta$) and utilizing the equilibrium expression

$$k_1\alpha_e\beta_e - k_{-1}\gamma_e = 0 \quad (50)$$

gives the rate equation as

$$-d\alpha/dt = k_1(\alpha_e\Delta\beta + \beta_e\Delta\alpha) - k_{-1}\Delta\gamma \quad (51)$$

since $\Delta\alpha = \Delta\beta = -\Delta\gamma$ and since $d\alpha/dt = d(\Delta\alpha)/dt$, Eq. (51) becomes

$$-d(\Delta\alpha)/dt = \Delta\alpha[k_1(\alpha_e + \beta_e) + k_{-1}] \quad (52)$$

This equation is readily integrated since ($\alpha_e + \beta_e$) is independent of time:

$$\ln(\Delta\alpha_0/\Delta\alpha) = [k_1(\alpha_e + \beta_e) + k_{-1}]t \quad (53)$$

Hence, a plot of $\ln(\Delta\alpha_0/\Delta\alpha)$ versus t has a slope equal to $[k_1(\alpha_e + \beta_e) + k_{-1}]$. By obtaining different values of this slope from experiments at different values of ($\alpha_e + \beta_e$), k_1 and k_{-1} can be found as the slope and intercept, respectively of a plot of $[k_1(\alpha_e + \beta_e) + k_{-1}]$ versus ($\alpha_e + \beta_e$) $\equiv (F_{f,e} + F_{b,max} - F_{b,e})$.

The relaxation time, τ, is that time at which $\Delta\alpha/\Delta\alpha_0 = 1/e$.

$$1/\tau = k_1(\alpha_e + \beta_e) + k_{-1} \quad (54)$$

This relaxation time unlike that of a simple first-order reaction is a function of the forward and backward rate constants and of the concentrations.

To relate $\Delta\alpha$ to the polarization, Eq. (9) is utilized:

$$\Delta F_f = \frac{MQ_b(p_b - p)}{Q_f(p - p_f) + Q_b(p_b - p)} - \frac{MQ_b(p_b - p_e)}{Q_f(p_e - p_f) + Q_b(p_b - p_e)} \quad (55)$$

if Δp is small, then $p \approx p_e$ and

$$\Delta F_f \equiv \Delta\alpha = \frac{-MQ_bQ_f(p_b - p_f)4\Delta p}{[Q_f(p_e - p_f) + Q_b(p_b - p_e)]^2} \quad (56)$$

Hence, $\ln|\Delta F_f| \equiv \ln|\Delta\alpha| = \ln|\Delta p| + $ constant and

$$\ln(\Delta F_{f,0}/\Delta F_f) \equiv \ln(\Delta\alpha_0/\Delta\alpha) = \ln(\Delta p_0/\Delta p) \quad (57)$$

Thus a plot of $\ln(\Delta p_0/\Delta p)$ versus t also has a slope equal to $[k_1(\alpha_e + \beta_e) + k_{-1}]$.

Instrumentation

Equilibrium and Slow Kinetic Measurements

Digital, Photon-Counting Fluorescence Polarometer

This instrument evolved from the need for precise equilibrium measurements at very low fluorescence levels together with a minimal illumination of the sample. To attain these goals, the following design features were incorporated: (1) a detector utilizing only one photomultiplier tube; (2) use of the detecting photomultiplier only as a photon counter with pulse height discrimination; (3) digital processing and readout of the photon count; (4) a continuously rotating polaroid analyzer making the measurements of p less susceptible to time-dependent variations in the signal, detector sensitivity, and excitation intensity; (5) a light source consisting of a tungsten lamp, the output of which is adjustable to $\pm 1\%$ over a 500-fold range, and stabilized by feedback control; (6) isolation of the excitation wavelengths and rejection of these wavelengths by a series of interference and absorption filters; (7) thermostatting of the cell holder ($\pm 0.2°C$) by means of a thermistor-controlled Peltier cooler.

A photograph of the instrument is shown in Fig. 1 and the arrangement

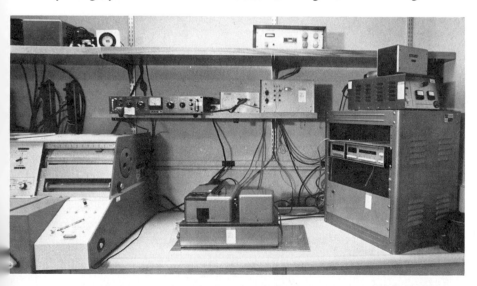

FIG. 1. Digital, photon-counting fluorescence polarometer. In the foreground the optical unit with lamphouse and cell compartment can be seen. The circuits for controlling light intensity and temperature are situated on the shelf above the optical unit. The digital computer is rack-mounted to the right. The lamp power supply rests on top of the rack unit.

of optical components is given by Fig. 2. A tungsten lamp was chosen as the light source, since the high intensities achievable with arc sources are somewhat undesirable from the viewpoint of local heating and possible photochemical changes but, more important, tungsten sources are not subject to erratic movement and stability of output is readily achieved. The lamp and blower are housed separately to avoid overheating interference filters F_2 and F_3. Interference filters used in tandem with each other or with nonfluorescent glass absorbing filters can achieve very high rejection of the incident wavelength, which is necessary to attain low blanks. In the incident beam, for example, an interference filter transmitting 62% at 520 nm, transmits about 0.01% at 490 nm (ratio = 6200). If two such filters are used in series, the ratio is 3.8×10^7. The same considerations apply to the fluorescent beam. Optimal filter combinations are best chosen empirically by comparing the instrument response to a solution of the fluorescent dye to that of a suspension of Ludox (colloidal silica from E. I. du Pont de Nemours, Wilmington, Delaware) or other material that scatters but shows negligible fluorescence. The ratio of fluorescence reading/scattering readings should be maximized, while retaining sufficient signal for easy detection. The plate, P (Fig. 2), is a 2-inch × 2-inch square of Pyrex glass (Corning Glassworks, Corning, New York, Glass No. 7740) which transmits down to about 300 nm. Low-fluorescence quartz would be somewhat preferable. A simple vacuum phototube was chosen to monitor the incident intensity because of its high stability and insensitivity to supply voltage. Thermostatting of the cell by thermoelectric cooling is quite precise and avoids the practical difficulties associated with circulating liquids. Control of the temperature is accomplished by feedback from a thermistor in the cell holder, C. Details of similar arrangements are well-known, and have not been included in the description below. Lenses used in the optical system are glass, but should be examined and chosen for low fluorescence. The cuvette should be of low-fluorescence quartz and may be either mirrored or painted black on two of the outside surfaces. While the mirrored cells provide an advantage in intensity of about a factor of 3 over the blackened cells, the latter are far more reproducible and stable, both over time and from cell to cell. Polarizing filters F_4 and F_7 (Fig. 2) are from Polaroid HN-22 film (Polaroid Corp., Cambridge, Massachusetts). For measurement below about 400 nm, other polarizers should be used (e.g., Polacoat Corp., Cincinnati, Ohio).

The components of the detector module are shown in Fig. 3. The photomultiplier feeds pulses originating from single photons into the SSR 1120 amplifier/discriminator. Only those pulses exceeding a predetermined amplitude are passed by the discriminator. In this way, "dark

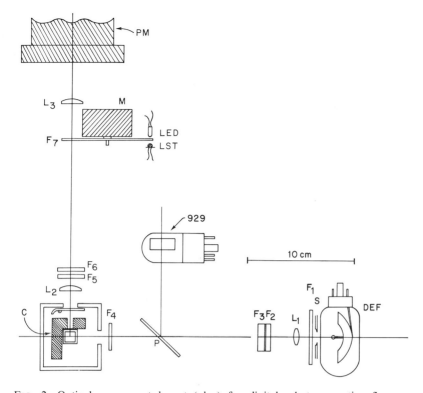

FIG. 2. Optical component layout (plan) for digital, photon-counting fluorescence polarometer. Light from the projection lamp (Sylvania DEF rated at 150 W, 21.5 V) passes through opening, S (10 mm in diameter), heat filter F_1 (Corning 1-69), and into lens L_1, which focuses an image of the source into the fluorescence cuvette held in a copper block, C, which is thermostatted by a Peltier cooler. The entire cuvette compartment is fitted with a light-tight lid. To protect the photomultiplier from ambient light, a shutter drops into the closed position when the lid is lifted and opens when the lid is replaced. Interference filters F_2 and F_3 isolate the desired wavelength (490 nm if fluorescein is to be used) and polarizer F_4 (Polaroid film HN-22 or HN-38) transmits light with the electric vector perpendicular to the plane of the drawing. Quartz or glass plate, P, reflects a few percent of the incident beam onto the photocathode of a vacuum phototube (RCA type 929), which furnishes the reference signal for the lamp control circuit. Fluorescent light from the solution in the fluorescence cuvette is focused by lens, L_2, onto the rotating analyzer, F_7. The fluorescence cuvette is of standard size (1 cm² × 5 cm high) made of low fluorescence quartz and painted black on two of the outside surfaces. The incident wavelength is removed by interference filter F_5 and colored glass F_6 (chosen for very low fluorescence). If fluorescein is to be used, F_5 is peaked at 520 nm and F_6 is a selected piece of Corning glass 3-69. Rotating polarizer, F_7, is made from a Polaroid sheet and is rotated at 90 rpm by a synchronous motor, M. The light-emitting diode, light-sensitive transistor (LED, LST) assembly, in conjunction with the light-chopping mask attached to the rotating polarizer furnishes a signal for phase discrimination in the photon-counting circuit. Light from the fluorescence solution passing through the rotating polarizer central to the mask mentioned above is focused by a lens, L_3, near the plane of the photocathode of the photomultiplier tube (EMI 9635QB), which is housed in the photomultiplier housing PM (SSR Model 1151 from SSR Instruments, Santa Monica, California).

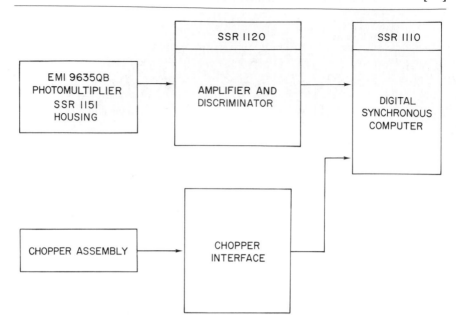

DETECTOR MODULE

FIG. 3. Block diagram of the detector module. The SSR 1110 computer, the SSR 1120 amplifier/discriminator, as well as the photomultiplier housing (SSR 1151) were units purchased from SSR Instruments Co., Santa Monica, California. The chopper signal is furnished by a light-sensitive transistor viewing a light-emitting diode through the rotating polarizer and its associated mask. The SSR 1110 displays, inter alia, two quantities, DIFF and SUM, which are the difference and sum, respectively, of photon counts over a selectable number of chopper cycles from the vertically polarized (v) and horizontally polarized (h) components in the fluorescent light. The polarization, p, is calculated from DIFF and SUM for the blank and sample according to Eq. (58).

current'' pulses are largely eliminated, since they are usually of smaller amplitude than those originating from photons. In addition, the photomultiplier is also fitted with a magnetic adapter which reduces the effective photocathode diameter from 1 cm to 2 mm, with a corresponding reduction in dark current. However, this reduction makes optical alignment more difficult and also decreases the amount of light collected. Use of a smaller photomultiplier possibly could achieve the same benefits as our arrangement. Use of the photomultiplier only as a photon counter also makes the resulting detection less sensitive to time-dependent changes in photomultiplier gain or in power supply voltage. The latter is normally set at 1150 V and is supplied by a Model 412B high-voltage power supply

(John Fluke Mfg. Co., Inc., Seattle, Washington). The detector module also interfaces the chopper signal from the rotating analyzer to the SSR 1110 by means of the circuit shown in Fig. 4. The chopper signal directs the pulses from the SSR 1120 into one of the two high-speed accumulating registers of the SSR 1110. At the end of each measurement, the SSR 1110 computes and displays the sum and difference of these registers, to three significant digits. The excitor module (Fig. 5) accepts photocurrent from the excitation monitor (929 phototube of Fig. 2) and compares it with an operator-selected reference voltage. The resulting control signal is used to drive the excitation lamp to deliver that intensity at which the excitation monitor signal equals the selected reference value. In addition, the module limits the maximum lamp current to 8 A. Three light-emitting diodes provide visible indication in the event that the excitation energy is less than or greater than the reference, and in the event the lamp current limit is activated. Schematics for the several parts of this module are given in Figs. 6–9. The excitor module is powered by a regulated ± 15 V supply based on the Motorola MC1539 and MC1561 integrated regulators (Motorola Semiconductor Products, Inc., Phoenix, Arizona).

Measurements of fluorescence polarization and intensity are obtained as follows: The instrument is warmed up for about an hour. During this time, the power supply for the lamp and power to the computer itself need not be on. The fluorescence cuvette and solution or the glass standard is then put into place and allowed to come to constant temperature. A light source intensity is selected to give perhaps up to 10^5 or 10^6 counts per second, and measurements are begun. The quantities read off the computer are SUM and DIFF, which are proportional respectively to $v + h$ and to $(v - h)$, and N, the number of chopper cycles. The most simplified

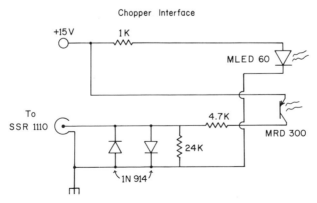

FIG. 4. Schematic of the interface between the chopper (LED, LST of Fig. 2) and the digital synchronous computer (see Fig. 3).

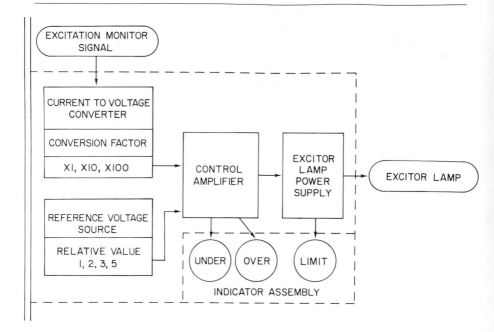

EXCITOR MODULE

Fɪɢ. 5. Block diagram of the excitor module. A signal from the excitation monitor (the 929 phototube of Fig. 2) is continuously compared to a standard reference voltage. The difference signal automatically maintains the lamp output constant at any one of 12 levels selectable over a 500-fold range of intensity.

procedure for handling these data is as follows. Obtain, from the computer registers, SUM and DIFF for both sample and blank, keeping the value of SUM below 10^5 counts per second so that deadtime corrections can be neglected. Take the differences, solution minus blank, to obtain $\Delta(\text{SUM})$ and $\Delta(\text{DIFF})$. The excess fluorescent intensity, $\Delta(v + h)$ is proportional to $\Delta(\text{SUM})$, and the polarization is

$$p = \frac{1}{\eta} \frac{\Delta(\text{DIFF})}{\Delta(\text{SUM})} \tag{58}$$

Values of η can be calculated from the equation

$$\eta = \frac{\sin (\omega t_\text{g} + 2\delta) \sin \omega t_\text{g}}{\omega t_\text{g}} \tag{59}$$

In Eq. (59), ω is the angular velocity of the rotating polarizer, t_g is the chopper gate time interval selected for the SSR 1110 computer and the angle, δ, is a measure of the alignment of the transmission axis of the rotating polarizer and the chopper and of the position of the light beam between the LED and LST (Fig. 2). If both the transmission axis and the light beam simultaneously pass through the center of one of the light areas of the optical chopper, then $\delta = 0$. Measurements made at different num-

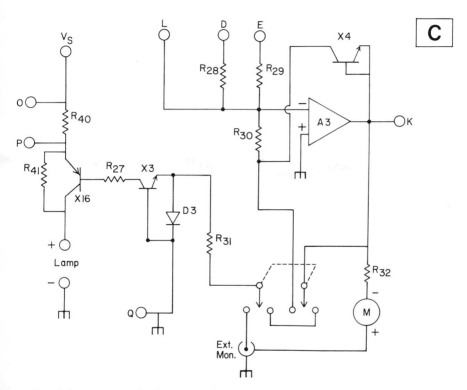

FIG. 6. Lamp control circuit (Module C). Control is accomplished by the "current limit" circuit (Fig. 7) connecting to points O and P and by the control signal feeding through R27 which controls the conductance of X16. This signal originates from the circuit shown in Fig. 9 and enters at points D and E of Fig. 6. A signal feeds out through point K to the circuit of Fig. 8, which supplies the "over"/"under" indication. The power supply (Model CA-150, Illumination Industries Inc., Sunnyvale, California) for the lamp feeds in at V_s and is capable of delivering up to about 10 A at 30 V dc. A variable transformer on the input of this supply is used to adjust V_s to the correct range where control will ensue as indicated by both the "over" and "under" indicators (Fig. 8) being off. The parts list includes: A3—741 op amp Fairchild U6A7741; D3—silicon diode, 1N914; M—1-mA panel meter, edge reading; R27, R31—1-KΩ, 10%, Ohmite RC20; R28-R30—49.9-KΩ, 1%, Corning RN60C; R32—10-KΩ, 50%, Ohmite; R40—0.5-Ω, 200-W (two 1-Ω, 100-W, w.w, in parallel), Ohmite; R41—2.0-Ω, 100-W, w.w, Ohmite; X3, X4—NPN transistor, Motorola 2N3904; and X16—PNP, Power Darlington transistor array, Motorola MJ2500.

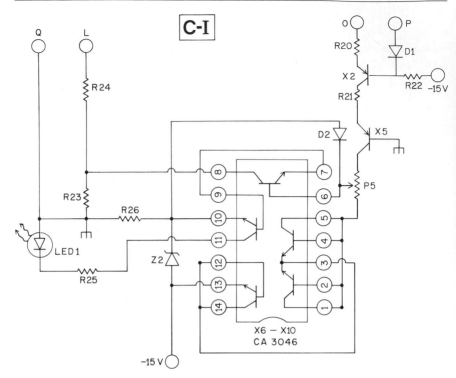

FIG. 7. Current limit circuit (Module C-1). This circuit limits the voltage drop across R40 (Fig. 6) and thus limits the lamp current. LED1 goes "on" if the current limit circuit is, at any moment, actually limiting the lamp current instead of its being controlled by the circuit of Fig. 6. If LED1 is "on," the voltage from the variable transformer (legend, Fig. 6) must be decreased. The parts list includes: D1, D2—silicon diode, 1N914; LED1—light-emitting diode, Monsanto MV5020; P5—10-KΩ, Bourns Trimpot 3006P-1-103; R20—3.01-KΩ, 1%, Corning RN60C; R21—6.8-KΩ, 10%, Ohmite RC20; R22—100-KΩ, 10%, Ohmite RC20; R23—1-MΩ, 10%, Ohmite RC20; R24—10-KΩ, 2%, Corning RL07S; R25—220-Ω, 10% Ohmite RC20; R26—1-KΩ, 10%, Ohmite RC20; X2, X5—PNP transistor, Motorola 2N3906; X6-X10—Linear transistor array, RCA CA3046; and Z2—Zener diode, 9-V, Motorola 1N4739.

bers (N) of chopper cycles and different gate times (t_g) may be normalized to counts per second by dividing the observed number of counts for the interval taken by $2Nt_g$. In addition, measurements may be normalized to a unit value of lamp intensity factor to facilitate comparison of data obtained at different incident intensities.

A series of intensity and polarization readings at a variety of lamp intensities and gate times were taken on a fluorescent glass standard which has a fluorescence emission intensity equivalent to $8 \times 10^{-11} M$ fluorescein at pH 7. The mean intensity reading was 1.838 with a standard error of

0.6%, and the mean polarization reading was 0.1795 with a standard error of 0.4%. Intensity readings on fluorescein solutions (pH 7) ranging in concentration from 8×10^{-13} M up to 2×10^{-10} M fitted the Eq. $\Delta SUM = 2.021 \times 10^{15}$ C + 29 with a regression coefficient of 1.000. Experimental results obtained with this instrument are to be found in Dandliker *et al.*[5] and Kelly *et al.*[4].

Analog, Direct-Reading Fluorescence Polarometer

This instrument represents a compromise between the ultimate in precision and the speed and ease of operation. Both intensity and polarization

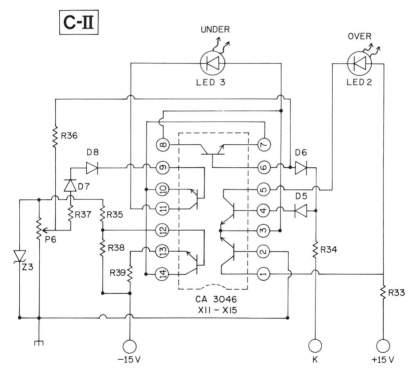

FIG. 8. "Over-Under" indicating circuit (Module C-II). This circuit indicates whether or not the variable transformer feeding the lamp power supply (cf. legend of Fig. 6) is set at a point where control will ensue. For this condition, both LED2 and LED3 must be "off." The parts list includes: D5–D8—silicon diode, 1N914; LED2, LED3—light-emitting diode, Monsanto MV2050; P6—2-KΩ Bourns Trimpot 3006P-1-202; R33—470-Ω, 10%, Ohmite RC20; R34—4.7-KΩ, 10%, Ohmite RC20; R35—91-Ω, 5%, Ohmite RC20; R36, R37—2.2-KΩ, 10%, Ohmite RC20; R38—510-Ω, 5% Ohmite RC20; R39—220-Ω, 5%, Ohmite RC20; X11-X15—linear transistor array, RCA CA3046; and Z3—Zener diode, 9-V, Motorola 1N4739.

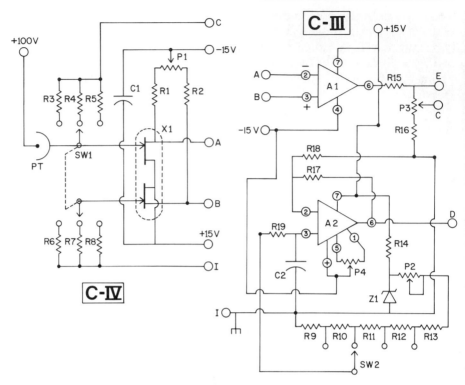

FIG. 9. Signal amplifier and voltage reference generator and amplifier (Modules C-III and C-IV). The parts list includes: A1, A2—741 op. amp., Fairchild U5B7741393; C1—0.1-μf, 25-V, Centralab UK25-104; C2—0.01 μf, disc capacitor, Centralab DD103; P1, P4—10-KΩ, Bourns Trimpot 3389P-1-103; P2, P3—2-KΩ, Bourns trimpot 3006P-1-202, front screwdriver adjustment; PT—Vacuum phototube, RCA type 929; R1, R2—100-KΩ, TI, Corning RL07S; R3, R6—10.0-MΩ, selected RC20; R4—1.00-MΩ, TI, selected Corning RL07S; R5—100-KΩ, TI, selected Corning RL07S; R7—1.0-MΩ, RC20; R8—100-KΩ, RC20; R9-R11—1.00-KΩ, TI, selected Corning R107S; R12—2.00-KΩ, TI, selected Corning RL07S; R13—10-Ω, selected, RC20; R14—1.5-KΩ, carbon, RC20; R15—10-KΩ, 2%, TI, Corning RL07S; R16—1-KΩ, 2%, TI, Corning RL07S; R17, R18—100-KΩ, 2%, TI, Corning RL07S, R19—47-KΩ, carbon RC20; SW1—Switch, 2P, 2-6 position, nonshorting, Centralab PA2003; SW2—Switch, 2P, 2-6 position, shorting, Centralab PA2002; X1—matched dual field-effect transistor, National Semiconductor FM3956; Z1—6.4-V zener, Motorola 1N4560.

are read out directly and blank subtraction is automatic. It is very well adapted to making rapid equilibrium measurements and also kinetic measurements with a time resolution of about 1 sec.

A photograph of the instrument is shown in Fig. 10, and the details of construction and operation are shown in Figs. 11–13. Measurements of the polarization of a glass standard with an intensity equivalent to $6 \times 10^{-11} M$ fluorescein were made over two successive 1-hr periods. The means and standard errors for about thirty readings in each group were

FIG. 10. Manual, direct-reading analog fluorescence polarometer. The top cover has been removed to show positioning of the optical components. The light source is located at the right rear of the unit and the cell compartment with a tubulation leading to the cooling or heating reservoir can be seen at the front right. A thermistor readout giving the temperature of the cell compartment is located just in front of the reservoir. Near the center the rotating polarizer and driving motor can be seen. The photomultiplier tube is located at the front left position, in front of the nullmeter. All controls are on the front panel.

0.3914 ± 0.0052 and 0.3919 ± 0.0031, respectively. Intensity measurements with this instrument are subject to erratic fluctuations due to wandering of the arc. A tungsten source eliminates this problem, but lowers the sensitivity of the instrument.

Experimental data on a number of systems have been obtained with this polarometer (see Dandliker and Levison,[2] Levison *et al.*,[14] Levison and Dandliker,[3] Levison *et al.*,[15] Dandliker,[11] Levison *et al.*,[16] Portmann *et al.*,[17] Levison *et al.*,[18] Portmann *et al.*,[19] Levison *et al.*[20]).

[14] S. A. Levison, A. N. Jancsi, and W. B. Dandliker, *Biochem. Biophys. Res. Commun.* **33**, 942 (1968).

[15] S. A. Levison, F. Kierszenbaum, and W. B. Dandliker, *Biochemistry* **9**, 322 (1970).

[16] S. A. Levison, A. J. Portmann, F. Kierszenbaum, and W. B. Dandliker, *Biochem. Biophys. Res. Commun.* **43**, 258 (1971).

[17] A. J. Portmann, S. A. Levison, and W. B. Dandliker, *Biochem. Biophys. Res. Commun.* **43**, 207 (1971).

[18] S. A. Levison, A. N. Hicks, A. J. Portmann, and W. B. Dandliker, *Biochemistry* **14**, 3778 (1975).

[19] A. J. Portmann, S. A. Levison, and W. B. Dandliker, *Immunochemistry* **12**, 461 (1975).

[20] S. A. Levison, W. B. Dandliker, R. J. Brawn, and W. P. VanderLaan, *Endocrinology* **99**, 1129 (1976).

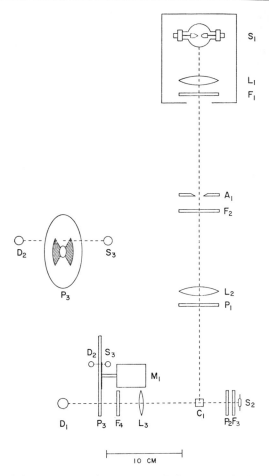

FIG. 11. Manual fluorescence polarometer. This instrument is designed to provide a direct analog readout of both fluorescence intensity and polarization. Helipots operated manually serve to null an indicating meter; the balance points provide the separate readout for intensity and polarization. An image of the source (S_1) is focused by lens (L_1) onto aperture (A_1) which acts as a secondary source of controllable size and shape. An image of A_1 is then focused into the cuvette (C_1) by lens L_2. The temperature of C_1 is controlled by a stream of air passing through the cuvette compartment. The temperature is sensed by a thermistor in the compartment and can be read externally. Heat filter (F_1) protects the interference bandpass filter F_2 from overheating, and polarizer (P_1) polarizes the beam in an azimuth perpendicular to the plane of the paper. Standard source (S_2) in conjunction with opal glass diffusor (F_3) and polarizer (P_2) supplies a polarized beam of constant but adjustable intensity to act both as a standard source for intensity measurements and as a completely polarized beam for polarization standardization. When the light from S_2 is being used for standardization the cuvette is removed and the beam from S_1 is cut off by a shutter between A_1 and F_2. When measurements on solutions are being made S_2 is automatically shut off by

Stopped-Flow Fluorescence Polarometer Design and Construction

This instrument is designed to give direct readouts of both fluorescence intensity and polarization as functions of time. It was constructed utilizing the mechanical components and flow system of the Durrum–Gibson stopped-flow apparatus (Durrum Instrument Corporation, Palo Alto, California). This apparatus is rugged, provides rapid mixing, has a low deadtime, and incorporates Gibson's hydraulic stopping mechanism, which eliminates cavitation due to stopping. In addition, the mechanism driving the syringes is operated on compressed gas, thus giving reproducible flow and mixing conditions. The essential changes and additions that were made include: (1) installation of a high-intensity mercury arc source above the instrument thus allowing an image of the arc to be focused directly into the flow cell; (2) use of two photomultiplier tubes, one viewing each end of the flow cell; (3) triggering both the oscilloscope and the solenoid controlling the driving mechanism by two separate signals generated by the triggering and control circuits; both of these signals are generated in an adjustable pattern subsequent to pressing the start button; (4) use of the stopping syringe to produce a mechanical stop only; no signal is taken from this syringe; (5) design and construction of detection and analog signal processing to directly read out intensity and polarization as functions of time on a storage oscilloscope. A photograph of the entire

---

the main selector switch and the shutter between A_1 and F_2 is in the open position except for the dark current setting. The fluorescent beam emerging from C_1 is focussed near the photocathode of D_1 by lens L_3. Interference filter (F_4) removes the incident wavelength, and P_3 the rotating polarizer allows D_1 to view a sinusoidally varying intensity the amplitude of which depends upon the difference between v and h in the fluorescent beam. Motor (M_1) is synchronous and rotates at 300 rpm. Source and detector assembly (S_3) and (D_2) shown on a magnified scale in the inset provide a phase reference signal for the lock-in (synchronous) detector (see Fig. 13). The parts list includes: S_1—200 W Hg arc lamp, Illumination Industries Type 202 mounted in an LH-350 housing driven by a CA-200 power supply; L_1 is part of the housing and focuses the arc image on A_1; F_1—Corning heat absorbing glass filter; A_1—rectangular aperture 6×20 mm, the image of which is focused on C_1 by biconvex lens L_2; F_2—50 mm diameter interference bandpass filter (IR Industries Waltham, Massachusetts); P_1, P_2—25 mm diameter 105UVwMR Polacoat Polarizing Filter; C_1—10 mm square quartz cuvette; F_3—25 mm diameter opal glass; L_3—25 mm diameter quartz biconvex lens, which focuses the image of C_1 on D_1; S_2—pinlite 60–20 lamp, which provides a calibration reference when the cuvette is removed (REFAC Electronics, Arkamsted, Ct.); F_4—50 mm diameter interference bandpass filter (IR Industries, Waltham, Massachusetts); P_3—polarizer/chopper wheel consisting of a circular piece of HN-38 (Polaroid Corp. Cambridge, Massachusetts) mounted between glass plates; D_1—RCA 1P 21 photomultiplier tube; D_2—1N2175 photodiode; S_3—No. 46 miniature lamp; M_1—KYC-22 300 RPM motor (Bodine Corp. Chicago, Illinois).

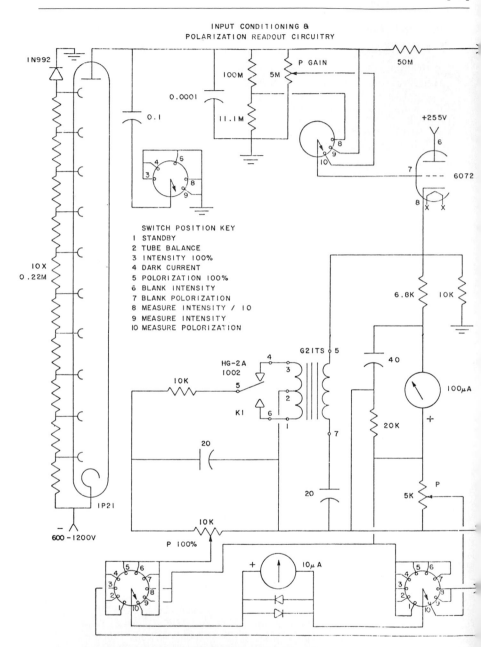

FIG. 12. Manual fluorescence polarometer—input conditioning and polarization readout circuit. The principle of this instrument involves the generation of a sinusoidal output from the photomultiplier as it views the fluorescence from the solution cuvette through a rotating

apparatus is shown in Fig. 14. Details of the optical layout and the electronics are given in Figs. 15–19.

Performance. Performance criteria for stopped-flow instruments have been discussed by Roughton and Chance,[21] Gibson,[22] and Sturtevant,[23] as well as in the technical notes from Durrum Instrument Co. The various factors that require special attention in stopped-flow instruments generally are as follows:

1. Mixing, which may be characterized by the time required to achieve a given degree, e.g., 99.5%, of mixing. The speed and adequacy of mixing can be determined by mixing solutions of widely differing refractive index, e.g., 5 M NaCl and water. The turbidity resulting from residual local differences in refractive index furnishes an empirical measure of the departure from complete mixing. For the instrument being described the point of 99.5% mixing is achieved in 2 msec, according to measurements by the manufacturer.

2. Time resolution; this is a complex composite parameter depending largely upon the time constants in the signal detecting and processing circuits and to a minor extent upon the difference in age of different portions of solution lying within the field of observation at any one time. Time resolution can be assessed by measuring the apparent rate constants

[21] F. J. W. Roughton and B. Chance, "Technique of Organic Chemistry" (S. L. Friess, E. S. Lewis, and A. Weissberger, eds.), Vol. VIII, p. 703. Wiley (Interscience) New York, 1963.
[22] Q. H. Gibson and L. Milnes, *Biochem. J.* **91,** 161 (1964).
[23] J. M. Sturtevant, *in* "Rapid Mixing and Sampling Techniques in Biochemistry" (B. Chance, Q. H. Gibson, R. H. Eisenhardt, and K. K. Lonberg-Holm, eds.), p. 89. Academic Press, New York, 1964.

polarizer. The resulting signal is of the form $(v - h)/2 \cos \omega t + (v + h)/2$. In the "Intensity" mode a capacitor is used to smooth out the ac term giving $(v + h)/2$. In the "Polarization" mode, a phase synchonous detector (lock-in detector) is used to extract the ac term. The ac component of the signal is separated from the dc component by capacitor coupling. The ac portion is rectified by relay K_1 driven by the phase reference signal. The average value of the rectified signal is then measured by the potential across the 20 μfd integrating capacitor. In order to obtain the polarization the dc component is applied to the entire length of the readout helipot and the rectified ac term is applied through the null meter to the slider which senses the current flow. The parts lists includes: K_1—Mercury wetted relay, H6-2A-1002 (Clare Corp. Chicago, Illinois); chopper input geoformer, G-21TS (TRIAD-UTRAD, Huntington, Indiana); P—10-turn 5K panel mount potentiometer with a 10-turn calibrated dial; P100%, P GAIN—single turn panel mount potentiometers; 10 μA meter—5-0-5 dc microammeter with 1N914 diodes for protection; the rotary switch sections are all ganged together with those in Fig. 13 to form a single 6-pole 10-position switch; 100, 50, and 11.1 Megohm resistors—1% tolerance; all other resistors are ½ W 10% tolerance with values in ohms; capacitors—400 V with values in μf; connections 1 through 4—see Fig. 13.

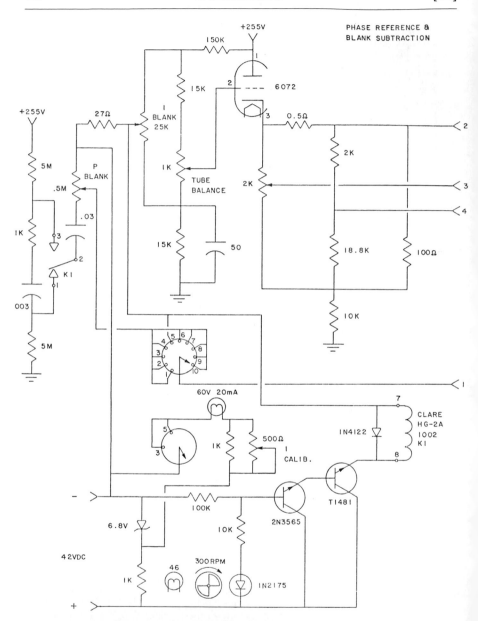

FIG. 13. Manual fluorescence polarometer. Blank generation, phase reference, and "intensity" readout potentiometer. The phase reference is generated by the No. 46 lamp, the rotating polarizer, and the 1N2175 assembly. The resultant signal drives the lock-in detector and also generates a phase-locked ac signal which is used to compensate that from the polarization blank. The intensity blank is generated by a simple dc takeoff from the 250 V

Fig. 14. Overall view of the stopped-flow fluorescence polarometer.

of a first-order or pseudo-first-order reaction at a series of different concentrations so as to obtain a series of higher and higher rates. A falloff in the value of the "constants" so obtained indicates the range of rates in which the time resolution of the instrument becomes inadequate. This test presupposes that a suitable test reaction is available and that it remains first order over all ranges of interest. On the other hand, such a reaction probably must be studied by flow techniques to determine the order in the first place. A partial solution to this dilemma is to investigate the order of reaction with varying instrument response times and to thus delineate the response time settings appropriate to rates of a given magnitude. In de-

supply for the 6072 tube. The blank compensating signals are subtracted from the output of the photomultiplier by current summing. In operation the selector switch is used to select between various calibrations and the polarization and intensity measurement modes. The parts list includes: K_1—see Fig. 12; S_2—see Fig. 11; P blank, I blank, and tube balance—single-turn panel-mount potentiometers; I and I Calibration—10-turn panel-mount potentiometers with calibrated 10-turn dials; S_3 is connected to a continuous source of 6.3 VAC; capacitors—400 VDC; values in μf; resistors—rated at $\pm10\%$, ½ W with values in ohms; switch sections—see Fig. 12; connections 1 through 4—see Fig. 12.

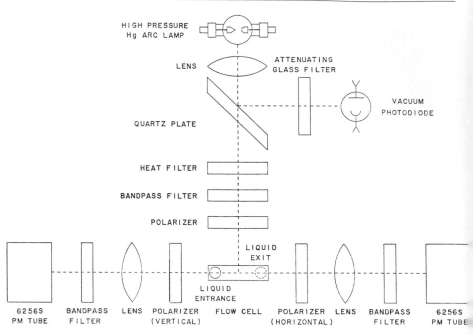

HIGH PRESSURE
Hg ARC LAMP

LENS   ATTENUATING
                        GLASS FILTER

QUARTZ PLATE   VACUUM
                                         PHOTODIODE

HEAT FILTER

BANDPASS FILTER

POLARIZER

                                    LIQUID
                                     EXIT

                                    LIQUID
                                   ENTRANCE

6256S   BANDPASS   LENS   POLARIZER   FLOW CELL   POLARIZER   LENS   BANDPASS   6256S
PM TUBE   FILTER   (VERTICAL)   (HORIZONTAL)   FILTER   PM TUBE

FIG. 15. Optical layout for the stopped-flow fluorescence polarometer. The excitation beam is focused into the flow cell by a single quartz lens. Before reaching the cell the light is filtered and polarized with the electric vector perpendicular to the long axis of the flow cell. A sample of the incident beam is diverted by a quartz plate and after attenuation falls onto a vacuum photodiode, which supplies a reference signal to the blanking circuit (compensator of Fig. 16). Ideally the filtering of the reference beam should be identical to that of the main excitation beam to allow for time-dependent changes in the spectral output of the lamp. The polarizers are UV transmitting 105 UVWRMR (Polacoat Corp. Cincinnati, Ohio). The two fluorescence beams exiting from the end windows of the flow cell are focused after polarization and filtering near the photocathodes of the two photomultiplier tubes. The lenses here are also quartz to allow measurements in the UV or the visible region. The two polarizers in the emission beams are oriented mutually perpendicularly. The parts list includes: EMI 62565 photomultiplier and dynode chains for dc and for frequency measurements (Gencom Division Emitronics, Inc. Plainview, New York); lamp—a type 112 (Illumination Industries, Sunnyvale, California) 100-Watt Hg high-pressure arc lamp mounted in an LH 350 lamp housing with integral focusing lens; the lamp is driven by a standard 100 W arc lamp power supply; attenuating glass filter—welding glass; glass heat absorbing filter—Corning; interference bandpass filter—25 mm diameter (IR Industries, Waltham, Massachusetts); exit lenses—biconvex quartz; flow cell—stopped-flow system modified for the additional over-head light path and dual exit paths (Durrum, Palo Alto, California).

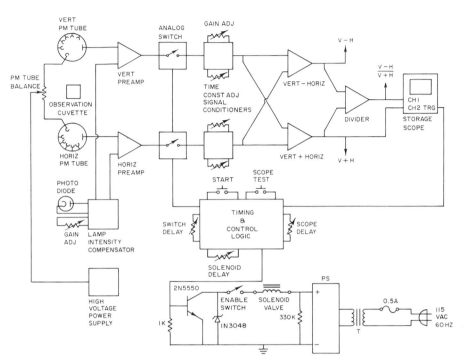

FIG. 16. Block diagram of the electronics for the stopped-flow fluorescence polarometer. Signals are received by the two photomultiplier tubes and conditioned, the sum and difference taken, and the ratio of the sum and difference as well as the sum are displayed by a storage oscilloscope. The conditioning includes subtraction of blanks with compensation for lamp intensity, manually adjustable time constant selection and suppression in the analog switch of any large transients possibly occurring before a small signal. The sequence of events is initiated by manually releasing (opening) the start switch. This initiates independently adjustable delays for the solenoid valve, analog switch, and scope trigger. Ancillary adjustments are made to control the relative gain of the two photomultiplier channels to allow for differences in the tubes and associated circuitry. The adjustment involves setting the output polarization to a known value for a calibration material contained in the cuvette. The $v + h$ output is linear regardless of the accuracy of the setting of the polarization output. The parts lists includes: high voltage power supply—Model 413C (Fluke, Seattle, Washington) Tektronix storage Oscilloscope 564 B with 3A3 preamplifier and 2B67 time base; T—Triad N-49X isolation transformer; PS—Durrum solenoid valve power supply; solenoid valve—standard Durrum unit; enable switch—toggle switch to prevent any possibility of accidental triggering during adjustments; resistors—¼ W 5% with values in ohms.

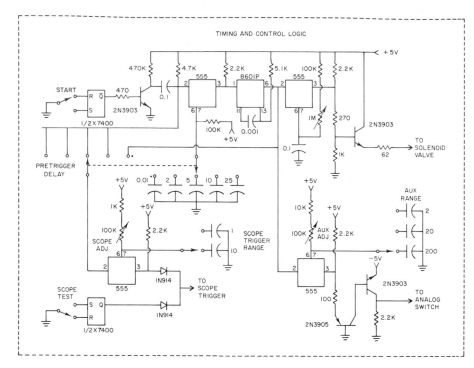

FIG. 17. Stopped-flow fluorescence polarometer timing and control circuitry. Releasing the start switch initiates the timing sequence and provides a scope trigger signal with adjustable delay. In addition it provides a signal for the activation of the solenoid valve and of the analog switch (activated at the same time). Thus the scope can be triggered before or after triggering the solenoid valve and analog switch. The parts list includes: 8601 P—monostable multivibrator (Motorola Semiconductor Products, Phoenix, Arizona); 555—timer (Signetics Corp., Sunnyvale, California); 1 Megohm potentiometer—Bourns 3059 P; other potentiometers—10-turn panel-mount units; capacitors—25 VDC, values are in μf; resistors ±5%, ¼ W, values in ohms; standard connections necessary for the operation of the 555, 8601P and 7400 are not shown here—see Signetics Data Book (1974) Signetics Corp.; Microelectronic Data Book (1969) Motorola Semiconductor Products Inc.

signing the electronics of the stopped-flow apparatus provision was made to vary the time constant manually from 10^{-6} to 0.5 sec.

3. Deadtime, which may be thought of as the "age" of the reaction mixture when observations can first be made, consists of contributions both from the time of transport of solution from the mixing chamber to the "point" of observation and also from time spent in the mixing chamber itself. The variation of the age of solution through the region of observa-

tion is often included in deadtime, but it has seemed somewhat more logical to consider this factor as a special kind of limit on time resolution. The deadtime may be assessed by measurement of the fraction of the total reaction that is observed at different flow rates. As an internal check, the extrapolation to infinite flow rate should give the same total change of the measured parameter as is obtained from a static experiment. Know-

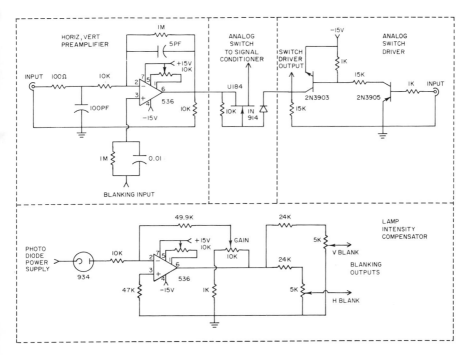

FIG. 18. Stopped-flow fluorescence polarometer. Schematics for the initial conditioning of the photomultiplier outputs. Each photomultiplier has associated with it a preamplifier and analog switch, as can be seen in Fig. 16. Both channels are driven by the same analog switch driver, which converts the level of the timing pulse to reliably drive each analog switch. The lamp intensity compensator provides two dc signals independently adjustable and each proportional to lamp intensity. Each of these signals is subtracted from that of one of the photomultiplier channels. The adjustments are made so that the signals from the two channels are nulled out (by dc balancing) when only buffer is present in the cell. The parts list includes: V blank and H blank—10-turn panel-mount potentiometers; all other potentiometers are Bourns 3059P PC mount; the photo diode power supply provides 90V dc ±2% at 1 mA max; 536 (Signetics); U184—FET (Siliconix, Santa Clara, California); 934—vacuum photodiode (RCA); capacitors—μf unless otherwise noted 50 VDC; resistors—$\frac{1}{2}$ W ±2% except for the 49.9K, which is 1% RN60C, values in ohms; standard power supply bypassing is not shown.

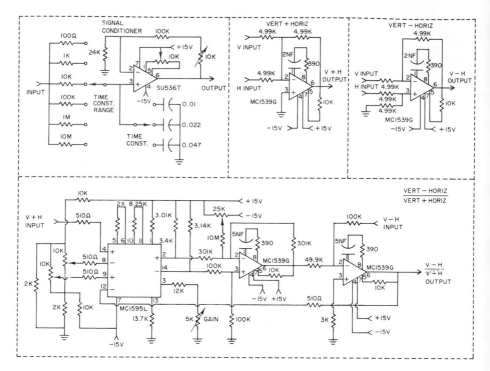

FIG. 19. Stopped-flow fluorescence polarometer. Schematics for the latter part of the signal conditioning and for analog computation. Each channel requires one signal conditioner which provides an adjustable time constant and a gain trim adjustment. The two conditioned signals (v and h) feed separately into both the "vert + horiz" and "vert-horiz" modules. The sum and difference are fed into the inputs of the divide module. The parts list includes: potentiometers—Bourns 3059P PC mount; all resistors in feedback or gain setting paths are ±1%; all other resistors are ±2%; all resistors are ¼ or ½ W; MC1539G, MC1595L, (Motorola); SU536T—(Signetics); capacitors—50VDC, are in μf unless otherwise noted.

ing the fraction of reaction observed at a particular flow rate for a first-order test reaction of known rate constant determines the deadtime at that flow rate. Obviously, the deadtime must be kept small compared to the half-time for reaction. The deadtime automatically eliminates the initial part of a reaction from study. For first-order reactions the loss of the initial phases of reaction are not so important as for second- or higher-order reactions, since a first-order rate constant can be obtained from measurements on any phase of the reaction. For second- or higher-order reactions where the measurement of initial rates gives valuable information, it is important that the deadtime be as short as possible. If in the study of a reaction initial rates are found to be consistent with second-order kinetics, for example, over a range of concentrations, and if at higher concentrations the observed rates are less than expected, it may be that the deadtime is the source of the deviation. Another useful way to assess the importance of deadtime is to make initial rate measurements at varying flow velocities. Measurements should be reliable if the initial rates are found to be independent of flow rate over the range used. The sensitivity of this stopped-flow system is such that useful measurements can be obtained down to about $10^{-10} M$ fluorescein or its equivalent with the time constant of the instrument set at 100 msec. Experimental results utilizing this stopped-flow system have been reported by Levison et al.[16,18,20]

[19] Quasi-Elastic Laser Light Scattering

By VICTOR A. BLOOMFIELD and TONG K. LIM

Quasi-elastic light scattering (QLS) is used in enzymology, and in other areas of molecular biology and macromolecular chemistry, to study the dynamics of macromolecules. Its predominant use has been to measure translational diffusion coefficients, which can be interpreted in terms of Stokes radii and combined with sedimentation coefficients in the Svedberg equation to yield molecular weights. Other applications are experimentally more difficult and therefore less common, though they may be promising in favorable cases. These include measurement of rotational diffusion coefficients of nonspherical scatterers, conformational changes, e.g., helix-coil transitions, kinetics of biomolecular reactions, measurements of intramolecular flexibility, molecular motion in gels, bacterial motility, muscular contraction, electrophoretic mobility, protoplasmic streaming, conduction of nerve impulses, and flow of blood and other biological fluids.

The frequency dependence of scattered light on dynamic molecular processes was predicted and observed in the 1920s in such manifestations as Raman and Brillouin scattering, and in 1926 Mandel'shtam[1] pointed out that translational diffusion coefficients of polymers could be obtained from the spectrum of their scattered light. Implementation of this idea had to wait, however, until the development in the early 1960s of lasers that emit highly monochromatic, coherent light. Application to macromolecular

[1] L. I. Mandel'shtam, Zh. Russ. Fiz.-Khim. Obshchest. 58, 381 (1926).

problems really dates from 1964, when Pecora[2] developed a theory to explain the spectrum of scattered light in terms of the motion of the dissolved macromolecular scatterers, and Cummins et al.[3] measured the diffusion coefficient of polystyrene latex spheres from the spectral broadening of the scattered light. In 1967, Dubin et al.[4] made the first application of QLS to the measurement of diffusion coefficients of proteins. Since then, there has been an explosive growth in the subject and a corresponding explosion in the number of reviews written about it. For a broad survey of applications, and references to earlier reviews, the reader is referred to Carlson.[5] Books about QLS have been written or edited by Chu,[6] Cummins and Pike,[7] and Berne and Pecora.[8] Other major reviews are those by Cummins and Swinney,[9] Benedek,[10] and Dubin.[11]

Introductory Survey

Because the principles of QLS are relatively complex and unfamiliar, it is well to preface a detailed, formal treatment with a more intuitive discussion of basic ideas.

Physical Basis of QLS

QLS is perhaps most easily understood as a form of Doppler shift spectroscopy. Doppler shifts in the frequency of scattered light arise because the scatterers are not stationary. Instead, they move with velocity related to diffusion, electrophoresis, internal flexibility, or other modes of motion. Although rigorous treatments of these cases will be given below, a "back-of-the-envelope" calculation for the most important and common case, translational diffusion, is useful at this point.

Let the frequency of the incident light be ν_0 and the Doppler shift be $\Delta\nu$, corresponding to velocity v of a scatterer. If the refractive index of the

[2] R. Pecora, *J. Chem. Phys.* **40**, 1604 (1964).
[3] H. Z. Cummins, N. Knable, and Y. Yeh, *Phys. Rev. Lett.* **12**, 150 (1964).
[4] S. B. Dubin, J. H. Lunacek, and G. B. Benedek, *Proc. Natl. Acad. Sci. U.S.A.* **57**, 1164 (1967).
[5] F. D. Carlson, *Annu. Rev. Biophys. Bioeng.* **4**, 243 (1975).
[6] B. Chu, "Laser Light Scattering." Academic Press, New York, 1974.
[7] H. Z. Cummins and E. R. Pike, "Photon Correlation and Light Beating Spectroscopy." Plenum, New York, 1974.
[8] B. J. Berne and R. Pecora, "Dynamic Light Scattering with Application to Chemistry, Biology, and Physics." Wiley (Interscience), New York, 1976.
[9] H. Z. Cummins and H. L. Swinney, *Prog. Optics* **8**, 133 (1970).
[10] G. B. Benedek, *in* "Brandeis Lecture in Theoretical Physics, 1966" (M. Chrétien, E. P. Gross, and S. Deser, eds.), Vol. 2, p. 1. Gordon & Breach, New York, 1968.
[11] S. B. Dubin, this series Vol. 26 [7].

solvent medium is n, the speed of light in the medium is c/n, where c is the speed of light *in vacuo*. Then

$$\Delta\nu/\nu_0 = \nu n/c \tag{1}$$

For a diffusing particle, ν cannot be defined precisely, but is on the order of

$$\nu \approx \langle x^2 \rangle^{1/2}/\tau \tag{2}$$

where $\langle x^2 \rangle^{1/2}$ is the rms displacement of a Brownian particle in time τ. According to the familiar Einstein relation for one-dimensional diffusion,

$$\langle x^2 \rangle = 2D\tau \tag{3}$$

where D is the translational diffusion coefficient. The frequency shift is detected as a "beat" between the unshifted and shifted light. In order for at least one beat to be observed, the particle must diffuse for a time

$$\tau \approx 1/\Delta\nu \tag{4}$$

Combining Eqs. (1)–(4) and rearranging, we find

$$\Delta\nu \approx 2(n/\lambda_0)^2 D \tag{5}$$

since λ_0, the wavelength of the incident light *in vacuo*, equals c/ν_0.

According to Eq. (5), the Doppler shift is proportional to the translational diffusion coefficient, and therefore D may be determined by measuring $\Delta\nu$. To make an order of magnitude estimate of the spectral shift to be expected, we let $D = 6 \times 10^{-7}$ cm²/sec, typical of a moderate-sized enzyme, $\lambda_0 = 632.8$ nm for helium–neon laser light, and $n = 1.33$ for aqueous solutions. Then $\Delta\nu \approx 500$ Hz, so that spectral shifts for diffusing biopolymers are in the kHz range, and can be detected by audiofrequency electronics. The spectral resolution of this experiment is, with quantities expressed in cgs units,

$$\nu_0/\Delta\nu = (c/\lambda_0)/\Delta\nu \approx (3 \times 10^{10}/6.328 \times 10^{-5})/5 \times 10^2 \approx 10^{12}$$

The special instrumentation required to achieve this enormous resolution is described below.

Another way to look at this experiment is to combine Eqs. (4) and (5) to obtain

$$(\lambda_0/n)^2 = 2D\tau \tag{6}$$

and compare this with Eq. (3). This shows that the mean distance traveled by the particle during the time required for the spectral shift to be manifested as a beat is on the order of the wavelength of light. This is also the distance that the particle must move in order for the light scattered from it to change its phase significantly. That is, again considering only a one-dimensional case for simplicity, the electric field E_s scattered from a particle located at position x in response to incident radiation of frequency ν_0 is

$$E_s = A \exp\{2\pi i[\nu_0 t - nx/\lambda_0]\} \qquad (7)$$

where A is a scattering amplitude. The phase of E_s changes by π (that is, the electromagnetic wave goes from a peak to a trough) when x changes by λ_0/n. We shall see below that the time dependence of this phase change can be measured directly by time correlation analysis of the photocurrent produced when the scattered light falls on the surface of a photomultiplier tube. The correlation time, that is, the time required for the phase of the scattered light to decay from its initial value, is inversely proportional to D, as can be seen by comparison of Eqs. (4) and (5).

These two physical pictures of the effect of molecular motion on scattered light—frequency shifts due to the Doppler effect, or loss of phase correlation due to molecular displacement—are complementary. They lead to two different experimental approaches, spectrum analysis and time correlation analysis, which are related to each other in much the same way as are conventional frequency-scanning infrared (IR) or nuclear magnetic resonance (NMR) spectroscopy with the corresponding Fourier transform spectroscopy. Although considerations of instrumental availability, signal-to-noise ratio, rapidity of data acquisition, or dynamic range may favor one or the other of these approaches, the information obtainable is in principle equivalent and interconvertible.

In either case, the foregoing analysis has exposed one of the major advantages of QLS, relative to traditional synthetic boundary methods, for measuring translational diffusion coefficients. In QLS, a typical molecule must move a distance only on the order of the wavelength of light; while in synthetic boundary experiments the boundary must broaden, or a typical molecule within it must move, a macroscopic distance (ca. 0.1 cm) for a reliable measurement of D to be made. From Eq. (3) we see that the ratio of the times is the ratio of the squares of the distances in these two measurements. With $\langle x^2 \rangle \approx (10^{-5} \text{ cm})^2$ in the QLS experiment, and $\langle x^2 \rangle \approx (10^{-1} \text{ cm})^2$ in the synthetic boundary experiment, the ratio of the characteristic times is on the order of 10^8. Even though extensive signal averaging, typically on the order of 10^5 to 10^6 accumulations of the beat frequency spectrum or correlation function, is required to obtain adequate signal-to-noise ratio in the QLS experiment, the time advantage is still several orders of magnitude. Thus QLS determination of D requires only seconds or minutes, compared with hours or days for synthetic boundary techniques.

Equation (5), or its time-domain counterpart, is only an order of magnitude estimate. It neglects the facts that only the component of particle velocity in the direction of the scattering vector will contribute to the observed Doppler shift, and that each particle in the solution will have a different instantaneous velocity so that a spectrum or distribution of Doppler shifts will be observed. A more complete and rigorous treatment will be presented below.

General Scope of Applicability

Perhaps the two most important questions that must be answered before embarking on QLS study of a macromolecular system are: Over what particle size range can useful data be obtained? How much material and time will the experiment take to achieve a satisfactory signal-to-noise ratio? We anticipate later results to give general answers to these questions here.

We shall show below that in translational diffusion coefficient determinations using homodyne analysis, the beat frequency spectrum has a Lorentzian shape with half-width at half-height

$$\Delta\nu_{1/2} = q^2 D/\pi \tag{8}$$

where

$$q = (4\pi n/\lambda_0)\,\sin(\theta/2) \tag{9}$$

and θ is the scattering angle. Correspondingly, the homodyne photocurrent autocorrelation function is a decaying exponential with correlation time

$$\tau = (2q^2 D)^{-1} \tag{10}$$

Present-day spectrum analyzers have useful spectral ranges between about 50 Hz and 500 kHz; correlation analyzers allow measurement of correlation times roughly the reciprocal of these frequencies. Taking θ to lie between 30° and 150°, which can be achieved without undue attention to optical problems in the scattering path; $\lambda_0 = 632.8$ nm for He–Ne laser light; and $n = 1.33$ for aqueous solutions, q^2 ranges between 4.67×10^9 cm^{-2} and 6.51×10^{10} cm^{-2}. Thus the measurable range of D (scattering intensity considerations are neglected for the moment) extends from 3.3×10^{-4} cm^2/sec at the high-frequency, low-angle limit to 2.4×10^{-9} cm^2/sec at the low-frequency, high-angle limit. Assuming spherical particles, for which D is related to the hydrodynamic radius R_h by the Stokes–Einstein equation

$$D = kT/6\pi\eta_0 R_h \tag{11}$$

where k is Boltzmann's constant, T the absolute temperature, and η_0 the solvent viscosity, these values of D correspond at 20° in water to $R_h = 6.5 \times 10^{-10}$ cm and 8.9×10^{-5} cm. That is, the size range potentially encompassed extends from small solvent molecules up to particles about 1 μm large, such as small bacterial cells. This range easily includes all known enzymes and most multisubunit complexes of finite dimensions. Indeed, the list of biological macromolecules whose translational diffusion coefficients have been measured by QLS extends from lysozyme and ribonuclease to F-actin and T4 bacteriophage.

For applications of QLS other than translational diffusion, the pertinent time scale also extends from about 10^{-5} to 10^{-1} sec.

The protein concentration required to obtain adequate precision in a QLS measurement of D depends in a complicated way on the detection scheme, laser power, time of signal accumulation, particle molecular weight, and other factors which will be discussed in detail later. Here we may give examples of the results achievable in current practice. With digital autocorrelation of singly clipped photon counting fluctuations from protein solutions illuminated with a 30 mW He–Ne laser, Foord et al.[12] obtained 1% precision in D for hemocyanin from *Murex trunculus* ($M_r = 9.2 \times 10^6$) at a protein concentration of 0.5 mg/ml in 30-sec measurement time. With the much smaller protein lysozyme ($M_r = 1.4 \times 10^4$) at 3 mg/ml, 2 hr were required to obtain D within ±4%. In homodyne scattering experiments such as these, in which the scattered light is mixed with itself on the surface of the photomultiplier tube, the amplitude of the autocorrelation function depends on the square of the molecular weight and the square of the weight concentration. In general, it may be stated that, to obtain diffusion coefficients of moderate-sized proteins ($M_r \leq 10^5$) accurate to ±2–4%, concentrations on the order of 10–30 mg/ml and signal accumulation times on the order of 10–60 min will be required using autocorrelation or real-time spectral analysis. For larger particles ($M_r > 10^7$), ±1 to 2% accuracy can be obtained with 0.5 mg/ml solutions in under 1 min.

Measurement of translational diffusion coefficients is by far the most common and reliable application of QLS, and we have emphasized this in this chapter. Among the other uses listed at the beginning of this article, electrophoresis is perhaps of broadest interest. The signal-to-noise considerations for this type of experiment are similar to those for conventional QLS. Problems in cell design are being overcome, as discussed below, and electrophoretic light scattering is emerging as a useful way to

[12] R. Foord, E. Jakeman, C. J. Oliver, E. R. Pike, R. J. Blagrove, E. Wood, and A. R. Peacocke, *Nature (London)* **227**, 242 (1970).

determine simultaneously the electrophoretic mobilities and diffusion coefficients of biological macromolecules.[13]

Rotational diffusion coefficients can be measured for elongated or optically anisotropic molecules by either polarized or depolarized QLS. The latter is generally superior, though plagued by low signal intensities. Details are discussed below, and in the previous review on this topic by Dubin.[11] It seems fair to say, however, that unless an investigator already has a functioning QLS laboratory, measurements of rotational diffusion are in most cases more handily made by other techniques, such as fluorescence depolarization or electric or flow birefringence.

Although QLS can, in principle, give useful information on reacting systems and internal motions in flexible polymers, in practice these applications have been fraught with difficulty, due to problems in dissecting exponential or Lorentzian signals with multiple components. The theoretical bases for interpretation of scattering from these types of systems are given below and in several reviews.[8,14,15]

Some interesting results have been obtained from observation of contributions of number fluctuations to the scattering spectrum.[16,17] These become important when the mean number $\langle N \rangle$ of scattering particles in the scattering volume is not too much greater than unity (say $\langle N \rangle \lesssim 10^4$), so that the relative fluctuation $\delta N / \langle N \rangle = \langle N \rangle^{-1/2}$ is measurable. Applications of number fluctuation spectroscopy have been made, for example in studies of bacterial motility.[18] Although the motility of cells is not directly a problem in enzymology, the underlying molecular mechanisms of motility and chemotaxis are certainly enzymic, and measurements at the level of the intact, functioning cell may shed light on the basic molecular processes. Therefore, the inclusion of some theory and methodology for number fluctuation experiments has been deemed appropriate for this chapter.

Theoretical Principles of Quasi-Elastic Laser Light Scattering

In this theoretical section, we examine the basic principles of QLS, obtaining equations by which particular types of experiments may be analyzed. The theoretical treatment is substantially more extensive than in most articles in this series, both because the theory is more complex

[13] B. R. Ware, *Adv. Colloid Interface Sci.* **4**, 1 (1974).
[14] Y. Yeh and R. N. Keeler, *Quart. Rev. Biophys.* **2**, 315 (1969).
[15] V. A. Bloomfield and J. A. Benbasat, *Macromolecules* **4**, 609 (1971).
[16] D. W. Schaefer and B. J. Berne, *Phys. Rev. Lett.* **28**, 475 (1972).
[17] E. L. Elson and W. W. Webb, *Annu. Rev. Biophys. Bioeng.* **4**, 311 (1975).
[18] D. W. Schaefer, *Science* **180**, 1293 (1973).

and less familiar than for most biochemical techniques, and because there are many different kinds of experiments that fall under the general rubric of QLS. In many respects, our treatment follows that of Cummins and Swinney.[9]

General Relations between Scattered Light and Photocurrent

In a QLS experiment, coherent laser light illuminates a volume in a scattering cell. The scattered light, and in some cases a portion of the unscattered laser light as well, falls on the surface of a photomultiplier tube, giving rise to a photocurrent. The behavior of the photocurrent is analyzed in the time or frequency domain and from this the properties of the scattering solution are inferred.

If the electric field of the light falling on the photocathode at time t is $E(t)$, then the intensity is

$$I(t) = E^*(t)E(t) = |E(t)|^2 \tag{12}$$

and the probability per unit time of photoelectron emission from the photocathode is

$$W^{(1)}(t) = \sigma I(t) \tag{13}$$

where σ is the quantum efficiency of the photodetector. The resulting instantaneous photocurrent is

$$i(t) = e\gamma W^{(1)}(t) = e\gamma\sigma I(t) \tag{14}$$

where e is the electron charge and γ is the gain of the photomultiplier tube. The joint probability per unit time that one photoelectron will be emitted at t and another at $t + \tau$ is

$$W^{(2)}(t,t + \tau) = \sigma^2 E^*(t)E(t)E^*(t + \tau)E(t + \tau) = \sigma^2 I(t)I(t + \tau) \tag{15}$$

If the field is stationary, i.e., on the average has the same behavior regardless of the time of measurement (this will be the case in all situations we describe), then the time-average photocurrent is

$$\langle i(t) \rangle = \langle i \rangle = e\gamma\sigma\langle I \rangle \tag{16}$$

and the average of $W^{(2)}(t,t + \tau)$ is

$$\langle W^{(2)}(t,t + \tau) \rangle = \sigma^2\langle I \rangle^2 g^{(2)}(\tau) \tag{17}$$

where the normalized second-order correlation function is

$$g^{(2)}(\tau) = \langle I(t)I(t + \tau) \rangle / \langle I \rangle^2 \tag{18}$$

The autocorrelation function for the photocurrent is the quantity actually measured. It is

$$C_i(\tau) = \langle i(t)i(t + \tau) \rangle \tag{19}$$

The photocurrent consists of a series of very narrow pulses. Therefore, there are two contributions to $C_i(\tau)$, one arising from different photoelectrons at t and $t + \tau$, and the other from the same electron at t and $t + \tau$. This latter, of course, can occur only when $\tau = 0$, and represents shot noise from randomly emitted electrons. Therefore,

$$C_i(\tau) = e\langle i \rangle\delta(\tau) + \langle i \rangle^2 g^{(2)}(\tau). \tag{20}$$

$\delta(\tau)$ is the delta function, equal to 1 when $\tau = 0$ and zero otherwise.

The electric field is itself characterized by an autocorrelation function

$$C_E(\tau) = \langle E^*(t)E(t + \tau) \rangle = \langle I \rangle g^{(1)}(\tau) \tag{21}$$

The normalized first-order correlation function $g^{(1)}(\tau)$ gives information on the dynamics of the scattering medium. We are interested in determining it from $g^{(2)}(\tau)$.

Two kinds of detection may be employed: homodyne or heterodyne. In homodyne detection, only scattered light falls on the phototube. It beats against itself, and homodyne detection is therefore sometimes called self-beat detection. If E is a Gaussian random process, as will be the case, for example, for a dilute solution of independent scattering macromolecules, then the relation between the normalized first-order and second-order correlation functions may be shown to be[19]

$$g^{(2)}(\tau) = 1 + \left| g^{(1)}(\tau) \right|^2 \tag{22}$$

For fields that do not obey Gaussian statistics, such as those arising from scattering solutions in which particle motions are strongly correlated, or in which there are only a few scatterers so that the central limit theorem is not valid, Eq. (22) will not hold. However, for the translational diffusion measurements which form the bulk of applications of QLS in biochemistry, Eq. (22) is quite satisfactory. Then the photocurrent autocorrelation function is the sum of three terms

$$C_i^{\text{homo}}(\tau) = e\langle i \rangle\delta(\tau) + \langle i \rangle^2 + \langle i \rangle^2 \left| g^{(1)}(\tau) \right|^2 \tag{23}$$

The first is the shot noise, the second is the constant, dc component, and the third contains the dynamic information of interest regarding the scattering system.

In heterodyne detection, the scattered light is mixed with a constant-intensity, coherent local oscillator signal, which is usually a portion of the

[19] L. Mandel, *Prog. Optics* 2, 181 (1963).

irradiating laser light. Then the electric field at the surface of the photo-cathode is

$$E(t) = E_S(t) + E_{LO}(t) \tag{24}$$

where S and LO refer to the scattered light and local oscillator. E_{LO} may be adjusted so it is much greater than E_S. Then it is found that the photo-current autocorrelation function is, neglecting terms of order $\langle i_S \rangle^2$,

$$C_i^{\text{het}}(\tau) = ei_{LO}\,\dot{\delta}(\tau) + i_{LO}^2 + i_{LO}\langle i_S \rangle \, [e^{i\omega_{LO}\tau}g^{(1)}(\tau) + e^{-i\omega_{LO}\tau}g^{(1)*}(\tau)] \tag{25}$$

Here i_{LO} is the photocurrent due to the local oscillator alone, $\langle i_S \rangle$ is the average photocurrent due to the scattered light alone, ω_{LO} is the frequency (rad/sec) of the local oscillator, and $g^{(1)}(\tau)$ is the normalized autocorrelation function of the scattered electric field. Again $C_i(\tau)$ contains three terms: shot noise, dc, and that related to scattering dynamics. However, we see that in heterodyne detection, $g^{(1)}(\tau)$ is obtained directly, rather than $g^{(2)}(\tau)$. Therefore, no assumptions are required about the statistics of the scattering sample.

If experiments are performed in the frequency domain using a spectrum analyzer, rather than in the time domain using a correlator, then the quantity measured is the power spectrum $P_i(\omega)$ of the photocurrent. That is, $P_i(\omega)d\omega$ is the power contained in the photocurrent output in the spectral range between ω and $\omega + d\omega$. According to the Wiener–Khintchine theorem,[20] the power spectrum and current autocorrelation function are related by the Fourier transformation

$$P_i(\omega) = 1/2\pi \int_0^\infty C_i(\tau)e^{i\omega\tau}\, d\tau \tag{26}$$

We shall see below that the normalized first-order correlation functions of concern in QLS experiments generally have the form

$$g^{(1)}(\tau) = e^{i\omega_0\tau}e^{-\Gamma|\tau|} \tag{27}$$

Therefore, the power spectrum in a homodyne QLS experiment will have the form

$$P_i^{\text{homo}}(\omega) = \frac{1}{2\pi}\, e\langle i \rangle + \langle i \rangle^2 \delta(\omega) + \frac{1}{\pi}\,\langle i \rangle^2\, \frac{2\Gamma}{\omega^2 + (2\Gamma)^2} \tag{28}$$

As in Eq. (23), the first term is due to shot noise, which contributes a constant, frequency-independent "white noise" background to the power spectrum. The second term is the dc component, with magnitude $\langle i \rangle^2$ at zero frequency and zero magnitude elsewhere. The third term is a Lorentzian with half-width at half-height of Γ, centered at $\omega = 0$.

[20] See, e.g., F. Reif, "Fundamentals of Statistical and Thermal Physics," p. 585. McGraw-Hill, New York, 1965.

Likewise, in a heterodyne QLS experiment,

$$P_i^{het}(\omega) = \frac{ei_{LO}}{\pi} + i_{LO}^2\delta'(\omega) + \frac{2}{\pi} i_{LO}\langle i_s\rangle \frac{\Gamma}{(\omega - |\omega_0 - \omega_{LO}|)^2 + \Gamma^2} \quad (29)$$

If $\omega_{LO} = \omega_0$, as will generally be the case, then the last term should be doubled for $\omega \geq 0$ and set equal to zero for $\omega < 0$. Likewise, the modified delta-function $\delta'(\omega)$ is twice as large for $\omega \geq 0$ and is zero for $\omega < 0$.

Light Scattering from Dilute Solutions

We now specialize our previous discussion of the photocurrent autocorrelation function or power spectrum, produced by an arbitrary optical field, to fields produced by solutions of macromolecular scatterers. We suppose that the scattering volume V contains N identical scattering molecules. The volume is illuminated by a laser light source whose electric field is $E = E_0e^{-i\omega_0 t}$ and which is polarized perpendicular to the scattering plane. The light scattered at an angle θ in that plane is observed by a detector a distance R_0 away (Fig. 1). It is assumed here that the detector is small enough that the scattered light is coherent over its surface. The effect of detector area will be treated later in our discussion of collection optics.

The electric field at the photodetector due to scattering from the jth particle located at $\mathbf{r}_j$ at time t is

$$E_j(t) = A_j(t)e^{i[\mathbf{q}\cdot\mathbf{r}_j(t)-\omega_0 t]} \quad (30)$$

$A_j(t)$ is the scattering amplitude, which may depend on orientation of the particle for nonspherical scatterers. The scattering vector $\mathbf{q}$ is defined

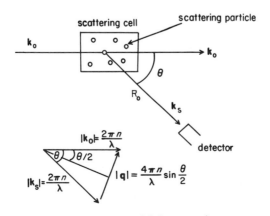

FIG. 1. Schematic diagram of light-scattering geometry.

as

$$\mathbf{q} = \mathbf{k}_0 - \mathbf{k}_s \tag{31}$$

where $\mathbf{k}_0$ and $\mathbf{k}_s$ are the wave vectors of the incident and scattered light, respectively. They have magnitudes of $2\pi n/\lambda_0$. The magnitude of $\mathbf{q}$ is

$$|\mathbf{q}| = q = \frac{4\pi n}{\lambda_0} \sin \frac{\theta}{2} \tag{32}$$

as may be seen from the geometrical construction in Fig. 1.

The total scattered field is the sum of contributions from each particle:

$$E_S(t) = \sum_{j=1}^{N} A_j(t) e^{i[\mathbf{q} \cdot \mathbf{r}_j(t) - \omega_0 t]} \tag{33}$$

The total intensity scattered from the volume is

$$\begin{aligned} I_S &= E_S^*(t) E_S(t) \\ &= \sum_{j=1}^{N} \sum_{k=1}^{N} A_j^*(t) A_k(t) \\ &= N |A(t)|^2 \end{aligned} \tag{34}$$

if it is assumed that all the scattering particles are identical (so that $A_j = A_k = A$) and that they are independent, so that the scattering amplitudes of j and k are uncorrelated if $j \neq k$. It is shown in standard references[21] that the total intensity of polarized light scattered from a macromolecular solution, assumed to be thermodynamically ideal, is

$$\begin{aligned} I_s &= \frac{I_0 \sin^2 \phi k_0^4 (\partial n/\partial c)^2 c V M P(q)}{4\pi^2 R_0^2 n^2 N_A} \\ &= R'Mc(\partial n/\partial c)^2 P(q) \end{aligned} \tag{35}$$

where I_0 is the intensity of the incident irradiation, ϕ is the angle between $\mathbf{k}_s$ and $\mathbf{E}_s$ (90° in the geometry discussed here), N_A is Avogadro's number, c the polymer concentration in g/ml, M the polymer molecular weight, and $P(q)$ the intramolecular scattering form factor. $(\partial n/\partial c)$ is the refractive index increment. Since $N = cN_A V/M$, comparison of Eqs. (34) and (35) shows that the rms scattering amplitude per particle is

$$(|A(t)|^2)^{1/2} = \frac{I_0^{1/2} \sin \phi k_0^2 (\partial n/\partial c) M P^{1/2}(q)}{2\pi R_0 n N_A} \tag{36}$$

The autocorrelation function of the scattered light is, from Eqs. (21) and (33), if we again invoke the identity and independence of individual

[21] C. Tanford, "Physical Chemistry of Macromolecules," Chap. 5. Wiley, New York, 1961.

scatterers, and assume that position $\mathbf{r}_j(t)$ and orientation [reflected in $A_j(t)$] are independent variables,

$$C_E(\tau) = Ne^{-i\omega_0\tau}\langle A^*(t)A(t+\tau)\rangle\langle e^{i\mathbf{q}\cdot[\mathbf{r}(t-\tau)-\mathbf{r}(t)]}\rangle \tag{37}$$
$$= Ne^{-i\omega_0\tau}C_A(\tau)C_\phi(\tau)$$

Translational Diffusion of Spherical Scatterers

When the scattering molecules are essentially spherical, so that the scattering amplitude does not depend on time, the amplitude autocorrelation function $C_A(\tau) = \langle A^*(t)A(t+\tau)\rangle$ becomes a constant, $|A|^2$. $C_E(\tau)$ is then determined by the phase autocorrelation function $C_\phi(\tau) = \langle\exp\{i\mathbf{q}\cdot[\mathbf{r}(t+\tau)-\mathbf{r}(t)]\}\rangle$, which equals $\langle\exp\{i\mathbf{q}\cdot[\mathbf{r}(\tau)-\mathbf{r}(0)]\}\rangle$ for a stationary process. The probability $G_s(\mathbf{r},\tau)$ that a particle undergoing translational diffusion will be at $\mathbf{r}(\tau)$ at time τ if it was at $\mathbf{r}(0)$ initially is governed by Fick's second law of diffusion

$$\frac{\partial G_s(\mathbf{r},\tau)}{\partial\tau} = D\nabla^2 G_s(\mathbf{r},\tau) \tag{38}$$

with the boundary condition $G(\mathbf{r},0) = \delta[\mathbf{r}-\mathbf{r}(0)]$. This has the well-known solution

$$G_s(\mathbf{r},\tau) = \frac{1}{(4\pi D\tau)^{3/2}} e^{-[\mathbf{r}(\tau)-\mathbf{r}(0)]^2/4D\tau} \tag{39}$$

When the autocorrelation function is calculated by

$$\langle e^{i\mathbf{q}\cdot[\mathbf{r}(\tau)-\mathbf{r}(0)]}\rangle = \int G_s(\mathbf{r},\tau)e^{i\mathbf{q}\cdot[\mathbf{r}(\tau)-\mathbf{r}(0)]}d^3\mathbf{r}(\tau)$$

the result is $e^{-Dq^2|\tau|}$. Thus the field autocorrelation function for N identical, spherical particles undergoing translational diffusion is

$$C_E(q,\tau) = N|A|^2 e^{-i\omega_0\tau}e^{-q^2D|\tau|} \tag{40}$$

This is of the form of Eq. (27) for the normalized first-order correlation function $g^{(1)}(\tau)$, with

$$\Gamma = q^2 D \tag{41}$$

Thus the homodyne photocurrent autocorrelation function $C_i(\tau)$ is, from Eqs. (16), (23), and (34)

$$C_i^{\text{homo}}(\tau) = e\langle i_s\rangle\delta(\tau) + \langle i_s\rangle^2 + \langle i_s\rangle^2 e^{-2q^2D\tau} \tag{42}$$

where $\langle i_s\rangle$ is related to molecular and detector parameters by Eqs. (16), (34), and (35). Likewise, from Eqs. (16), (25), and (34), with $\omega_0 = \omega_{\text{LO}}$, the heterodyne photocurrent autocorrelation function is

$$C_i^{\text{het}}(\tau) = ei_{\text{LO}}\delta(\tau) + i^2_{\text{LO}} + 2i_{\text{LO}}\langle i_s\rangle e^{-q^2D\tau} \tag{43}$$

The correlation functions decay exponentially, with decay rates $2q^2D$ and q^2D in the homodyne and heterodyne cases, respectively. The corresponding power spectra are, from Eqs. (28), (29), and (41)

$$P_i^{\text{homo}}(\omega) = \frac{1}{2\pi} e \langle i_s \rangle + \langle i_s \rangle^2 \delta(\omega) + \frac{\langle i_s \rangle^2}{\pi} \frac{2q^2D}{\omega^2 + (2q^2D)^2} \qquad (44)$$

and

$$P_i^{\text{het}}(\omega) = \frac{ei_{\text{LO}}}{\pi} + i_{\text{LO}}^2 \delta'(\omega) + \frac{4}{\pi} i_{\text{LO}} \langle i_s \rangle \frac{q^2D}{\omega^2 + (q^2D)^2} \qquad (45)$$

The last terms in each of these expressions are Lorentzian line shapes with half-width at half-height $2q^2D$ and q^2D, respectively. In the frequency units of cycles per second or Hz, in which instrumental readout is actually presented, the half-widths are ($\Delta \nu_{1/2} = \Delta \omega_{1/2}/2\pi$)

$$\Delta \nu_{1/2}^{\text{homo}} = q^2D/\pi \text{ Hz} \qquad (46)$$
$$\Delta \nu_{1/2}^{\text{het}} = q^2D/2\pi \text{ Hz} \qquad (47)$$

Comparison of these expressions with Eq. (5) and the definition of q confirms the earlier assertion, based on simple physical Doppler shift arguments, that $\Delta\nu$ is proportional to D and inversely proportional to λ^2.

Examples of experimentally determined homodyne correlation functions and power spectra are shown in Fig. 2.

Many biopolymer solutions contain more than one scattering component. If there are two independent macromolecular components, Eq. (40) becomes

$$C_E(\tau) = [N_1 |A_1|^2 e^{-q^2D\tau} + N_2 |A_2|^2 e^{-q^2D_2\tau}]e^{-i\omega_0\tau} \qquad (48)$$

and the total average scattering intensity is, from Eqs. (34) and (35)

$$I_s = N_1 |A_1|^2 + N_2 |A_2|^2$$
$$= R'[M_1 c_1 (\partial n/\partial c_1)^2 P_1(q) + M_2 c_2 (\partial n/\partial c_2)^2 P_2(q)] \qquad (49)$$

so that the normalized first-order correlation function is

$$g^{(1)}(\tau) = \frac{M_1 c_1 (\partial n/\partial c_1)^2 P_1(q) e^{-i\omega_0\tau} e^{-q^2D_1\tau} + M_2 c_2 (\partial n/\partial c_2)^2 P_2(q) e^{-i\omega_0\tau} e^{-q^2D_2\tau}}{M_1 c_1 (\partial n/\partial c_1)^2 P_1(q) + M_2 c_2 (\partial n/\partial c_2)^2 P_2(q)}$$
$$(50)$$

Then the photocurrent autocorrelation functions are

$$C_i^{\text{homo}}(\tau) = e \langle i_s \rangle \delta(\tau) + \langle i_s \rangle^2 + \langle i_s \rangle^2 [X_1^2 e^{-2q^2D_1\tau} + X_1 X_2 e^{-q^2(D_1+D_2)\tau} + X_2^2 e^{-2q^2D_2\tau}] \qquad (51)$$

and

$$C_i^{\text{het}}(\tau) = ei_{\text{LO}} \delta(\tau) + i_{\text{LO}}^2 + 2i_{\text{LO}} \langle i_s \rangle [X_1 e^{-q^2D_1\tau} + X_2 e^{-q^2D_2\tau}] \qquad (52)$$

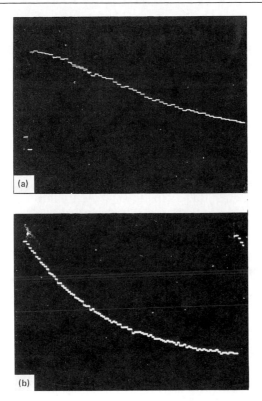

FIG. 2. (a) Photocurrent spectrum for light scattered from 910 Å diameter polystyrene latex spheres in water, $\theta = 50°$. The total spectral width is 1 kHz. (b) Time correlation function of photocurrent under same scattering conditions, obtained in a 100-channel correlator with delay time of 20 μsec/channel.

where

$$X_i = \frac{M_i c_i (\partial n/\partial c_i)^2 P_i(q)}{\sum_{j=1}^{2} M_j c_i (\partial n/\partial c_j)^2 P_j(q)} \quad (i = 1,2) \tag{53}$$

is the fraction of the total scattering intensity due to component i. Two important things should be noted about Eqs. (51) and (52). First, the homodyne photocurrent autocorrelation function in a solution with two scattering species is the sum of *three* exponential decays: two "self" terms and a cross term. Second, each component contributes to the homodyne correlation function in proportion to the *square* of its contribution to the total scattered intensity. Thus strongly particles will influence the homodyne scattering relatively more than the heterodyne scattering.

In the general case of n different scattering species, the correlation functions become

$$C_i^{homo}(\tau) = e\langle i_s \rangle \delta(\tau) + \langle i_s \rangle^2 + \langle i_s \rangle^2 \sum_{i=1}^{n} \sum_{j=1}^{n} X_i X_j e^{-q^2(D_i + D_j)\tau} \qquad (54)$$

and

$$C_i^{het}(\tau) = ei_{LO}\delta(\tau) + i_{LO}^2 + 2i_{LO}\langle i_s \rangle \sum_{i=1}^{n} X_i e^{-q^2 D_i \tau} \qquad (55)$$

where X_i and X_j are defined as in Eq. (53), but with the indexes running from 1 to n. The corresponding power spectra are

$$P_i^{homo}(\omega) = \frac{1}{2\pi} e\langle i_s \rangle + \langle i_s \rangle^2 \delta(\omega) + \frac{\langle i_s \rangle^2}{\pi} \sum_{i=1}^{n} \sum_{j=1}^{n} \frac{X_i X_j q^2 (D_i + D_j)}{\omega^2 + [q^2(D_i + D_j)]^2} \qquad (56)$$

and

$$P_i^{het}(\omega) = \frac{ei_{LO}}{\pi} + i_{LO}^2 \delta'(\omega) + \frac{4}{\pi} i_{LO}\langle i_s \rangle \sum_{i=1}^{n} \frac{X_i q^2 D_i}{\omega^2 + (q^2 D_i)^2} \qquad (57)$$

Procedures for analyzing experimental data from polydisperse systems are discussed later.

Uniform Translational Motion

When a particle moves in a straight line with constant velocity $\mathbf{v}$ for a distance long compared to λ (more precisely, compared to q^{-1}) the distance traveled is $\mathbf{r}(\tau) - \mathbf{r}(0) = \mathbf{v}\tau$. If all particles are moving with the same Eq. (37) becomes

$$C_E(\tau) = N|A|^2 e^{-i(\omega_0 - \mathbf{q} \cdot \mathbf{v})\tau}$$

or

$$g^{(1)}(\tau) = e^{-i(\omega_0 - \mathbf{q} \cdot \mathbf{v})\tau} \qquad (58)$$

so the heterodyne photocurrent autocorrelation function oscillates with frequency $\nu_0 - \mathbf{q} \cdot \mathbf{v}/2\pi$. Applying the Wiener–Khintchine theorem, Eq. (26), to Eq. (25) with $\omega_{LO} = \omega_0$, we see that the corresponding photocurrent power spectrum is

$$P_i^{het}(\omega) = (ei_{LO})/\pi + i_{LO}^2 \delta(\omega) + \frac{4}{\pi} i_{LO}\langle i_s \rangle \delta(\omega - \mathbf{q} \cdot \mathbf{v}) \qquad (59)$$

The last term corresponds to a delta-function spike in frequency at the

Doppler-shift frequency

$$\Delta \nu = \mathbf{q} \cdot \mathbf{v}/2\pi \tag{60}$$

The homodyne autocorrelation function is, from Eqs. (21) and (58), simply

$$g^{(2)}(\tau) = 2$$

That is, $g^{(2)}(\tau)$ is simply a constant, and correspondingly the photocurrent power spectrum $P_i^{\text{homo}}(\omega)$ is independent of frequency (except for the dc spike). Homodyne QLS is useless to determine phase information relating to systematic translational velocity, so long as Eq. (21) is obeyed, and heterodyne QLS must be employed to measure such velocities.

In practice, correlation functions such as Eq. (58) are encountered only with translational motion of very large particles, such as cells, for which diffusional Brownian motion is unimportant. In this case, QLS serves as a laser Doppler velocimeter, with Doppler shift proportional to the speed $|\mathbf{v}|$ and to the angle between $\mathbf{q}$ and $\mathbf{v}$. More generally, particularly with small macromolecules, the Doppler-shifted line is broadened by diffusion. This is treated in the next section.

Electrophoretic Light Scattering[13]

When an electric field $\mathbf{E}$ is applied to a solution containing charged molecules with electrophoretic mobility μ, the electrophoretic drift velocity produced is

$$\mathbf{v}_d = \mu \mathbf{E} \tag{61}$$

The particles will in addition still be undergoing random diffusive motion. The proper generalization of Eq. (38) to account for the motion under the influence of the applied field in the x-direction is

$$[\partial G_s(\mathbf{r},\tau)]/\partial \tau = D\nabla^2 G_s(\mathbf{r},\tau) \pm \mathbf{v}_d[\partial G_s(\mathbf{r},\tau)]/\partial x \tag{62}$$

The sign of the second term depends on the sign of μ and the direction of $\mathbf{E}$, which determine the direction of $\mathbf{v}_d$.

The solution of Eq. (62), corresponding to Eq. (39) for simple diffusion, is

$$G_s(r,\tau) = [1/(4\pi D\tau)]^{3/2} e^{-[(x \pm v_d\tau)^2 + y^2 + z^2]/4D\tau} \tag{63}$$

Then evaluation of the field autocorrelation, assuming N independent spherical particles gives

$$C_E(q,\tau) = N|A|^2 e^{-i\omega_0\tau_e \pm i\mathbf{q}\cdot\mathbf{v}_d\tau_e} e^{-q^2 D\tau} \tag{64}$$

and the heterodyne photocurrent autocorrelation function is

$$C_i^{het}(\tau) = ei_{LO}\delta(\tau) + i_{LO}^2 + 2i_{LO}\langle i_s\rangle e^{\pm i\mathbf{q}\cdot\mathbf{v}_d\tau}e^{-q^2D\tau} \qquad (65)$$

Comparison of Eqs. (43) and (65) shows that the exponentially decaying part of the photocurrent autocorrelation function is modulated by a sinusoidal frequency $\mathbf{q}\cdot\mathbf{v}_d$. This is shown in Fig. 3.

The power spectrum corresponding to Eq. (65) is

$$P_i^{het}(\omega) = \frac{ei_{LO}}{\pi} + i_{LO}^2\delta'(\omega) + \frac{4}{\pi}i_{LO}\langle i_s\rangle\frac{q^2D}{(\omega \pm \mathbf{q}\cdot\mathbf{v}_d)^2 + (q^2D)^2} \qquad (66)$$

The last term, which is the one of interest, corresponds to a Lorentzian spectrum of half-width at half-height $\Delta\nu_{\frac{1}{2}} = q^2D/2\pi$, whose maximum is displaced from zero by

$$\Delta\nu_{shift} = \pm\mathbf{q}\cdot\mathbf{v}_d/2\pi = \pm\mu\mathbf{q}\cdot\mathbf{E}/2\pi. \qquad (67)$$

An example is given in Fig. 3. Thus an electrophoretic light-scattering experiment can yield simultaneous measurements of the transport parameters D and μ, from the line width and Doppler shift, respectively.

The advantages of electrophoretic light scattering over conventional electrophoresis experiments are summarized by Ware.[13] They include simultaneous measurement of μ and D, greater rapidity of measurement (by as much as 10^3 to 10^4), potentially greater resolution, the absence of macroscopic concentration gradients or boundaries, applicability over a wide range of ionic strengths (particularly very low ionic strengths) and to very large particles, and the possibility of studying reaction dynamics (see below).

Some disadvantages, on the other hand, are the complexity and expense of the equipment required to do electrophoretic light scattering (ELS), the loss of preparative capability, interferences by dust and large-scattering particles, and some difficulties in reproducible heterodyning and other aspects of scattering-cell design. In many cases, however, ELS represents a notable advance in electrophoresis technique.

To gain further insight into the feasibility of ELS, we consider the resolution R, defined as the ratio of the Doppler shift to the diffusion broadening:

$$R = \Delta\nu_{shift}/\Delta\nu_{\frac{1}{2}} = \mathbf{q}\cdot\mathbf{v}_d/q^2D \qquad (68)$$

This is optimized when working at very low scattering angles, where $\mathbf{E}$ and hence $\mathbf{v}_d$ is almost parallel to $\mathbf{q}$, and when $\mathbf{q}$ is nearly perpendicular to $\mathbf{k}_0$, the wavevector of the incident light. Then, as seen from Fig. 4,

$$R = \frac{q\mu E\cos(\theta/2)}{q^2D} = \frac{\mu E\cos(\theta/2)}{\left(\frac{4\pi n}{\lambda_0}\right)D\sin(\theta/2)} \qquad (69)$$

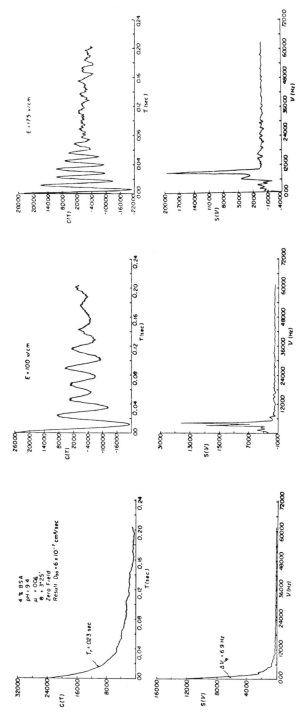

FIG. 3. Heterodyne spectrum and correlation function for electrophoretic light scattering from a solution of bovine serum albumin at electric fields of 0 (left), 100 (center), and 175 (right) V/cm. From B. R. Ware and W. H. Flygare, *J. Colloid Interface Sci.* **39**, 670 (1972).

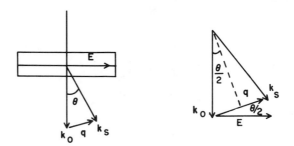

FIG. 4. Geometry of electrophoretic light-scattering experiment.

At low angles, $\cos(\theta/2) \simeq 1$, $\sin(\theta/2) \approx \theta/2$, so

$$R = (\mu E \lambda_0/(2\pi n D \theta)) \tag{70}$$

The resolution improves with increasing E and decreasing θ.

From the relations[22]

$$\mu = (Ze\phi)/f \tag{71}$$
$$D = kT/f \tag{72}$$

where Z is the macromolecule charge in units of the proton charge e, f is the translational frictional coefficient, and ϕ is a function of ionic strength through the inverse thickness κ of the Debye–Hückel ion atmosphere $\kappa = [(8\pi N_A e^2)/(1000\epsilon\, kT)]^{1/2}I^{1/2}$, and of particle size and shape, we see that

$$\mu/D = Ze\phi/k_BT \tag{73}$$
$$R = (E\lambda_0/2\pi n\theta)(Ze\phi/k_BT) \tag{74}$$

so the major molecular parameter on which the resolution depends is the charge Z.

Likewise, simultaneous measurement of μ and D enables determination of $Z\phi$ from Eq. (73). When the ionic strength is very low and/or the particle has a small hydrodynamic radius R_h, so that $\kappa R_h \ll 1$, then the particle behaves like a bare charge Ze, and $\phi = 1$. For larger sizes and ionic strengths, matters become more complicated. A first-order approximation based on Debye–Hückel theory applied to solid spheres gives the Henry equation[23]

$$\phi(\kappa R_h) = X_1(\kappa R_h)/(1 + \kappa R_h) \tag{75}$$

where $X_1(\kappa R_h)$ is a function that goes from 1 at $\kappa R_h \ll 1$ to 3/2 at $\kappa R_h \ll 1$. Measurement of D gives R_h, and knowledge of ionic strength

[22] Tanford,[21] Sections 21, 24.
[23] D. C. Henry, *Proc. R. Soc. London, Ser. A* **133**, 106 (1931).

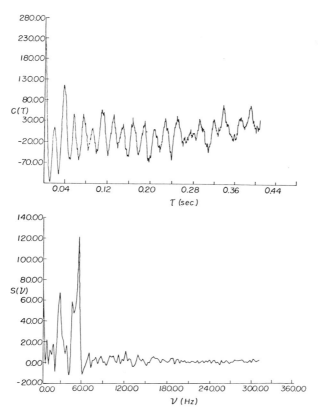

FIG. 5. Electrophoretic light-scattering correlation function and spectrum of a mixture of 4% bovine serum albumin (including some dimers) and 1% fibrinogen at pH 9.55; $\mu = 0.005$, E = 125 V/cm, and $\theta = 3° 37'$. From B. R. Ware and W. H. Flygare, *J. Colloid Interface Sci.* **39**, 670 (1972).

gives κ, so $\phi(\kappa R_h)$ can be determined. A full discussion of the interpretation of electrophoretic mobilities is beyond the scope of this chapter. See the volumes edited by Bier.[24]

Electrophoretic mobilities and diffusion coefficients of individual components in a mixture of proteins may be determined in a single experiment if the components have sufficiently different mobilities. An example is shown in Fig. 5, for a mixture of bovine serum albumin and fibrinogen.

It is also possible to combine conventional band electrophoresis in stabilizing density gradients with measurement of D by QLS.[25] This tech-

[24] M. Bier, ed. "Electrophoresis: Theory, Methods, and Applications." Academic Press, New York, Vol. I, 1959; Vol. II, 1967.

[25] T. K. Lim, G. Baran, and V. A. Bloomfield, *Biopolymers* **16**, 1473 (1977).

nique will be discussed in detail below, in the section on experimental analysis of polydisperse systems.

Rotational Motion

If the scattering molecules are not spherical, but instead are anisotropic in shape or polarizability, the possibility arises to use QLS to measure rotational diffusion coefficients. This may be done by analysis of either the polarized or the depolarized scattering.

Isotropic Scattering from Cylindrical Rods. We consider first the scattering from a long rigid rod that is optically isotropic. Since the polarizability of the rod is the same along and perpendicular to its long axis, the scattered light will retain its original polarization. If the length of the rod is L and its radius is b, the rotational diffusion coefficient obtained from the Perrin[26] equation for prolate ellipsoids of semimajor axis $L/2$ and semiminor axis b, in the limit $L/2 \gg b$, is

$$\Theta = \frac{3kT[2 \ln (L/b) - 1]}{16\pi\eta_0(L/2)^3} \tag{76}$$

If L is not much less than q^{-1}, then rotational diffusion will contribute to the time dependence of the scattering through the amplitude autocorrelation function, $C_A(\tau)$, in Eq. (37).

The total intensity of scattering from a solution of rodlike molecules is given by Eq. (35), where the form factor is[27]

$$P(q) = \frac{2}{qL} Si (qL) - \left[\frac{\sin (qL/2)}{qL/2} \right]^2 \tag{77}$$

The sine integral is a tabulated[28] function defined as

$$Si(X) = \int_0^x \frac{\sin y}{y} dy \tag{78}$$

It can be shown[2,29] that Eq. (77) results from two contributions: a "spherical" part

$$P(q)_{sph} = [Si(qL/2)/(qL/2)]^2 \tag{79}$$

arising only from translational motion of the rod, and a "modified" part

$$P(q)_{mod} = Si(qL)/(qL/2) - [\sin (qL/2)/(qL/2)]^2 - [Si(qL/2)/(qL/2)]^2 \tag{80}$$

arising from both translational and rotational motions.

[26] F. Perrin, *J. Phys. Radium* [7] **5**, 497 (1934).
[27] T. Neugebauer, *Ann. Phys. (Leipzig)* [5], **42**, 509 (1943).
[28] M. Abramowitz and I. A. Stegun, eds. "Handbook of Mathematical Functions," p. 231. National Bureau of Standards, Washington, D.C., 1965.
[29] R. Pecora, *J. Chem. Phys.* **48**, 4126 (1968).

To obtain the autocorrelation function or spectrum of light scattered from the solution of rods, it is necessary to solve the equations of translational and rotational diffusion for the time evolution of the distribution functions, and to use these in Eq. (37) to evaluate the correlation functions. This was done by Pecora,[2,29] who showed that

$$C_A(\tau) = |A_0|^2 \sum_{\substack{l=0 \\ \text{even}}}^{\infty} B_l e^{-l(l+1)\Theta\tau} \tag{81}$$

while $C_\phi(\tau)$ is still given by $e^{-q^2 D\tau}$. In Eq. (81), $|A_0|$ is $\langle |A(t)| \rangle^{1/2}$, Eq. (36), divided by $P^{1/2}(q)$, and

$$B_l = (2l + 1) \left[\frac{2}{qL} \int_0^\infty \frac{J_l(X)}{X} \, dx \right]^2 \tag{82}$$

where $J_l(X)$ is the Bessel function of order l.

The total field autocorrelation function from a solution of rods undergoing both translational and rotational diffusion is therefore

$$C_E(\tau) = N|A_0|^2 e^{-i\omega_0\tau} \sum_{l=0}^{\infty} B_l e^{-[q^2 D + l(l+1)\Theta]\tau} \tag{83}$$

This is an infinite series of even exponentials, with relaxation times $\tau_0 = (q^2 D)^{-1}$, $\tau_2 = (q^2 D + 6\Theta)^{-1}$, $\tau_4 = (q^2 D + 20\Theta)^{-1}$, . . . and corresponding relative weights B_0, B_2, B_4 . . . . These weights are computed as a function of the reduced variable qL in Fig. 6. In the range of accessible angles, only B_0 and B_2 contribute significantly.

To obtain the power spectrum of scattered light, we Fourier transform Eq. (83) according to the Wiener–Khintchine theorem, obtaining

$$I_s(\omega) = \frac{N|A_0|^2}{\pi} \sum_{\substack{l=0 \\ \text{even}}}^{\infty} B_l \frac{q^2 D + l(l + 1)\Theta}{(\omega - \omega_0)^2 + [q^2 D + l(l + 1)\Theta]^2} \tag{84}$$

The spectrum is a sum of Lorentzians centered at $\omega = \omega_0$, weighted by B_l with half-widths at half-heights $\Delta\omega_{1/2,l} = q^2 D + l(l + 1)\Theta$. Integration of this equation over all frequencies yields the total scattered intensity. The first, $l = 0$, term gives rise to $P(q)_{\text{sph}}$, Eq. (79), since it depends only on translation; while the remaining terms in aggregate give rise to $P(q)_{\text{mod}}$, Eq. (80), which depends on both translation and rotation.

The dependence of spectral half-width on scattering angle for TMV, measured both by heterodyne and homodyne techniques, is shown in Fig. 7. For an autocorrelation function such as Eq. (83), which is the sum of exponential decays, analysis of the homodyne scattering is more complicated than that of the heterodyne scattering, since cross-terms arise. That is, considering only the $l = 0$ and $l = 2$ terms, the normalized

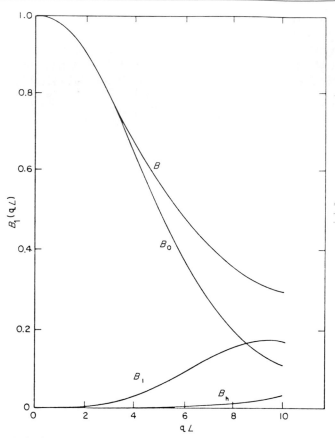

FIG. 6. Relative integrated intensities of light scattered from optically isotropic rigid rods. B is the total relative integrated intensity, B_0 the intensity of the pure translational part, B_1 the first nonzero term whose spectral width contains the rotational diffusion coefficient, and B_h the sum of intensities of all other terms. From B. J. Berne and R. Pecora, "Dynamic Light Scattering with Application to Chemistry, Biology, and Physics." Wiley (Interscience), New York, 1976.

field autocorrelation function is

$$g^{(1)}(\tau) = \frac{B_0}{B_0 + B_2} e^{-q^2 D \tau} + \frac{B_2}{B_0 + B_2} e^{-(q^2 D + 6\Theta)\tau} \tag{85}$$

while the corresponding homodyne correlation function is

$$g^{(2)}(\tau) = 1 + \left(\frac{B_0}{B_0 + B_2}\right)^2 e^{-2q^2 D \tau} + \left(\frac{B_2}{B_0 + B_2}\right)^2 e^{-2(q^2 D + 6\Theta)\tau}$$

$$+ \frac{2 B_0 B_2}{B_0 + B_2} e^{-(2q^2 D + 6\Theta)\tau} \tag{86}$$

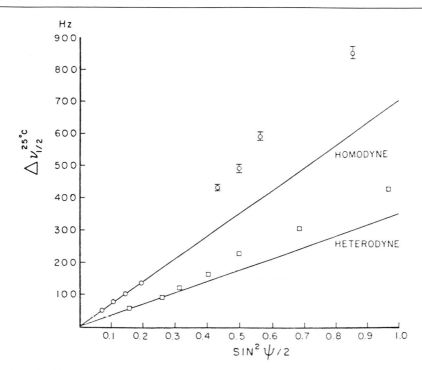

FIG. 7. Single Lorentzian half-widths at 25° as a function of $\sin^2 \psi/2$ obtained from homodyne and heterodyne spectra of TMV; ○, homodyne spectra; □, heterodyne spectra; solid lines correspond to best least-squares linear fit to data for $\psi < 60°$. From H. Z. Cummins, F. D. Carlson, T. J. Herbert, and G. Woods, *Biophys. J.* **9**, 518 (1969).

Likewise, the corresponding homodyne power spectrum will be the sum of three Lorentzians, with half-widths $2q^2D$, $2q^2D + 12\Theta$, and $2q^2D + 6\Theta$.

From Fig. 6 it is evident that the relative weight of B_2 and higher l contributions to $I(\omega)$ go to zero at zero scattering angle. Therefore, the low-angle scattering in Fig. 7 is due only to translation. The upward curvature at higher angles is characteristic of rotational diffusion. By subtracting the pure translational part, extrapolated from low angles, from the higher-angle scattering, Cummins *et al.*[30] were able to determine $D_{20,w}^{\circ} = 2.80 \pm 0.6 \times 10^{-8}$ cm²/sec, and $\Theta_{20,w}^{\circ} = 320 \pm 18$ sec⁻¹, in good agreement with results from other methods.

Rotational contribution to polarized scattering of the type discussed above will be significant only for highly elongated macromolecules whose length is comparable to the wavelength of light. To use QLS for determi-

[30] H. Z. Cummins, F. D. Carlson, T. J. Herbert, and G. Woods, *Biophys. J.* **9**, 518 (1969).

nation of Θ of smaller, less asymmetric proteins, it is necessary to use depolarized scattering.

Depolarized Scattering from Anisotropic Molecules. A molecule is optically anisotropic when it has different polarizabilities along different molecular axes. When a laser beam with electric field polarized along the Z axis impinges upon such a molecule, it will induce an oscillating molecular dipole moment not only in the Z-direction, but also in the x and y directions. This will in turn give rise to scattered light with the electric vector $\mathbf{E}_s$ having components along the x and y axes as well as the Z axis. This light is said to be depolarized. Since the molecules are continuously reorienting by rotational diffusion, the depolarized and polarized components fluctuate with time. The dynamics of these fluctuations thus give information on the rotational diffusion coefficient.

We shall consider here only the results for cylindrically symmetric molecules. These have polarizability $\alpha_\parallel$ alone the cylinder axis, and $\alpha_\perp$ perpendicular to it. The theory on which these results are based, and the extension to more complicated anisotropic molecules, is developed in detail in Chapter 7 of the book by Berne and Pecora.[8]

Assuming translational and rotational motion to be uncorrelated, we need to calculate the influence of depolarized scattering on the orientational correlation function $C_A(\tau)$, Eq. (37). There are two standard polarization arrangements in light scattering. By convention, the scattering plane, in which $\mathbf{k}_0$ and $\mathbf{k}_s$ lie is the x-y plane, and $\mathbf{k}_0$ is along the $+x$ direction (Fig. 1). The incident laser light is polarized vertically, along the Z axis. Then in so-called VV (vertical–vertical) scattering, the scattered light is detected after passage through a vertically oriented analyzer. In VH (vertical–horizontal) scattering, the analyzer is rotated 90° into the x-y plane. Thus we have two orientational correlation functions, $C_{A,VV}(\tau) = \langle A_{VV}^*(0)A_{VV}(\tau)\rangle$ and $C_{A,VH}(\tau) = \langle A_{VH}^*(0)A_{VH}(\tau)\rangle$. When these are evaluated using the rotational diffusion equation, the corresponding field autocorrelation functions are found to be

$$C_{E,VV}(\tau) = N|A'|^2 e^{-i\omega_0\tau_e - q^2 D\tau}\left[\alpha^2 + \frac{4}{45}\beta^2 e^{-6\Theta\tau}\right] \tag{87}$$

and

$$C_{E,VH}(\tau) = N|A'|^2 e^{-i\omega_0\tau_e - q^2 D\tau} \times \frac{1}{15}\beta^2 e^{-6\Theta\tau} \tag{88}$$

In these equations α is the average, isotropic polarizability

$$\alpha = \frac{1}{3}(\alpha_\parallel + 2\alpha_\perp) \tag{89}$$

which is related to measurable solution properties[31] by

$$\alpha = Mn(\partial n/\partial c)/2\pi N_A \tag{90}$$

Comparison of Eqs. (36) and (90) shows that

$$A' = I_0^{1/2} \sin \phi k_0^2 P^{1/2}(q)/R_0 n^2 \tag{91}$$

The quantity β is the optical anisotropy,

$$\beta = \alpha_{\parallel} - \alpha_{\perp} \tag{92}$$

From Eqs. (40), (87), and (88) we see that

$$C_{E,VV}(\tau) = C_E^{iso}(\tau) + 4/3\, C_{E,VH}(\tau) \tag{93}$$

where $C_E^{iso}(\tau)$, the isotropic part of the autocorrelation function, arises from purely translational diffusion with fluctuations governed by the average polarizability α.

The heterodyne power spectra corresponding to Eqs. (87) and (88) are

$$P_{VV}^{het}(q,\omega) = \frac{N|A'|^2}{15\pi}\left[\alpha^2\,\frac{q^2D}{\omega^2+(q^2D)^2} + \frac{4}{45}\,\beta^2\,\frac{q^2D+6\Theta}{\omega^2+(q^2D+6\Theta)^2}\right] \tag{94}$$

$$P_{VH}^{het}(q,\omega) = \frac{N|A'|^2}{15\pi}\,\beta^2\left(\frac{q^2D+6\Theta}{\omega^2+(q^2D+6\Theta)^2}\right) \tag{95}$$

Integrating these overall frequencies (this is equivalent to taking the $\tau = 0$ limit of $C_E(\tau)$), we find the total scattered intensities

$$I_{VV} = I_{ISO} + \frac{4}{3}\,I_{VH}$$

$$I_{VH} = \frac{1}{15}\,\beta^2 N|A'|^2 \tag{96}$$

$$I_{ISO} = N|A'|^2\alpha^2$$

The depolarization ratio is then

$$\frac{I_{VH}}{I_{ISO}} = \frac{1}{15}\,\frac{\beta^2}{\alpha^2} = \frac{3}{5}\left[\frac{\alpha_{\parallel} - \alpha_{\perp}}{\alpha_{\parallel} + 2\alpha_{\perp}}\right]^2 \tag{97}$$

This ratio is usually very small, since $|\alpha_{\parallel} - \alpha_{\perp}| \ll \alpha$. This means that depolarized scattering is generally difficult to measure, since it is weak and since, if the polarizers are not of highest quality, strongly scattered polarized light may "leak" through the horizontal analyzer and contribute spuriously to the depolarized component.

[31] Tanford,[21] p. 278.

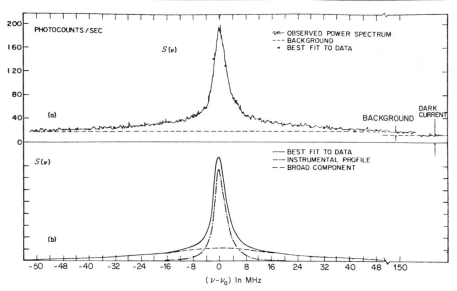

FIG. 8. (a) Power spectrum of the depolarized light scattered by a 15% solution of lysozyme in a 0.1 M sodium acetate–acetic acid buffer at 24°. The solid circles indicate the best least-squares fit to the spectrum: A convolution of the experimental profile $I(\nu)$ with a Lorentzian of $\Delta\nu_{1/2}$ of 18.1 mHz and a delta function of area two-thirds as large as the Lorentzian. These components rest in a 5-count/sec background. This experimental trace was swept out in 15 min. (b) Breakdown of the measured spectrum in terms of the component having the instrumental profile and the broad component from the lysozyme. The measured spectrum is the sum of these components. From S. B. Dubin, N. A. Clark, and G. B. Benedek, *J. Chem. Phys.* **54**, 5158 (1971).

A virtuoso illustration of the use of depolarized scattering to measure rotational diffusion of a small protein was provided by Dubin *et al.*,[32] who worked with lysozyme. Because of the small size and weak scattering of the protein, they worked at 15% concentration. Also, because Θ is so large, $(16.7 \pm 0.8) \times 10^6 \text{ sec}^{-1}$, the spectral width is beyond the capacity of available spectrum analyzers, so Fabry–Perot interferometry was required. The resulting signal is shown in Fig. 8. For details of instrumentation and analysis, the reader is referred to the original paper[32] and to Dubin's previous article in this series.[11]

It will be noted that, at $q = 0$, $C_{E,VH}(\tau)$ contains time-dependence only due to rotational motion. Therefore, measurement of very low-angle depolarized scattering is in some cases a useful way to observe rotation without interference from translational diffusion.

[32] S. B. Dubin, N. A. Clark, and G. B. Benedek, *J. Chem. Phys.* **54**, 5158 (1971).

Scattering from Flexible-Chain Polymers

We saw in the preceding section that rotational contributions to the scattering autocorrelation function were significant only if the molecular size were comparable to the wavelength, $qL > 1$, so that intraparticle interference effects could be manifested. Likewise, internal motions in flexible polymers will contribute to the scattering dynamics only if they have a scale comparable to λ or q^{-1}. The theory has been worked out only for flexible chains that obey Gaussian conformational statistics. It was shown by Debye[33,34] that the form factor in the total scattered intensity for Gaussian chains is

$$P(q) = 2y^{-2}(e^{-y} - 1 + y) \tag{98}$$

where $y = q^2 R_G^2$ and R_G, the rms radius, or "radius of gyration" of the chain is

$$R_G^2 = nl^2/6 \tag{99}$$

for a chain of n freely jointed subunits, each of length l.

The dynamics of this model for flexible polymers has been worked out by Rouse[35] and by Zimm,[36] who showed that the chain motions could be expressed as a sum of normal modes. In the Rouse theory, which is simpler but neglects hydrodynamic interaction between the subunits, the relaxation time of the kth normal mode is

$$\tau_k = (l^2 n^2 \zeta)/(3k_B T \pi^2 k^2) \tag{100}$$

where ζ is the Stokes frictional resistance of a subunit ($= 3\pi\eta_0 l$ if the subunits are spheres of diameter l). Pecora[37] used the Rouse theory to evaluate the time-dependence of $P(q,\tau)$:

$$P(q,\tau) = P_0(y)e^{-q^2 D\tau} + P_2(y)e^{-(q^2 D + 2/\tau_1)\tau} + \cdots \tag{101}$$

Scattering is dominated by the longest and slowest modes. The dependence of P and its components P_0, P_2, and $P_n = P - P_0 - P_2$ on y are shown in Fig. 9. We note that the terms involving internal relaxations $\tau_1, \ldots$ will be important only for very long chains. In this case, a plot of observed autocorrelation time vs q^2 will have a finite intercept at $q = 0$, given approximately by $\tau_1/2$.

The depolarized scattering from flexible chain molecules has also been analyzed theoretically.[38] Perhaps the most interesting conclusion is that at

[33] P. Debye, *J. Appl. Phys.* **15**, 338 (1944).
[34] P. Debye, *J. Phys. Colloid Chem.* **51**, 18 (1947).
[35] P. E. Rouse, Jr., *J. Chem. Phys.* **21**, 1272 (1953).
[36] B. H. Zimm, *J. Chem. Phys.* **24**, 269 (1956).
[37] R. Pecora, *J. Chem. Phys.* **49**, 1032 (1968).
[38] K. Ono and K. Okano, *Jpn. J. Appl. Phys.* **9**, 1356 (1971).

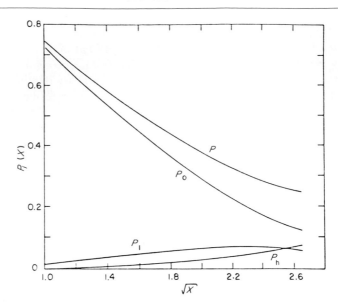

FIG. 9. Relative integrated intensities of light scattered from Gaussian coils vs $\sqrt{x} = qR_G$. P is the total relative intensity, P_0 the intensity of the pure translational part, P_1 the first significant term whose time correlation function depends on intramolecular relaxation times, and $P_h = P - (P_0 + P_1)$. From B. J. Berne and R. Pecora, "Dynamic Light Scattering with Applications to Chemistry, Biology, and Physics." Wiley (Interscience), New York, 1976.

zero scattering angle, the observed correlation function is a sum of contributions from all normal modes of the chain. Each mode is weighted equally, instead of the longest wavelength modes being weighted more strongly, as is the case for isotropic scattering. Schmitz and Schurr[39] have used forward depolarized scattering to study internal motions in DNA.

Reaction Kinetics

One of the most intriguing potential applications of QLS is in the determination of chemical reaction rates. Two cases may be distinguished, one in which the reaction half-life is long compared to the time required to determine the diffusion coefficient, the other in which the reaction is at dynamic equilibrium with half-life comparable to the characteristic diffusion time scale $(q^2D)^{-1}$.

Reaction Slow Compared to Measurement Time. When the reaction proceeds to its final equilibrium position with a half-time $t_{\frac{1}{2}}$ substantially longer than the time required to measure the diffusion coefficient of the

[39] K. S. Schmitz and J. M. Schurr, *Biopolymers* **12**, 1543 (1973).

scattering species in the system, then the reaction kinetics can be elucidated from the time-dependence of the average diffusion coefficient. The requirement is that the average D not change appreciably (by more than the standard deviation in the error of its measurement) during the time of measurement. If the data are analyzed by the cumulant expansion procedure (see below), then the relevant average is the Z-average diffusion coefficient $\bar{D}_Z$.

Consider for example the irreversible unimolecular isomerization A $\xrightarrow{k_f}$ B. The Z-average D, which also equals the number- or weight-average D if $M_A = M_B$, is

$$\bar{D} = f_A D_A + (1 - f_A)D_B \qquad (102)$$

If the reaction starts with pure A, then the fraction remaining as A is

$$f_A = e^{-k_f t} \qquad (103)$$

If the measurement of $\bar{D}$ requires a time T, so that it is initiated at reaction time t and completed at $t + T$, then the relative change in $\bar{D}$ is

$$\frac{\bar{D}(t + T) - \bar{D}(t)}{\bar{D}(t)} = \frac{(1 - \beta)e^{-k_f t}[e^{-k_f T} - 1]}{\beta + (1 - \beta)e^{-k_f t}} \qquad (104)$$

where $\beta = D_B/D_A$. We choose for simplicity $t = t_{1/2} = 0.693/k_f$. Then one can show that, if the fractional standard deviation in D_A or D_B is $\pm\delta$, with $\delta \ll 1$, then one requires $T \leq \delta/k_f$ if $\bar{D}$ is not to change by more than δ during T. For example, if $\delta = 0.03$, and $T = 1$ min, then $k_f \leq 0.03$ min^{-1} or $t_{1/2} \gtrsim 23$ min. If the measurement were done at early times, $t \approx 0$, then the requirement would be $T \leq \delta/|1 - \beta|k_f$. With $\beta = 1.3$, then $k_f \leq 0.1$ min^{-1} or $t_{1/2} \gtrsim 6.93$ min.

Another limitation on this type of experiment is that $t_{1/2}$ be long compared to the time required for convective mixing disturbances to die out. This is typically on the order of 2 min. Finally, the difference in diffusion coefficients of reactants and products must be larger than the error in determination of either, $|D_A - D_B| > \delta$.

An example of a case in which these requirements were met is the attachment of T4 bacteriophage heads (H) to tails (T), giving tailfiberless virus particles (HT).[40] The time course of the average diffusion coefficient is shown in Fig. 10. In this case the reaction is H + T $\xrightarrow{k_f}$ HT, but since the head is much more massive than the tail (M_r 180 $\times$ 10^6 and 18 $\times$ 10^6, respectively), from a light-scattering point of view the scattering species are just H (reactant) and HT (product). At any reaction time t the homodyne autocorrelation function $C_{it}^{homo}(\tau)$ is (neglecting dc and shot

[40] J. A. Benbasat and V. A. Bloomfield, J. Mol. Biol. **95**, 335 (1975).

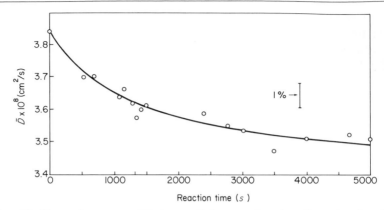

FIG. 10. Change in average diffusion coefficient as a function of reaction time for T4 phage assembly reaction H + T → HT at 22°. Initial concentrations were $[H]_0 = [T]_0 = 4.4 \times 10^{10}$/ml, corrected for fraction of inactive heads as determined by electron microscopy. From J. A. Benbasat and V. A. Bloomfield, *J. Mol. Biol.* **95**, 335 (1975).

noise components and constants of proportionality)

$$C_{it}^{homo}(\tau) = [f_H(t)e^{-q^2D_H\tau} + (1 - f_H(t))e^{-q^2D_{HT}\tau}]^2 \qquad (105)$$

since $M_{HT} \approx M_H$. Then taking square roots and rearranging,

$$[C_{it}^{homo}(\tau)]^{1/2} - e^{-q^2D_{HT}\tau} = f_H(t)[e^{-q^2D_H\tau} - e^{-q^2D_{HT}\tau}] \qquad (106)$$

From previous measurements on pure heads and fiberless particles, D_H and D_{HT} were known. Therefore, the left-hand side of Eq. (106) can be plotted against the coefficient of $f_H(t)$ on the rhs to obtain $f_H(t)$, as shown in Fig. 11. This, when plotted in a manner suitable for a second-order reaction, yielded k_f.

The success of this experiment is based largely on the fact that viruses are large, strongly scattering particles. This permitted the measurement time T to be as short as 40 sec, while the concentrations of H and T were low enough that $t_{1/2} \approx 500$ sec. Whether such an experiment could succeed with smaller enzyme molecules, with molecular weights on the order of 10^5 instead of 10^8, is doubtful given present instrumental capabilities.

Rapid Reactions in Dynamic Equilibrium. When the reaction is in dynamic equilibrium, with forward and reverse reaction rates comparable to q^2D, then the reaction process is reflected in the autocorrelation function itself. This is because concentration fluctuations are produced and dissipated not only by diffusion, but also by chemical reaction. The simplest case is the reversible unimolecular isomerization

$$A \underset{k_b}{\overset{k_f}{\rightleftharpoons}} B$$

When the concentrations are written as sums of equilibrium and fluctuation terms, $c_A(r,t) = \bar{c}_A + \delta c_A(r,t)$, $c_B = \bar{c}_B + \delta c_B(r,t)$, then the equations governing the time and space dependence have both diffusion and reaction terms

$$[\partial \delta c_A(r,t)]/\partial t = D_A \nabla^2 \delta c_A(r,t) - k_f \delta c_A(r,t) + k_b \delta c_B(r,t)$$
$$[\partial \delta c_B(r,t)]/\partial t = D_B \nabla^2 \delta c_B(r,t) + k_f \delta c_A(r,t) - k_b \delta c_B(r,t) \tag{107}$$

It can be shown[15] that the scattered field autocorrelation function for this system is the sum of two exponential relaxations:

$$C_E(\tau) = B_- e^{-\Gamma_- \tau} + B_+ e^{-\Gamma_+ \tau} \tag{108}$$

The inverse relaxation times are

$$\Gamma_\pm = -\frac{1}{2} \left\{ -q^2(D_A + D_B) - \tau_c^{-1} \right.$$
$$\left. \pm \tau_c^{-1} \left[1 + 2 \left(\frac{K_e' - 1}{K_e' + 1} \right) \left(\frac{q^2 \Delta}{\tau_c^{-1}} \right) + \left(\frac{q^2 \Delta}{\tau_c^{-1}} \right)^2 \right]^{1/2} \right\} \tag{109}$$

where $\tau_c = (k_f + k_b)^{-1}$ is the chemical relaxation time, $K_e' = k_f/k_b = \bar{c}_B/\bar{c}_A$, and $\Delta - D_A - D_B$.

When the diffusion coefficients of reactants and products are the same, $D_A = D_B = D$, the reaction can be detected by scattering only if the refractive index increments $(\partial n/\partial c_A)$ and $(\partial n/\partial c_B)$ are different.[15]

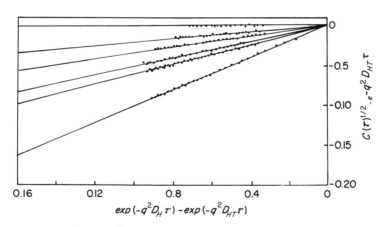

FIG. 11. Plot of $[C_{it}^{homo}(\tau)]^{1/2} - e^{-q^2 D_{HT}\tau}$ vs $(e^{-q^2 D_H\tau} - e^{-q^2 D_{HT}\tau})$ for the T4 phage assembly reaction H + T → HT. The reaction times and percentage of phage parts reacted, reading from bottom to top curve, are 0 min, 0%; 625 sec, 39.5%; 1070 sec, 48.2%; 1276 sec, 65.4%; 2762 sec, 79.2%; 5000 sec, 98.1%. From J. A. Benbasat and V. A. Bloomfield, *J. Mol. Biol.* **95**, 335 (1975).

In this case we find

$$\Gamma_- = (\tau_c^{-1} + q^2 D)$$
$$\Gamma_+ = q^2 D \tag{110}$$

and the relative weights of the exponential decays are, letting $(\partial n/\partial c_B) = \zeta(\partial n/\partial c_A)$

$$B_- = \left(\frac{\partial n}{\partial c_A}\right)^2 M_A c_A \frac{K_e'(1 - \zeta)^2}{1 + K_e'} \tag{111}$$

$$B^+ = \left(\frac{\partial n}{\partial c_A}\right)^2 M_A c_A \frac{(1 + \zeta K_e')^2}{1 + K_e'}$$

If $(\partial n/\partial c_A) = (\partial n/\partial c_B)$, $\zeta = 1$ and only a single exponential decay, $\Gamma_+ = q^2 D$, will appear in the autocorrelation function since $B_- = 0$. If $\zeta \neq 1$, two exponentials will appear, and as the scattering angle is decreased, eventually $\tau_c^{-1} \gg q^2 D$, and the time dependence of $C_E(\tau)$ is dominated by the chemical reaction rate. Unfortunately, polarizabilities or refractive index increments rarely change very much during chemical processes. This means that ζ is very close to 1, and the B_- will be very small compared to B_+. If $K_e' = 1$, $\zeta = 1.06$ (corresponding to a change in $\partial n/\partial c$ from 0.18 to 0.19 on going from A to B), then $B_-/B_+ = 7.3 \times 10^{-4}$ so the chemical relaxation term will contribute only 0.007% of the total scattering intensity. This would not normally be detectable.

The other extreme case, more likely for macromolecular reactions, occurs when the refractive index increments remain constant $= \partial n/\partial c$, but the diffusion coefficient changes during the reaction. Then the decay rates are given by the full equations (109), and the relative weights by

$$B_- = -\frac{M_A \bar{c}_A (\partial n/\partial c)^2}{\Gamma_+ - \Gamma_-} [q^2(D_A + K_e' D_B) - \Gamma_+(1 + K_e')] \tag{112}$$

$$B_+ = -\frac{M_A \bar{c}_A (\partial n/\partial c)^2}{\Gamma_- - \Gamma_+} [q^2(D_A + K_e' D_B) - \Gamma_-(1 + K_e')]$$

In the limit of very fast reactions, $\tau_c^{-1} \gg q^2 \Delta$, these expressions simplify to

$$\Gamma_- = \tau_c^{-1} \tag{113}$$
$$\Gamma_+ = q^2[D_A + (K_e' D_B)]/(1 + K_e')$$

and

$$B_- = M_A c_A (\partial n/\partial c)^2 \left(\frac{K_e'}{1 + K_e'}\right)\left(\frac{q^2 \Delta}{\tau_c^{-1}}\right)^2 \tag{114}$$

$$B_+ = M_A c_A (\partial n/\partial c)^2 (1 + K_e')$$

In this limit the relaxation times are similar to those in Eq. (110), with D replaced by an average diffusion coefficient weighted by the equilibrium fractions of the two components. However, in contrast to the previous case, the intensity B_- associated with the chemical relaxation mode decreases rapidly, as q^4, with decreasing scattering angle. In this case nothing is gained by working at low scattering angles to dissect out the contribution of reaction to the autocorrelation function, and again the intensity of the reaction mode is much less than that of the diffusion mode.

If experiments are carried out under conditions where $\bar{c}_A = \bar{c}_B$, so that $K'_e = 1$, then the field autocorrelation function can be written in the form

$$\frac{C_E(\tau)}{M_A\bar{c}_A(\partial n/\partial c)^2} = \left(1 - \frac{1}{\sigma}\right)\exp\left[-\left(1 + \frac{x}{2} + \frac{x\sigma}{2}\right)q^2\bar{D}\tau\right]$$
$$+ \left(1 + \frac{1}{\sigma}\right)\exp\left[-\left(1 + \frac{x}{2} - \frac{x\sigma}{2}\right)q^2\bar{D}\tau\right] \quad (115)$$

where $x = \tau_c^{-1}/q^2\bar{D}$, $\bar{D}$ is the average diffusion coefficient $(D_A + D_B)/2$, and $\sigma = [1 + (q^2\Delta/\tau_c^{-1})^2]^{1/2} = [1 + (\Delta/x\bar{D})^2]^{1/2}$. It can be seen from these expressions that in order for reaction to be observable, τ_c^{-1} must be $\gtrsim q^2\bar{D}$ and $\Delta/\bar{D}$ cannot be too much less than 1. These criteria are difficult to satisfy simultaneously in practice, since few protein reactions are known that are in rapid dynamic equilibrium while occasioning large changes in molecular volume. Some potentially favorable cases are described by Benbasat and Bloomfield.[41]

The same equations are applicable to the reaction $2A \underset{k_b}{\overset{k_f}{\rightleftharpoons}} A_2$, if the parameters are redefined as $\tau^{-1} = 4k_f\bar{c}_A/M_A + k_b$ and $K'_e = 4\bar{c}_Ak_f/M_Ak_b$.

Reaction Kinetics from Electrophoretic Light Scattering (ELS). The difficulty in studying rapid reactions at equilibrium by QLS lends special interest to the possibility of doing so by ELS, despite the additional technical complications. The advantage of ELS in this application is spectral separation conferred on the Lorentzian signals from reactants and products, so long as they have different electrophoretic mobilities.

Figure 12 shows simulated ELS spectra for the unimolecular isomerization $A \underset{k_b}{\overset{k_f}{\rightleftharpoons}} B$ at different exchange rates, with transport parameters typical of small globular proteins.[13] When the rate constants are zero, the spectrum is simply that for a mixture, consisting of two Lorentzians. That for A has Doppler shift $\omega_A = \mu_A\vec{q}\cdot\vec{E}$ and half-width $\Delta\omega_A = q^2D_A$; and that for B has $\omega_B = \mu_B\vec{q}\cdot\vec{E}$ and $\Delta\omega_B = q^2D_B$. In the limit of very fast kinetics, when $k_f,k_b \gg \omega_A,\omega_B$ and q^2D_A,q^2D_B, the two peaks coalesce into a single peak at an intermediate frequency. In the intermediate regime, where

[41] J. A. Benbasat and V. A. Bloomfield, *Macromolecules* 6, 593 (1973).

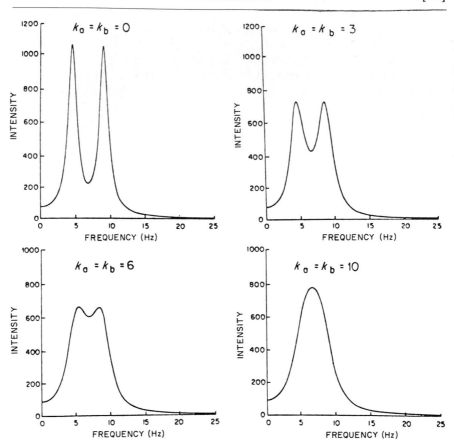

FIG. 12. Predicted electrophoretic light-scattering spectra for the system A $\rightleftarrows$ B. Experimental conditions assumed were $D_A = D_B = 6 \times 10^{-7}$ cm^2/sec; $\mu_A = 1 \times 10^{-4}$ cm^2/sec V; $\mu_B = 2 \times 10^{-4}$ cm/sec V; E = 100 V/cm; $\theta = 1°$; $\lambda_0 = 5.145 \times 10^{-5}$ cm. From B. R. Ware, Adv. Colloid Interface Sci. **4**, 1 (1974).

reaction rate is comparable to ω_A and ω_B, the two separate lines broaden and move toward each other. The situation is analogous to chemical exchange studies by NMR spectroscopy. Sample curves for the dimerization reaction $2A \rightleftarrows A_2$ have been computed by Berne and Gininger.[42] They are qualitatively similar to those in Fig. 12. The complete equations describing these curves are very complicated, except in the slow- and fast-exchange limits. The reader is referred to the original literature[13, 42] for these results.

[42] B. J. Berne and R. Giniger, *Biopolymers* **12**, 1161 (1973).

Number Fluctuations in Very Dilute Solutions

When the average number $\langle N \rangle$ of particles in the scattering volume V is sufficiently small, so that the relative fluctuation $\delta N/\langle N \rangle = \langle N \rangle^{-\frac{1}{2}}$ is significant compared to unity, homodyne QLS can give information on the dynamics of the occupation number fluctuations $\delta N(t)$. The theory has been developed by Schaefer and Berne.[16] They showed that the normalized second-order autocorrelation function for scattering from a very dilute solution of identical, spherical, independent particles is

$$g^{(2)}(\tau) = 1 + e^{-2q^2D\tau} + \langle \delta N(0)\delta N(\tau) \rangle/\langle N \rangle^2 \qquad (116)$$

The first two terms will be recognized as arising from the familiar Siegert relation based on the Gaussian approximation. The last term is new, and arises from a Poisson distribution of occupation number fluctuations.

There are two relaxation times associated with Eq. (116): the diffusional relaxation time $(2q^2D)^{-1}$ and the characteristic time for a particle to traverse the scattering volume. Generally the latter is much longer than the former. Therefore $g^{(2)}(\tau)$ will initially decay rapidly due to diffusion, and then more slowly due to number fluctuations. The latter decay may appear as a time-dependent baseline.

Laser beams have Gaussian intensity profiles. If we call the standard deviation in intensity across the beam σ_1, and assume that the scattered light is observed through collection optics defining a Gaussian profile of width σ_2 at right angles to the direction of propagation, then

$$\langle \delta N(0)\delta N(\tau) \rangle = \langle N \rangle \left[1 + \frac{2\langle \Delta r^2(\tau) \rangle}{3\sigma_1^2} \right]^{-1} \left[1 + \frac{2\langle \Delta r^2(\tau) \rangle}{3\sigma_2^2} \right]^{-1/2} \qquad (117)$$

where $\langle \Delta r^2(\tau) \rangle$ is the mean-square displacement of a particle in time τ. This equation has been used by Schaefer[18] to interpret number fluctuation dynamics on suspensions of motile and dead *Escherichia coli* (Fig. 13). The solid curve is calculated for a particle that swims in a straight line with speed V for a distance (mean free path) $\Lambda \gg \sigma_1, \sigma_2$ so that $\langle \Delta r^2(\tau) \rangle = \langle V^2 \rangle \tau^2$. The dot–dash curve is calculated for a diffusing particle, whose mean free-path is $\ll \sigma_1, \sigma_2$, and for which $\langle \Delta r^2(\tau) \rangle = 6D\tau$, where $D = \frac{1}{3}\Lambda V$. Dead bacteria fit the diffusion equation, though they are so large that D is very small and relaxation is hardly measurable. Motile bacteria fit neither the diffusing nor free-particle models, but instead obey random-walk statistics.

Further applications of QLS to bacterial motility and chemotaxis are discussed in the literature.[43]

Number fluctuation measurements of the type described here are but

[43] See, e.g., G. B. Stock, *Biophys. J.* **16**, 535 (1976) and references cited therein.

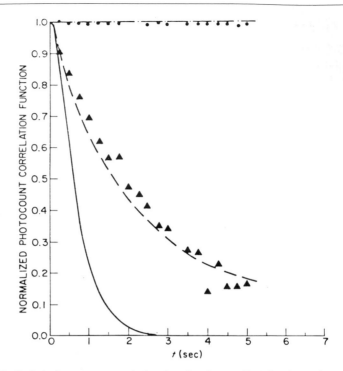

FIG. 13. Scaled photocount correlation function for motile (triangles and squares) and nonmotile (circles) *Escherichia coli* bacteria. The time-independent background has been subtracted and the data are normalized to unity at $t \to 0$. The scattering angles are 15° (squares) and 90° (triangles and circles). Solid curve, calculated curve for a free particle. Dot–dash curve, calculated curve for diffusive motion. Dashed curve, approximate curve for random-walk motion. The parameters used for the theoretical curves are $\langle V^2 \rangle^{1/2} = 39$ μm/sec, $\langle L \rangle = 17$ μm, $\sigma_1 = 56$ μm, $\sigma_2 = 15.5$ μm, and $D = 4 \times 10^{-9}$ cm^2/sec. From D. W. Schaefer, *Science* **180**, 1293 (1973).

one example of the general field of fluctuation correlation spectroscopy. Detection of these fluctuations does not depend on the specific coherence properties of laser scattering, but instead can be based on any sensitive technique for measuring concentrations, such as fluorescence or ionic conductivity. For a general review of this field, see Elson and Webb.[17]

Instrumentation and Experimental Procedures

Optics

The optical part of a self-beating (homodyne) light-scattering apparatus consists of a light source, focusing optics, sample cell, and collect-

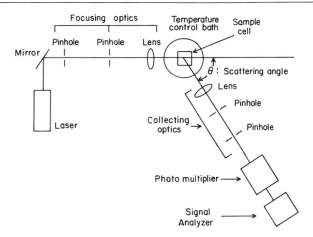

FIG. 14. Block diagram of quasi-elastic light-scattering apparatus.

ing optics. A block diagram of the arrangement of these components is given in Fig. 14. The components should be mounted on a vibration isolation table to insulate them from building vibrations. Light from the laser light source is reflected by a mirror into the focusing optics. By adjusting the vertical and horizontal orientations of the mirror, the direction and position of the beam is determined. (While the laser beam may be focused directly on the scattering cell, use of the mirror simplifies optical alignment). The focusing optics focus the beam onto the sample. The light scattered by the sample is collected by the collecting optics and directed to the photocathode of the photomultiplier tube. The scattering angle is determined by the two pinholes in the collecting optics. We now discuss the role and specifications of each component in more detail, starting with the laser.

Laser Light Source. A light source for light-scattering spectroscopy must have narrow linewidth (<1 Hz), spatial coherence, and high power density (1–100 W/cm^2). Although there exist conventional arc sources that satisfy the power density requirement, only lasers have adequately narrow lines and spatial coherence.[44] A continuous wave gas laser is used in QLS. In choosing such a laser, one must consider both output power and frequency.

For a given length of data accumulation time, the signal-to-noise (S/N) ratio depends strongly on the scattered intensity. According to Eq. (35), the scattered intensity depends on the product Mc, since the refractive

---

[44] See, however, E. Jakeman, P. N. Pusey, and J. M. Vaughan, *Opt. Commun.* **17**, 305 (1976) for an example of QLS from a liquid crystal using a conventional Hg arc source.

index increment $(\partial n/\partial c)$ is very nearly the same for all proteins. For example, a 1 mW laser provides adequate power for signal accumulation in 10 min with a sample of $M_r = 5 \times 10^6$ at $c = 1$ mg/ml. Then for the same wavelength, scattering angle, and signal accumulation time, a 100 mW laser will be necessary for a protein with $M = 5 \times 10^4$ at the same concentration.

The stability of the power output must also be considered. All lasers undergo some small long-time (seconds or minutes) drift in their output. In the analog mode of operation of an autocorrelation or spectrum analyzer, a low-frequency filter will eliminate this fluctuation, so long as it occurs on a time scale much longer than the relevant correlation time for the process under study. However, in the photon counting mode the drift contributes to a deviation in the base line from its true value. A well-stabilized laser should have an rms instability no larger than 0.2%. It is desirable to monitor the laser output stability with a light power meter frequently when photon counting is being done, and to adjust the laser so as to minimize output fluctuations.

The frequency or wavelength of the laser chosen has several consequences. It can affect the scattering intensity, through the Rayleigh λ^{-4} dependence in Eq. (35). It influences the range of scattering vector employed, according to $q = (4\pi n/\lambda) \sin (\theta/2)$. That is, a high value of q, which might be desirable for example in QLS study of rotation or internal motion, can be achieved either by working at high θ or at low λ. Finally, the wavelength should generally not be chosen to fall in an absorption band of a molecule in solution, because of laser-induced heating of the solution.[45] Wavelengths of commonly available laser lines, and typical power associated with each line, are listed in the table.

The output of a laser operating at a given frequency is composed of several longitudinal modes separated by frequency $c/2L$ where c is the speed of light and L is the length of the laser cavity. In each longitudinal mode there are several transverse modes, corresponding to possible solutions of the differential equation for the electromagnetic field with boundary conditions imposed by the geometry of the plasma tube.[46] The separation between the longitudinal modes is of the order of 100 MHz, and between the transverse modes, 5 MHz. When the frequency range to be studied is much less than $c/2L$ (this will be the case in nearly all macromolecular applications), then the beat frequencies between the longitudinal modes can easily be integrated out electronically.

---

[45] However, great enhancement of the depolarized scattering intensity of diphenylpolyenes has been achieved by working near a molecular absorption frequency. D. R. Bauer, B. Hudson, and R. Pecora, *J. Chem. Phys.* **63**, 588 (1976).

[46] B. Lengyel, "Introduction to Laser Physics," Chap. 2. Wiley, New York, 1966.

WAVELENGTHS AND POWERS (mW) OF TYPICAL CURRENTLY AVAILABLE CW GAS
LASERS USEFUL IN QLS EXPERIMENTS[a]

Wavelength (nm)	He–Ne	Argon	Krypton	Argon/krypton
799.3			30	
793.1			10	
752.5			100	
676.4			120	20
647.1			500	200
640.1	50			
632.8	50			
611.8	50			
568.2			150	80
530.9			200	20
520.8			70	200
514.5		800		20
501.7		140		50
496.5		300		200
488.0		700		10
482.5		300	30	60
476.5				
476.2			50	
472.7		60		
465.8		50		
457.9		150		
454.5				
351.1 + 363.8				
350.7 + 356.4				

[a] Lasers may be available that have higher power outputs.

However, beats between transverse modes can pose difficulties in determining correlation functions of rapid processes (time scales shorter than 5 μsec). Also, any correlation between amplitudes of different modes, or mode hopping, will give additional noise. For example, Oliver[47] found that S/N is a factor of 2–4 worse in a QLS experiment employing an Ar$^+$ laser operating in multiple transverse modes then when the same laser operates in a single mode. Single-mode operation is obtained by introducing an etalon into the laser cavity. Such etalons may be purchased from laser manufacturers. To select a single mode by tuning the etalon, one also needs a Fabry–Perot interferometer which reveals the mode structure of the laser. When the laser is not tuned properly and has several longitudinal modes, the output from the interferometer will show several peaks, as in

[47] C. J. Oliver, in "Photon Correlation and Light Beating Spectroscopy" (H. Z. Cummins and E. R. Pike, eds.), p. 151. Plenum, New York, 1973.

Fig. 15a. Adjusting the etalon eliminates these peaks, and a single peak will be observed, as in Fig. 15b.

For optimal spatial coherence of the laser output one selects the TEM_{00} transverse mode. (TEM stands for *Transverse Electro Magnetic*; the subscripts denote the number of nodes in the standing electromagnetic waves orthogonal to the laser cavity axis.). The TEM_{00} mode has a Gaussian intensity distribution of the beam cross-section. This can be monitored by scanning a small (1–5 μm diameter) pinhole across the beam, or by observing the reflected beam, after the beam profile is expanded 10–100 times by a lens.

Focusing Optics. The mirror in Fig. 14 is used to direct the laser beam through the two pinholes that define the zero (forward) angle of scattering. The mirror should be on a mount that allows it to be rotated horizontally and vertically. To avoid destroying the optical coherency of the laser

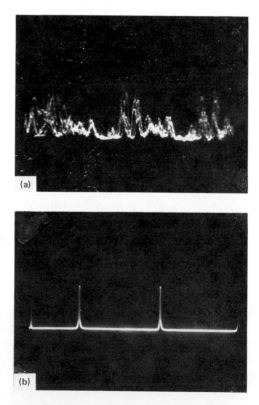

FIG. 15. Mode pattern of Ar^+ laser, analyzed by a Fabry–Perot interferometer. (a) without etalon in the laser cavity; (b) with a well-tuned etalon.

beam, the mirror must be flat to within one-half wavelength over its surface. It is also desirable that the mirror have greater than 98% reflectivity.

The two pinholes in the focusing optics determine the zero angle and
clean the beam profile of the laser. As stated above, the theoretical cross-
sectional intensity distribution of the TEM_{00} mode is Gaussian. However,
there may be stray light accompanying this Gaussian beam. The stray
light is spatially incoherent and should be eliminated to improve the signal
to background ratio. To accomplish this, the first pinhole should have a
diameter less than that of the stray light but greater than the $1/e$ point of
the laser intensity profile. This point, given in the laser manual, is typically 1–2 mm.

To define the zero angle accurately, the two pinholes should be as
widely separated as possible. The angular uncertainty due to the finite
sizes of the pinholes is

$$\delta\theta = (r_1 + r_2)/d \tag{118}$$

where r_1 and r_2 are the pinhole radii and d is the distance between them.

The focusing lens serves to focus the incident beam onto the sample so
that a smaller scattering volume and higher power density can be
achieved. As we shall see below, a small scattering volume is necessary to
obtain a small number of coherence areas at the phototube surface. In a
homodyne experiment with a 15–30-cm focal length lens, the beam is
focused to a diameter of 100–200 μm.

The focused beam at the focal point is not a geometrical point, but
rather has a finite cross section due to diffraction and imperfections in the
focusing lens. When a collimated beam of diameter D_f is focused by a lens
of focal length f, the beam will have a diameter at the focal point of

$$d_0 = 1.22\lambda f/D_f \tag{119}$$

in the diffraction-limited case.[6] The beam will have a cross section close to
this near the focal point. As the beam propagates from the focal point, its
diameter increases according to

$$d(x)^2 = d_0^2[1 + (4\lambda x/\pi \, d_0)^2] \tag{120}$$

If we take the focal depth x as the point where the power density is 90% of
its value at the focal point, $x = 0$, so $d(x)/d_0 = (1/.9)^{\frac{1}{2}} = 1.054$, we find

$$x = \pi \, d_0^2/12\lambda \tag{121}$$

For a lens of focal length 15 cm and beam diameter $D_f = 2$ mm with
$\lambda = 632.8$ nm, $d_0 = 5.79 \times 10^{-3}$ cm, and $x = 0.14$ cm.

In addition to these diffraction effects, imperfections in the lens will
produce further broadening of the focused beam profile. For applications

that depend critically on a definition of the scattering volume, the beam profile at each point should be measured as described in the previous section.

Another factor affecting the choice of the focusing lens is the angular divergence of the light produced by the focusing. This is $\delta\theta = D_f/2f$, which can be important in small-angle scattering, since the correlation function or spectrum will be averaged over $\delta\theta$. With $f = 15$ cm, $D_f = 2$ mm, $\delta\theta = 6.7$ milliradians $= 0.38$ degree, which is a 5% or greater angular uncertainty when the scattering angle is 7 degrees or less. Obviously, the angular uncertainty can be reduced by increasing f.

Sample Cell and Temperature Control. In choosing a sample cell, one must consider its geometrical shape, the optical quality of the cell walls, and the total sample volume required to fill the cell. Either a cylindrical cell or a rectangular one, such as a fluorescence cuvette, can be used. The cell walls should be optically polished to avoid intense light scattering from scratches on these surfaces, particularly when homodyne QLS is measured at low scattering angles. It is found that even polished cells give problems when θ is less than 15 deg. The intensity of light scattered from the walls can be estimated from the signal obtained when the macromolecular sample is replaced with filtered buffer in the cell.

Several special designs to avoid scattering from cell walls are discussed by Chu,[6] and are shown in Fig. 16. A long-path rectangular cell is most easily obtained. The scattering arrangement is depicted in Fig. 16a.

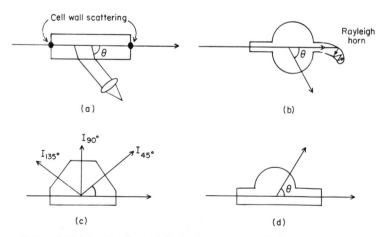

FIG. 16. Special light scattering cell designs: (a) long path length rectangular cell, (b) cell with flat entrance window and Rayleigh horn at the exit, (c) hemioctagonal cell, (d) cell with flat entrance and exit windows. From B. Chu, "Laser Light Scattering." Academic Press, New York, 1974.

The basic idea underlying all these designs is to locate the cell walls as far as possible from the scattering volume observed by the collecting optics.

Since the intensity of scattered light is proportional to the square of the molecular weight for a given number of particles, QLS as well as total intensity light scattering is very sensitive to the presence of dust or larger aggregates. The sample should always be centrifuged or filtered thoroughly to remove large particles before it is placed in the cell. In many cases a standard glass centrifuge tube may be used as the scattering cell, and the solution centrifuged in the cell itself. For more sophisticated development of this idea, see Haas et al.[48] The scattering cell should also be cleaned thoroughly with cleaning solution and then flushed with filtered water for 2–5 min. If the cell has been properly cleaned, the scattered light intensity from the cell filled with water or buffer should be less than 2% of the scattered intensity from the macromolecular sample. The cell surface should also be cleaned thoroughly to minimize scattering from the walls and avoid destruction of the coherence of incident and scattered light. Further details of solution and cell cleaning are given in previous reviews in this series.[11,49]

It is generally desirable to thermostat the scattering cell by surrounding it with a bath of fluid that may be circulated. If the walls of the cell are thick, it may be desirable that the circulating fluid have a refractive index closely matching that of the glass, to reduce refraction effects at the interface. Otherwise, water may be used. Particularly for low-angle scattering experiments, it is important to filter the circulating fluid. The windows for the incident and scattered light should be maintained scrupulously clean.

When the refractive index of the scattering solution is the same as that of the liquid in the thermostatting bath, and the scattering cell wall is less than 1 mm thick, no correction of the scattering angle need be made. However, if the refractive indexes are different, a correction is required for a rectangular cell. The correction is based on Snell's law,

$$n_1 \sin \theta_1 = n_0 \sin \theta_0 \tag{122}$$

where n_1 and n_0 are the refractive indexes inside and outside the cell. If the scattered light is observed through the forward face of the cell in Fig. 17a, then the true scattering angle is

$$\theta_1 = \sin^{-1}\left[(n_0/n_1) \sin \theta_0\right] \tag{123}$$

[48] D. D. Haas, R. V. Mustacich, B. A. Smith, and B. R. Ware, *Biochem. Biophys. Res. Commun.* **59**, 174 (1974).
[49] E. P. Pittz, J. C. Lee, B. Bablouzian, R. Townend, and S. N. Timasheff, this series Vol. 27 [10].

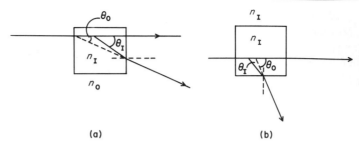

(a)   (b)

FIG. 17. Relationship between the true scattering angle θ_1 and the apparent scattering angle θ_0 for a rectangular cell oriented perpendicular to the incident beam. (a) Low-angle scattering. (B) Large-angle scattering.

If scattering is observed through the side face (Fig. 17b), then

$$\theta_1 = 90° - \sin^{-1}[(n_0/n_1)\sin(90° - \theta_0)] \tag{124}$$

When a cylindrical cell is used and the scattering volume is located at the rotation axis of the collecting optics, the scattered light always exits normal to the cell wall and no correction is required.

Collecting Optics. The lens and pinholes of the collecting optics serve to define the scattering angle and the scattering volume, to collect and send scattered light to the phototube, and to determine the number of coherence areas on the photocathode surface. These components are mounted on a rotating arm to change the scattering angle. The angular uncertainty is again given by Eq. (118), where the quantities now refer to the collecting pinholes.

The scattering volume is that volume of the sample from which scattered light is selected and analyzed by the instrument. From Fig. 18, we

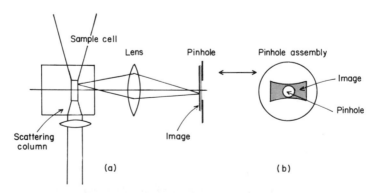

FIG. 18. Schematic diagram of collecting optics. An image of the scattering volume is formed at the surface of the pinhole. The pinhole should be positioned to select the desired portion of image.

see that the collecting lens forms an image of the scattering column on the surface of the first pinhole. A small portion of this image is passed through to the second pinhole. By adjusting the position of the first pinhole with respect to the image, one can select the desired portion of the scattering column for analysis. This enables one to avoid detecting stray light scattered from the cell walls.

The concept of spatial coherence and the number of coherence areas on the phototube surface are well described by Cummins and Swinney[9] and Lastovka.[50] We follow the argument of the former authors. Consider the homodyne photocurrent spectrum given by Eq. (28). The light beating signal-to-shot noise ratio at $\omega = 0$ is

$$P_S/P_{SN} = \langle i \rangle / e\Gamma \tag{125}$$

which is the mean photoelectron emission rate divided by the half-width of the optical spectrum. The photocurrent generated by a portion of the photocathode surface is proportional to its area, ΔA. The total current generated by the phototube with surface area A_d is $A_d/\Delta A$ times the current generated by ΔA, so long as the optical field is coherent over the phototube surface. Therefore, the signal (and dc) contributions to Eq. (28) will increase like A^2 or $\langle i \rangle^2$, while the shot noise contribution will vary only like A or $\langle i \rangle$, and P_S/P_{SN} increases linearly with A or $\langle i \rangle$.

However, if the optical field is not spatially coherent over the phototube surface, then the currents from different areas add in an uncoordinated, random walk fashion and P_S will increase linearly with A or $\langle i \rangle$. When this condition is reached, P_S/P_{SN} becomes a constant, independent of detector area. In this case[51]

$$P_S/P_{SN} = \langle i \rangle / n_c e\Gamma \tag{126}$$

where n_c is the number of coherence areas on the photocathode surface.

The dependence of the homodyne autocorrelation function on the number of coherence areas is more complicated, but Jakeman et al.[52] have shown numerically that

$$C_i^{\text{homo}}(\tau) = C(\infty)[1 + f(n_c)|g^{(1)}(\tau)|^2] \tag{127}$$

When $n_c \gg 1$, $f(n_c) = 1/n_c$. However, a typical correlation measurement is done in the range $n_c \sim 1$ to 2. Then $f(n_c)$ should be obtained from Fig. 19. In subsequent equations where no ambiguity will result, we suppress the superscript "homo" or "het" on $C_i(\tau)$.

[50] J. Lastovka, Ph.D. thesis, Massachusetts Institute of Technology, 1967.
[51] A. T. Forrester, R. A. Gudmundsen, and P. O. Johnson, *Phys. Rev.* **99**, 1691 (1955).
[52] E. Jakeman, C. J. Oliver, and E. R. Pike, *J. Phys. A: Gen. Phys.* **3**, L 45 (1970).

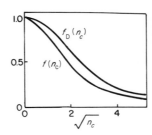

We now show how the coherence area or n_c can be estimated from the optical properties of the system. When light from an incoherently emitting source of finite size (the scattering volume) is detected by a detector of finite size, there will exist a partial coherency of the light falling on different points of the detector. The solid angle over which the finite size source, of area A_S, gives coherence is[53]

$$\Omega_c = \lambda^2/A_S \qquad (128)$$

When the detector is a distance d from the source, the area A_c of the source over which coherence persists is

$$A_c = \Omega_c d^2 \qquad (129)$$

Then the number of coherence areas on the detector surface of area A_d is

$$n_c = A_d/A_c = A_d A_S/\lambda^2 d^2 \qquad (130)$$

When the radius of the detector is 0.1 cm, the radius of the source is 0.01 cm, $d = 30$ cm, and $\lambda = 632.8$ nm, $n_c = 2.7$. This falls in the range $n_c = 2$ to 4 desired for optimal utilization of the signal.[47]

Signal Detection and Processing

Once the scattered light falls on the surface of the photomultiplier tube, it produces an electrical signal that is processed and analyzed to obtain the desired autocorrelation function or spectrum. In this section we describe the various steps in this process. A block diagram of the signal detection and processing electronics is given in Fig. 20.

Photomultiplier. Since the efficiency of data collection depends critically on the choice of the photomultiplier tube and optimization of condi-

53 M. Born and E. Wolf, "Principle of Optics." Pergamon, Oxford (1970).

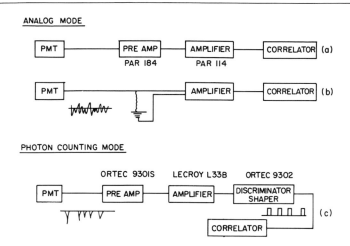

FIG. 20. Block diagrams of electronics for correlation analysis of photocurrent. (a) Analog mode operation with relatively low current output; (b) analog mode operation with relatively high current output; (c) photon counting mode.

tions for its use, we describe these factors here in some detail. A useful general source of information on photomultipliers has been prepared by RCA.[54]

The main considerations in choosing a photomultiplier tube are quantum efficiency, spectral response, gain, response time, and dark current level. To understand quantum efficiency, we recognize that although we have treated the scattered light in a classical way as a continuous field, it in fact consists of a series of discrete particles, or photons. Each photon arriving at the photocathode will have a probability of ejecting an electron, which will be multiplied in passing through the following dynode chain in the tube. The probability that an incoming photon will eject an electron is the quantum efficiency of the tube. This depends on the wavelength of the incident light. The spectral response, which depends on the material used for the photocathode and the phototube window, is specified by the manufacturer as S-1, S-20, etc. For the red light of a He–Ne laser, a long wavelength sensitivity is required. Although all phototubes have rather low quantum efficiencies in this region, the S-20 surface is the best.

The multiplication factor of the dynode chain is called the gain, G. At the output stage, or anode, there will be a train of electric pulses whose average charge $\langle Q \rangle$ is equal to eG where e is the electron charge. The actual charge Q of a specific pulse has a nearly Poisson distribution about

[54] RCA, "Photomultiplier Manual," Harrison, N. J. (1970).

$\langle Q \rangle$ due to the statistical nature of the photoemission and multiplication processes. Normally the quantum efficiency is in the range 0.01 to 0.3, and the gain is 10^5 to 10^7.

The response time τ_{RC} of a phototube is an important consideration when photon counting measurements are to be made. It is given by $\tau_{RC} = RC,$ where C is the tube capacitance and R is either the output resistance of the tube or the input resistance of the instrument into which the phototube output is fed, whichever is larger. Normally, C is about 3×10^{-11} farad, so R less than 3×10^5 ohms should be used to get τ_{RC} less than 1 μsec. The response time is roughly the width or duration of the output pulse from the tube. In the photon counting mode, it is desirable that successive pulses do not overlap.

Dark current is the most important factor in phototube noise. The dark current is the output current of the tube when there is no signal entering it. It has temperature-dependent and independent components. The main source of temperature-dependent dark current is thermionic emission at the cathode and gain noise of the dynode chain. To reduce noise from this source, it is customary to cool the tube with dry ice or liquid nitrogen. However, this also decreases the quantum efficiency. The relative importance of these effects depends upon the counting rate. Harker et al.[55] have shown that S/N increases upon cooling when an ITT FW130 tube (S-20) operates at low counting rates, below 8000/sec, but decreases at higher rates. When the temperature-dependent part of the dark current is reduced to a low level (a few hundred counts per second), the nonthermal sources begin to dominate. These result from a variety of sources, such as photoemission in the tube window due to light generated by cosmic rays, radioactivity, and ionization of residual gas in the tube, impurities in the photocathode surface, and dirt in the dynode chain. More details on dark current and optimal operating temperatures are available from the manufacturers and in the RCA manual.[54] Particularly in experiments where low scattering intensities are anticipated, such as small proteins and high dilutions, careful attention to minimizing dark current should be paid before ordering a particular tube.

Another important factor affecting S/N is the voltage across the photomultiplier dynode chain. Although the optimal operating voltage is generally specified by the tube manufacturer, there are many situations when one will wish to work at a higher voltage than optimal, particularly when the duration of the experiment is a limiting factor, as in irreversible kinetics studies. Then the signal to dark current ratio can be sacrificed to obtain data in a shorter time span. Since the dark current increases exponentially and the signal linearly as a function of applied voltage, a point is

[55] Y. D. Harker, J. D. Masso, and D. F. Edwards, *Appl. Opt.* **8**, 2563 (1969).

eventually reached where further voltage increase leads to a decrease in S/N. This occurs when the equivalent noise input (ENI) exceeds the signal level. The ENI is defined as the equivalent amount of incident light which, when modulated as a square wave by a chopper, produces an rms output current equal to the rms noise current within the bandwidth Δf of the electronics. It is given by[54]

$$\text{ENI} = \pi(ei_T\Delta f)^{\frac{1}{2}}/S \tag{131}$$

where i_T is the dark current at the photocathode, S is the sensitivity of the tube, that is, the current at the photocathode generated per lumen of incident light, and the other symbols have been defined previously. In terms of more easily measurable quantities, the relation between the current i_s of the signal at the output stage of the tube and the dark current at the anode $i_d = i_T\gamma$ should be such that

$$i_s \gg \text{ENI} \times S \times \gamma = \pi(ei_d\Delta f\gamma)^{\frac{1}{2}} \tag{132}$$

The output from a photomultiplier tube is a train of pulses of different charges, as shown in Fig. 21. These pulses appear to be randomly distributed in time. However, it is the analysis of this seemingly random pulse train by an autocorrelator or spectrum analyzer that leads to the correlation functions or spectra that provide information on dynamic processes in solution. After the output is properly amplified and processed, by procedures to be discussed in the next sections, the signal is sampled for a duration ΔT by the autocorrelator. There are two modes of sampling, the analog mode and the photon-counting mode. In the photon-counting mode the individual pulses are counted during ΔT and this number, $n(t)$ is fed

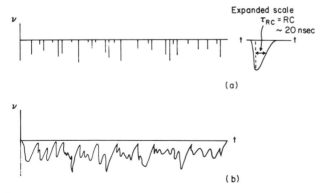

(a)

(b)

FIG. 21. Train of output pulses from photomultiplier. (a) Low counting rate case with fast response photomultiplier, shown with expanded pulse shape; (b) high counting rate where pulses start to overlap each other with slow response photomultiplier.

into the correlator which forms the correlation function defined as

$$C(\tau) \equiv \langle n(0)n(\tau) \rangle \tag{133}$$

In the analog mode, the total charge Q in the pulses received during ΔT is integrated and input to the correlator, which forms the correlation function

$$C(\tau) \equiv \langle Q(0)\, Q(\tau) \rangle \tag{134}$$

Signal Amplification: Analog Mode. To process signals properly in the analog mode, one must understand the shape of the phototube pulse and the role of the bandwidth of the electronics used for amplification. The shape of each pulse is an exponentially decaying function $q(t) = \exp(-t/\tau_{RC})$. In the high-counting rate situations where the analog mode is normally used, the pulses overlap each other and the signal appears as a randomly fluctuating charge

$$Q(t) = \sum_{t'=-\infty}^{t} q(t')\exp[-(t - t')/\tau_{RC}] \tag{135}$$

or current $i(t) = dQ/dt$.

Even in the high counting rate conditions characteristic of the analog mode, the current output from the phototube is very low, in the range 10^{-7} amp for homodyne scattering. Therefore, it must be amplified before processing by the autocorrelator. Normally a preamplifier of gain 10–100 should be used. The output from the preamplifier is then input to a high-gain (50 to 500 ×) amplifier. Both the preamplifier and amplifier should be characterized by low noise and wide bandwidth (adjustable up to about 1 MHz). Canberra, Ortec, and PAR are among the companies that provide suitable instruments.

If the current output from the photomultiplier tube is high enough (this will rarely be the case for scattering from enzyme solutions, but may pertain for scattering from larger particles or in heterodyne scattering), the preamplifier may be eliminated. Then the output from the photomultiplier is passed through a resistor to ground. The voltage across the resistor is then amplified and fed to the autocorrelator. This voltage is calculated from Ohm's law, $V = iR$. Since i is about 10^{-7} amp, R must be about 1 MΩ to get a voltage of 100 mV suitable for subsequent amplification. However, this resistor, combined with the internal capacitance C of the photomultiplier tube, determines the time constant τ_{RC}. Since C is about $30\,pF$, τ_{RC} will be about 30 μsec. However, the time constant should be less than the sample time ΔT, so one is limited in using high resistances.

The time constant τ_{RC} can be measured by letting an uncorrelated signal, such as light from a flashlight bulb (care should be taken to keep the light level very low, so as not to damage the phototube) fall on the photocathode surface and be processed by the electronics. The only correlation function that should be detected is $\exp(-t/\tau_{RC})$ due to the RC integration. The effect of RC on the correlation function is discussed in detail by King and Lee.[56]

A wide-bandwidth preamplifier is desirable to follow the rapid time response of the photomultiplier (commonly about 20 MHz). However, for certain purposes it is desirable to put high and low frequency cutoff filters in the amplifier. Introducing a high frequency (ω_c) filter is equivalent to integrating the signal for a time $1/\omega_c$, thereby improving S/N. Also, from Eq. (131), ENI is proportional to the square root of the bandwidth, so the smaller ω_c, the lower the noise contributed by the dark current i_d. However, if $1/\omega_c > \Delta T$, an undesired term $\exp(-\omega_c t)$ will appear in the correlation function. For example, an amplifier with a cutoff frequency at 30 kHz will introduce an additional correlation time of 30 μsec. This effect would easily be observed in a correlator set at $\Delta T = 5$ μsec: the first six channels would be significantly distorted.

A low-frequency cutoff filter has two roles. It eliminates low-frequency noise such as line voltage fluctuations and mechanical vibrations, and, more important, it cuts off the long-time background part of the correlation function, thereby increasing the ease of data analysis. The autocorrelation function obtained in homodyne QLS has been shown to have the form of Eq. (127). It is difficult in the analog mode to determine the background term $C(\infty)$. However, since this term is time independent, a low-frequency filter automatically eliminates it, leaving only the time-dependent term to be analyzed. Care must be taken, of course, to choose the low-frequency cutoff ω_L so as not to deform the correlation function. That is, one requires $1/\omega_L \gg M\,\Delta T$, where M is the total number of channels in the correlator. The equivalent condition is that ω_L be substantially lower than the frequency of the slowest process of interest in the experiment. Since this is not always known beforehand, care must be taken in setting ω_L and ω_c.

Signal Amplification: Photon-Counting Mode. Photon counting is useful when the input signal level is low enough that individual pulses from the photomultiplier do not overlap each other. For photon counting one must use a photomultiplier tube which is fast enough to give a narrow pulse for each incident photon. The ITT FW30, the RCA 8850, and Channeltron photo tubes are adequate for this purpose. The output pulses from these photomultipliers have widths less than 20 nsec. The height of such a

[56] T. A. King and W. I. Lee, *J. Phys. E* **5**, 1091 (1972).

pulse in amperes can be calculated from

$$I = Q/\Delta T = \gamma e/\Delta T \tag{136}$$

where ΔT is the pulse width. For a gain of 5×10^6 and a width of 20 nsec, $I = 4.5 \times 10^{-5}$ amp. When this pulse is passed through a 50-Ω resistor, the standard input impedance of most rapid response electronics, it gives a pulse height on the order of 2 mV. The distribution of pulse heights is normally Poisson but to a certain extent depends on the phototube characteristics.

As in the case of analog mode processing a preamplifier is used to amplify this low level signal and the output of the preamp is then fed into another amplifier before the signal enters a discriminator. Both the preamplifier and amplifier should have very wide band widths, larger than 20 mHz, and should be dc coupled. LeCroy, Princeton Applied Research, and Ortec supply amplifiers and preamplifiers adequate for these specifications.

The output pulses from the amplifier consist not only of the pulses generated by photons, but also of pulses due to other origins, such as afterpulses and electronic noise in the photo tube. The electronic noise level is high in these low-level photon-counting experiments and therefore should be eliminated to the extent possible. This is made possible by the different pulse height distribution of the electronic noise and signal as depicted in Fig. 22. For this purpose one uses a discriminator set at a level between 30 mV to 1 V, such as those manufactured by Ortec, Canberra, and Princeton Applied Research. These discriminators have as output a standardized square wave pulse. These standardized pulses are more easily handled by most autocorrelators and are free of base-line fluctuations and other on-line noise.

Although the electronic noise and the signal have different pulse height distributions, they overlap to an extent that makes it difficult to set the

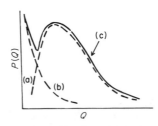

FIG. 22. Typical pulse height distributions $P(Q)$ as a function of charge/pulse Q in a photomultiplier used for photon counting. Curve (a) represents the signal, (b) the electronic noise; (c) is the resultant.

discriminator level at a point that eliminates the electronic noise without overly affecting the signal. The most desirable way to set the discriminator level is to experimentally measure the pulse height distribution for electronic noise and dark current. In a phototube suitable for photon counting experiments, there is a dip between these two distributions, as shown in Fig. 22. One sets the discriminating level at the dip to get rid of the electronic noise. It is always necessary to measure this pulse-height distribution at the same high voltage across the phototube as that to be used in the actual light-scattering experiment. The pulse height distribution would normally be obtained using a multichannel analyzer. If such an instrument is not available, one can obtain a crude distribution function by measuring the number of pulses of a given height from an oscilloscope display, while keeping the counting rate low.

If a pulse-height distribution cannot be measured directly, the following procedure should be employed to set the discriminator level. First, measure the counting rate of the output from the amplifier in the absence of an incoming light signal. Adjust the discriminator level so that the counting rate decreases to the dark current level specified for the phototube (this normally is provided by the manufacturer).

Autocorrelator Operation. In this section we discuss how an autocorrelator samples incoming signals and processes them to form the autocorrelation function. We follow closely the description given by Oliver.[47]

The incoming signals, in the form of randomly distributed pulses, are divided into time periods ΔT as shown in Fig. 23a. The number of counts in the ith time division are recorded in the ith shift register of the correlator as shown in Fig. 23b. At the same time $n_j(0)$ will be recorded in the common line of the autocorrelator. Multiplication is performed between the number in this common line and the numbers in the shift register line. The results of these multiplications are sent to memory where each number is added to the previously stored number in the ith register of the memory line. After the multiplication operation, each number in the shift register is shifted by ΔT, or the processing time T_p if T_p is greater than ΔT, and a new sample is taken as shown in Fig. 23b. The processing time is the time required for the correlator to do all the multiplications, storing, and shifting of numbers among registers. After a number N_s of these operations the number stored in the ith channel of memory will be $\sum_{k=0}^{N_s} n_k(0)$ $n_k(-i\ \Delta T)$. Since the autocorrelation function is an even function of time, then

$$C(i\ \Delta T) = \sum_{k=0}^{N_s} n_K(0)n_K(i\ \Delta T) \tag{137}$$

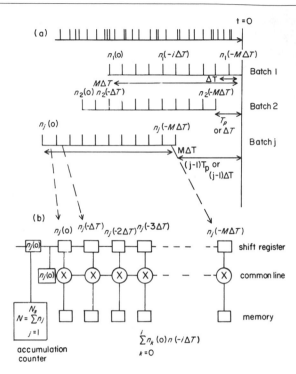

FIG. 23. Schematic diagram of (a) the photon-counting sequential-sampling scheme for a (b) correlator of M channels with sampling time T and processing time T_p. The correlator has a counter for the incoming signal at the front of the first channel.

When the processing time T_p is very small compared to the sampling time ΔT, there will be no delay due to processing and the next sampling batch is simply shifted by ΔT in time. However, when T_p is greater than ΔT the correlator has to wait T_p sec until necessary calculations are completed. During T_p the correlator will lose information, thereby decreasing its efficiency.

To avoid this problem some correlators use a procedure called clipping. Clipping may be applied to either or both of the shift register and the common line. In clipping mode, the counter is set at a level n_k. If $n_j > n_k$, the counter sends 1 to the shift register line; if $n_j < n_k$, the counter sends 0. In this way one substitutes addition for multiplication, thus saving processing time. The signal can be clipped either in just one of the shift register or common lines (single clip mode), or in both lines (double clipping). In the clipping modes the measured correlation function deviates to some extent from the original functional form.[57] In the single clip mode

[57] E. Jakeman and E. R. Pike, *J. Phys. A: Gen. Phys.* **2**, 411 (1969).

with clipping level set at k, the correlation function is given by[47]

$$g_k^{(2)}(\tau) = 1 + \left(\frac{1+k}{1+\langle n \rangle} \right) f(n_c) |g^{(1)}(\tau)|^2 + O(1/N_s) \qquad (138a)$$

where $\langle n \rangle$ is the average count rate per ΔT. Therefore it is best to set the clipping level $k = \langle n \rangle$ for best signal to background ratio. In the double clipping mode the form of the correlation function has been derived only for the case where both clipping levels are set at zero[57]:

$$g_{00}^{(2)}(\tau) = \left\{ 1 + f(n_c) \frac{1-\langle n \rangle}{1+\langle n \rangle} |g^{(1)}(\tau)|^2 \right\} \left\{ 1 - \left(\frac{\langle n \rangle}{1+\langle n \rangle} \right)^2 |g^{(1)}(\tau)|^2 \right\}^{-1}$$
$$(138b)$$

It is evident from these equations that clipping introduces some distortion into $g^{(2)}(\tau)$, but this is usually not significant.

Once the autocorrelation function has been accumulated in memory, it may be displayed on an oscilloscope screen or plotted in analog form, or may be fed into a computer for numerical processing. The manufacturer should be consulted about the nature of the output of the correlator to assure adequate interfacing with subsequent processing instruments.

The optimal utilization of a correlator, to obtain the best value of the line width Γ and to minimize the uncertainty $\Delta \Gamma / \Gamma$, involves selecting the number of channels M and the sweeping time range $M \Delta T$. Several companies make correlators suitable for QLS experiments. Honeywell-Saicor makes 100-channel and 400-channel correlators that are 100% efficient down to $\Delta T = 200$ μsec. They become progressively less efficient, however, as ΔT decreases below this time. Precision Instruments and Devices, Ltd., Malvern, England manufactures a correlator that is 100% efficient down to smaller channel times ΔT, but has only 50 channels. Correlators are currently undergoing rapid development, and improved instruments may be anticipated by the time this chapter appears in print.

Although the autocorrelation function apparently can be determined with greater statistical precision if there are more channels, and therefore more experimental points, this advantage is somewhat illusory, because more channels require a longer processing time T_p. This leads to the decreased inefficiency mentioned above. Oliver[47] has shown that in the low counting rate limit ($r \ll 1$, where r is the number of counts per correlation time τ_c), the time T_β required to attain the same accuracy in estimation of τ_c as would be achieved by a perfectly efficient correlator in time $T_{\beta=1}$ is

$$T_\beta = T_{\beta=1}/\beta \qquad (139)$$

where $\beta = \Delta T/T_p$ is the fractional efficiency. In the high counting rate limit, $r \gg 1$,

$$T\beta = (\tau_c/T_p)T_{\beta=1} \tag{140}$$

if $T_p > \tau_c$. Generally a 50- or 100-channel correlator will be adequate.

Once the total number M of channels has been determined by purchase of a correlator, and the correlation time $\tau_c = \Gamma^{-1}$ is known approximately, one must choose the optimal time base ΔT to minimize the relative variance of Γ. When the correlation function is of the form $C(\tau) = C(\infty)[1 + B \exp(-2\Gamma\tau)]$ one can get Γ either by a three-parameter fit to $C(\infty)$, B, and Γ, or by a two-parameter fit to B and Γ with the background $C(\infty)$ known from separate information. A computer simulation of the reduced standard deviation $(\mathrm{Var}\ \Gamma/\Gamma^2)^{1/2}$ as a function of sweep time $M\Delta T/\tau_c$ for a 60-channel correlator is shown in Fig. 24.[58] The range $M\Delta T = 2\tau_c$ to $5\tau_c$ leads to minimum error in the estimate of τ_c.

The signal-to-noise ratios for analog and photon counting have been calculated by Oliver.[47] In the photon-counting mode

$$(\mathrm{S/N})^{-2} = \frac{1}{N_s}\left[\frac{1}{\langle n \rangle} + \frac{1}{\langle n \rangle^2}\right] \tag{141}$$

In the analog mode

$$(\mathrm{S/N})^{-2} = \frac{1}{N_s}\left[\frac{1}{\langle m \rangle \tau_{RC}} + \frac{1}{\langle m \rangle^2 \tau^2_{RC}}\right] \tag{142}$$

where $\langle m \rangle$ is the average number of counts per second. N_s as before is the total number of summations or batchings performed.

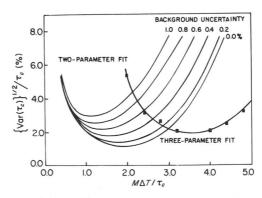

FIG. 24. The relative variance of the correlation time is plotted as a function of W for data generated with a computer. The curves are drawn through the results for the variance in the correlation time by fitting the computer-synthesized data; for example, the results for the variance for the case of the three-parameter fit are shown by the squares. [$C(\infty)/w = 5000$].

[58] T. K. Lim, Ph.D. thesis, Johns Hopkins University, Baltimore, Maryland, 1973.

At low counting rates when $\langle n \rangle \ll 1$ or $\langle m \rangle \tau_{RC} \ll 1$, the second terms in these equations are dominant and S/N is proportional to the square of the counting rate. At high counting rates, when $\langle n \rangle \gg 1$ and $\langle m \rangle \tau_{RC} \gg 1$, the first terms are dominant and S/N is proportional to the first power of the counting rate. To compare signal to noise ratios in analog and digital processing, we note that τ_{RC} must be less than ΔT. Therefore S/N is always better in digital processing by a factor of $\Delta T / \tau_{RC}$ for high counting rates and $(\Delta T / \tau_{RC})^2$ for low counting rates. However, if the counting rates are too high, the pulses begin to overlap each other. In this case, the analog mode must be used. For rapid photomultipliers used in photon counting, the pulse width is about 20 nsec. Therefore the maximum counting rate at which digital processing can be employed is 10^7 counts per second or less. Since the pulses are randomly distributed, the actual limits should be even less than this, say 2×10^6 pulses per second. Under typical conditions with $\Delta T = 10^{-5}$ sec and a counting rate of 5×10^5 counts per second, 10^5 sums are required to obtain S/N = 1.5×10^3.

In the analog mode, when only the current can be measured, Eq. (140) can be modified as follows. Since each pulse has an average charge $\langle q \rangle = e\gamma$ then

$$\langle m \rangle \tau_{RC} = \frac{\langle m \rangle \langle Q \rangle \tau_{RC}}{\langle Q \rangle} = \langle i \rangle \tau_{RC}/e\gamma \tag{143}$$

and

$$(\text{S/N})^{-2} = \frac{1}{N_s} \left[\frac{1}{\langle i \rangle (\tau_{RC}/e\gamma)} + \frac{1}{\langle i \rangle^2 (\tau_{RC}/e\gamma)^2} \right] \tag{144}$$

Here $\langle i \rangle$ is the output current from the phototube anode.

Similar considerations lead to the expression for the relative variance of the decay rate in the photon counting mode[47,59,60]

$$\text{Var}(\Gamma)/\Gamma^2 = (2\Gamma\Delta T)/C(\infty)B \tag{145}$$

The relative variance in the analog mode is twice this under optimal conditions.

Data Analysis

The homodyne photocurrent autocorrelation function is of the form

$$C_i(\tau) = C(\infty)[1 + B|g^{(1)}(\tau)|^2] \tag{146}$$

When there is a single type of relaxation process (normally the trans-

[59] E. Jakeman, *J. Phys. A: Gen. Phys.* **5**, 249 (1972).
[60] E. Jakeman, E. R. Pike, and S. Swain, *J. Phys. A: Gen. Phys.* **4**, 517 (1971).

lational diffusion of a single macromolecular species) contributing to the scattering, $g^{(1)}(\tau) = \exp(-\Gamma\tau)$. When there are several relaxation processes, the normalized scattering autocorrelation function will be of the form

$$g^{(1)}(\tau) = \sum_{j=1}^{r} G_j e^{-\Gamma_j \tau} \tag{147}$$

or

$$g^{(1)}(\tau) = \int_0^\infty G(\Gamma) e^{-\Gamma\tau} d\Gamma \tag{148}$$

where the G's are weighting functions. In this section we discuss procedures to determine the relaxation rates and weights, and to test whether the autocorrelation function is adequately accounted for by a single relaxation process.

Background. Although the background term $C(\infty)$ may be obtained along with the other parameters in Eq. (146) by nonlinear least-squares fitting procedures, it is generally preferable to determine it separately. This may be seen from Fig. 24. If the background can be determined to better than $\pm 0.4\%$, which is usually easy in practice, Γ and B can be more accurately determined from a linear two-parameter fit than from a nonlinear three-parameter fit.

In the analog correlation mode, as discussed above, a low-frequency cutoff filter eliminates the dc term $C(\infty)$. In the photon counting mode, $C(\infty)$ may be determined directly by the following procedure. The autocorrelation function is output from the correlator as

$$C(\tau) = \sum_{k=0}^{N_s} n_k(\tau) n_k(0) \tag{149}$$

where k denotes the batch number and N_s the number of batches. In the limit $\tau \gg \tau_c = \Gamma^{-1}$, $n_k(\tau)$ and $n_k(0)$ are completely uncorrelated, so

$$C(\tau \gg \tau_c) = C(\infty) = N_s \langle n \rangle^2 \tag{150}$$

Most commercially available photon counting correlators give as output N_s, and also the total number of counts in the accumulation counter (Fig. 23)

$$N_T = \sum_{k=0}^{N_s} n_k(\tau) = N_s \langle n \rangle \tag{151}$$

Note that $\langle n \rangle$ does not depend on τ. Thus one obtains

$$C(\infty) = N_T^2 / N_s \tag{152}$$

This procedure can be subject to error due to timing of the counter and shift register and other electronic uncertainties. It should therefore initially be checked by using a dim light bulb as a light source. Since this is an uncorrelated source, $C(\tau)$ should be flat and equal to $C(\infty)$ in each channel within random error $(S/N)^{-1}$. Assuming Poisson statistics, $S/N = \sqrt{C(\infty)}$. In a typical experiment $\langle n \rangle \sim 0.05$, $N_s \sim 5 \times 10^7$, $N_T \sim 2.5 \times 10^6$, $C(\infty) \sim 1.25 \times 10^5$, and $S/N \sim 3.5 \times 10^2$, so $(S/N)^{-1} \sim 2.8 \times 10^{-3}$.

Single Exponential Behavior. With $C(\infty)$ determined as above, and $g^{(1)}(\tau) = \exp(-\Gamma\tau)$, Eq. (146) indicates that a plot of

$$y_j = \ln\{[C_i(\tau_j) - C(\infty)]/C_i(0)\} = a + b\tau \qquad (153)$$

vs τ_j should be linear, with slope $b = -2\Gamma$. The actual data will be somewhat scattered, and a weighted linear least-squares analysis should be performed to obtain the best value of Γ.

The weighting factor for each point (τ_j, y_j) should be inversely proportional to the standard deviation $\sigma_y(\tau_j)$ of that point. Since $C_i(\tau_j)$ will be Poisson-distributed about its mean value, its standard deviation is

$$\sigma_y(\tau_j) = [C_i(\tau_j)]^{1/2} \qquad (154)$$

This propagates to $\sigma_y(\tau_j)$ as

$$\sigma_y(\tau_j) = \sigma_c(\tau_j)/\{[C_j(\tau_j) - C(\infty)]/C_i(0)\}$$
$$= C_i(0)[C_i(\tau_j)]^{1/2}/[C_j(\tau_j) - C(\infty)] \qquad (155)$$

With this as the reciprocal of the weighting factor, one obtains from standard linear least-squares analysis[61]

$$a = \frac{1}{\Delta}\left(\sum \frac{x_i^2}{\sigma_i^2} \sum \frac{y_i}{\sigma_i^2} - \sum \frac{x_i}{\sigma_i^2} \sum \frac{x_i y_i}{\sigma_i^2}\right) \qquad (156)$$

$$b = \frac{1}{\Delta}\left(\sum \frac{1}{\sigma_i^2} \sum \frac{x_i y_i}{\sigma_i^2} - \sum \frac{x_i}{\sigma_i^2} \sum \frac{y_i}{\sigma_i^2}\right) \qquad (157)$$

$$\Delta = \sum \frac{1}{\sigma_i^2} \sum \frac{x_i^2}{\sigma_i^2} - \left(\sum \frac{x_i}{\sigma_i^2}\right)^2 \qquad (158)$$

The intercept a should equal unity within statistical error.

To test whether the correlation function truly obeys single-exponential behavior, it is common to fit y to a second-order polynomial in τ (see below) and to check whether the coefficient of the τ^2 term is suitably small compared to the coefficient of τ. This test should be accompanied by a plot of the residuals $(y_j - \hat{y}_j)$, where $\hat{y}_j$ is calculated from the least-squares values of Γ and b, against channel number j. If the residuals scatter randomly on either side of zero, this is a strong indication of single-exponential behavior.

[61] P. R. Bevington, "Data Reduction and Error Analysis for the Physical Sciences." McGraw-Hill, New York, 1969.

Multiple Exponential Behavior. If the scattering solution is polydisperse, or if relaxation processes other than translational diffusion are occurring, the scattering autocorrelation function will have the form of Eqs. (54) or (86). In this case a plot of y_j vs τ will be curved. Koppel[62] has shown that a cumulant analysis of the normalized autocorrelation function leads to well-defined averages. If one writes

$$y(\tau) = K_0 + K_1(-\tau) + K_2(-\tau)^2 \tag{159}$$

then the limiting slope at $\tau = 0$, or first cumulant, is

$$K_1 = \langle \Gamma \rangle = \int_0^\infty \Gamma G(\Gamma) d\Gamma \Big/ \int_0^\infty G(\Gamma) d\Gamma \tag{160}$$

and the normalized limiting curvature or second cumulant is

$$K_2/K_1^2 = \langle (\Gamma - \langle \Gamma \rangle)^2 \rangle / \langle \Gamma \rangle^2 \tag{161}$$

In the most common case of a polydisperse polymer solution, where for species k, $\Gamma_k = q^2 D_k$, the angular brackets denote a Z average:

$$\langle \Gamma \rangle = q^2 \langle D \rangle_Z \tag{162}$$
$$= q^2 \left(\sum_k M_k C_k D_k \Big/ \sum_k M_k C_k \right)$$

If the scattering particles are not small compared to the wavelength of light, then the form factors $P_k(q)$ must also be included in Eq. (162). In this case the average diffusion coefficient will vary with scattering angle. In general, $\langle D \rangle_Z$ will increase with increasing q, because the form factors of the larger particles (those with smaller D) decrease more rapidly with increasing angle, thereby decreasing the relative contribution of these particles to $\langle D_Z \rangle$. Analysis of data in this circumstance has been discussed by Brehm and Bloomfield.[63]

The normalized second cumulant K_2/K_1^2 reflects both the sample size distribution through $G(\Gamma)$ and noise in the scattering data. To check whether polydispersity is a factor, graphical as well as numerical inspection of the data is important. As in the single-exponential case described above, it is useful to plot the residuals $(y_j - \hat{y}_j)$ vs. channel number, where $\hat{y}_j$ is computed from the linear approximation with the best values of K_0 and K_1. Systematic deviation of the residuals to the positive or negative side indicates that polydispersity or some other relaxation process is important.

The equations for the determination of the coefficients K_i in a cumulant

[62] D. E. Koppel, *J. Chem. Phys.* **57**, 4814 (1972).
[63] G. A. Brehm and V. A. Bloomfield, *Macromolecules* **8**, 663 (1975).

expansion, determined from weighted quadratic least squares regression analysis, are given in determinantal form as[61]

$$K_0 = \frac{1}{\Delta} \begin{vmatrix} \sum \frac{y_i}{\sigma_i^2} & \sum \frac{x_i}{\sigma_i^2} & \sum \frac{x_i^2}{\sigma_i^2} \\ \sum \frac{x_i y_i}{\sigma_i^2} & \sum \frac{x_i^2}{\sigma_i^2} & \sum \frac{x_i^3}{\sigma_i^2} \\ \sum \frac{x_i^2 y_i}{\sigma_i^2} & \sum \frac{x_i^3}{\sigma_i^2} & \sum \frac{x_i^4}{\sigma_i^2} \end{vmatrix} \tag{163}$$

$$K_1 = \frac{1}{\Delta} \begin{vmatrix} \sum \frac{1}{\sigma_i^2} & \sum \frac{y_i}{\sigma_i^2} & \sum \frac{x_i^2}{\sigma_i^2} \\ \sum \frac{x_i}{\sigma_i^2} & \sum \frac{x_i y_i}{\sigma_i^2} & \sum \frac{x_i^3}{\sigma_i^2} \\ \sum \frac{x_i^2}{\sigma_i^2} & \sum \frac{x_i^2 y_i}{\sigma_i^2} & \sum \frac{x_i^4}{\sigma_i^2} \end{vmatrix} \tag{164}$$

$$K_2 = \frac{1}{\Delta} \begin{vmatrix} \sum \frac{1}{\sigma_i^2} & \sum \frac{x_i}{\sigma_i^2} & \sum \frac{y_i}{\sigma_i^2} \\ \sum \frac{x_i}{\sigma_i^2} & \sum \frac{x_i^2}{\sigma_i^2} & \sum \frac{x_i y_i}{\sigma_i^2} \\ \sum \frac{x_i^2}{\sigma_i^2} & \sum \frac{x_i^3}{\sigma_i^2} & \sum \frac{x_i^2 y_i}{\sigma_i^2} \end{vmatrix} \tag{165}$$

$$\Delta = \begin{vmatrix} \sum \frac{1}{\sigma_i^2} & \sum \frac{x_i}{\sigma_i^2} & \sum \frac{x_i^2}{\sigma_i^2} \\ \sum \frac{x_i}{\delta_i^2} & \sum \frac{x_i^2}{\sigma_i^2} & \sum \frac{x_i^3}{\sigma_i^2} \\ \sum \frac{x_i^2}{\sigma_i^2} & \sum \frac{x_i^3}{\sigma_i^2} & \sum \frac{x_i^4}{\sigma_i^2} \end{vmatrix} \tag{166}$$

If there are only a few species or relaxation processes that account for the scattering, then the autocorrelation function may be represented according to Eq. (54) as a sum of exponentials. Although nonlinear least squares regression analysis may be used to estimate the G_j's and Γ_j's, results are often uncertain even for data of high accuracy, unless the approximate values are known in advance. Standard references to these problems in numerical analysis are cited in footnotes 64–66. Generally, computer programs will have to be written to deal with specific cases. The investigator faced with these problems is urged to consult the local computer center or numerical analysis expert.

[64] S. W. Provencher, *J. Chem. Phys.* **64,** 2772 (1976).
[65] K. Tsuji, H. Watanabe, and K. Yoshioka, *Adv. Mol. Relax. Proc.* **8,** 49 (1976).
[66] M. L. Johnson and T. M. Schuster, *Biophys. Chem.* **2,** 32 (1974).

One case in which a correlation function that is the sum of several exponentials can be satisfactorily dissected is that where the relaxation rates differ by almost an order of magnitude. The procedure is familiar from the analysis of the decay of radioactive isotopes. If, for example, $g^{(1)}(\tau) = G_1 \exp(-\Gamma_1\tau) + G_2 \exp(-\Gamma_2\tau)$, with $\Gamma_1 > 5\Gamma_2$, then a plot of $y(\tau)$ vs τ will show two well-defined regions of different slopes. The slope of the long-τ region gives Γ_2, and extrapolation of this line back to $\tau = 0$ gives G_2. Then G_1 and Γ_1 are obtained from a semilog plot of $[C_i(\tau) - C(\infty) - G_2 \exp(-\Gamma_2\tau)]$ vs τ.

When relaxation processes in addition to translational diffusion, such as rotation or chemical reaction, are suspected, analysis of the dependence of the correlation function on scattering angle or q will provide valuable information. In many such cases only a single Γ can be extracted, with statistical significance, from the data at a given angle. If only diffusion of a single macromolecular component is important, a plot of Γ vs q^2 should be linear and pass through the origin, with slope D. Perhaps a more sensitive plot for the recognition of deviations from this simplest case is one of Γ/q^2 vs q^2, which should be horizontal with ordinate D. If rotation, internal chain motion, or reaction (q-independent relaxation processes) are important, a plot of Γ vs q^2 will have a positive intercept, and a plot of Γ/q^2 vs q^2 will curve upward as q^2 approaches zero. If polydispersity is important, particularly for large particles with significant angle-dependent form factors, a plot of Γ vs q^2 will be concave upward, and may have an apparent negative intercept if measurements are not performed at very low angles. The plot of Γ/q^2 vs q^2 will be horizontal at low q^2 and curve upward at high q^2.

Distortion of the Correlation Function. Several factors may lead to distortion of the correlation function from its true form. Chief among these are scattering by dust and extraneous light.

Although the number concentration of dust particles is small, their contribution may be significant since, for a given number concentration, the light scattering intensity is proportional to the square of the particle mass. If the mean counting rates per channel time ΔT for sample and dust separately are $\langle n \rangle_s$ and $\langle n \rangle_d$, Eq. (51) indicates that the observed homodyne autocorrelation function will be

$$g^{(2)}(\tau) = 1 + \frac{2\langle n \rangle_s \langle n \rangle_d}{(\langle n \rangle_s + \langle n \rangle_d)^2} g_s^{(1)}(\tau) g_d^{(1)}(\tau) + \left(\frac{\langle n \rangle_s}{\langle n \rangle_s + \langle n \rangle_d}\right)^2 |g_s^{(1)}(\tau)|^2$$
$$+ \left(\frac{\langle n \rangle_d}{\langle n \rangle_s + \langle n \rangle_d}\right)^2 |g_d^{(1)}(\tau)|^2 \quad (167)$$

Since the diffusional relaxation time for the dust will be much longer than that for the macromolecules of interest, $g_d^{(1)}(\tau) \approx 1$ over the range of τ's where $g_s^{(1)}(\tau)$ changes appreciably. Therefore the last term on the rhs of Eq. (167) will give rise to an increased base line, while the second term (which arises owing to partial heterodyning) will distort the apparent relaxation rate. The apparent rate will be[47]

$$\Gamma_{app} \approx \Gamma_s(1 + \langle n \rangle_d / \langle n \rangle_s) \tag{168}$$

Since the dust concentration will generally be very low, non-Gaussian number fluctuations[16] may also contribute to the correlation function. These will generally result in a rise in the base line.

Extraneous light coming from outside the scattering volume can have two effects. If it comes from a light leak, incoherent with the signal, no interference will occur and the effect will simply be an increase in background level. If the background is determined as described above, there will be no effect on Γ. However, if the extraneous light is coherent with the signal (e.g., light scattered from the cell walls) then the correlation function will have both heterodyne and homodyne terms[47]:

$$g^{(2)}(\tau) = 1 + \frac{2 \langle n \rangle_s \langle n_0 \rangle}{(\langle n \rangle_s + \langle n \rangle_0)^2} g_s^{(1)}(\tau) + \left(\frac{\langle n \rangle_s}{\langle n \rangle_s + \langle n \rangle_0} \right)^2 |g_s^{(1)}(\tau)|^2 \tag{169}$$

where $\langle n \rangle_0$ is the mean number of local oscillator counts in time ΔT. Force-fitting this homodyne correlation function to a single decay process will result in a relaxation rate Γ_{app} given by Eq. (168) where $\langle n \rangle_d$ is replaced by $\langle n \rangle_0$.

Obviously, it is vastly preferable to eliminate these sources of signal distortion ahead of time, by careful cleaning of solutions and cells, than to attempt to unscramble them later.

Spectrum Analysis and Heterodyne Spectroscopy

The power spectrum and the autocorrelation function in principle yield equivalent information, since they are Fourier transforms of each other according to the Wiener–Khintchine theorem. In practice, however, one or the other may be substantially preferable in a given case. For example, the optical field autocorrelation function of a system with two relaxation processes is

$$g^{(1)}(\tau) = A_1 e^{-\Gamma_1 \tau} + A_2 e^{-\Gamma_2 \tau} \tag{170}$$

This leads to the current autocorrelation function $g^{(2)}(\tau) = 1 + |g^{(1)}(\tau)|^2$, which has three time-dependent terms to be analyzed. However, one can take the square root of $[g^{(2)}(\tau) - 1]$ and regain a two-exponential function.

The spectra corresponding to $g^{(1)}(\tau)$ and $g^{(2)}(\tau)$ are

$$P^{(1)}(\omega) = \frac{A_1 \Gamma_1/\pi}{\omega^2 + \Gamma_1^2} + \frac{A_2 \Gamma_2/\pi}{\omega^2 + \Gamma_2^2} \tag{171}$$

and

$$P^{(2)}(\omega) = A_1^2 \frac{(2\Gamma_1)/\pi}{\omega^2 + (2\Gamma_1)^2} + \frac{2A_1 A_2 (\Gamma_1 + \Gamma_2)/\pi}{\omega^2 + (\Gamma_1 + \Gamma_2)^2} + \frac{A_2^2 (2\Gamma_2)/\pi}{\omega^2 + (2\Gamma_2)^2} \tag{172}$$

so the three-term $P^{(2)}(\omega)$ is not simply related to the two-term $P^{(1)}(\omega)$.

On the other hand, the autocorrelation function and power spectrum of a particle undergoing translation with velocity v_0 and diffusion are

$$g^{(1)}(\tau) = e^{-q^2 D \tau_e - i\mathbf{q} \cdot \mathbf{v} \tau} \tag{173}$$

and

$$P^{(1)}(\omega) = \frac{q^2 D/\pi}{(\omega - \mathbf{q} \cdot \mathbf{v})^2 + (q^2 D)^2} \tag{174}$$

$g^{(1)}(\tau)$ represents an exponentially damped oscillation, while $P^{(1)}(\omega)$ is simply a Lorentzian with half-width $q^2 D$ whose maximum is displaced from zero frequency by $\mathbf{q} \cdot \mathbf{v}$. In this case, $P^{(1)}(\omega)$ is substantially easier to analyze.

Computer Fourier transformation of the autocorrelation function to the power spectrum introduces a truncation error, since the correlation function is only measured over a time $M\Delta T$. Therefore the computed

$$[P(\omega)]_{\text{calc}} = \int_0^{M\Delta T} C(\tau) e^{i\omega\tau} d\tau \tag{175}$$

and true

$$P[\omega] = \int_0^\infty C(\tau) e^{i\omega\tau} d\tau \tag{176}$$

power spectra will differ by

$$\int_{M\Delta T}^\infty C(\tau) e^{i\omega\tau} d\tau \tag{177}$$

This affects mainly the low-frequency part of $P(\omega)$, but the entire spectrum will be distorted to some extent.

Spectrum Analyzers. QLS experiments were first performed[3,4] with a conventional autofrequency sweeping single-channel spectrum analyzer. In this apparatus, a bandpass filter of bandwidth $\Delta\nu$ centered at ν_c selects a frequency window $\nu_c - \frac{1}{2} \Delta\nu < \nu < \nu_c + \frac{1}{2} \Delta\nu$. The amplitude of the signal at this range is recorded, and the filter then shifts to another ν_c'. At any given time, frequency components other than those in the range under examination are discarded. Thus the single-channel spectrum analyzer is markedly inefficient and slow. A multiband pass filter analyzer could be built that would be 100% efficient, but this would be very costly.

Recently, 100% efficient real time spectrum analyzers have been developed, by Honeywell-Saicor, Federal Scientific, and Princeton Applied Research, which operate by a time-compression technique. This is based on the fundamental principle (the sampling theorem) that a signal must be processed for a time no less than $\Delta T = 1/\Delta\nu$ if the desired resolution (width of bandpass filter) is $\Delta\nu$. If the total frequency range to be analyzed is $\nu_T = M \; \Delta\nu$ where M is the number of channels, then it will require $M \; \Delta T = M/\Delta\nu$ sec for a single analysis of the entire spectrum. In the time-compression procedure, the signal during a period T is read into memory, and is then read out during a shorter time T_a. The time compression ratio is T/T_a. This has the effect of increasing the total frequency range to $(T/T_a)\nu_T$. Therefore the analysis time per channel is $M/[(T/T_a)\nu_T]$ and the total analysis time is $M^2/[(T/T_a)\nu_T] = M/[(T/T_a)\Delta\nu]$, less than that for a conventional swept frequency analyzer by the factor T/T_a. Real time analysis is achieved when $MT_a = T$, so no incoming signal is lost during processing. Then the total analysis time is $1/\Delta\nu$ sec.

For example, in a conventional $M = 200$ channel spectrum analyzer, if the desired frequency range is 2 KHz, the resolution is 10 Hz and the time per spectrum analysis is $200 \times (1/10) = 20$ sec. However, in the Honeywell-Saicor 200-channel real-time spectrum analyzer, the processed signal length is $T = 0.1$ sec, corresponding to $T_a = 5 \times 10^{-4}$ sec. Thus an increase in speed of $(20/0.1) = 200$ is achieved. Depending on the particular instrument, spectral widths from 25 Hz to 1 MHz can be measured by these techniques.

As is the case with time autocorrelation analysis, signal averaging must also be applied to spectrum analysis to achieve adequate S/N. Generally, (S/N) improves as $N^{\frac{1}{2}}$, where N is the number of repetitive spectral scans. Since N will generally be large ($\gtrsim 10^4$), this emphasizes the importance of the time saved by use of a real-time, time-compression analyzer.

Data Analysis. The most common case is homodyne analysis of scattering from a monodisperse solution of scatterers. The power spectrum for this situation is given by Eq. (44). Normally the spectrum analyzer does not pass the DC component of the spectrum, so one can write

$$P_i^{\text{homo}}(\omega) = A + B/[\omega^2 + (2\Gamma)^2] \qquad (178)$$

where $A = e \langle i_s \rangle / 2\pi$ is the shot noise term, $B = 2\Gamma \langle i_s \rangle^2 / \pi$, and $\Gamma = q^2 D$. These constants may be determined by nonlinear least-squares fitting to a three-parameter equation, or the shot noise may be determined independently.

This may be done by measuring $P_i^{\text{homo}}(\omega)$ at high frequencies $\omega \gg \Gamma$, so that the Lorentzian contribution to the signal is negligible. However, rolloff at high frequencies may render this approach somewhat unreliable. Perhaps a better procedure is to substitute a flashlight bulb for the scattering volume, and to adjust the intensity of the light to produce an average photocurrent equal to $\langle i_s \rangle$ produced by the sample. Since the light from the bulb is totally incoherent, it will give the same shot noise as the sample without yielding any signal. Then $P_i^{\text{homo}}(\omega)$ should be a flat function of frequency, with magnitude A.

Since spectrum analyzers may not be accurately linear, particularly at high frequencies or at very low frequencies near the dc cutoff, it is important to calibrate an analyzer with a periodic signal generator of known output voltage. The instrumental profile so determined should be subtracted from the measured spectrum before data analysis.

Once A has been determined, Γ and B in Eq. (178) can be determined either by nonlinear fitting to a two-parameter equation, or by a linear plot such as $[P_i^{\text{homo}}(\omega) - A]^{-1}$ vs ω^2, which yields slope of $1/B$ and intercept of $(2\Gamma)^2/B$. As with autocorrelation analysis, the angular variation of Γ yields D. The constancy of D with angle, and the passing of the plot of Γ vs q^2 through the origin, are indications that the translational diffusion of a single macromolecular species is being observed.

The signal-to-noise ratio in spectrum analysis has been discussed by Cummins and Swinney.[9] They define S/N as the ratio of the signal amplitude at frequency ω to the rms fluctuation in signal plus shot noise at ω. If the signal is a Lorentzian centered at $\omega = 0$, the maximum signal and S/N are at that frequency. Then

$$\left(\frac{S}{N}\right)_{\text{max}} = \left[\frac{\mu \langle i \rangle_s / n_c e \Gamma}{1 + (\mu \langle i \rangle_s / n_c e \Gamma)}\right] (1 + \delta T_{\text{RC}})^{1/2} \qquad (179)$$

where $\mu = 1$ for homodyning, $\mu = 2$ for heterodyning with $\omega_{\text{LO}} > \Gamma$, and $\mu = 4$ for heterodyning with $\omega_{\text{LO}} = 0$. n_c is the number of coherence areas subtended by the detector, $\delta = 2\pi\Delta\nu$ is the filter bandwidth, and T_{RC} is the time constant of the RC integrator in the spectrum analyzer. In the usual case $\delta T_{\text{RC}} \gg 1$. Equivalently, if signal averaging is performed not with an RC integrator but by repetitive digital averaging, δT_{RC} should be replaced by N, the number of accumulations, in Eq. (179).

Heterodyne Spectroscopy. One of the major uses of spectrum analysis rather than autocorrelation analysis in QLS arises in heterodyne spectroscopy. The local oscillator produces such an intense signal that in the time domain the signal of interest is just a small contribution resting on a large dc base line, which tends to saturate the correlator. In the spectral domain, however, this dc component is filtered out and the signal of interest is exhibited plainly.

In the homodyne technique discussed up to this point, different frequency components of the scattered light mix with one another on the photocathode surface to give a beat spectrum or correlation function. The heterodyne technique differs from this only in that a reference monochromatic beam is combined with the scattered light to produce the beat spectrum. The reference light must have a definite, stable relationship in phase and amplitude with respect to the incident beam. Although it is possible to use a separate laser light source, even one of different frequency,[61] as the reference, it is technically easier and more common to use a portion of the incident beam itself. We shall describe several ways to obtain this reference beam and to adjust its intensity relative to the scattered light so as to optimize data quality.

Except for some minor differences in the processing electronics, all the description of focusing and collecting optics and data processing given above for homodyning applies to heterodyning as well.

Three methods to obtain a reference beam from the incident beam are diagrammed in Fig. 25. In the first method, Fig. 25a, part of the incident light is scattered from the cell wall. When a sharp image is formed on the first collecting pinhole by the collecting lens, bright spots from cell wall reflections can be seen at both ends of the column defined by the pencil of incident light (Fig. 25a). The collecting pinhole should be adjusted to

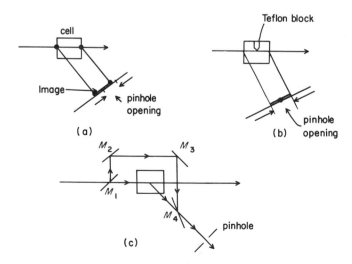

FIG. 25. Some optical arrangements for heterodyne scattering. The scattered light is mixed with a local oscillator obtained by reflecting part of the incident beam from (a) the cell walls, (b) a Teflon block, and (c) mirrors in a separate optical path.

include both the bright reference spot and the column corresponding to the image of the scattered light. The ratio of intensities can be regulated by adjusting the position of the pinhole to include more or less of the bright spot. This technique has the advantage of being insensitive to mechanical vibrations, the major cause of phase and amplitude instabilities that destroy the heterodyning relationship. However, it is difficult to adjust the intensity mixing ratio over a wide range, and scattered light is collected only near the cell wall. This is undesirable in electrophoretic light scattering when electroosmotic effects are significant.

A second way to produce heterodyning is to reflect the incident beam from some stationary object within the scattering volume (Fig. 25b). The strength of the reference beam can be adjusted by adjusting the position of the object. A Teflon wedge has been used for this purpose,[30] and in an electrophoretic light-scattering experiment, Uzgiris[67] reflected the beam from an electrode in the cell. These procedures are difficult, however, because vibrations of the stationary scatterer relative to the rest of the light-scattering assembly may lead to spurious frequency shifts.

The third procedure is to deflect part of the incident light around the scattering cell and then recombine it with the scattered light, as shown in Fig. 25c. This is done with half-transparent mirrors M_1 and M_4, while mirrors M_2 and M_3 control the deflected beam direction. The intensity of the deflected light is controlled by a neutral density filter in the path between M_2 and M_3. For this procedure to succeed, the mirrors must be immobile with respect to the sample. This is achieved by mounting them in solid blocks that are bolted to the same metal plate on which the cell is mounted. So long as this setup is vibration-free, it is preferable to the others described because of greater versatility in mixing ratio and range of scattering angles.

It is important in this third procedure to keep the difference in the two optical path lengths less than the coherence length l_c of the laser. If the frequency difference between adjacent modes is $\Delta\omega$, $l_c = c/\Delta\omega$. A He–Ne laser without a stabilizing etalon will have numerous longitudinal modes within the gain profile of width about 1.5×10^9 Hz. Thus l_c, is calculated to be about 20 cm, and may be smaller in practice. In an Ar$^+$ laser with internal etalon, l_c is much larger. To test whether the optical path is less than l_c, one collects the beams from the two paths (with the sample blanked with solvent) on the phototube surface. The photocurrent should contain no beat signal.

An important consideration in heterodyne spectroscopy is the mixing ratio, that is, the ratio of scattered light to reference light intensities. Equation (25) gives the heterodyne photocurrent autocorrelation function

[67] E. E. Uzgiris, *Opt. Commun.* **6**, 55 (1972).

$C_i^{het}(\tau)$ in terms of the average photocurrent $\langle i_s \rangle$ due to scattering and the reference local oscillator photocurrent i_{LO}. This equation was derived assuming $\langle i_s \rangle \ll i_{LO}$. Since the neglected terms in Eq. (25) are of order $\langle i_s \rangle^2$, and the time-dependent part of the signal itself is of order $\langle i_s \rangle i_{LO}$, $\langle i_s \rangle / i_{LO}$ must be less than 1/30 if the error in the correlation function is to be less than 3%. However, $\langle i_s \rangle / i_{LO}$ cannot be too small if adequate S/N is to be maintained. Generally it is best to keep $\langle i_s \rangle / i_{LO}$ between 0.01 and 0.03.[9,47]

Electrophoretic Light Scattering

One of the major applications of heterodyne spectroscopy is electrophoretic light scattering. The apparatus for this experiment consists of a QLS apparatus in which the scattering cell is an electrophoretic cell of standard Tiselius or similar design. The theory of electrophoretic light scattering has already been presented, as have the description of the QLS apparatus and procedures for heterodyne spectroscopy. Therefore we confine our attention here to discussion of cell design. Pioneering work in electrophoretic light scattering (ELS) was done by Ware and Flygare[13] and Uzgiris[67] who made major contributions in developing effective ELS cells.

An ELS cell is composed of a sample chamber, 2 electrodes and a power supply to produce the electric field. A block diagram of a prototype cell is given in Fig. 26. The sample chamber may be rectangular or cylindrical. It should have uniform cross section in the region in which the electrophoretic motion and scattering is being detected, so that the electric field is uniform and parallel to the windows at the sides of the cells.

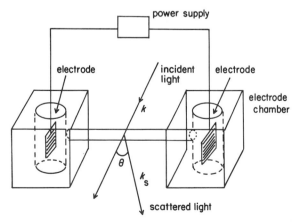

FIG. 26. Schematic diagram of an electrophoretic light-scattering cell with electrodes and power supply. The center tube should be optically transparent.

The laser light is incident on the sample and is collected after scattering through these windows. The electrode chambers must be in electrical contact with the sample chamber through the buffer, but it is preferable that they be separated from the sample chamber by polyacrylamide gel plugs, which prevent contaminating electrode products from entering the sample volume. The entire cell and electrode assembly should be replaced in a circulating coolant bath to control temperature.

Problems in cell design arise from such features as limitations of power supplies, flow characteristics of electrophoresing solutions, Joule heating, and contamination by electrode products. We shall discuss these in turn.

Power Supply. The electrophoretic Doppler shift is proportional to the applied electric field E across the cell. For highest resolution it is desirable to have the highest possible E; however, this must be reconciled with the limited output current capacity of power supplies. If the cross-sectional area of the cell is A and the specific conductivity of the buffer is κ, the total current across the cell which must be provided by the power supply is given by Ohm's law, in the form

$$i = A\kappa E \qquad (180)$$

In most experiments E will be about 100 V/cm and κ is in the range 2×10^{-3} ohm^{-1} cm^{-1} for standard buffers employed in ELS experiments at about $10^{-2} M$ ionic strength. For a power supply with a current capacity of 20 mA, A must be 0.1 cm^2. Obviously the smaller the cross-sectional area, the less the current-carrying demands on the power supply. Also a cell of large A will require more sample for a given experiment. However if A is too small, it will be difficult to select the desired portion of the scattering column to be analyzed as is necessary to control the mixing ratio in heterodyne spectroscopy. These considerations lead to a compromise cross section which typically is about 2 mm $\times$ 3 mm.

Electroosmosis. ELS measures the velocity of macromolecules with respect to a cell-wall fixed-coordinate system, while the molecules themselves are characterized by their electrophoretic mobility relative to solvent. Therefore any movement of the solution relative to the cell walls will lead to an incorrect electrophoretic mobility and will also distort the spectrum if the solution movement is inhomogeneous across the cell cross section. Perhaps the major source of solution movement is electroosmosis, which has been a major problem in developing ELS experiments since their inception.

Electroosmosis results when the cell surface adsorbs specific ions from the solution.[68,69] The charged surface and its associated counterions

[68] J. S. Newman, "Electrochemical Systems." Prentice-Hall, Englewood Cliffs, New Jersey, 1973.
[69] M. Bier, "Electrophoresis," Vol. 2. Academic Press, New York, 1967.

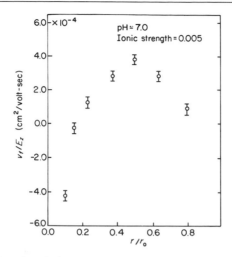

FIG. 27. Electrophoretic velocity v_f of polystyrene latex spheres divided by field strength E_z as a function of relative distance r/r_0 from the wall of a cylindrical capillary of radius r_0. The parabolic flow profile is due to electroosmosis. From T. K. Lim, and G. Flynn, B. J. Berne, Columbia University, unpublished work.

form a diffuse double layer. An external electric field tangent to the cell surface exerts forces on the charges in the diffuse layer and thus causes motion of the solution. As the solution moves in one direction there must be a return flow if the total solution volume is confined in a closed cell. The flow patterns that develop are shown in Fig. 27. The electroosmotic velocity pattern inside a cylindrical cell has been calculated by several authors[68,69] to be parabolic. This result has been confirmed by experiment.[70] The ELS spectrum in the presence of this parabolic flow profile has been calculated as shown in Fig. 28. As confirmed by experiment, this figure clearly indicates that the electroosmotic effect can result in extra peaks in the ELS spectrum. It is, therefore, very important to eliminate electroosmosis to the fullest possible extent.

Electroosmosis can best be minimized by reducing the surface charge on the cell walls. Lim et al.[25] have used Siliclad (Clay Adams, division of Becton, Dickinson and Company, Parsippany, New Jersey 07054) to coat the glass surfaces and found that this treatment reduced the electroosmotic effect to 0.05×10^{-4} cm²/V-sec over the pH range 4.5 to 9.5. The coating must be kept thin to minimize scattering from the walls.

Joule Heating. When a voltage is applied across a conducting solution, the temperature of the solution will increase owing to Joule heating by the electric current produced. At sufficiently high electric fields and temper-

[70] T. K. Lim, B. J. Berne, and G. Flynn, unpublished work, Columbia University.

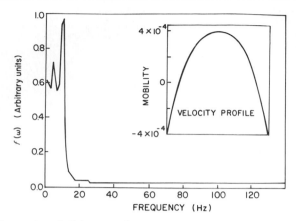

FIG. 28. Electrophoretic light scattering spectrum calculated for the flow pattern given in the inset, which is similar to that in Fig. 27. μ is taken to be 4×10^{-4} cm²/V-sec, and D such that the linewidth is 1 Hz. From T. K. Lim, B. J. Berne and G. Flynn. Columbia University, unpublished work.

atures, turbulence due to convective instabilities will develop and destroy the spectrum one is attempting to study. This effect sets a limit to the maximum useful electric field and therefore to the maximum resolution of the experiment. In addition, the temperature should be maintained essentially constant because for a molecular interpretation of electrophoretic mobility one must know the temperature in order to know the viscosity.

Some order of magnitude estimates of Joule heating may be useful. Consider the solution contained between two surfaces of area A, separated by distance d. The power input/cm³ by Joule heating is $iE/A = i^2/A\kappa = E^2\kappa$ W/cm³. Without any mechanism for heat dissipation, this will produce a temperature rise of $dT/dt = E^2\kappa/c$ deg/sec, where c is the specific heat. With $E = 100$ V/cm, $\kappa = 2 \times 10^{-3}$ ohm^{-1} cm^{-1}, and $c \approx 4.2$ joules/deg-cm³ for aqueous solutions, one finds $dT/dt = 5$ deg/sec. To a certain extent this temperature rise is counteracted by heat flow out through the cell walls. If the walls are maintained at temperature T_0, and the center of the cell is at $T = T_0 + \Delta T$, and if one assumes a linear gradient from the center to each wall (this approximation will not introduce serious inaccuracy in our final result), then the heat flow out is $2KA\Delta T/(d/2)$ joules/sec where K is the thermal conductivity of the solution. This is balanced in the steady state by the heat flow in, $E^2\kappa Ad$. Thus $\Delta T \approx E^2 d^2 \kappa/4K$ deg. For water, K is about 10^{-2} joule/cm-deg-sec. Therefore, to achieve essentially constant temperature throughout the cell, say $\Delta T \leq 0.1$ deg, one requires a spacing $d \leq 0.014$ cm. Needless to say, such a narrow gap between cell walls or electrodes is very difficult to work with, owing to problems with bubbles and aggregated material near the electrode surface. Uzgiris[71] has used electrode gaps of 0.022 cm and 0.1 cm, and has found severe turbulence with the latter when κE^2 was greater than 30.

To further reduce Joule heating, Ware and Flygare[72] developed a pulsing technique in which the voltage is applied for a time Δt and then turned off for a time t_{off}. The polarity of E is reversed in each pulse to prevent concentration polarization; that is, the accumulation of charge carriers at one or another of the electrodes. The pulse width is limited by the requirement $\Delta t \geq 1/\Delta\omega$ where $\Delta\omega$ is the resolution (bandwidth) of the spectrum analyzer. A shorter pulse will give a distorted spectrum. t_{off} should be long enough to allow for adequate heat dissipation. For a solid of density ρ bounded by two parallel planes kept at constant temperature, Carslaw and Jaeger[73] showed that the temperature distribution as a function of time is an infinite series of decaying exponentials with relaxation times $\tau_n = d^2\rho c/\pi^2 K n^2$. With parameters appropriate for aqueous solutions, the longest ($n = 1$) relaxation time is about 50 d^2 sec, or about 5 sec with an 0.3 cm gap. Under typical operating conditions, $\Delta t = 1$ sec and $t_{off} = 20$ sec.

Electrodes. In designing electrodes one must consider such problems as overvoltage, concentration polarization, and bubble formation.

When electrodes are immersed in an aqueous solution, a voltage drop occurs between the electrode material and the liquid. This is termed overvoltage.[74,75] Unless the opposing voltage is greater than the overvoltage, there will be no current flowing through the interface. The overvoltage depends on the electrode material and on the solution itself but is generally not more than 1 V and normally amounts to about 100 mV. If the total voltage drop between the electrodes is in the range 10–100 V, the overvoltage can be neglected. However, for a small electrode gap in which the potential drop is less than 5 V, overvoltage may be a significant problem.

Concentration polarization arises when, owing to current flow, charged carriers move from one electrode to the other. This concentration gradient generates an extra electrical potential at the electrode–solution interface. For significant currents, the potential may drop by as much as 30% between the electrodes. However, in the alternating pulse mode of

[71] E. E. Uzgiris, *Rev. Sci. Instrum.* **45**, 74 (1974).

[72] B. R. Ware and W. H. Flygare, *J. Coll. Interface Sci.* **39**, 670 (1972).

[73] H. S. Carslaw and J. S. Jaeger, "Conduction of Heat in Solid." Oxford Univ. Press, London and New York, 1959.

[74] E. C. Potter, "Electrochemistry, Principle and Applications." Cleaver Hume, London, 1956.

[75] S. Glasstone, "An Introduction to Electrochemistry." Van Nostrand-Reinhold, Princeton, New Jersey, 1946.

operation described above, concentration polarization will automatically be eliminated.

At high electrode current densities one may observe many gas bubbles generated at the electrode surfaces. These bubbles may enter the scattering volume and severely distort the ELS spectrum. Bubbles also change the field pattern. By choosing the proper electrode material one can minimize the problem of bubble formation. Uzgiris[71] tested several electrode materials including Ag, Ag/AgCl, Pt, and platinized Pt. He found that although Ag/AgCl gives the fewest bubbles, it produces other contaminating particulate matter. Uzgiris concluded that platinized Pt works best. It has the largest effective surface area and therefore the lowest current density. A Pt electrode can be platinized by immersing it in chloroplatinic acid (H_2PtCl_2). Pt is electrodeposited on each electrode by running anodic current through it and a third Pt electrode, typically at 300 mA/cm² for about 3 min. The Pt electrode becomes black, corresponding to an enormous increase in surface area. Even with platinized Pt electrodes, gas bubbles may be formed at high current densities. These may be removed by circulating the buffer in the electrode chamber from the bottom to the top as shown in Fig. 29. This technique has been effectively employed in the apparatus devised by Lim et al.[25] In this apparatus it was also found that gel plugs inserted between the electrode chambers and the sample volume prevented contaminants from the electrode from entering the scattering cell.

Haas and Ware[76] have constructed a cell of ingenious design, which has properties of small solution volume, short thermal relaxation time, and uniform electrode current density (Fig. 30). The Ag/AgCl electrodes are hemicylindrical and separated by dielectric (Teflon) inserts that form a narrow rectangular gap in the middle. The isopotential lines are concentrated in the gap, where light scattering is observed. Therefore, the field at the electrode surfaces can be kept low, less than 10 V/cm, minimizing problems with bubble and aggregate formation and electrode irreversibility, while the field in the scattering volume is about 15-fold higher and substantially uniform. The entire chamber is designed to be inserted into a preparative ultracentrifuge tube, allowing high-speed sedimentation to remove large particles.

Although this cell holds promise of predictable, satisfactory performance, we are unable confidently to recommend it or any other ELS cell design, until more experience has been gained.

Data Analysis. In ELS experiments the photocurrent can be analyzed either by an autocorrelator or a real-time spectrum analyzer. In theory, as noted in the preceding section, these procedures should give the same results. However, each apparatus has only a finite number of channels,

[76] D. D. Haas and B. R. Ware, *Anal. Biochem.* **74**, 175 (1976).

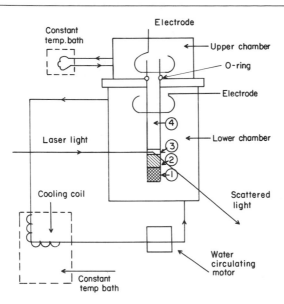

FIG. 29. Diagram of electrophoresis scattering cell and temperature-control system. The lower chamber has windows to pass the incident and scattered light. The solvent circulation inlets are located at the bottom of each chamber. The two chambers and the cell are mounted on a vertical translation stage attached to this center plate. The cell is a glass tube of inner diameter 5 ml, 11 cm long, with 1 mm-thick walls. The bottom of the cell (region 1) is plugged by a 3% polyacrylamide gel, which is supported by a piece of cheesecloth to prevent the gel plug from slipping out of the bottom. A buffered sucrose solution, 38% by weight, is layered on the gel plug (region 2). The sample itself is in a buffered 25% sucrose solution and is layered as a thin band about 1 mm thick over the underlying supporting solution (region 3). On top of the sample (region 4) is a buffered linear sucrose gradient of 3% per inch with 16% by weight sucrose at the junction between regions 3 and 4. The polarity of the electrode is chosen so that when the electric field is applied, the sample band moves upward. From T. K. Lim, G. J. Baran, and V. A. Bloomfield, *Biopolymers* **16**, 1473 (1977).

thereby giving a correlation function over the finite time $M\Delta T$ or spectrum of width $M\Delta\nu$. This truncation introduces problems in the correlation function analysis. It is possible to fit a single Lorentzian spectral peak by recursive graphical analysis using a plotter driven by a desk-top calculator of the type commonly available. However, for most accurate fitting, or when multiple peaks are involved, nonlinear least-square computer fitting is essential.

QLS Analysis of Polydisperse Systems in Density Gradients

We have previously discussed methods for analyzing QLS of macromolecular systems that are polydisperse with respect to size, molecular weight, or charge. The autocorrelation function or power spectral density

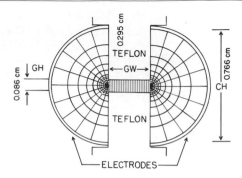

FIG. 30. A computer-generated electric field map for hemicylindrical electrodes separated by a pair of dielectric inserts froming a narrow gap. The isopotential lines are vertical lines in the gap and become arcs of circles concentric with the electrodes in the cavities. They are drawn at equal increments of potential. The electric field lines normal to the isopotential lines indicate the path of current flow in the conducting medium, and the spacing between the electric field lines is proportional to the current density. The important design features indicated by this map are the greatly increased electric field strength in the gap and the uniform current density at the electrodes. The abbreviations are GW = gap width, GH = gap height, and CH = cavity height. From D. D. Haas and B. R. Ware, *Anal. Biochem.* **74,** 175 (1976).

of the scattered light from these systems can be analyzed by several procedures to obtain averages for the system. However, these analysis procedures require highly accurate data to obtain more than one or two moments of the distribution, and they often presuppose some form for the distribution. It is often more satisfactory to study the distribution directly. This may be done either by preparatively separating the components and studying each by itself or, in a group of techniques recently developed, by combining band sedimentation or electrophoresis with QLS and carrying out the analysis directly in a density gradient column.

Sedimentation. Density-gradient centrifugation, usually in sucrose, has been widely employed for both preparative and analytical purposes. Koppel[77] has combined this technique with QLS so that the diffusion coefficient as well as the sedimentation coefficient can be measured in the preparative ultracentrifuge tube. Combination of sedimentation and diffusion coefficients yields the molecular weight distribution at each point in the tube.

The sample is layered onto a preformed sucrose gradient in the appropriate buffer. In a typical experiment the sucrose concentration ranges from 30% by weight at the bottom of the centrifuge tube to 10% at the top in a total cell length of 3.5 inches. Particles of different sedimentation

[77] D. E. Koppel, *Biochemistry* **13,** 2712 (1974).

coefficients separate from each other during centrifugation into distanct bands. These bands are stabilized by the sucrose gradient. Upon attainment of adequate band separation, the tube is removed from the ultracentrifuge and placed in a vertical translational stage for scanning by the QLS laser beam. Two scans are made. During the first, the average intensity of the scattered light is measured as a function of height in the tube. This permits location of the bands, which scatter most strongly. The distance moved by each band, divided by the time of centrifugation and the angular acceleration $\omega^2 r$, gives the sedimentation coefficient. The magnitude of the scattered intensity gives a measure of the particle concentration at each point, according to Eq. (35). To obtain the solvent refractive index n_0 required in this equation, the correlation between sucrose concentration and n_0 in the ISCO tables[78] is needed.

The second scan along the tube measures the autocorrelation function at each point, enabling the determination of the diffusion coefficient of the macromolecules in each band by procedures with which we have already become familiar. Since both the diffusion coefficient and sedimentation coefficient are inversely proportional to the solvent viscosity, an accurate calibration of the sucrose gradient is needed if results are to be adjusted to standard $s_{20,w}^0$ and $D_{20,w}^0$. The density is also required for standardization to $s_{20,w}^0$. The correlation of sucrose concentration with refractive index, viscosity, and density is available in the ISCO tables. Two methods can be employed to characterize the gradient. The refractive index of the solution collected from several positions in the cell can be measured by differential refractometry. Alternatively, the diffusion coefficient of a standard scattering sample, such as polystyrene latex spheres (Dow Chemicals) 91.0 nm in diameter, can be measured at varying heights in the sucrose gradient. From the measured diffusion coefficient and known radius, the viscosity of the sucrose solution at each height can be determined and thereby the solution density from the ISCO tables. Once the sedimentation and diffusion coefficients and the density have been determined, the molecular weight of material in each band can be evaluated after proper extrapolation to zero concentration if necessary according to the Svedburg equation. Koppel[77] has used this procedure satisfactorily to characterize *E. coli* ribosomes and their 30 S and 50 S subunits. Lowenstein and Birnboim[79] have used a similar procedure, involving sedimentation in capillary tubes, to measure the sedimentation coefficients, diffusion coefficients, and molecular weights of R17 bacteriophage and 28 S ribosomal rat liver RNA.

[78] L. H. Newburn, ed., "Iscotables," 6th ed. Instrument Specialties Co., Lincoln, Nebraska, 1975.
[79] M. A. Loewenstein and M. H. Birnboim, *Biopolymers* 14, 419 (1975).

Electrophoresis. Lim *et al.*[71] have recently pursued a similar strategy in combining band electrophoresis in sucrose gradients with QLS to measure the electrophoretic mobilities and diffusion coefficients of biological macromolecules. As described in our previous discussion of electrophoretic light scattering, combination of electrophoretic mobilities and diffusion coefficients leads to information on molecular charge as well as size. The electrophoresis cell is diagrammed in Fig. 29. It consists of an upright cylindrical glass chamber, 2 electrodes, upper and lower cooling chambers, a circulation pump, and a power supply. The bottom of the cell (region 1 in Fig. 29) is plugged by a 3% polyacrylamide gel supported by a piece of cheesecloth. A buffered 38% (w/w) sucrose solution is layed on the gel plug (region 2 in Fig. 29). The sample itself is in a buffered 25% sucrose solution and is layered as a band about 1 ml thick over the underlying supporting solution (region 3 in Fig. 29). On top of the sample (region 4) is a buffered linear sucrose gradient, 3% per inch, with 16% sucrose at the junctions of regions 3 and 4. The cylindrical electrophoresis cell is attached to upper and lower buffer chambers, in which the platinum electrodes are immersed. The buffer chambers are isolated from each other so that the current flows only through the electrophoresis chamber. An electric field of about 10 V/cm is applied for 60 sec and then turned off for 80 sec, to allow heat dissipation. Temperature is controlled by circulating buffers in both chambers. The temperature rise during the 60 sec on cycle is not more than 2° as measured by a telethermometer probe. To eliminate electroosmosis, the cell is treated with Siliclad as discussed previously. The electrical potential at each point along the cell is measured by an electric wire probe of small surface area. Electric field strength (V/cm) is obtained from the slope of this curve, thereby automatically eliminating the junction potential at the interface of buffer and probe. After 12 on–off cycles of the voltage, a light-scattering scan of the cell is made to determine positions of the migrating bands. After each following 12 on–off cycles, the column is scanned again by total intensity light scattering, thus generating a series of band positions for electrophoretic mobility measurements. A plot of band displacement divided by field strength as a function of total time of field application should be linear with slope equal to the electrophoretic mobility. Once the bands have been located, they are characterized with respect to their diffusion coefficients by QLS in exactly the same manner as described above for the band sedimentation case.

Author Index

Numbers in parentheses are footnote reference numbers and indicate that an author's work is referred to although his name is not cited in the text.

Subject Index

A

F

Fabry-Perot interferometer, 455
Ferredoxin, partial specific volume, 29
Ferric thiocyanate complex, pressure-jump studies, 310
Fibrinogen, electrophoretic light scattering properties, 435
Filter, low-frequency cutoff, 467
Flavin adenine dinucleotide, fluorescent acceptor, 378
Flotation method, density determinations and, 23–29
Fluorescein
  emission spectrum, 369
  intensity readings, on digital, photon-counting fluorescence polarometer, 401
Fluorescence
  definition, 348–349
  donor-acceptor pairs, 362–363
  donor quenching, 354–355, 360, 364–365
  dye choice, 359
  emission, polarization of, 350–352
  influence of local environment on, 348–350
  for measurement of reaction equilibria and kinetics, 380–415
    advantages and disadvantages, 381
  molar
    definition, 384
    evaluation, 385
  nonradiative energy transfer
    efficiency, 354–356
    measurement of, 354–356
    optimal conditions for monitoring of, 360
    random labeling case, data analysis, 373–375
    rate of, 353–354
    sample preparation, 363–364
    specific-site labeling case, data analysis, 368–373
    theory, 352–355
  parameters of, species concentration and, 384–385
  polarization, evaluation, 385
  reaction stoichiometry, 382–383
  sensitized emission, 355, 361
  time scale of, 348
  types of labeling, 356–359, 364

Fluorescence polarometer
  analog, direct reading, 401–404, 406, 408
    sensitivity, 402–403
  digital, photon-counting, 393–401
    detector module components, 394, 396–397
    excitor module components, 397–402
    operating procedure, 397–401
    optical components, 394–395
    sensitivity, 400–401
    stopped-flow, 405, 407, 409–415
    sensitivity, 415
Fluorescence spectroscopy, 348–356
  for donor lifetime quenching technique, 367–368
  for donor quenching technique, 364–365
  ligand binding and, 307
  for sensitized emission technique, 365–367
  single photon counting data, analysis of, 367
Flux equations, for asymptotic boundary profiles, 196–201
Formaldehyde, protein labeling and, 332
Formate buffer, protein elution, 343
Form factor, *see* Intramolecular scattering form factor
Förster critical distance, 354, 356
  range of, calculation, 369–373
  values, for assembled systems, 378–379
Fractionation experiment, 245
Frictional coefficient, lack of knowledge of, 4
Fringe displacement
  absolute, 170
  numerical integration of, 183–184
FTC
  fluorescent acceptor and donor, 362–363
  spectral properties, 362
  structural formula, 362
Fusidate, sodium salt, partial specific volume, 21

G

Gel-chromatographic procedure of Hummel and Dreyer, 307
Gel electrophoresis
  in absence of detergents, 4
  molecular weight determinations and, 3–10